Lecture Notes in Computer Science 16106

Advanced Research in Computing and Software Science

Subline of Lecture Notes in Computer Science

More information about this series at https://link.springer.com/bookseries/558

Artur Jeż · Jan Otop
Editors

Fundamentals of Computation Theory

25th International Symposium, FCT 2025
Wrocław, Poland, September 15–17, 2025
Proceedings

Springer

Editors
Artur Jeż
University of Wrocław
Wrocław, Poland

Jan Otop
University of Wrocław
Wrocław, Poland

ISSN 0302-9743 ISSN 1611-3349 (electronic)
Lecture Notes in Computer Science
ISBN 978-3-032-04699-4 ISBN 978-3-032-04700-7 (eBook)
https://doi.org/10.1007/978-3-032-04700-7

This Springer imprint is published by the registered company Springer Nature Switzerland AG
The registered company address is: Gewerbestrasse 11, 6330 Cham, Switzerland

Preface

The conference series Fundamentals of Computation Theory (FCT) was established in 1977 as a forum for researchers from all branches of theoretical computer science. The first edition was in Poznań, Poland and from its inception FCT maintained strong ties to the Polish theoretical computer science community. FCT has been held biennially since its founding. The 25th International Symposium on Fundamentals of Computation Theory (FCT 2025) was hosted by the University of Wrocław, Poland, on September 15–17th, 2025; this was the second time FCT was held in Wrocław; the first time was in 2009 when it was hosted by the Technical University of Wrocław.

The Program Committee of FCT 2025 consisted of 32 scientists from 14 countries and was co-chaired by Artur Jeż and Jan Otop (University of Wrocław, Poland).

FCT 2025 received 50 submissions in total. Each submission was reviewed by at least three Program Committee members, sometimes with the help of external reviewers. The review process was single-blind. Based on the reviews and discussions, 32 contributions were chosen for presentation during FCT 2025 and inclusion in this volume. The decision was based on quality, originality, and relevance to the symposium.

The Program Committee selected the contribution:

Computing Diverse and Nice Triangulations
Waldo Gálvez, Mayank Goswami, Arturo Merino, GiBeom Park, Meng-Tsung Tsai

for the Best Paper Award; the award was sponsored by Springer, the publisher of this volume.

Three invited talks were given at FCT 2025:

- Bernhard Haeupler (ETH & INSAIT Sofia University)
  Graph Decompositions and Length-Constrained Expanders
- Sławomir Lasota (University of Warsaw)
  Reachability in extensions of Petri nets
- Inbal Talgam-Cohen (Technion)
  Algorithmic Contract Design: Recent Progress

The PC chairs (and main local organizers) would like to thank the following, in chronological order:

- the FCT Steering Committee for entrusting us with organizing FCT 2025;
- Springer, for publishing the proceedings and sponsoring the Best Paper Award;
- authors of all submissions, for their willingness to contribute and be part of this event;
- Program Committee members and subreviewers, for their hard work;
- the Springer technical team, for facilitating the publication process;

– the invited speakers, for making this event more attractive and interesting;
– the Municipality of Wrocław, for kindly providing the City Hall for the conference banquet.

July 2025

Artur Jeż
Jan Otop

Organization

Program Committee Chairs

Jeż, Artur	University of Wrocław, Poland
Otop, Jan	University of Wrocław, Poland

Steering Committee

Chlebus, Bogdan	Augusta University, USA
Karpiński, Marek (Chair)	University of Bonn, Germany
Lingas, Andrzej	Lund University, Sweden
Santha, Miklos	CNRS and Université Paris Cité, France
Upfal, Eli	Brown University, USA

Program Committee Members

Belov, Alexander	University of Latvia, Latvia
Björklund, Johanna	Umeå University, Sweden
Bérczi, Kristóf	Eötvös Loránd University, Hungary
Chen, Hubie	King's College London, UK
Czerwiński, Wojciech	University of Warsaw, Poland
Dürr, Christoph	CNRS & Sorbonne University, France
Faliszewski, Piotr	AGH University of Science and Technology, Poland
Fernandez, Maribel	King's College London, UK
Fici, Gabriele	University of Palermo, Italy
Figueiredo, Celina	Federal University of Rio de Janeiro, Brazil
Fijalkow, Joanna	CNRS & University of Bordeaux, France
Ganardi, Moses	Max Planck Institute for Software Systems, Germany
Golovach, Petr	University of Bergen, Norway
Gąsieniec, Leszek	University of Liverpool, UK
He, Meng	Dalhousie University, Canada
Hirvensalo, Mika	University of Turku, Finland
Hliněný, Petr	Masaryk University, Czech Republic
I, Tomohiro	Kyushu Institute of Technology, Japan

Ibsen-Jensen, Rasmus	University of Liverpool, UK
Jeż, Artur (Co-chair)	University of Wrocław, Poland
Klonowski, Marek	Wrocław University of Science and Technology, Poland
Mukherjee, Anish	University of Liverpool, UK
Otop, Jan (Co-chair)	University of Wrocław, Poland
Petrov, Tatjana	University of Trieste, Italy
Podolskii, Vladimir	Tufts University, USA
Shirmohammadi, Mahsa	CNRS & IRIF, France
Skibski, Oskar	University of Warsaw, Poland
Sornat, Krzysztof	AGH University of Science and Technology, Poland
Vaccaro, Ugo	University of Salerno, Italy
Weiß, Armin	University of Stuttgart, Germany
Wiese, Andreas	Technical University of Munich, Germany
Wojtczak, Dominik	University of Liverpool, UK

Additional Reviewers

Abbasi, Fateme
Balabán, Jakub
Banerjee, Sandip
Berglund, Martin
Björklund, Henrik
Bojko, Dominik
Bosch-Calvo, Miguel
Carneiro, Alan Diego
Chauhan, Archit
Cho, Kyungjin
Cordasco, Gennaro
Costa, Marcio
Csáji, Gergely
Czyżewska, Jadwiga
D'Alessandro, Flavio
Dey, Sanjana
Doerr, Benjamin
Dumas, Mael
Fonseca, Guilherme Dias
Fournier, Hervé
Gargano, Luisa
Geniet, Colin
Guttenberg, Roland
Jacobs, Dale
Jain, Rhea
Jeang, John
Jedelský, Jan
Jonsson, Anna
Khan, Asif
Klein, Julia
Kloks, Ton
Kobayashi, Yusuke
Le, Hoang-Oanh
Levet, Michael
Lohrey, Markus
Löding, Christof
Marcais, Guillaume
Mascle, Corto
Meeks, Kitty
Milovanov, Alexey
Nadara, Wojciech
Nogueira, Loana
Paul, Soumyajit
Pekárková, Kristýna
Pierczyński, Grzegorz
Prior, Morgan
Rescigno, Adele
Roy, Shivesh K.

Silva, Ana
Slizkov, Alexey
Stober, Florian
Terziadis, Soeren
Thielen, Clemens
Thierauf, Thomas
Van Santvliet, Hannah
Varonka, Anton
Waleń, Tomasz
Waniek, Marcin
Welker, Janette
Wu, Kaiyu
Wąs, Tomasz
Węgrzycki, Karol
Yamakami, Tomoyuki

Contents

On the Parameterized Complexity of Connected Cluster Vertex Deletion

Ankit Abhinav[1], Sriram Bhyravarapu[2(✉)], A. Mohanapriya[3], and Saket Saurabh[3,4]

[1] National Institute of Science Education and Research, An OCC of Homi Bhabha National Institute, Bhubaneswar, India

[2] Indian Institute of Technology Guwahati, Guwahati, India
sriramb@iitg.ac.in

[3] The Institute of Mathematical Sciences, HBNI, Chennai, India
{mohana,saket}@imsc.res.in

[4] University of Bergen, Bergen, Norway

Abstract. In the Connected Cluster Vertex Deletion problem, the input is a graph G and an integer k, and the task is to determine whether there exists a vertex set $S \subseteq V(G)$ of size at most k such that $G[S]$ is connected and the graph $G - S$ is a cluster graph—that is, every connected component of $G - S$ is a clique. The problem is known to be NP-complete and has been previously studied from the viewpoint of classical computational complexity.

In this paper, we initiate the study of Connected Cluster Vertex Deletion from the perspective of parameterized complexity. We show that the problem is fixed-parameter tractable (FPT) when parameterized by the solution size k, by designing an algorithm with running time $\mathcal{O}(5.2^k \cdot n^{\mathcal{O}(1)})$. Furthermore, we show that Connected Cluster Vertex Deletion does not admit a polynomial kernel unless NP $\subseteq$ coNP/Poly. To cope with this kernelization lower bound, we complement our result by designing a polynomial-size α-approximate kernel for any fixed $\alpha > 1$.

1 Introduction

Clustering in graph theory refers to the task of partitioning a graph into groups or clusters, where nodes within each cluster are more densely connected to each other than to nodes in different clusters. This technique is widely used to reveal the underlying structure of complex networks, such as identifying social communities, detecting functional groups in biological networks, or grouping related documents in web mining [10,14,19]. By grouping similar nodes together, clustering helps uncover meaningful patterns and relationships, facilitating applications such as community detection, recommendation systems, and image segmentation.

One of the central problems in this area is the Cluster Vertex Deletion problem, which has been extensively studied in the parameterized complexity

A. Jeż and J. Otop (Eds.): FCT 2025, LNCS 16106, pp. 1–14, 2026.
https://doi.org/10.1007/978-3-032-04700-7_1

literature. In Cluster Vertex Deletion, the goal is to delete at most k vertices from a graph so that the resulting graph is a cluster graph, i.e., a disjoint union of cliques. Cai [4] initiated the study of its parameterized complexity and gave an $\mathcal{O}^*(3^k)$-time algorithm.[1] Over the years, this has been improved, with the current best known running time being $\mathcal{O}^*(1.811^k)$ due to Tsur [18].

Recently, Le and Le [12] initiated the study of the connectivity-constrained variant of Cluster Vertex Deletion, known as Connected Cluster Vertex Deletion (CCVD). In Connected Cluster Vertex Deletion, the goal is to delete a set $S \subseteq V(G)$ of at most k vertices such that the induced subgraph $G[S]$ is connected and the remaining graph $G - S$ is a cluster graph—that is, every connected component of $G - S$ is a clique. Le and Le [12] observed that, by leveraging results from Komusiewicz [11], one can conclude that Connected Cluster Vertex Deletion is NP-complete and does not admit an algorithm with running time $2^{o(n+m)}n^{\mathcal{O}(1)}$, unless the Exponential Time Hypothesis (ETH) fails. Furthermore, they established a classical complexity dichotomy for Connected Cluster Vertex Deletion on graph families that exclude a fixed graph H as an induced subgraph.

While this added connectivity constraint makes the problem theoretically more challenging, it also makes it more relevant in applications such as network design, community detection, and security [1,17], where isolating a cohesive and connected substructure while simplifying the remainder of the network is often desirable. In addition to its practical motivations, Connected Cluster Vertex Deletion fits into a broader and increasingly studied class of connectivity-constrained graph problems, where the goal is to find a small connected solution satisfying a classical graph property. Such variants have been investigated for fundamental problems like Vertex Cover [5], Feedback Vertex Set [15], Odd Cycle Transversal [3], and Dominating Set [8], leading to both algorithmic advances and new complexity insights. Motivated by these developments, we undertake a parameterized complexity study of Connected Cluster Vertex Deletion, aiming to better understand its algorithmic behavior and computational boundaries. We start by defining the problem formally.

Connected Cluster Vertex Deletion (CCVD) **Parameter:** k
Input: An undirected graph G and an integer k.
Question: Does there exist a set $S \subseteq V(G)$ of size at most k such that $G[S]$ is connected and $G - S$ is a cluster graph?

Our Results. A naive algorithm for several connectivity problems proceeds by enumerating all minimal solutions for unconnected version of the problem whose size is at most k, and then solving the corresponding instance of the Steiner Tree problem where the vertices in the minimal solutions are treated as terminals.

[1] Throughout this paper, we use the $\mathcal{O}^*$ notation to suppress factors polynomial in the input size.

Recall that a graph is a *cluster graph* if and only if it does not contain any induced P_3 (a path on three vertices). Therefore, the CLUSTER VERTEX DELETION (CVD) problem asks for a set of at most k vertices whose removal destroys all induced P_3's in the graph. A simple branching algorithm shows that there are at most 3^k minimal cluster deletion sets of size at most k. Applying the known $\mathcal{O}^*(2^k)$ algorithm for the STEINER TREE problem [16] on each such set leads to an overall $\mathcal{O}^*(6^k)$ time algorithm for the CCVD problem.

Our first result improves upon this naive strategy. Specifically, we give a deterministic algorithm that solves CCVD in time $\mathcal{O}^*(5.236^k)$. This leads to the following theorem.

Theorem 1. CONNECTED CLUSTER VERTEX DELETION *can be solved in* $\mathcal{O}^*(5.236^k)$ *time.*

It is known that the CONNECTED VERTEX COVER (CVC) problem does not admit a polynomial kernel unless $\mathsf{NP} \subseteq \mathsf{coNP/Poly}$ [7]. This kernelization lower bound is inherited by the CCVD problem due to a parameter-preserving reduction, implying that CCVD shares the same kernelization hardness. To formalize this, we state the following theorem.

Theorem 2. CONNECTED CLUSTER VERTEX DELETION *does not admit a polynomial kernel unless* $\mathsf{NP} \subseteq \mathsf{coNP/Poly}$.

Theorem 2 establishes that CCVD does not admit a polynomial kernel unless the polynomial hierarchy collapses, ruling out the possibility of an efficient preprocessing that reduces the instance size to a function polynomial in the solution size while preserving exact solvability. To overcome this limitation, we adopt the framework of *lossy kernelization* [13]. A lossy kernel is a polynomial-time reduction that produces a smaller instance whose size is polynomial in both the solution size and an accuracy parameter α, and which preserves approximate solvability.

In this work, we obtain a lossy kernel for CCVD by combining the Sunflower Lemma with the known lossy kernel for the STEINER TREE problem. This yields an efficient preprocessing routine for CCVD that guarantees an α-approximate solution. We formalize this in the following theorem.

Theorem 3. *For any* $\alpha > 1$, CONNECTED CLUSTER VERTEX DELETION *parameterized by the solution size* k *has a time efficient polynomial-size approximate kernelization scheme with at most* $k^{f(\alpha)}$ *vertices where* f *is some computable function.*

Preliminaries: In this paper, we consider finite, undirected and connected graphs. Given a graph $G = (V, E)$, we use $V(G)$ and $E(G)$ to denote the vertex and the edge sets of G. For a vertex $v \in V(G)$, we use $N_G(v)$ (open neighborhood of v) to denote the set of neighbors of v in G. The closed neighborhood of v is denoted by $N_G[v] = N_G(v) \cup \{v\}$. Let S be a connected cluster vertex deletion set. It is clear that $G - S$ is a collection of cliques. When we say a partition of a clique $C \in G-S$, we mean that set of vertices in the same partition of C having same neighbors in S. Observe that a clique $C \in G-S$ can have at most $2^{|S|}$ many partitions.

2 FPT Algorithm for CCVD Using Iterative Compression

In this section, we prove Theorem 1 by using iterative compression technique [6]. In Sect. 2.1, we prove that CCVD exhibits a natural algorithm that runs in $\mathcal{O}^*(6^k)$. The compression version in Sect. 2.1 uses branching. We show that by employing a faster branching strategy without enumeration of all possible intersections of the size k during the compression version. Our approach involves thoroughly examining the graph's structure and finding the best solution based on a vertex that has been initially identified as not part of the solution. By using a careful bounded search tree in the compression version, we improve the run time of CCVD from $\mathcal{O}^*(6^k)$ to $\mathcal{O}^*(5.236^k)$, we exhibit this in Sect. 2.2.

2.1 Algorithms for Connected Cluster Vertex Deletion

Now we present two algorithms for CCVD using iterative compression technique. Using this technique, we first show an algorithm running in time $\mathcal{O}^*(6^k)$ in Lemma 3. Then we do a bit of hard work to improve the running time and the algorithm is presented in Sect. 2.2, Theorem 4. We define a compression problem where the input is an undirected graph G, an integer k and a connected cluster vertex deletion set of size at most $k+2$, and the question is to find a connected cluster vertex deletion set of size at most k. We formally state the problem below.

> Compression Connected Cluster Vertex Deletion(Comp-CCVD)
> **Input:** An unweighted undirected graph G, an integer k and a connected cluster vertex deletion set $Z \subseteq V(G)$ of G of size at most $k+2$.
> **Output:** Find a connected cluster vertex deletion set of size at most k.

We initialize $G^* = G$ with the input graph. Let (G, k, Z) be an instance of Comp-CCVD. Let $S \subseteq V(G)$ be a hypothetical connected cluster vertex deletion set in the graph G. We start by guessing the intersection of S with Z. Let $Y = Z \cap S$ and $N = Z \setminus Y$. Notice that each of the connected components in $G[N]$ is a clique, otherwise there exists a P_3 in $G[N]$ and the guessed set Y is not valid and thus we ignore it. Therefore we assume that $G[N]$ is a cluster graph. Further, we update the solution size by including Y in the solution and the resulting instance is $(G, k-|Y|, Z)$ of Comp-CCVD. From now, we look for a set of vertices from $G-Z$ of size at most $k-|Y|$ which together with Y will constitute the solution to the original instance of Comp-CCVD. Since a connected cluster vertex deletion set is also a cluster vertex deletion set, the objective is to kill all the P_3's in $G-Y$. Depending on the edges between N and $G-N$, we classify P_3's into the following four types. Throughout the section, we use C_i, C_j to refer to maximal cliques in $G-Z$, and C'_a, C'_b to refer to maximal cliques in $G[N]$.

Type 1: There exists a vertex $u \in C_i$, where C_i is a maximal clique in $G - Z$ and there are two vertices in $x, y \in G[N]$ such that $x, y \in C'_a$, where C'_a is a maximal clique in $G[N]$, and $u \in N_G(x)$ and $u \in N_G(y)$.

Type 2: There exists a vertex $u \in C_i$, where C_i is a maximal clique in $G - Z$ and there are two vertices in $x, y \in G[N]$ such that $x \in C'_a$, $y \in C'_b$ where C'_a, C'_b, $a \neq b$ are maximal cliques in N, and $u \in N_G(x)$ and $u \notin N_G(y)$.

Type 3: There exists a vertex $x \in G[N]$, there are two vertices u, v such that $u \in C_i$, $v \in C_j$, $j \neq i$, C_i, C_j are maximal cliques in $G - Z$, and $u, v \in N_G(x)$.

Type 4: There are at least two vertices $u, v \in C_i$, where C_i is amaximal clique in $G - Z$ and there is a vertex in $x \in G[N]$ such that $x \in C'_a$ where C'_a is a maximal clique in $G[N]$, and $u \in N_G(x)$ and $v \notin N_G(x)$ (Fig. 1).

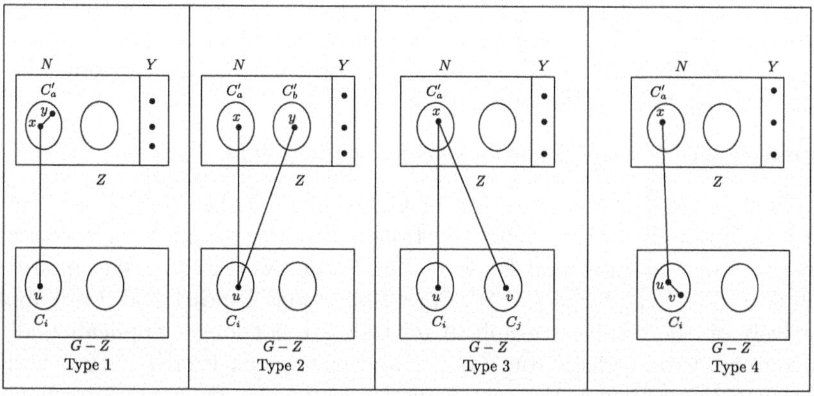

Fig. 1. An illustration for different representative types.

We say a tuple (u, x, v) *corresponds* to Type 3 if there exist two cliques C_i, C_j in $G - N$ such that $u \in C_i$, $v \in C_j$, $x \in N$, and $u - x - v$ forms a P_3. Similarly, we say a tuple (u, v, x) *corresponds* to Type 4 if there exists a clique C_i in $G - N$ such that $u, v \in C_j$, $x \in N$, and $v - u - x$ forms a P_3. We now look at a few reduction rules.

Reduction Rule 1. *Let* (y, x, u) *be a tuple corresponding to* Type 1 *in* (G, k, Z). *Then the resulting instance is* $(G - \{u\}, k - 1, Z)$.

Lemma 1. *Reduction Rule 1 is safe.*

Proof. In order to prove the safeness, it is sufficient to show: (G, k, Z) is a Yes-instance of COMP-CCVD if and only if $(G - \{u\}, k - 1, Z)$ is a Yes-instance of COMP-CCVD.

Let S be a solution to the instance (G, k, Z). Since there exists a P_3, $y - x - u$ of Type 1 in G and $x, y \notin S$, u must be in S. Thus, we get the instance $(G - \{u\}, k - 1, Z)$ and $S \setminus \{u\}$ is a solution to $(G - \{u\}, k - 1, Z)$.

Let S' be a solution to $(G' = G - \{u\}, k-1, Z)$. Then we show that $S' \cup \{u\}$ is a solution to (G, k, Z). Suppose not, there exists a P_3 in $G - (S' \cup \{u\})$. Then such a P_3 also exists in $G' - S'$ as $G' - S'$ is a subgraph of $G - S'$, a contradiction to the fact that S' is a solution to $(G', k-1, Z)$. □

Reduction Rule 2. *Let (x, u, y) be a tuple corresponding to* Type 2 *in (G, k, Z). Then the resulting instance is $(G - \{u\}, k-1, Z)$.*

Lemma 2. *Reduction Rule 2 is safe.*

Proof. In order to prove the safeness, it is sufficient to show: (G, k, Z) is a Yes-instance of COMP-CCVD if and only if $(G - \{u\}, k-1, Z)$ is a Yes-instance of COMP-CCVD.

Let S be a solution to the instance (G, k, Z). Since there exists a P_3, $x - u - y$ of Type 1 in G and $x, y \notin S$, u must be in S. Thus, we get the instance $(G - \{u\}, k-1, Z)$ and $S \setminus \{u\}$ is a solution to $(G - \{u\}, k-1, Z)$.

Let S' be a solution to $(G' = G - \{u\}, k-1, Z)$. Then we show that $S' \cup \{u\}$ is a solution to (G, k, Z). Suppose not, there exists a P_3 in $G - (S' \cup \{u\})$. Then such a P_3 also exists in $G' - S'$ as $G' - S'$ is a subgraph of $G - S'$, a contradiction to the fact that S' is a solution to $(G', k-1, Z)$. □

Lemma 3. COMP-CCVD *can be solved in $\mathcal{O}^*(6^k)$ time.*

Proof. Let (G, k, Z) be an instance of COMP-CCVD. Let $S \subseteq Z$ be a hypothetical connected cluster vertex deletion set in the graph G. We guess the intersection of S with Z and let $Y = Z \cap S$ and $N = Z \setminus Y$. We update the graph $G = G - Y$ and $k = k - |Y|$. The Reductions Rules 1 and 2 are applied recursively on the resultant graph. After they are not longer applicable, all the P_3's contains two vertices from $G - N$ and one vertex from N. That is, they are of Type 3 and Type 4. For each tuple (u, x, v) (resp. (u, v, x)) corresponding to Type 3 (resp. Type 4), we create two subproblems including either $u \in S$ or $v \in S$, and dropping the value of k by 1. We discard a branch if there exists a P_3 of one of the types Type 3 and Type 4 and $k \leq 0$. As a consequence, there are at most 2^k subproblems. For each of the subproblem, to achieve the connectedness property, we ask for a Steiner tree of the resultant graph with S^* as the terminal set where S^* which is the set of vertices included in the solution so far leading to this branch. To find a connected cluster vertex deletion set of size at most k, we rely on STEINER TREE parameterized by the number of terminals, with the input instance $(G, S^*, |S^*|)$. Since STEINER TREE is FPT when parameterized by the number of terminals, we obtain a $\mathcal{O}^*(2^k)$ time algorithm to find a connected cluster deletion set. We go over all the subproblems and pick a Steiner tree with the least number of Steiner vertices. If such has at most k vertices then we output that there exists a connected cluster vertex deletion set of size at most k. If for all the choices, of Y there is no set Steiner tree, then we conclude that the given instance is a No-instance of COMP-CCVD.

The running time of the algorithm is $(\sum_{i=0}^{k} \binom{k}{i} 2^{k-i}) 2^k = \mathcal{O}^*(6^k)$. □

2.2 A Faster Algorithm for the Compression Version

In this section, we present an efficient algorithm for COMP-CCVD.

Theorem 4. COMP-CCVD *can be solved in* $\mathcal{O}(5.236^k)$ *time.*

Proof. Let (G, k, Z) be an instance of COMP-CCVD. Let $S \subseteq V(G)$ be a hypothetical connected cluster vertex deletion set in the graph G. We guess the intersection of S with Z and let $Y = Z \cap S$ and $N = Z \setminus Y$. We update the graph $G = G - Y$ and $k = k - |Y|$. The Reductions Rules 1 and 2 are applied recursively on the resultant graph. After they are not longer applicable, all the P_3's contains two vertices from $G - N$ and one vertex from N. That is, they are of Type 3 and Type 4. The aim of this section is to enhance the branching algorithm outlined in Lemma 3 by eliminating the use of the branching vector (1,1), which led to $2^{(k-i)}$ factor in the running time. We introduce a more efficient branching algorithm by employing various branching steps. The algorithm always applies the first applicable rule, in the order presented below, at each stage.

Reduction Rule 3. *Let* (u, x, v) *be a tuple corresponding to* Type 3 *in* (G, k, Z) *and* $w \in N_G(x)$ *be a vertex in* $C_i - \{u\}$. *Then the resulting instance is* $(G - \{v\}, k - 1, Z)$ *or* $(G - \{u, w\}, k - 2, Z)$.

Lemma 4. *Reduction Rule 3 is safe.*

Proof. To prove the correctness of the lemma, it is sufficient to show the following: (G, k, Z) is a Yes-instance of COMP-CCVD if and only if $(G - \{v\}, k - 1, Z)$ or $(G - \{u, w\}, k - 2, Z)$ is a Yes-instance of COMP-CCVD.

Let S be a solution to the instance (G, k, Z). Since there exists a P_3, $u - x - v$ of Type 3 in G and $x \notin S$, u or v belongs to S. Observe that there exists a P_3, $w - x - v$ in G where $w \in C_i - \{u\}$. In order to kill these P_3's, $u - x - v$ and $w - x - v$, it must be the case that either $v \in S$ or $u, w \in S$. In the former case, we get the instance $(G - \{v\}, k - 1, Z)$ and $S \setminus \{v\}$ is a solution to $(G - \{v\}, k - 1, Z)$. For the latter case, we get the instance $(G - \{u, w\}, k - 2, Z)$ and $S \setminus \{u, w\}$ is a solution to $(G - \{u, w\}, k - 2, Z)$.

Suppose that at least one of the instances $(G' = G - \{v\}, k - 1, Z)$ or $(G' = G - \{u, w\}, k - 2, Z)$ is a Yes-instance of COMP-CCVD. Let S' be a solution to $(G' = G - \{v\}, k - 1, Z)$. Then we show that $S' \cup \{v\}$ is a solution to (G, k, Z). Suppose not, there exists a P_3 in $G - (S' \cup \{v\})$. Then such a P_3 also exists in $G' - S'$ as $G' - S'$ is a subgraph of $G - S'$, a contradiction to the fact that S' is a solution to $(G', k - 1, Z)$. Suppose that S' is a solution to $(G' = G - \{u, w\}, k - 2, Z)$. Then we show that $S' \cup \{u, w\}$ is a solution to (G, k, Z). Suppose not, then there exists a P_3 in $G - (S' \cup \{u, w\})$ and such a P_3 also exists in $G' - S'$, a contradiction to the fact that S' is a solution to $(G', k - 2, Z)$. □

Reduction Rule 4. *Let* (u, x, v) *be a tuple corresponding to* Type 3 *in* (G, k, Z) *and* $w \notin N_G(x)$ *be a vertex in* C_i. *Then the resulting instance is* $(G - \{u\}, k - 1, Z)$ *or* $(G - \{v, w\}, k - 2, Z)$.

Reduction Rule 5. *Let (u, x, v) be a tuple corresponding to* Type 3 *in (G, k, Z) and $w \in N_G(x)$ be a vertex in C_ℓ, where $\ell \notin \{i, j\}$. Then the resulting instance is $(G - \{u, w\}, k - 2, Z)$ or $(G - \{v, w\}, k - 2, Z)$ or $(G - \{u, v\}, k - 2, Z)$.*

Reduction Rule 6. *Let (u, v, x) be a tuple corresponding to* Type 4 *in (G, k, Z) and $w \in N_G(x)$ be a vertex in $C_i - v$. Then the resulting instance is $(G - \{u\}, k - 1, Z)$ or $(G - \{v, w\}, k - 2, Z)$.*

Reduction Rule 7. *Let (u, v, x) be a tuple corresponding to* Type 4 *in (G, k, Z) and $w \notin N_G(x)$ be a vertex in $C_i - \{u\}$. Then the resulting instance is $(G - \{v\}, k - 1, Z)$ or $(G - \{u, w\}, k - 2, Z)$.*

Lemma 5 (♠).[2] *Reduction Rules 4, 5, 6 and 7 are safe.*

We apply the Reduction Rules 3, 4, 5, 6 and 7 whenever applicable. When none of these reduction rules are applicable, the resultant graph G contains P_3's either of Type 3 or Type 4. Let $\mathcal{Q} = \{(u, x, v) \mid (u, x, v)$ corresponds to Type 3$\}$ and $\mathcal{R} = \{(u, v, x) \mid (u, v, x)$ corresponds to Type 4$\}$.

Observation 1. *Let $(u, x, v) \in \mathcal{Q}$. Then $|C_i| = 1$ and $|C_j| = 1$. Moreover, there is no clique $C' \in G - Z$, where $C' \notin \{C_i, C_j\}$, that contains a vertex adjacent to x. Consequently, $|N_G(x) \cap (V(G) \setminus Z)| = 2$.*

Proof. Since Reduction Rules 3 and 4 are not applicable, we have that $|C_i| = 1$ and $|C_j| = 1$. Furthermore, since Reduction Rule 5 is not applicable, there is no clique $C' \in G - Z$, where $C' \neq C_i$, that contains a vertex adjacent to x. □

Observation 2. *Let $(u, v, x) \in \mathcal{R}$. Then $|C_i| = 2$. Consequently, $|N_G(x) \cap (V(G) \setminus Z)| = 1$.*

Proof. Since Reduction Rule 6 is not applicable, we have that $\{w \in C_i \mid w \notin N_G(x)\} = \{v\}$. Furthermore, since Reduction Rule 7 is not applicable, we have that $\{w \in C_i \mid w \in N_G(x)\} = \{u\}$. Therefore we have that $|C_i| = 2$ and $C_i = \{u, v\}$. □

Let us consider a branch with the resultant instance (G, k, Z). Let the cliques of $G - Z$ be $C_1, C_2, \ldots, C_r$. Let Q_ℓ represent the ℓth tuple of $\mathcal{Q}$ and R_d represent the dth tuple of $\mathcal{R}$, where $1 \leq \ell \leq |\mathcal{Q}|$ and $1 \leq d \leq |\mathcal{R}|$. Without loss of generality, let C_{2p+1} and C_{2p+2} be the cliques corresponding to the tuple Q_{p+1}, where $0 \leq p < |\mathcal{Q}|$. Similarly, let $C_{2|\mathcal{Q}|+d}$ be the clique corresponding to the tuple R_d, where $1 \leq d \leq |\mathcal{R}|$. For the instance (G, k, Z), we now construct a graph H and formulate an instance of Steiner Tree on H. The vertex set of H is $V(H) = (V(G) \setminus N) \cup \mathcal{X}_V \cup B \cup W \cup S'$ where $B = \{b_i \mid 1 \leq i \leq |\mathcal{Q}|\}$, $W = \{w_i \mid 1 \leq i \leq |\mathcal{R}|\}$, $\mathcal{X} = \{X_i \mid X_i$ is a path on n vertices, where $1 \leq i \leq 2|\mathcal{Q}|\}$, $\mathcal{X}_V$ is the union of vertices in each path of $\mathcal{X}$, and S' is the set of vertices included in the solution till we reach the instance (G, k, Z). Let the two endpoints of each $X_i \in \mathcal{X}$ be x_i^1 and x_i^n. The edge set $E(H)$ is defined as follows:

[2] Results labeled ♠ can be found in the extended version of this paper.

$$E(H) = E(\mathcal{X}) \cup \{ux^1_{2p+1}, vx^1_{2p+2} \mid Q_{p+1} = (u, x, v), 0 \leq p < |\mathcal{Q}|\}$$
$$\cup \{yy' \in E(G^*) \mid y, y' \in Y \cup S'\} \cup \{b_{p+1}x^n_{2p+1}, b_{p+1}x^n_{2p+2} \mid 0 \leq p < |\mathcal{Q}|\}$$
$$\cup \{uw_p, vw_p \mid R_p = (u, v, x), 2|\mathcal{Q}| < p \leq 2|\mathcal{Q}| + |\mathcal{R}|\}$$
$$\cup \{y\ell \in E(G^*) \mid y \in Y \cup S', \ell \in C_i, 1 \leq i \leq r\}$$
$$\cup \{y\ell \in E(G^*) \mid y, \ell \in C_i, 2|\mathcal{Q}| + |\mathcal{R}| < i \leq r\}.$$

where $E(\mathcal{X})$ represents the edges of the paths in $\mathcal{X}$ and G^* is the initial input graph. The construction of H is complete. We consider the set $T = Y \cup B \cup W \cup S'$ as our terminal set and (H, T, k') as an instance of Steiner Tree where $k' = |T| \leq k$. We now show the following result (Fig. 3).

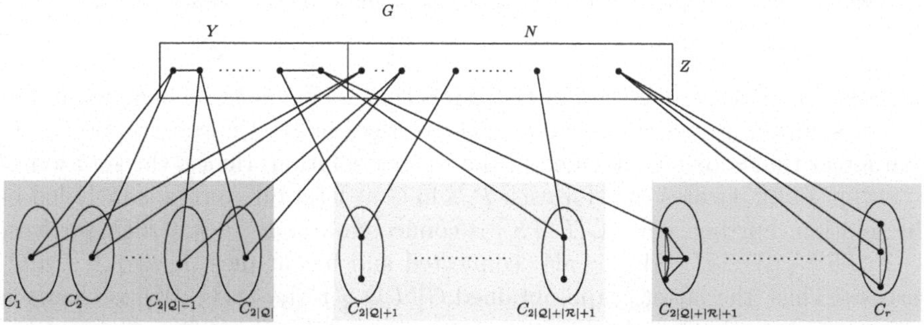

Fig. 2. An illustration of a graph G.

Lemma 6. *G has a* Comp-CCVD *of size at most k if and only if* Steiner Tree *(H, T, k') uses at most $k + n|\mathcal{Q}|$ Steiner vertices.*

Proof. Let S be the solution of Comp-CCVD of size at most k and S' be the set of vertices included upon till now by the current branch. Observe that $S' \subseteq T$ and $Y \subseteq T$. Further, for each tuple in $\mathcal{Q}$, u or v must be in S, so we use the corresponding path adjacent to u or v in H to connect b_{p+1}, where $0 \leq p < |\mathcal{Q}|$. Furthermore, for each tuple in $\mathcal{R}$, u or v must be in S, this ensures that w_p is connected, where where $2|\mathcal{Q}| \leq p < 2|\mathcal{Q}| + |\mathcal{R}|$ Since $G[S]$ is connected, we have a Steiner tree for H with T terminals that uses at most $k+n|\mathcal{Q}|$ Steiner vertices.

Let S' be a solution to (H, T, k) of Steiner Tree that uses at most $k+n|\mathcal{Q}|$ Steiner vertices. Observe that in order to ensure connectivity for each vertex $b \in B$, exactly one path of length n must be chosen from $\mathcal{X}$. Suppose that there exists a vertex $b \in B$ such that it uses both paths of length n adjacent to it. Then, the Steiner tree has at least $n|\mathcal{Q}| + n$ Steiner vertices which is more than $k + n|\mathcal{Q}|$, which implies that it is not possible to get a connected subgraph by using vertices from $H - \mathcal{X}$. Thus, we need $n|\mathcal{Q}|$ vertices to connect

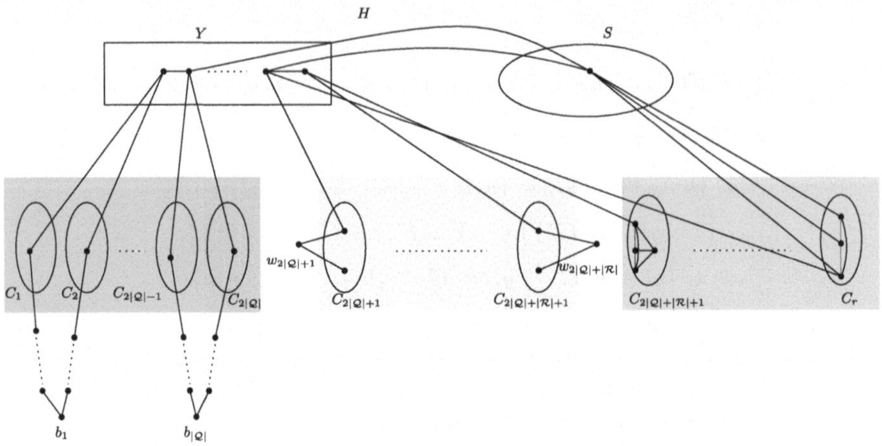

Fig. 3. An illustration of Graph H corresponding to the graph G in Fig. 2.

vertices in B. Further, vertices in B ensures that u or v must be included in the Steiner solution. Furthermore, to connect vertices in W, the construction of H guarantees that u or v is included in the Steiner solution. Hence, the solution of Steiner Tree ensures that for each P_3's in G at least one vertex is included in the solution. Further, since $G[T \cup S']$ is connected and $|S'| \leq k + n|\mathcal{Q}|$ vertices, $G[(T \cup S') \setminus (\mathcal{X} \cup B \cup W)]$ is also connected and has at most $k + n|\mathcal{Q}| - n|\mathcal{Q}|$ vertices. Thus, the latter graph obtained $G[(T \cup S') \setminus (\mathcal{X} \cup B \cup W)]$ is clearly a connected cluster vertex deletion set of size at most k. □

Analysis of the Algorithm. We now analyze the algorithm presented in Theorem 4, by considering the branching vectors of the reduction rules and the construction of H and invoking Steiner Tree algorithms. Notice that it is safe to assume that the branching vectors produce the upper bound on the number of leaves in recursion tree. The branching vectors for the reductions rules are as follows:

- (1) for the Reduction Rules 1 and 2,
- (1, 2) for the Reduction Rules 3, 4, 6 and 7, and
- (2, 2, 2) for the Reduction Rule 5.

After the recursive application of Reduction Rules 1, 2, 3, 4, 5, 6 and 7, the number of leaves in the recursion tree is upper bounded by 1.618^k. Further, we apply Lemma 6 for each branch, which runs in time $\mathcal{O}^*(2^k)$. Given an instance (G, k, Z) of Comp-CCVD, we go over all $Y \subseteq Z$ that belongs to the solution from Z. Since the number of such sets is $\sum_{i=0}^{k} \binom{k}{i}$, and the number of leaves in the recursion corresponding to a set Y is at most 1.618^{k-i}, the runtime of Comp-CCVD is $(\sum_{i=0}^{k}\binom{k}{i}1.618^{k-i})2^k = \mathcal{O}^*(5.236^k)$. □

We now present the proof of Theorem 1.

Proof (Proof of Theorem 1). To start with, we consider an ordering of the vertices in $V(G)$, say $\sigma = v_1, v_2, \ldots, v_n$ such that for each $i > 1$, the vertex v_i has a neighbor, say v_j where $j < i$. Such an ordering exists since we assumed that G is connected and an ordering can be found in linear time by arbitrarily choosing a vertex and include vertices in a breadth first search order. Let G_i denote the graph induced by the vertex set $\{v_1, v_2, \ldots, v_i\}$. Notice that for all i, G_i is connected. To start with, consider the set $Z = \{v_1, v_2, \ldots, v_{k+2}\}$ which is a connected cluster vertex deletion set of G_{k+2}. We apply the algorithm of Comp-CCVD on the instance $(G := G_{k+2}, k, Z)$ using Theorem 4. We add next the vertex v in the ordering σ to Z. If $G[Z]$ is connected, then create another instance of Comp-CCVD. Otherwise, we choose a $N(v)$ from the resultant graph of previous iteration, then create another instance of Comp-CCVD. We keep on doing this until we call the instance on the graph G^*. Since we invoke Comp-CCVD at most n times, the running time follows. □

3 CCVD Does Not Admit a Polynomial Kernel

It is known that Connected Vertex Cover does not admit a poynomial kernel unless the polynomial hierarchy collapses to $\sum_p^3$ [7]. To show that CCVD does not admit a polynomial kernel does not admit a polynomial kernel unless $\mathsf{NP} \subseteq \mathsf{coNP/Poly}$, we give a polynomial time reduction from Connected Vertex Cover. Let (G, k) be an instance of Connected Vertex Cover. We create a graph G' from G as follows: $V(G') = \{u_i, u'_i \mid v_i \in V(G)\}$, and $E(G') = \{u_r u_s \mid v_r v_s \in E(G)\} \cup \{u_i u'_i \mid v_i \in V(G)\}$. The number of vertices and edges in G' is $2n$ and $m + n$ respectively where n and m denote $|V(G)|$ and $|E(G)|$.

Claim 1. *G has connected vertex cover of size at most k if and only if G' has a connected cluster vertex deletion set of size at most k.*

Proof. Let $S = \{v_1, \ldots, v_k\}$ be a connected vertex cover of G. Then $G - S$ is an edgeless graph. Consider the set $S' = \{u_1, \ldots, u_k\}$. Since G is a subgraph of G' by construction, and $G[S]$ is connected in G, we have that $G[S']$ is connected in G'. Moreover, $G' - S'$ is a graph where every component has exactly one edge where the edges of $G' - S'$ is the set $\{u_{k+1}u'_{k+1}, \ldots, u_n u'_n\}$. Thus S' is a connected cluster vertex deletion set.

Let $S' \subseteq V(G')$ be a connected cluster vertex deletion set of size at most k. We know that $G' - S'$ is a cluster graph. By construction of G', we know that each vertex $u_i \in V(G')$ has a pendant vertex u'_i adjacent to it. Further, for each $u_r u_s \in E(G')$, we have two P_3's $u'_r - u_r - u_s$ and $u'_s - u_s - u_r$. Thus, for each edge we need to choose at least one vertex. Since, $|S'| \leq k$, we choose the corresponding v_i in S for G. Suppose that $u'_s \in S'$, for some s, then we choose the corresponding $u_s \in S$ for G. If S' is a connected cluster vertex deletion set of size at most k, then S is a connected vertex vertex cover of size at most k. □

This concludes the proof of Theorem 2.

4 Lossy Kernel

In this section, we give an overview of the proof of Theorem 3. Given an instance (G, k) of CCVD, we first show that there exists a subset $Z \subseteq V(G)$ of size at most $\mathcal{O}(k^3)$ such that all minimal solutions of size at most k (a set X that is inclusion minimal and $G - X$ is a cluster graph) are contained in Z. We call such a set Z as *k-minimal solution cover.* To find the set Z, we use the sunflower lemma from [9].

Theorem 5 (Sunflower lemma *[9]***).** *Let F be a family of sets (without duplicates) over a universe U, such that each set in F has cardinality at most d. If $|F| > d \cdot d!(k-1)^d$, then F contains a sunflower with k petals and such a sunflower can be computed in time polynomial in $|F|$, $|U|$, and k.*

Lemma 7. *Let (G, k) be an instance of* Cluster Vertex Deletion. *Then in polynomial time we can compute a k-minimal solution cover of size at most $\mathcal{O}(k^3)$.*

Proof. Towards the proof, we first cast Cluster Vertex Deletion as an instance of 3-Hitting Set, where universe U is $V(G)$ and the set family $\mathcal{F}$ consists of all 3-sized vertex subsets of $V(G)$ such that the corresponding subset forms an induced path. It is well-known that G is a cluster graph if and only if G does not have an induced P_3 (an induced path on 3 vertices). Thus, a set $S \subseteq V(G)$ of size at most k is a minimal cluster vertex deletion set of G if and only if S is a minimal 3-hitting set of the instance $(U, \mathcal{F}, k)$.

By applying the sunflower lemma (Theorem 5), if $|\mathcal{F}| > 18(k+1)^3$, we can obtain a sunflower with $k+2$ petals. If the core of the sunflower is empty, the instance is a NO-instance because hitting all $k+2$ petals would require a cluster vertex deletion set of size $k+2$, exceeding the allowed budget of k. Therefore, we can assume that the core is non-empty. Next, we arbitrarily delete one of the petals, say Q. □

Claim 2. *G has a minimal cluster vertex deletion set of size at most k if and only if $G - V_Q$ has a minimal cluster vertex deletion set of size at most k, where V_Q is the set of vertices in the petal Q.*

Proof. Suppose G has a minimal cluster vertex deletion set S of size at most k. Without loss of generality, assume $S = \{v_1, \ldots, v_k\}$. If S does not contain any vertex from P_i, then we are done, as S is also a cluster vertex deletion set for $G \setminus P_i$. Otherwise, S contains a vertex $v_i \in P_i$. In this case, $G \setminus P_i$ does not include v_i, but it shares the core vertices, which together induce a P_3 with v_i

in G. We can arbitrarily remove v_i from S and replace it with another vertex $v_j \in S$, forming a new solution S' for the cluster vertex deletion set of $G \setminus P_i$ of size k.

Now, we prove the reverse direction. Let S be a cluster vertex deletion set for the graph $G - V_Q$ of size k. The graph G contains the vertices from the petal Q, which together with the core induces a P_3. Additionally, the core induces a P_3 with the other petals that remain unremoved. Therefore, to hit all the unremoved P_3's in $G - V_Q$, we are required to select at least one vertex from the core. Otherwise, the budget would exceed k. Thus, the core vertices must be included in the solution S, which also covers the vertices in the petal Q. Hence, S is also a cluster vertex deletion set for G. □

We continue this process until the family size is bounded above by $18(k+1)^3$, which is $\mathcal{O}(k^3)$.

Claim 3. *Let Z be the set of elements in the reduced instance where the sunflower lemma can no longer be applied. Then Z contains all minimal cluster vertex deletion sets of size at most k.*

Proof. Since Z is the reduced instance where the sunflower lemma can no longer be applied, and as shown in the previous claim, removing a petal from the $k+2$ petals of the sunflower does not affect the minimal solution for G, it follows that $G[Z]$ contains all cluster vertex deletion sets of size at most k. □

Observation 3. *Any connected cluster vertex deletion set of size at most k contains a minimal cluster vertex deletion set of size at most k.*

The proof of the above observation is easy to see.

The objective of Theorem 3 is to find a *polynomial size approximate kernelization scheme* (PSAKS) for CONNECTED CLUSTER VERTEX DELETION. Towards this, we utilize the PSAKS algorithm for STEINER TREE in [13]. By using Lemma 7, we obtain k-minimal solution cover, Z, of size at most $\mathcal{O}(k^3)$. Further, from Observation 3, we get that any connected cluster vertex deletion set S of size at most k, contains a set $S' \subseteq S$ such that $S' \subseteq Z$. Observe that once we know the set S', we can obtain S by treating S' as our terminal set in STEINER TREE (G, T, p), where $p = |T|$. We present the lossy kernel and the proof of correctness in the extended version of the paper.

References

1. Bergmann, S., et al.: Characterization of the yeast transcriptional network. Genome Res. **14**(8), 1743–1754 (2004)
2. Björklund, A., Husfeldt, T., Kaski, P., Koivisto, M.: Fourier meets möbius: fast subset convolution. In: Proceedings of the Thirty-Ninth Annual ACM Symposium on Theory of Computing, STOC '07, pp. 67–74, New York, NY, USA, 2007. Association for Computing Machinery
3. Bojikian, N., Kratsch, S.: Tight algorithm for connected odd cycle transversal parameterized by clique-width. CoRR, abs/2402.08046, 2024. arXiv:2402.08046

4. Cai, L., et al.: Fixed parameter tractability of graph modification problems for hereditary properties. SIAM J. Discret. Math. **9**(4), 520–536 (1996)
5. Cygan, M.: Deterministic parameterized connected vertex cover. In: Fomin, F.V., Kaski, P. (eds.) SWAT 2012. LNCS, vol. 7357, pp. 95–106. Springer, Heidelberg (2012). https://doi.org/10.1007/978-3-642-31155-0_9
6. Cygan, M., et al.: Lower bounds for kernelization. In: Parameterized Algorithms, pp. 523–555. Springer, Cham (2015). https://doi.org/10.1007/978-3-319-21275-3_15
7. Dom, M., Lokshatanov, D., Saurabh, S.: Kernelization lower bounds through colors and ids. ACM Trans. Algorithms **11**(2), 13:1–13:20 (2014)
8. Fomin, F.V., Lokshtanov, D., Saurabh, S., Thilikos, D.M.: Kernels for (connected) dominating set on graphs with excluded topological minors. ACM Trans. Algorithms **14**(1), 6:1–6:31 (2018)
9. Fomin, F.V., Lokshtanov, D., Saurabh, S., Zehavi, M.: Theory of Parameterized Preprocessing. Cambridge University Press, Kernelization (2019)
10. Fortunato, S.: Community detection in graphs. Phys. Rep. **486**(3–5), 75–174 (2010)
11. Komusiewicz, C.: Tight running time lower bounds for vertex deletion problems. ACM Trans. Comput. Theory **10**(2), 6:1–6:18 (2018)
12. Le, H.-O., Le, V.B.: Complexity of the (connected) cluster vertex deletion problem on h-free graphs. Theor. Comp. Sys. **68**(2), 250–270 (2024)
13. Lokshtanov, D., Panolan, F., Ramanujan, M.S., Saurabh, S.: Lossy kernelization. In: Proceedings of the 49th Annual ACM SIGACT Symposium on Theory of Computing, STOC 2017, pp. 224–237, New York, NY, USA, 2017. Association for Computing Machinery
14. Ma, H., et al.: A survey of computational methods for protein-protein interaction network analysis. J. Biomed. Inform. **42**(3), 460–473 (2009)
15. Misra, N., Philip, G., Raman, V., Saurabh, S., Sikdar, S.: FPT algorithms for connected feedback vertex set. J. Comb. Optim. **24**, 131–146 (2012). https://doi.org/10.1007/s10878-011-9394-2
16. Nederlof, J.: Fast polynomial-space algorithms using Möbius inversion: improving on Steiner tree and related problems. In: Albers, S., Marchetti-Spaccamela, A., Matias, Y., Nikoletseas, S., Thomas, W. (eds.) Automata, Languages and Programming. ICALP 2009. LNCS, vol. 5555, pp. 713–725. Springer, Berlin, Heidelberg (2009). https://doi.org/10.1007/978-3-642-02927-1_59
17. Newman, M.E.J.: Modularity and community structure in networks. Proc. Natl. Acad. Sci. **103**(23), 8577–8582 (2006)
18. Tsur, D., et al.: Faster parameterized algorithms for cluster vertex deletion. Theory Comput. Syst. **51**(3), 409–425 (2012)
19. Zeng, D., et al.: A survey on web page clustering. Int. J. Web Grid Serv. **2**(3), 268–292 (2006)

Tight Bounds for the Number of Absent Subsequences

Duncan Adamson[1], Pamela Fleischmann[2], Annika Huch[2(✉)], Florin Manea[3], Paul Sarnighausen-Cahn[3], and Max Wiedenhöft[2]

[1] University of St Andrews, St Andrews, UK
duncan.adamson@st-andrews.ac.uk
[2] Department of Computer Science, Kiel University, Kiel, Germany
{fpa,ahu,maw}@informatik.uni-kiel.de
[3] Department of Computer Science, University of Göttingen, Göttingen, Germany
{florin.manea,paul.cahn}@cs.uni-goettingen.de

Abstract. A *subsequence* of a word w is a word u that can be obtained by deleting some letters from w while maintaining the relative order of the remaining letters, e.g., `lala` is a subsequence of `alfalfa`. A word, over some alphabet Σ, which has all possible words of length ι over Σ as subsequences is called ι-universal, and the largest ι for which this holds is called the universality index of w, and denoted $\iota(w)$. Moreover, words that are not subsequences of w are called absent subsequences (AS) of w, and their investigation was started in (Kosche et al., 2022). In this paper, we present tight bounds on the number of AS of a given length k among all words with the same universality index ι. For both the lower and upper bound, we construct words that have, respectively, a minimal and maximal number of absent subsequences of the respective length k, and, in the case of the lower bound, we provide the exact number of missing subsequences as a closed form. Finally, we present efficient enumeration algorithms for the set of subsequences of given length of a word: we give a novel, optimal enumeration algorithm with output linear delay of this set of subsequences, with preprocessing time $O(|w|)$, which is further improved to an incremental enumeration algorithm with $O(1)$ delay of this set of subsequences, with preprocessing time $O(|w|)$.

1 Introduction

For a given word w, a *subsequence* (*scattered factor*) of w is a word obtained by deleting letters from w, i.e., formally a word u such that there exist indices $1 \leq i_1 < i_2 < \cdots < i_{|u|} \leq |w|$ with $u = w[i_1]w[i_2]\cdots w[i_{|u|}]$ and $|u|$ (resp. $|w|$) denotes the length of u (resp. w). For instance, `mega`, `meat`, and `gate` are subsequences of $w =$ `pomegranate` while `goat` is not, because its letters do not occur in the correct order in w. The relation between words and their subsequences has been studied in logics [20,33], as well as language and automata theory [1,22,23,42–44]. The set of subsequences of a word reveals information about existing and absent parts of the word, and thus are not only of theoretical but of practical

A. Jeż and J. Otop (Eds.): FCT 2025, LNCS 16106, pp. 15–29, 2026.
https://doi.org/10.1007/978-3-032-04700-7_2

interest, too. When examining discrete data, subsequences are often used to model corrupted data [12,18,36] related to the reconstruction problem. These problems appear in a wide variety of fields, such as formal software verification [20,44]. Subsequences have also been studied in the context of database theory [4,29,30,41] or with motivation coming from this field [3,11,35].

In terms of theoretical work, the study of subsequences is strongly related to the 1972-work of Simon [42], where the famous congruence relation, now known as *Simon's congruence*, regarding piecewise-testable languages is introduced. Two words u and v are called *Simon k-congruent* w.r.t. a natural number k if they have the same set of subsequences of length at most k. For example, aaba and abaa are Simon 2-congruent but not 3-congruent. For a more detailed overview of Simon's congruence, we refer to [34, Chapter 6] by Sakarovich and Simon, as well as the surveys [32,38]. Despite intense scrutiny [14,19] and several attempts toward giving the number of congruence classes, this problem remains open and continues to attract interest [16,17,26–28]. Current developments inspect both the shortest absent subsequence of words [31] and relations to universality index on the other hand [5,10]. This notion of universality serves as a measure of the containment of all subsequences of up to a given length and originates from two sources: Karandikar et al. [22] introduced the notion of k-richness in the context of piecewise testable languages [25,37] which coincides with the notion of k-universality. The language of k-universal words is extensively studied and characterised in [1,5,10,21]. One of the main insights concerning subsequence-universality is that the k-universality of words is tightly related to the *arch factorisation* by Hébrard [21]. One approach followed in [15,31] is to characterise absent subsequences of words in order to determine the index of Simon's congruence, i.e., parametrise the problem by the number of absent subsequences. Additionally, absent subsequences seem to naturally occur in rather practical scenarios such as in the context of reachability and avoidability problems [31]. In [31], the authors gave an implicit characterisation and a tree-based representation of all shortest absent subsequences of a given word.

Own Contribution. We extend the ideas of [15,16,31] to the analysis of absent subsequences, that are longer than the shortest ones. A first generalisation is to ask for the minimal and maximal number of absent subsequences of a word in relation to the number of arches ι and the size of the alphabet. Thus, we ask - given an alphabet Σ, and positive integers ι and k where $\iota < k$ - what is the structure of a word containing exactly ι arches with the greatest (resp. fewest) number of absent subsequences of length k? In this way, we build on the work of [1] in looking at the properties of k-universality within languages. While [1] considered the problem of deciding whether there exists a k-universal word in a given language, we look for the word that is closest (resp. furthest) to k-universality within the restricted language of ι-universal words. We give a tight characterisation for both. First, we describe the structure of words $w \in \Sigma^*$ that have the maximal number of existing subsequences, i.e., the minimal number of absent subsequences, of any given length $k \in \mathbb{N}$ among all words $w' \in \Sigma^*$ with the same number of arches (Sect. 3). Additionally, we characterise the words fulfilling the above property and have the shortest possible length. After that, we exhibit a class of words $w_{\min} \in \Sigma^*$ with a minimal number of existing subsequences, i.e., a maximal number of absent subsequences among all words $w' \in \Sigma^*$ with the same number of arches (Sect. 4). Finally, we address the problem of enumerating the set of subsequences of given length of an input word (Sect. 5). In particular, we give a novel, optimal enumeration algorithm with output linear delay of this set of subsequences with preprocessing time $O(|w|)$. This is then further improved to an incremental enumeration algorithm (where only the differences between two consecutive outputs are succinctly provided) with constant delay of this set of subsequences with preprocessing time $O(|w|)$.

An extended version can be found at arxiv.org/abs/2407.18599.

2 Preliminaries

Let $\mathbb{N}$ be the set of all natural numbers, $\mathbb{N}_0 = \mathbb{N} \cup \{0\}$, $[n] = \{1, \ldots, n\}$, $[i, j] = [j] \backslash [i-1]$, and $[n]_0 := [n] \cup \{0\}$. An *alphabet* Σ is a non empty, finite set whose elements are called *letters* or *symbols*. Given $\sigma \in \mathbb{N}$, we fix $\Sigma = \{\mathtt{a}_1, \ldots, \mathtt{a}_\sigma\}$. A *word* is a finite sequence of letters from Σ. Let Σ^* be the set of finite words over Σ and ε be the empty word. Set $\Sigma^+ := \Sigma^* \setminus \{\varepsilon\}$. Let $w \in \Sigma^*$. For $n \in \mathbb{N}_0$, inductively define $w^0 = \varepsilon$ and $w^n = ww^{n-1}$. The *length* of w is the number of its letters; so $|\varepsilon| = 0$. For all $k \in \mathbb{N}_0$ set $\Sigma^k := \{w \in \Sigma^* \mid |w| = k\}$ (define $\Sigma^{\leq k}$, $\Sigma^{\geq k}$ analogously). We denote w's i^{th} letter by $w[i]$ and $w[i : j] = w[i] \cdots w[j]$ if $i < j$, and ε if $i > j$ for all $i, j \in [|w|]$. Set $\text{alph}(w) = \{\mathtt{a} \in \Sigma \mid \exists i \in [|w|] : w[i] = \mathtt{a}\}$ as w's *alphabet* and for each $\mathtt{a} \in \Sigma$ set $|w|_\mathtt{a} = |\{i \in [|w|] \mid w[i] = \mathtt{a}\}|$. The word $u \in \Sigma^*$ is a *factor* (or infix) of w if there exist $x, y \in \Sigma^*$ such that $w = xuy$. We call u a *prefix* (*suffix*) of w if $x = \varepsilon$ ($y = \varepsilon$). We define the *reverse* of w by $w^R = w[|w|] \cdots w[1]$. Let $<$ be a total order on Σ, and assume that $\mathtt{a}_1 < \mathtt{a}_2 < \ldots < \mathtt{a}_\sigma$ are the letters of Σ; define $w_\Sigma \in \Sigma^\sigma$ as $w_\Sigma = \mathtt{a}_1 \cdots \mathtt{a}_\sigma$. For further definitions see [34].

Definition 1. *Let $w \in \Sigma^*$ and $n \in \mathbb{N}_0$. A word $u \in \Sigma^n$ is called a* subsequence *of w - denoted as $u \in \text{SubSeq}(w)$ - if there exist $v_1, \ldots, v_{n+1} \in \Sigma^*$ such that $w = v_1 u[1] v_2 u[2] \cdots v_n u[n]\ v_{n+1}$. Set $\text{SubSeq}_k(w) = \{u \in \text{SubSeq}(w) \mid |u| = k\}$.*

The words `cafe` and `ufo` are all subsequences of `cauliflower` but neither `flour` nor `row`. Subsequences are tightly related to *Simon's congruence*.

Definition 2. *Two words $w, v \in \Sigma^*$ are* Simon congruent *w.r.t. $k \in \mathbb{N}_0$ (denoted $w \sim_k v$) if $\text{SubSeq}_\ell(w) = \text{SubSeq}_\ell(v)$ for all $\ell \leq k$.*

Since $\text{SubSeq}_k(w) \subseteq \Sigma^k$ holds for all $k \in \mathbb{N}_0$, determining the index of Simon's congruence (i.e., the number of equivalence classes w.r.t. $\sim_k$) can be split into the parametrised problem on determining how many subsequence sets - or equivalently how many different words - exist with $|\text{SubSeq}_k(w)| = \sigma^k - m$ for all $m \in \mathbb{N}_0$ (cf. [5,10] for $m = 0$). We disregard all $w \in \Sigma^{<k}$ because they form singleton classes. For all $w \in \Sigma^{\geq k}$ it suffices to consider the set $\text{SubSeq}_k(w)$ instead of $\text{SubSeq}_\ell(w)$ for all $\ell \leq k$ because all shorter subsequences are contained in a subsequence of length k.

Definition 3. *For given* $k, m \in \mathbb{N}_0$*, a word* $w \in \Sigma^*$ *is called* m-nearly k-universal *if* $|\operatorname{SubSeq}_k(w)| = \sigma^k - m$*. If* $m = 0$*, the words are called* k*-universal and the set of all* k*-universal words over* Σ *is* $\operatorname{Univ}_{\Sigma,k}$*. The largest* $k \in \mathbb{N}_0$*, such that a word is* k*-universal, is called the* universality index $\iota(w)$.

For a unary alphabet $\{a\}$, all words are of the form $\mathtt{a}^\ell$ for $\ell \in \mathbb{N}_0$ and their subsequences can be determined easily. Thus, we only consider here alphabets with two or more letters. Moreover, we assume $\Sigma = \operatorname{alph}(w)$ for a given w, if not stated otherwise. One of the main tools used in the investigation of (m-nearly) k-universal words is the *arch factorisation*, introduced by Hébrard [21].

Definition 4. *For* $w \in \Sigma^*$ *the* arch factorisation *is* $w = \operatorname{ar}_1(w) \cdots \operatorname{ar}_k(w)\, \operatorname{r}(w)$ *for some* $k \in \mathbb{N}_0$ *with*

(a) $\iota(\operatorname{ar}_i(w)) = 1$ *for all* $i \in [k]$*,*
(b) $\operatorname{ar}_i(w)[|\operatorname{ar}_i(w)|] \notin \operatorname{alph}(\operatorname{ar}_i(w)[1 : |\operatorname{ar}_i(w)| - 1])$ *for all* $i \in [k]$*, and*
(c) $\operatorname{alph}(\operatorname{r}(w)) \subsetneq \Sigma$.

The words $\operatorname{ar}_i(w)$ *are called* arches *and* $\operatorname{r}(w)$ *is the* rest *of* w*. The* modus $\operatorname{m}(w)$ *is given by* $\operatorname{ar}_1(w)[|\operatorname{ar}_1(w)|] \cdots \operatorname{ar}_k(w)[|\operatorname{ar}_k(w)|]$*, i.e., it is the word containing the unique last letters of each arch. The* inner part of the ith arch of w*, denoted* $\operatorname{in}_i(w)$*, is defined as* $\operatorname{in}_i(w) = \operatorname{ar}_i(w)[1 : |\operatorname{ar}_i(w)| - 1]$.

To visualise the arch factorisation in explicit examples we use parenthesis, e.g., $(\mathtt{aab}) \cdot (\mathtt{bba}) \cdot \mathtt{a}$ has two arches and the rest $\mathtt{a}$. Here, the modus is $\mathtt{ba}$ and the two inner parts are $\mathtt{aa}$ and $\mathtt{bb}$ resp.

Definition 5. *We call a word* $w \in \Sigma^*$ perfect k-universal *if* $\iota(w) = k$ *and* $\operatorname{r}(w) = \varepsilon$*. The set of all these words with* $\operatorname{alph}(w) = \Sigma$ *is denoted by* $\operatorname{PUniv}_{\Sigma,k}$*. Additionally, we call* w minimal perfect k-universal *if* $w \in \operatorname{PUniv}_{\Sigma,k}$ *and* w *is of minimal length among all words in* $\operatorname{PUniv}_{\Sigma,k}$*. The set of those words over* Σ *is denoted by* $\operatorname{MinPUniv}_{\Sigma,k}$*.*

For the algorithmic results we use the standard computational model RAM with logarithmic word-size (see, e.g., [24]). We also follow a standard assumption from stringology, that if w is the input word for our algorithms, then we assume that w is over an alphabet $\Sigma = \{1, 2, \ldots, \sigma\}$ for some $\sigma \leq |w|$.

3 Lower Bound for the Number of Absent Subsequences

In this section, we consider the following problem: for a specific alphabet Σ, universality index $\iota \in \mathbb{N}$, and number k, what is the minimum number of absent subsequences of length k for any word w with $\iota(w) = \iota$? Our approach is to construct a word that contains as many subsequences as possible (and thus as few absent ones as possible) and compute its number of absent subsequences. Additionally, we are particularly interested in finding the shortest words fulfilling this property. We start by understanding which words are always absent subsequences from $\operatorname{SubSeq}_k(w)$ (for w with the above properties) and show that there actually exist words in which only these words are absent, providing a tight lower bound on the number of absent subsequences.

Proposition 1. *For $k \in \mathbb{N}$, $w \in \Sigma^*$ with $k > \iota(w)$, the next are equivalent:*

1) *For all $w' \in \Sigma^*$ with $\iota(w') = \iota(w)$, we have $|\operatorname{SubSeq}_k(w')| \leq |\operatorname{SubSeq}_k(w)|$ (i.e., w is a word with a maximal number of subsequences of length k, over all words with the same universality index).*
2) *There exists $\mathtt{a} \in \Sigma$ such that $\Sigma \setminus \operatorname{alph}(\operatorname{r}(w)) = \{\mathtt{a}\}$ and $\operatorname{SubSeq}_k(w) = \{v \in \Sigma^k \mid \operatorname{m}(w)\mathtt{a}$ is not a subsequence of $v\}$.*

Intuitively, for all words w that are n-universal for some $n \in \mathbb{N}$, we know that their rest $\operatorname{r}(w)$ always misses at least one letter $\mathtt{a}$. Additionally, we know that its modus $\operatorname{m}(w)$ occurs only once in w and ends right before $\operatorname{r}(w)$. Hence, all words v of length $k > |n| = |\operatorname{m}(w)|$ that have $\operatorname{m}(w)\mathtt{a}$ as a subsequence, can in fact not be a subsequence of w. In fact, in the framework of Proposition 1, we can even construct a word $w_{\max}$ in which only subsequences $v \in \Sigma^k$ with $\operatorname{m}(w_{\max})\mathtt{a} \in \operatorname{SubSeq}(v)$ are absent subsequences of length k, i.e., $v \notin \operatorname{SubSeq}_k(w_{\max})$. Assume $\Sigma = \{\mathtt{a}_1, \ldots, \mathtt{a}_\sigma\}$ with $\mathtt{a}_\sigma = \mathtt{a}$. Consider

$$w_{\max} = ((\mathtt{a}_1 \cdots \mathtt{a}_{\sigma-1})^{k-\iota(w)}\mathtt{a})^{\iota} \cdot (\mathtt{a}_1 \cdots \mathtt{a}_{\sigma-1})^{k-\iota(w)}.$$

By Proposition 1, $w_{\max}$ must have the maximal number of subsequences of length k, i.e., the minimal number of absent subsequences of length k, of all words $w \in \Sigma^*$ with $\iota(w) = \iota(w_{\max}) = n$. In general, assume w to be a word with a maximal number of subsequences of length k over all words with the same universality index $\iota(w)$. By Proposition 1 and the existence of $w_{\max}$, we know that all absent subsequences of w contain $\operatorname{m}(w)\mathtt{a}$ as a subsequence themselves. Hence, we can derive a tight bound on the minimum number of absent subsequences of w.

Proposition 2. *Let $k \in \mathbb{N}_0$ and $w \in \Sigma^*$ with $k > \iota(w)$ such that, for all $w' \in \Sigma^*$ with $\iota(w) = \iota(w')$, we have $|\operatorname{SubSeq}_k(w)| \geq |\operatorname{SubSeq}_k(w')|$. Then, $|\operatorname{SubSeq}_k(w)| = \sum_{j \in [0,\iota(w)]} \binom{k}{j}(\sigma-1)^{k-j}$.*

Corollary 1. *Let $k \in \mathbb{N}_0$, $w \in \Sigma^*$ with $k > \iota(w)$ such that, for all $w' \in \Sigma^*$ with $\iota(w) = \iota(w')$, $|\operatorname{SubSeq}_k(w)| \geq |\operatorname{SubSeq}_k(w')|$ holds. The number of absent subsequences of length k of w is $\sigma^k - \sum_{j \in [0,\iota(w)]} \binom{k}{j}(\sigma-1)^{k-j}$.*

Example 1. Let $\Sigma := \{\mathtt{a}, \mathtt{b}, \mathtt{c}\}$, $k = 4$, and $\iota = 2$. Then $w = (\mathtt{ababc}) \cdot (\mathtt{ababc}) \cdot \mathtt{abab}$ is a word with a minimal number of absent subsequences of length k among all words with universality index ι. Note that exactly those words of length 4 that are absent subsequences of w have $\mathtt{ccc}$ as a subsequence.

Let $\iota, k \in \mathbb{N}$ with $k > \iota$, and consider $w \in \Sigma^*$, with $\iota(w) = \iota$, such that for all $w' \in \Sigma^{\geq |w|}$ with $\iota(w) = \iota(w')$ we have $|\operatorname{SubSeq}_k(w)| \geq |\operatorname{SubSeq}_k(w')|$ and, for all $w' \in \Sigma^{<|w|}$ with $\iota(w) = \iota(w')$ we have $|\operatorname{SubSeq}_k(w)| > |\operatorname{SubSeq}_k(w')|$. Then w is the shortest word over Σ which has $\sum_{j \in [0,\iota(w)]} \binom{k}{j}(\sigma-1)^{k-j}$ (i.e., the maximum number of) subsequences of length k, among all words with universality index ι.

We can now determine the structure of these shortest words. We first show that in such a word w, the ith arch fulfils $\operatorname{in}_i(w) \in \operatorname{Univ}_{(\Sigma \setminus \{\operatorname{m}(w)[i]\}), k-\iota}$, for all $i \in [\iota]$, and also $\operatorname{r}(w) \in \operatorname{Univ}_{(\Sigma \setminus \{\mathtt{a}\}), k-\iota}$, for some $\mathtt{a} \in \Sigma$. From this, and the example-word $w_{\max}$ above, we can first derive their length.

Proposition 3. *For $\iota, k \in \mathbb{N}$ with $k > \iota$, let w be the shortest word over Σ which has a maximum number of subsequences of length k, among all words with universality index ι. Then, we have $|w| = (\iota+1)\cdot(\sigma-1)\cdot(k-\iota)+\iota$.*

Proof. Let $\Sigma^- = \Sigma\backslash\{\mathtt{a}_\sigma\}$ and $w = (w_{\Sigma^-}^{k-\iota}\mathtt{a}_\sigma)^\iota w_{\Sigma^-}^{k-\iota}$ as explained before. $|w| = (\iota+1)\cdot(\sigma-1)\cdot(k-\iota)+\iota$. Suppose there exists some $w' \in \Sigma^*$ with $\iota(w) = \iota(w')$ and $|w'| < |w|$ such that $|\operatorname{SubSeq}_k(w')| \geq |\operatorname{SubSeq}_k(w)|$. Then either $|\operatorname{r}(w')| < |\operatorname{r}(w)|$ or $|\operatorname{in}_i(w')| < |\operatorname{in}_i(w)|$ for some $i \in [\iota]$. By construction outlined before, $\operatorname{r}(w), \operatorname{in}(w) \in \operatorname{MinPUniv}_{\Sigma^-,k-\iota}$. So, either $\operatorname{r}(w') \notin \operatorname{Univ}_{\Sigma^-,k-\iota}$ or $\operatorname{in}_i(w') \notin \operatorname{Univ}_{\Sigma^-,k-\iota}$. Hence, by the above discussion on the form of the structure of every arch and the rest of words with a maximal amount of subsequences we get that $|\operatorname{SubSeq}_k(w')| < |\operatorname{SubSeq}_k(w)|$ holds, which is a contradiction. □

Proposition 3 gives a tight bound on the length of the shortest words with a minimal number of absent subsequences, of fixed universality, allowing us to also identify the precise structure of such words, in Proposition 4. Ultimately, this leads to the main result of this section, namely Theorem 1.

Proposition 4. *For $\iota, k \in \mathbb{N}$ with $k > \iota$, let w be a shortest word over Σ with a maximum number of subsequences of length k, among all words with universality index ι. Assume $\mathtt{a} \notin \operatorname{alph}(r(w))$ is the unique missing letter in the rest. Then, for all $i \in [\iota]$ and the corresponding arch $\operatorname{ar}_i(w)$, we have that its inner part $\operatorname{in}_i(w) \in \operatorname{MinPUniv}_{(\Sigma\backslash\{\operatorname{m}(w)[i]\}),k-\iota}$, and $\operatorname{r}(w) \in \operatorname{MinPUniv}_{(\Sigma\backslash\{\mathtt{a}\}),k-\iota}$.*

Theorem 1. *For $\iota, k \in \mathbb{N}$, $w \in \Sigma^*$ with $k > \iota = \iota(w)$, the next are equivalent:*

1) w is the shortest word over Σ with a maximum number of subsequences of length k, among all words with universality index ι.
2) there exists some $\mathtt{a} \in \Sigma$ such that $\operatorname{m}(w) = \mathtt{a}^\iota$ and, for all $i \in [\iota]$, we have $\operatorname{in}_i(w), \operatorname{r}(w) \in \operatorname{MinPUniv}_{\Sigma',k-\iota}$ for $\Sigma' = \Sigma \setminus \{\mathtt{a}\}$.

Proof. This first direction follows according to Proposition 4. The other direction follows simply by constructing such a word and seeing that we obtain the shortest possible length of words that fulfil all properties. □

Further, we are interested in the total number of shortest words w with a minimal number of absent subsequences of length k over all words that are $\iota(w)$-universal. First, consider the binary case $\Sigma = \{\mathtt{a}, \mathtt{b}\}$. Choose w.l.o.g. $\operatorname{m}(w) = \mathtt{a}^\iota$. Thus, the inner part of each arch and the rest are unary words, i.e., $w = (\mathtt{b}^{k-\iota}\mathtt{a})^\iota \mathtt{b}^{k-\iota}$. Since the choice of $\mathtt{a}$ fixes the remaining letters, for any (k,ι)-combination, we have exactly two shortest words with a minimal number of absent subsequences. For alphabets with σ letters, we have σ choices for the modus. The inner part of each arch and the rest need to be $(k-\iota)$-universal over $\Sigma\backslash\{\mathtt{a}\}$ if $\mathtt{a}$ is the *modus-letter*. This allows us to compute the number of shortest words having a minimal number of absent subsequences, of fixed universality.

Proposition 5. *For $k, \iota \in \mathbb{N}$, there exist $\sigma((\sigma-1)!)^{(\iota+1)(k-\iota)}$ different words $w \in \Sigma^*$ of minimal length $|w| = (\iota+1)\cdot(\sigma-1)\cdot(k-\iota)+\iota$, with universality index ι, each having a maximum number $\sum_{j\in[0,\iota(w)]}\binom{k}{j}(\sigma-1)^{k-j}$ of subsequences of length k.*

Proof. Let $\mathtt{a} \in \Sigma$ with $\mathrm{m}(w) = \mathtt{a}^{\iota}$. We have σ choices for a. For each such choice, we have ι-many inner parts and each inner part is the $(k-\iota)$ power of $w_{\Sigma \setminus \{a\}}$. Thus, we have $\iota(k-\iota)$ times the option to choose a permutation over $\Sigma \setminus \{\mathtt{a}\}$. Since in fact the rest behaves like an arch in the case of the minimality of absent subsequences, we have to count one additional arch. □

This concludes all results regarding the lower bound on the number of absent subsequences among words with a common universality index.

4 Upper Bound for the Number of Absent Subsequences

In this section, we aim to identify words, over an alphabet Σ of size σ, with fixed universality index ι, which have a maximum possible number of absent subsequences (of length k). We do so by constructing a ι-universal word $w_{\min}$ that, in fact, has the minimum number of existing subsequences for all possible lengths $k > \iota$. To this end, note that it is not enough to show, for some universality index ι, that $w_{\min}$ has the fewest subsequences for some fixed length $k > \iota$ to get that this holds for all other lengths $k' > \iota$. One can verify that this does not hold in general. Hence, to show our results, we need to have a more general approach and not focus on one specific value k, but consider all $k \in [\iota(w)+1, |w|]$.

First, we define the function $\mathrm{next}_w(i,s)$ taking, as input, some alphabet size $s \in [\sigma]$ and index $i \in [|\,\mathrm{w}|]$. The function returns the leftmost position j following after i in w such that $w[i+1:j]$ contains s distinct letters or -1 if no such position exists, i.e., if $\mathrm{alph}(w[i+1,|w|]) < s$.

Definition 6. *Let $w \in \Sigma^*$. We define* $\mathrm{next}_w : [|w|] \times [|\Sigma|] \to [|w|] \cup \{-1\}$ *by*

$$\mathrm{next}_w(i,s) = \begin{cases} j, & \text{if } |\,\mathrm{alph}(w[i+1:j])| = s \text{ and } |\,\mathrm{alph}(w[i+1:j-1])| < s, \\ -1, & \text{otherwise.} \end{cases}$$

As an important concept for this section, as well as for Sect. 5, we recall the *greedy embedding* of a subsequence $v \in \Sigma^k$ occurring in a word w, of length n. For $i \in [k]$, we define j_i as the leftmost position of w, such that $w[1:j_i]$ contains the subsequence $v[1:i]$. Then, $w[j_1]\cdots w[j_k] = v$ and the sequence $j_1,\ldots,j_k$ is the *greedy embedding* of v in w, and can be computed in linear time (see, e.g., [31]). Worth noting, this sequence $j_1,\ldots,j_k$ of indices satisfies $j_i \le j_i'$, for any $i \in [1,k]$ and sequence $1 \le j_1' < j_2' < \cdots < j_k' \le |w|$ with $w[j_1']w[j_2']\ldots w[j_k'] = v$.

Lemma 1. *For $w \in \Sigma^*$, $k \in [|w|]$, and $u \in \mathrm{SubSeq}_k(w)$, let $j_1 < \cdots < j_k$ be the greedy embedding of u in w. Then* $\mathrm{next}_w(j_i, |\,\mathrm{alph}(w[j_i+1, j_{i+1}])|) = j_{i+1}$.

Proof. Let $u \in \mathrm{SubSeq}_k(w)$ and assume $e = (j_1, ..., j_k)$ to be the greedy embedding of u in w. Let $i \in [k-1]$. Assume $s = |\,\mathrm{alph}(w[j_i : j_{i+1}])|$. We know that $w[j_{i+1}] \notin \mathrm{alph}(w[j_i : j_{i+1}-1])$, as otherwise e would not be a greedy embedding. Hence, $|\,\mathrm{alph}(w[j_i : j_{i+1}-1])| < s$. By the definition of next_w, we obtain $\mathrm{next}_w(j_i, |\,\mathrm{alph}(w[j_i : j_{i+1}])|) = j_{i+1}$.

Using the properties of greedy embeddings, we give the specific construction of words with the minimum number of subsequences, for a given universality index and subsequence length. In particular, we show that all words $w \in \Sigma^*$ that satisfy the structure $\mathrm{ar}_1(w) = \mathtt{a}_1 \cdots \mathtt{a}_\sigma$ and $\mathrm{ar}_i(w) = \mathrm{ar}_{i-1}(w)^R$ have the minimum number of subsequences amongst all words with universality index $\iota(w)$ over that alphabet $\Sigma = \{\mathtt{a}_1, \mathtt{a}_2, \ldots, \mathtt{a}_\sigma\}$. We assume, for the remainder of this section, that $w_{\min}$ has this structure, with $\mathrm{ar}_1(w_{\min}) = \mathtt{a}_1 \cdots \mathtt{a}_\sigma$, noting that the same construction can be applied for any permutation of Σ. Kosche et al. in [31] show that $w_{\min}$ has the fewest subsequences of length $\iota(w_{\min}) + 1$ amongst all words over $\mathrm{alph}(w_{\min})$ with universality index $\iota(w_{\min})$. We generalise this by showing that $w_{\min}$ has for all k, the fewest subsequences of length k, amongst the set of words $\{v \in \Sigma^* | \iota(v) = \iota(w_{\min})\}$, noting that for any $k \leq \iota(w_{\min})$, all words in this set contain every word in Σ^k as a subsequence.

Lemma 2. *Let $w \in \Sigma^*$. If $|w| > \iota(w) \cdot \sigma$, then there exists a word $w' \in \Sigma^*$ with $\iota(w') = \iota(w)$ and $|w'| = \iota(w) \cdot \sigma$ such that, for all $k \in [|w|]$, we have $|\operatorname{SubSeq}_k(w')| \leq |\operatorname{SubSeq}_k(w)|$.*

Proof. Let w satisfy $|w| > \iota(w) \cdot \sigma$. Now, there exists some index $i \in [|w|]$ such that $\iota(w[1 : i-1]w[i+1 : |w|]) = \iota(w)$. Further, given $u \in \operatorname{SubSeq}(w[1 : i-1]w[i+1 : |w|])$, $u \in \operatorname{SubSeq}(w)$ by construction. Thus $\operatorname{SubSeq}(w[1 : i-1]w[i+1 : |w|]) \subseteq \operatorname{SubSeq}(w)$, and by extension $|\operatorname{SubSeq}_k(w[1 : i-1]w[i+1 : |w|])| \leq |\operatorname{SubSeq}_k(w)|$ for all $k \in [|w|]$, completing the proof. □

Lemma 2 allows us to only consider words $w \in \Sigma^*$ with $|w| = \iota(w) \cdot \sigma$, which have the property that every arch is a permutation of the alphabet Σ. The next two lemmata give upper bounds on possible values of next_w when $|w| = \iota(w) \cdot \sigma$. Further, we show that the word $w_{\min}$ matches these bounds.

Lemma 3. *Let $w \in \Sigma^{\iota(w)\sigma}$. Then, for all $s \in [\sigma]$ and $j \in [|w|]$, $\mathrm{next}_w(j, s) \leq \lfloor \frac{j}{\sigma} \rfloor \cdot \sigma + \sigma + s$.*

Lemma 4. *Let $w \in \Sigma^{\iota(w)\sigma}$. For $s \in [\sigma]$ and $j \in [|w|]$, if either $j \bmod \sigma = 0$ or $j \bmod \sigma \neq 0$ and $s \leq \sigma - (j \bmod \sigma)$, then $\mathrm{next}_w(j, s) = j + s$ or $\mathrm{next}_w(j, s) = -1$.*

Lemma 5. *For the word $w_{\min}$, $s \in [\sigma]$, $j \in [|w|]$, and $r = j \bmod \sigma$, we have*

$$\mathrm{next}_{w_{\min}}(j, s) = \begin{cases} j + s, & \text{if } r = 0 \text{ and } |w| - j \geq \sigma, \text{ or } s \leq \sigma - r, \\ \lfloor \frac{j}{\sigma} \rfloor \cdot \sigma + \sigma + s, & \text{if } r \neq 0, \lfloor \frac{j}{\sigma} \rfloor + 1 < \iota(w_{\min}), \text{ and } s > \sigma - r, \\ -1, & \text{otherwise.} \end{cases}$$

We can now construct an injective mapping of subsequences from words structured like $w_{\min}$ to subsequences of any other word $w' \in \Sigma^{\iota(w_{min})\sigma}$ with the same universality index and the same size. The function is defined by mapping a subsequence $u \in \operatorname{SubSeq}(w_{min})$, with the greedy embedding $(j_1, ..., j_k)$, to the subsequence $v \in \operatorname{SubSeq}(w')$, with the greedy embedding $(j'_1, ..., j'_k)$, that can be obtained by continued application of the next function using the same sequence of numbers $(s_1, ..., s_n) \in [\sigma]^k$, starting in both cases in position 0. In particular, if we have $\text{next}_{w_{min}}(0, s_1) = j_1$, $\text{next}_{w_{min}}(j_i, s_{i+1}) = j_{i+1}$, then we define $(j'_1, ..., j'_k)$ such that $\text{next}_{w'}(0, s_1) = j'_1$, and $\text{next}_{w'}(j'_i, s_{i+1}) = j'_{i+1}$, for $i \in [k-1]$. Following the previous results, this mapping is total and injective.

Proposition 6. *For $w_{\min}$ and each $w' \in \Sigma^*$ with $\iota(w_{\min}) = \iota(w')$ and $|w_{\min}| = |w'|$, there exists an injective function $f : \text{SubSeq}_k(w_{\min}) \to \text{SubSeq}_k(w')$.*

Proof. Consider $w_{\min}$ as defined before and let $w' \in \Sigma^*$ with $\iota(w_{\min}) = \iota(w')$ and $|w_{\min}| = |w'|$. Let $k \in [|w_{\min}|]$. Let $f : \text{SubSeq}_k(w_{\min}) \to \text{SubSeq}_k(w')$ and define $f(u) = v$ if there exist $j_1, \ldots, j_k \in [|w_{\min}|]$, $j'_1, \ldots, j'_k \in [|w'|]$, and $s_1, \ldots, s_{k-1} \in [\sigma]$ such that $u = w_{\min}[j_1] \cdots w_{\min}[j_k]$, $v = w'[j'_1] \cdots w'[j'_k]$, $j_1 = j'_1$, $|w_{\min}[1 : j_1 - 1]|_{w_{\min}[j_1]} = 0$, $\text{next}_w(j_i, s_i) = j_{i+1}$, and $\text{next}_w(j'_i, s_i) = j'_{i+1}$. We now show that f is an injective function. Let $u \in \text{SubSeq}_k(w_{\min})$. By Lemma 1 we know that there exist $j_1, \ldots, j_k \in [|w_{\min}|]$ and $s_1, \ldots, s_{k-1} \in [\sigma]$ such that $|w_{\min}[1 : j_1 - 1]|_{w_{\min}[j_1]} = 0$ and $\text{next}_{w_{\min}}(j_i, s_i) = j_{i+1}$ for all $i \in [k-1]$. By Lemmas 3, 4 and 5 we know that for all $i \in [k-1]$, $\text{next}_{w_{\min}}(j_i, s_i)$ witnesses its upper bound over all words $w' \in \Sigma^{|w_{\min}|}$ with $\iota(w_{\min}) = \iota(w')$. Therefore, $\text{next}_{w_{\min}}(j_i, s_i) \geq \text{next}_{w'}(j'_i, s_i)$, for every $i \in [k]$, and thus the word constructed by f must be a subsequence of w'. To show that $f(u_1) \neq f(u_2)$, for any pair $u_1, u_2 \in \text{SubSeq}_k(w_{\min})$ where $u_1 \neq u_2$, let i be the index such that $u_1[1 : i-1] = u_2[1 : i-1]$, and $u_1[i] \neq u_2[i]$. Further, let $j_1, \ldots, j_k$ be the leftmost set of indices such that $u_1 = w_{\min}[j_1] \ldots w_{\min}[j_k]$, and $\ell_1, \ldots, \ell_k$ be the leftmost set of indices such that $u_2 = w_{\min}[\ell_1] \ldots w_{\min}[\ell_k]$. First, as $u_1[1 : i-1] = u_2[1 : i-1]$, we have by definition $j_1, \ldots, j_{i-1} = \ell_1, \ldots, \ell_{i-1}$. Further, as $u_1[i] \neq u_2[i]$, $j_i \neq \ell_i$, and thus $|\text{alph}(w_{\min}[j_{i-1} + 1 : j_i])| \neq |\text{alph}(w_{\min}[\ell_{i-1} + 1 : \ell_i])|$. Now, let $j'_1, \ldots, j'_k$ be the leftmost set of indices embedding $f(u_1)$ into w', and let $\ell'_1, \ldots, \ell'_k$ be the leftmost set of indices embedding $f(u_2)$ into w'. Again, we note that $j'_1, \ldots, j'_{i-1} = \ell'_1, \ldots, \ell'_{i-1}$. Now, we have $j'_i = \text{next}(j'_{i-1}, |\text{alph}(w_{\min}[j_{i-1} + 1 : j_i])|)$ and $\ell'_i = \text{next}_{w_{\min}}(j'_{i-1}, |\text{alph}(w_{\min}[\ell_{i-1} + 1 : \ell_i])|)$. As $|\text{alph}(w_{\min}[j_{i-1} + 1 : j_i])| \neq |\text{alph}(w_{\min}[\ell_{i-1} + 1 : \ell_i])|$, $w'[j'_i] \neq w'[\ell'_i]$, and thus $f(u_1)[i] \neq f(u_2)[i]$. Due to the shape of w', we have $f(u_1) \neq f(u_2)$. □

Using this construction, we can conclude the main result of this section.

Theorem 2. *Let $\Sigma := \{\mathtt{a}_1, \ldots, \mathtt{a}_\sigma\}$ for some $\sigma \in \mathbb{N}$. Let $w \in \Sigma^*$ such that w.l.o.g. $\text{ar}_1(w) = \mathtt{a}_1 \cdots \mathtt{a}_\sigma$ and $\text{ar}_i(w) = \text{ar}_{i-1}(w)^R$ for all $i \in [2, \iota(w)]$. Then, for all $w' \in \Sigma^*$ with $\iota(w) = \iota(w')$, we have $|\text{SubSeq}_k(w)| \leq |\text{SubSeq}_k(w')|$.*

Proof. Let w be given as in the statement and let $k \in [|w|]$. Let $w' \in \Sigma^*$ be any other word with $\iota(w) = \iota(w')$. By construction of w', $|w'| \geq \iota(w)\sigma$. If $|w'| > \iota(w') \cdot \sigma$ holds, we know by Lemma 2 there exists a word $w'' \in \Sigma^*$ with $\iota(w'') = \iota(w')$, $|w''| = \iota(w) \cdot \sigma$, and $|\text{SubSeq}_k(w'')| \leq |\text{SubSeq}_k(w')|$. Now, by Proposition 6, there exists an injective function $f : \text{SubSeq}_k(w) \to \text{SubSeq}_k(w'')$. By that, we know that $|\text{SubSeq}_k(w)| \leq |\text{SubSeq}_k(w'')|$ and as $|\text{SubSeq}_k(w'')| \leq |\text{SubSeq}_k(w')|$, we also have $|\text{SubSeq}_k(w)| \leq |\text{SubSeq}_k(w')|$. □

Note that the previous results do not exclude the existence of ι-universal words $w \in \Sigma^{\iota\sigma}$ that do not follow the structure of w_{min} but have a minimal number of existent subsequences over all $k \in [\iota(w)+1, |w|-1]$.

5 Enumeration Algorithms

So far, we focused on the combinatorial properties of the set of subsequences of length k of a word w (and their implications w.r.t. the set of missing subsequences). In this section, we consider the algorithmic problem of enumerating the elements of this set. In particular, we propose optimal enumeration algorithms for the set of subsequences of length k of a given word. Enumeration algorithms with output-linear delay for this set of subsequences can be obtained (as explained below) via subsequence-automata [9], while counting the length-k subsequences of w can be done in $O(k|w|)$ time by a dynamic programming algorithm [13] (which can also be extended to an enumeration algorithm). However, such enumeration algorithms are not optimal, and our results improve them significantly. We are not aware of any other results related to the efficient (incremental) enumeration of the subsequences of a word; recent related works approach, based on totally different ideas, the problems of designing efficient algorithms for the enumeration of the longest common subsequences of two strings (see [39,40] and the references therein) and for the enumeration of the maximal common subsequences of two strings (see [7,8] and the references therein).

Our results are based on two key ideas. The first is related to the notion of *greedy embedding* of a subsequence v, of length k, occurring in a word w, of length n, introduced in Sect. 4 in connection to Lemma 1. In the context of this section, we note that two distinct greedy embeddings correspond to two distinct subsequences, and every subsequence has a greedy embedding, we conclude that it is enough to enumerate the sets of greedy embeddings of subsequences of length k.

Secondly, we recall that one can construct in time $O(n\sigma)$ a deterministic finite automaton $A = (Q, \Sigma, q_0, F, \delta)$ with $L(A) = \{v | v \text{ is a subsequence of } w\}$; see, e.g., [9]. In this automaton, we have the states $Q = \{0, \dots, n, n+1\}$, $q_0 = 0$, $F = Q \setminus \{0, n+1\}$. Moreover, we define the transition function $\delta : Q \times \Sigma \to Q$ as follows. For $i \in [n]$ and $a \in \Sigma$, we set $\delta(i, a) = \min(\{n+1\} \cup \{j \mid i < j \leq n, w[j] = a\})$; further, $\delta(n+1, a) = n+1$ for all $a \in \Sigma$ (i.e., $n+1$ is an error-state). Importantly, for a subsequence v, of length m, occurring in w, the automaton A will accept v while going through the sequence of states $0, i_1, \dots, i_m$, where $i_1, \dots, i_m$ is the greedy embedding of v in w.

Thus, we may enumerate all subsequences of length k of w (more precisely, their greedy embeddings) after an $O(n\sigma)$ time preprocessing, with $O(k)$ delay, by way of a depth-first traversal over the length k paths in the graph corresponding to the automaton A. This can be improved, as long as we are interested in an incremental enumeration: we simply produce the first output (the subsequence $w[1:k]$, represented, e.g., as pair (i, k)), and then, in subsequent steps of our enumeration, we output a short sketch of the edits needed to be performed to obtain the current subsequence from the previous one. In particular, if these edits are "remove suffix of length ℓ, add $w[a+1 : a+\ell]$ as a new suffix to the current subsequence" (where in the last part, $w[a : b]$ is simply represented as the pair of positions a, b of w), then we can obtain, using the techniques from [2], an enumeration algorithm with $O(n\sigma)$ time preprocessing and $O(1)$ delay. This algorithm allows, at any moment, to output explicitly the currently enumerated subsequence.

We improve these results, and show that, both in the case of an explicit enumeration algorithm and in that of an incremental enumeration algorithm, the preprocessing time can be reduced to $O(n)$ and, hence, we obtain optimal explicit and, respectively, incremental enumeration algorithms.

The main obstacle in achieving $O(n)$ preprocessing time is the explicit construction of the DFA A, of size $O(n\sigma)$. However, when working with A, we only needed the states Q, and for each $i \in Q \backslash \{n, n+1\}$ the set of its relevant successor-states, $L_i = \{j \neq n+1 \mid \exists a \in \Sigma, \delta[i, a] = j\}$. The key observation is that we can identify the elements of L_i, for $i < n$, on the fly, and we do not need to precompute these sets. However, we need some additional data structures: the array samePrev[$\cdot$], of size n, where samePrev[i] $= \max\{j \mid w[j] = w[i], j < i\}$, and samePrev[$i$] $= 0$ if $w[i]$ does not occur in $w[1 : i-1]$; and range minimum query data structures for samePrev[$\cdot$] (denoted RMQ_{samePrev}). Now, back to the computation of L_i, we always have $i{+}1 \in L_i$ (for $i < n$). All the remaining elements of L_i, if any, are in the range $[i+2 : n]$; so we define a list I_i containing only this range. Then we iterate the following process (which simulates the traversal of the list L_i, in the case when it was precomputed), but on the fly, each time we need to identify a new state-successor of i. Until I_i becomes empty, we extract the range $[a : b]$ from I_i and set $j = RMQ_{\text{samePrev}}[a, b]$. If samePrev[$j$] $\leq i$, then $j \in L_i$ holds, and further elements of L_i can be potentially found in the ranges $[a, j-1]$ and $[j+1 : b]$, so we add them to I_i. If samePrev[j] $> i$, then there is no element of L_i in the range $[a : b]$, as all samePrev[ℓ] $> i$ for all $\ell \in [a : b]$. This process allows us to identify the successors of a state in A in $O(1)$ time per successor, with $O(n)$ preprocessing, and it immediately leads to an implementation of the previous enumeration algorithm with $O(n)$ preprocessing time and $O(k)$ delay. This is optimal when we want the outputs explicitly.

All the algorithms based on the exploration of A maintain a stack which contains the positions $i_1, \ldots, i_k$ of the greedy embedding of the currently enumerated subsequence. So, to get to $O(1)$ delay (for incremental enumeration, as described above), the novel idea is to maintain the greedy embedding of the currently enumerated subsequence in a more succinct way, which allows us to identify quicker the next subsequence to be enumerated, and still do the enumeration in the same order as in the previous algorithms. This succinct representation is actually relatively simple: as contiguous factors of the currently enumerated subsequence v can also be embedded, in the greedy embedding of v in w, into contiguous factors of w, we will not maintain this embedding of v as a sequence of separate positions of w, but as a sequence of maximal contiguous factors of w (that is, we represent compactly, as a range $[i : j]$, the infixes of the greedy embedding which correspond to the factor $w[i:j]$). This representation can be implemented in a way that allows us to identify in $O(1)$ the longest prefix (i.e., the rightmost position i_j) of the greedy embedding $i_1, \ldots, i_k$ of v in w such that the sequence of positions $i_1, \ldots, i_j$ can be extended to obtain a subsequence not enumerated before. A key idea here is that this extension can also be done in $O(1)$: we perform this extension with a sequence $i'_{j+1}, i'_{j+1}+1, \ldots, i'_{j+1}+k-1$ (represented succinctly as the range $[i'_{j+1} : i'_{j+1}+k-1]$, corresponding to the factor $w[i'_{j+1} : i'_{j+1}+k-1]$ of w). We choose i'_{j+1} exactly in the same way as in the RMQ-based algorithm, overviewed above, which will guarantee that we did not enumerate any subsequence whose greedy embedding starts with $i_1, \ldots, i_j, i'_{j+1}$ until now. Moreover, our algorithm enumerates then all subsequences whose greedy embedding starts with $i_1, \ldots, i_j, i'_{j+1}$, and then moves on and finds a new position i''_{j+1} and enumerate all subsequences whose greedy embedding starts with $i_1, \ldots, i_j, i''_{j+1}$. The result is an enumeration algorithm for the set of subsequences of length k of w, with $O(n)$ preprocessing time and $O(1)$ delay. The currently enumerated subsequence can be also output, in $O(k)$ time.

6 Conclusion

In this work, we gained a better understanding of the set of length k subsequences of a given word. Based on the notions of m-nearly k-universality and absent subsequences, we provided the minimal and maximal number of absent subsequences for a given alphabet Σ, subsequence length k and universality index ι. During the investigation we gave two tight bounds for m as well as constructions on ι-universal words over Σ that omit this number of absent subsequences of length k. With these results we obtained a further restriction on the index of Simon's Congruence. While absent subsequences were shown to be instrumental in gaining a better understanding in the set of subsequences of given length of a word from a combinatorial perspective (both here and in works such as [15,31]), it also makes sense to ask whether we can effectively produce the respective set of present subsequences, by an efficient algorithm. We prove that this set can be enumerated optimally, i.e., we give an algorithm with linear preprocessing and linear output delay, or with constant delay in the case, when we only output some initial subsequence and short sketches showing how the current subsequence can be produced from the previous one. While minimal and maximal bounds for the number of absent subsequences m are shown, for given universality ι and subsequence length k, it remains open to determine all possible values of m, the structure and count of corresponding words, and whether efficient algorithms for obtaining them exist.

References

1. Adamson, D., Fleischmann, P., Huch, A., Koß, T., Manea, F., Nowotka, D.: k-universality of regular languages. In: ISAAC. LIPIcs, vol. 283, pp. 4:1–4:21 (2023)
2. Adamson, D., Gawrychowski, P., Manea, F.: Enumerating m-length walks in directed graphs with constant delay. In: Soto, J.A., Wiese, A. (eds.) LATIN 2024: Theoretical Informatics. LATIN 2024. LNCS, vol. 14578, pp. 35–50. Springer, Cham (2024). https://doi.org/10.1007/978-3-031-55598-5_3
3. Adamson, D., Kosche, M., Koß, T., Manea, F., Siemer, S.: Longest common subsequence with gap constraints. In: WORDS, pp. 60–76 (2023)
4. Artikis, A., Margara, A., Ugarte, M., Vansummeren, S., Weidlich, M.: Complex event recognition languages: tutorial. In: DEBS, pp. 7–10 (2017)
5. Barker, L., Fleischmann, P., Harwardt, K., Manea, F., Nowotka, D.: Scattered factor-universality of words. In: Jonoska, N., Savchuk, D. (eds.) DLT 2020. LNCS, vol. 12086, pp. 14–28. Springer, Cham (2020). https://doi.org/10.1007/978-3-030-48516-0_2
6. Bender, M.A., Farach-Colton, M.: The level ancestor problem simplified. In: Rajsbaum, S. (ed.) LATIN 2002. LNCS, vol. 2286, pp. 508–515. Springer, Heidelberg (2002). https://doi.org/10.1007/3-540-45995-2_44
7. Buzzega, G., Conte, A., Kobayashi, Y., Kurita, K., Punzi, G.: The complexity of maximal common subsequence enumeration. Proc. ACM Manag. Data **3**(2) (2025)
8. Conte, A., Grossi, R., Punzi, G., Uno, T.: Enumeration of maximal common subsequences between two strings. Algorithmica **84**(3), 757–783 (2022)
9. Crochemore, M., Melichar, B., Tronícek, Z.: Directed acyclic subsequence graph - overview. J. Discret. Algorithms **1**(3–4), 255–280 (2003)
10. Day, J.D., Fleischmann, P., Kosche, M., Koß, T., Manea, F., Siemer, S.: The edit distance to k-subsequence universality. In: STACS. LIPIcs, vol. 187, pp. 25:1–25:19 (2021)
11. Day, J.D., Kosche, M., Manea, F., Schmid, M.L.: Subsequences with gap constraints: complexity bounds for matching and analysis problems. In: ISAAC. LIPIcs, vol. 248, pp. 64:1–64:18 (2022)
12. Dress, A.W.M., Erdős, P.L.: Reconstructing words from subwords in linear time. Ann. Comb. **8**, 457–462 (2005)
13. Elzinga, C., Rahmann, S., Wang, H.: Algorithms for subsequence combinatorics. TCS **409**(3), 394–404 (2008)
14. Fleischer, L., Kufleitner, M.: Testing Simon's congruence. In: MFCS. LIPIcs, vol. 117, pp. 62:1–62:13 (2018)
15. Fleischmann, P., Haschke, L., Höfer, J., Huch, A., Mayrock, A., Nowotka, D.: Nearly k-universal words - investigating a part of Simon's congruence. TCS **974**, 114113 (2023)
16. Fleischmann, P., Höfer, J., Huch, A., Nowotka, D.: α-β-factorization and the binary case of Simon's congruence. In: Fernau, H., Jansen, K. (eds.) Fundamentals of Computation Theory. FCT 2023. LNCS, vol. 14292, pp. 190–204. Springer, Cham (2023). https://doi.org/10.1007/978-3-031-43587-4_14
17. Fleischmann, P., et al.: Matching patterns with variables under Simon's congruence. In: Bournez, O., Formenti, E., Potapov, I. (eds.) Reachability Problems. RP 2023. LNCS, vol. 14235, pp. 155–170. Springer, Cham (2023). https://doi.org/10.1007/978-3-031-45286-4_12
18. Fleischmann, P., Lejeune, M., Manea, F., Nowotka, D., Rigo, M.: Reconstructing words from right-bounded-block words. Int. J. Found. Comput. Sci. **32**(6), 619–640 (2021)

19. Gawrychowski, P., Kosche, M., Koß, T., Manea, F., Siemer, S.: Efficiently testing Simon's congruence. In: STACS. LIPIcs, vol. 187, pp. 34:1–34:18 (2021)
20. Halfon, S., Schnoebelen, P., Zetzsche, G.: Decidability, complexity, and expressiveness of first-order logic over the subword ordering. In: LICS, pp. 1–12. IEEE Computer Society (2017)
21. Hébrard, J.: An algorithm for distinguishing efficiently bit-strings by their subsequences. TCS **82**(1), 35–49 (1991)
22. Karandikar, P., Kufleitner, M., Schnoebelen, P.: On the index of Simon's congruence for piecewise testability. Inf. Process. Lett. **115**(4), 515–519 (2015)
23. Karandikar, P., Schnoebelen, P.: The height of piecewise-testable languages and the complexity of the logic of subwords. Log. Methods Comput. Sci. **15**(2) (2019)
24. Kärkkäinen, J., Sanders, P., Burkhardt, S.: Linear work suffix array construction. J. ACM **53**(6), 918–936 (2006)
25. Kátai-Urbán, K., Pach, P.P., Pluhár, G., Pongrácz, A., Szabó, C.: On the word problem for syntactic monoids of piecewise testable languages. In: Semigroup Forum, vol. 84, pp. 323–332. Springer (2012)
26. Kim, S., Han, Y., Ko, S., Salomaa, K.: On Simon's congruence closure of a string. TCS **972**, 114078 (2023)
27. Kim, S., Han, Y.S., Ko, S.K., Salomaa, K.: On the Simon's congruence neighborhood of languages. In: Drewes, F., Volkov, M. (eds.) Developments in Language Theory. DLT 2023. LNCS, vol. 13911, pp. 168–181. Springer, Cham (2023). https://doi.org/10.1007/978-3-031-33264-7_14
28. Kim, S., Ko, S., Han, Y.: Simon's congruence pattern matching. TCS **994**, 114478 (2024)
29. Kleest-Meißner, S., Sattler, R., Schmid, M.L., Schweikardt, N., Weidlich, M.: Discovering event queries from traces: laying foundations for subsequence-queries with wildcards and gap-size constraints. In: ICDT. LIPIcs, vol. 220, pp. 18:1–18:21 (2022)
30. Kleest-Meißner, S., Sattler, R., Schmid, M.L., Schweikardt, N., Weidlich, M.: Discovering multi-dimensional subsequence queries from traces - from theory to practice. In: BTW. LNI, vol. P-331, pp. 511–533 (2023)
31. Kosche, M., Koß, T., Manea, F., Siemer, S.: Absent subsequences in words. Fundam. Inform. **189**(3–4), 199–240 (2022)
32. Kosche, M., Koß, T., Manea, F., Siemer, S.: Combinatorial algorithms for subsequence matching: a survey. In: NCMA, vol. 367, pp. 11–27 (2022)
33. Kuske, D., Zetzsche, G.: Languages ordered by the subword order. In: Bojańczyk, M., Simpson, A. (eds.) Foundations of Software Science and Computation Structures. FoSSaCS 2019. LNCS, vol. 11425, pp. 348–364. Springer, Cham (2019). https://doi.org/10.1007/978-3-030-17127-8_20
34. Lothaire, M.: Combinatorics on Words. Cambridge University Press, Cambridge Mathematical Library (1997)
35. Manea, F., Richardsen, J., Schmid, M.L.: Subsequences with generalised gap constraints: upper and lower complexity bounds. In: CPM. LIPIcs, vol. 296, pp. 22:1–22:17 (2024)
36. Manuch, J.: Characterization of a word by its subwords. In: DLT 1999, pp. 210–219. World Scientific (1999)
37. Pach, P.P.: Normal forms under Simon's congruence. In: Semigroup Forum, vol. 97, pp. 251–267. Springer (2018)
38. Pin, J.E.: The influence of Imre Simon's work in the theory of automata, languages and semigroups. In: Semigroup Forum, vol. 98, pp. 1–8. Springer (2019)

39. Sakai, Y.: Linear-space LCS enumeration for two strings. In: Bonizzoni, P., Mäkinen, V. (eds.) CPM. LIPIcs, vol. 331, pp. 2:1–2:14 (2025)
40. Sakai, Y.: Linear-space LCS enumeration with quadratic-time delay for two strings. CoRR **abs/2504.05742** (2025)
41. Sattler, R., Kleest-Meißner, S., Lange, S., Schmid, M.L., Schweikardt, N., Weidlich, M.: Disces: systematic discovery of event stream queries. Proc. ACM Manag. Data **3**(1) (2025)
42. Simon, I.: Hierarchies of events with dot-depth one. Ph.D. thesis, University of Waterloo (1972)
43. Simon, I.: Piecewise testable events. In: Brakhage, H. (ed.) GI-Fachtagung 1975. LNCS, vol. 33, pp. 214–222. Springer, Heidelberg (1975). https://doi.org/10.1007/3-540-07407-4_23
44. Zetzsche, G.: The complexity of downward closure comparisons. In: ICALP. LIPIcs, vol. 55, pp. 123:1–123:14 (2016)

On the Exact & Approximate Complexity of the Strongly Connected Steiner Subgraph Problem on Two Terminals with Demands

Kevin Kurien Alex(✉), Rajesh Chitnis, and Alex Tempest

School of Computer Science, University of Birmingham, Birmingham, UK
{kka362,axt826}@student.bham.ac.uk,rajeshchitnis@gmail.com

Abstract. The Strongly Connected Steiner Subgraph (SCSS) problem only demands strong connectivity between a set of terminals: nothing is specified about how many (disjoint) paths are required between each pair of terminals. Chitnis et al. [IPEC '14, Algorithmica '17] introduced a variant of SCSS (on two terminals) with demands for the number of paths between the two terminal pairs. Formally, the 2-SCSS-(k_1, k_2) problem is defined as follows: given an edge-weighted directed graph $G = (V, E)$ on n vertices with weight function $\omega : E \to \mathbb{R}^{\geq 0}$, two terminal vertices s, t, and integers k_1, k_2, p ; the question is whether there exists a set of k_1 paths $F_1, F_2, \ldots, F_{k_1}$ from $s \rightsquigarrow t$ and k_2 paths $B_1, B_2, \ldots, B_{k_2}$ from $t \rightsquigarrow s$ such that $\sum_{e \in E} \omega(e) \cdot \phi(e) \leq p$, where $\phi(e) = \max\Big\{|\{i \in [k_1] : e \in F_i\}| \ , \ |\{j \in [k_2] : e \in B_j\}|\Big\}$.

For each $k \geq 1$, Chitnis et al. [IPEC '14, Algorithmica '17] designed a $n^{O(k)}$ time algorithm for graphs with n vertices, and obtained a matching lower bound of $f(k) \cdot n^{o(k)}$ under the Exponential Time Hypothesis (ETH) for any computable function f. In this paper, we address the next natural questions for the 2-SCSS-$(k, 1)$ problem:

- `(Strong restriction)` **Finding exact solutions in planar graphs**: The 2-SCSS-$(k, 1)$ problem is W[1]-hard even on planar graphs, and moreover has no $f(k) \cdot n^{o(k)}$ algorithm under ETH.
- `(Weak restriction)` **Finding solutions which approximate planar solutions**: We consider an intermediate version called 2-SCSS-$(k, 1)_{\text{PLANAR}}$: here the graph is general, but we want to find a solution whose cost is at most p if there is a planar solution of weight at most p. We show that the 2-SCSS-$(k, 1)_{\text{PLANAR}}$ problem, and hence also 2-SCSS-$(k, 1)$ is $W[1]$-hard parameterized by $k + p$. Our reduction also implies lower bounds for parameterized approximation of the 2-SCSS-$(k, 1)_{\text{PLANAR}}$ problem:
  - For any computable function f, there is no $(1+\epsilon)$-approximation in $f(\epsilon, k, p) \cdot n^{O(1)}$ time unless FPT$= W[1]$.
  - For any computable function g, there is no $(1+\epsilon)$-approximation in $g(\epsilon, k, p) \cdot n^{o(k+\sqrt{p+1/\epsilon})}$ time under ETH.

A. Tempest—Work done as a student at the University of Birmingham.

A. Jeż and J. Otop (Eds.): FCT 2025, LNCS 16106, pp. 30–44, 2026.
https://doi.org/10.1007/978-3-032-04700-7_3

– (No restriction) **Finding approximate solutions in polynomial time on general graphs**: We design a polynomial time $(k+1)$-approximation algorithm for the 2-SCSS-$(k, 1)$ problem. This is complemented by a lower bound showing that our analysis is almost-tight: there are graphs where our algorithm does not produce a solution having cost smaller than k times that of an optimal solution.

As a consequence of our results, we also obtain W[1]-hardness of the 2-SCSS-$(k, 1)$ problem on general graphs parameterized by the cost p of the solution: this does not follow from the W[1]-hardness proof of Chitnis et al. [IPEC '14, Algorithmica '17] since the value of p in their reduction was a function of both k and n.

Keywords: Strongly connected steiner subgraph · Parameterized complexity · Planar graphs

1 Introduction

The Steiner Tree (ST) Problem is one of the most fundamental problems in the field of graph connectivity: given an undirected graph $G = (V, E)$ with edge weights given by $\omega : E \to \mathbb{R}^{\geq 0}$ and a set $T \subseteq V$ of terminals, the aim is to find a tree S of minimum weight $\omega(S) = \sum_{e \in S} \omega(e)$ that spans all the terminals. In the directed variant, known as the Directed Steiner Tree (DST) Problem, we are also given a root vertex r and the aim is to find the tree of minimum weight in the directed graph such that the root vertex r is connected to every vertex in the set of terminals T. Both the Steiner Tree and Directed Steiner Tree Problems are known to be NP-complete [24]. Even though it was studied purely from a theoretical interest, the Steiner Tree problem has recently found applications in areas like multicast routing, circuit design and network design [27]. In particular, the DST problem has applications when designing networks where the aim is to send multiple signals from a single source, such as satellite-based network routing [34].

However, since the DST problem only deals with sending signals from a root vertex, a natural generalization of this problem would be to consider the cost of sending signals throughout the network and not necessarily in one direction. Requiring two-way connectivity between a set of terminals leads to the ℓ-Strongly Connected Steiner Subgraph (SCSS) problem: given an edge-weighted digraph $G = (V, E)$ with weight function $\omega : E \to \mathbb{R}^{\geq 0}$ and ℓ many terminals $t_1, t_2, ..., t_\ell \in V$, the objective is to find a set of edges $X \subseteq E$ of minimum weight such that the induced graph $G[X]$ contains a $t_i \rightsquigarrow t_j$ path for every $1 \leq i \neq j \leq \ell$.

1.1 Prior Work on the Strongly Connected Steiner Subgraph Problem

The NP-hardness of the ℓ-SCSS problem means that unless P = NP, we cannot produce an algorithm that gives the exact solution in polynomial time with respect to the size of the input graph. Multiple algorithmic paradigms have been used to cope with the NP-hardness of the ℓ-SCSS problem:

- Parameterized Complexity: A natural parameter for the ℓ-SCSS problem is the number of terminals. Feldman and Ruhl [14] designed an XP algorithm that solves the ℓ-SCSS problem on graphs with n vertices in $n^{O(\ell)}$ time. This algorithm was shown to be almost-tight by Chitnis et al. [9] who proved that for any computable function f, the ℓ-SCSS problem on n vertices has no $f(\ell) \cdot n^{o(\ell/\log \ell)}$ time algorithm unless the Exponential Time Hypothesis (ETH) fails. Recall that the Exponential Time Hypothesis (ETH) states that n-variable m-clause 3-SAT cannot be solved in $2^{o(n)} \cdot (n+m)^{O(1)}$ time [21,22].
- Polynomial time Approximation: The best known approximation ratio in polynomial time for the ℓ-SCSS problem is ℓ^{ε} for any $\varepsilon > 0$ by Charikar et al. [3]. Moreover, Halperin and Krauthgamer [20] showed that SCSS has no $\Omega(\log^{2-\epsilon} n)$-approximation for any $\varepsilon > 0$, unless NP has quasi-polynomial Las Vegas algorithms.
- FPT Approximation: Chitnis et al. [10] showed that the ℓ-SCSS problem admits a 2-approximation in $3^{\ell} \cdot n^{O(1)}$ time. This ratio was later shown to be tight by Chitnis et al. [7] who proved that under Gap-ETH, for any $\varepsilon > 0$ and any computable function f, there is no $f(\ell) \cdot n^{O(1)}$ time algorithm that computes a $(2-\varepsilon)$-approximation for the ℓ-SCSS problem.

Another direction towards tractability of the ℓ-SCSS problem is to impose a restriction on either the structure of the input graph or (less severely) on the structure of an optimal solution.

- Finding optimal solutions on planar graphs: Chitnis et al. [9] showed that the PLANAR-ℓ-SCSS problem, i.e., the ℓ-SCSS problem restricted to planar graphs can be solved in time $2^{O(\ell)} \cdot n^{O(\sqrt{\ell})}$ time which is faster than the $n^{O(\ell)}$ algorithm of Feldmann and Ruhl on general graphs. Chitnis et al. [9] also showed a tight lower bound for the PLANAR-ℓ-SCSS problem: under ETH, there is no computable function f such that there exists a $f(\ell) \cdot n^{o(\sqrt{\ell})}$ algorithm. This is an example of the square-root phenomenon in planar graphs [29]: the planar version for several problems [1,11,16,17,25,26,28,30–32] is easier by (roughly) a square root factor in the exponent as compared to general graphs, and there are lower bounds indicating that this improvement is essentially the best possible.
- Finding solutions of cost at most that of a planar solution: A meet-in-the-middle option between no restrictions on the input graph versus requiring it to be planar is to ask for solutions which have cost comparable to a planar solution. More formally, the $\text{SCSS}_{\text{PLANAR}}$ problem defined by Chitnis et al. [7] has a general graph as the input and the question is to find a (not necessarily planar) solution of cost at most p if there exists a planar solution of size at most p. Chitnis and Feldmann [6] showed that the $\text{SCSS}_{\text{PLANAR}}$ problem is $W[1]$-hard parameterized by $p+k$.

1.2 Introducing Demands: The 2-SCSS-(k_1, k_2) Problem

The ℓ-SCSS problem only requires one path between every pair of terminals. This can be a drawback in certain networks as having possibly just one path between terminals does not guarantee network connectivity if some parts of the network fail. Hence, a natural generalization of these problems is to introduce a network redundancy requirement that the terminals are connected by several disjoint paths. This led to the SURVIVABLE NETWORK DESIGN PROBLEM (SNDP) problem [33]: given an (un)directed graph $G = (V, E)$ with weight function $\omega : E \rightarrow \mathbb{R}^{\geq 0}$, a set of terminals $T \subseteq V$ and a non-negative integer demand $d_{s,t}$ for each terminal pair $s, t \in T$; the objective is to find a subgraph $H \subseteq G$ of minimum cost containing T such that every pair of terminals $s, t \in T$ is connected by at least $d_{s,t}$ many edge-disjoint paths in H. There have been multiple results for the undirected version of SNDP problem. Jain [23] developed a factor 2 approximation algorithm and Feldmann et al. [15] worked on the parameterized complexity of the SNDP problem and obtained a FPT algorithm under certain constraints. However, there have been limited results for the directed version of the SNDP problem.

Chakrabarty et al. [2] and Goemans et al. [19] observed that the directed version of the SNDP problem becomes much easier to approximate if we are allowed to take multiple copies of the edges in the graph G: essentially this allows one to **"buy disjointness"** by paying for an edge multiple times. This can be applied to large real-world networks where it is more feasible to add parallel paths between terminals in the network as compared to the computationally intractable task of finding low-cost disjoint paths in the pre-existing network. With this intuition, Chitnis et al. [5] defined the following generalization of the 2-SCSS problem which they called the 2-SCSS-(k_1, k_2) problem:

2-SCSS-(k_1, k_2):

Input: An edge-weighted directed graph $G = (V, E)$ with weight function $\omega : E \rightarrow \mathbb{R}^{\geq 0}$, two terminal vertices s, t and positive integers k_1, k_2.
Question: Find a set of k_1 paths $F_1, F_2, \ldots, F_{k_1}$ from $s \rightsquigarrow t$ and k_2 paths $B_1, B_2, \ldots, B_{k_2}$ from $t \rightsquigarrow s$ such that the total weight $\sum_{e \in E} \omega(e) \cdot \phi(e)$ is minimized, where $\phi(e) = \max\Big\{ |\{i \in [k_1] : e \in F_i\}| \ , \ |\{j \in [k_2] : e \in B_j\}| \Big\}$.

Observe that the 2-SCSS-$(1, 1)$ problem is the same as the 2-SCSS problem. Chitnis et al. [5] completely resolved the parameterized complexity of the 2-SCSS-$(k, 1)$ problem parameterized by k on general graphs: they designed a $n^{O(k)}$ time algorithm and proved that it does not have a $f(k) \cdot n^{o(k)}$ time algorithm for any computable function f unless the Exponential Time Hypothesis (ETH) fails. Chitnis et al. [5] also showed that their algorithm does not work for 2-SCSS-(k_1, k_2) when $\min\{k_1, k_2\} \geq 2$: it is open whether 2-SCSS-$(2, 2)$ is NP-hard or solvable in polynomial time.

1.3 Our Results for the 2-SCSS-$(k, 1)$ Problem

In this paper, we complement the results of Chitnis et al. [5] on 2-SCSS-$(k, 1)$ by asking the following three questions: does the problem get easier on planar graphs, what is the complexity of finding solutions that approximate planar solutions and what is the best approximation ratio that we can obtain in polynomial time?

Obtaining a Tight Bound on Planar Graphs: Our first result answers the question of whether the 2-SCSS-$(k, 1)$ problem becomes easier to solve on planar graphs as compared to general graphs.

Theorem 1. *The* PLANAR-2-SCSS-$(k, 1)$ *problem is* $W[1]$*-hard parameterized by* k*. Moreover, under the Exponential Time Hypothesis (ETH), the* PLANAR-2-SCSS-$(k, 1)$ *problem on graphs with* n *vertices cannot be solved in* $f(k) \cdot n^{o(k)}$ *time for any computable function* f*.*

Theorem 1 shows that the $n^{O(k)}$ time algorithm of [5] for 2-SCSS-$(k, 1)$ cannot be asymptotically improved even if the input graph is restricted to be planar. The proof of Theorem 1 is a modification of the W[1]-hardness proof of 2-SCSS-$(k, 1)$ on general graphs from [5]: in a sense, we planarize the graph obtained in [5] while also simplifying some of the arguments.

An interesting consequence of Theorem 1 is that this shows that 2-SCSS-$(k, 1)$ is a "genuinely planar" problem which is known to not admit the square-root phenomenon [29] on planar graphs. The term "genuinely planar" was coined by [30] to distinguish from problems [4,8,9] that are known to not admit the square-root phenomenon on planar graphs but encode non-planarity as part of the problem definition. To the best of our knowledge, we are aware of only two [18,30] other genuinely planar problems that are known to not admit the square-root phenomenon in planar graphs.

Ruling Out Solutions Which Approximate Planar Solutions. Next, we consider an intermediate version of the 2-SCSS-$(k, 1)$ problem where the input graph is general, but we want to find a solution with cost at most p if there is a planar solution of cost at most p. We obtain the following lower bounds on exact and approximation algorithms for this version:

Theorem 2. [$\star$] *The* 2-SCSS-$(k, 1)_{\text{PLANAR}}$ *problem is* $W[1]$*-hard parameterized by* $p + k$*. Moreover, under the Exponential Time Hypothesis (ETH), for any computable function* f*:*

- *there is no* $f(k, p) \cdot n^{o(k+\sqrt{p})}$ *time algorithm for* 2-SCSS-$(k, 1)_{\text{PLANAR}}$
- *for every* $\epsilon > 0$*, there is no* $f(k, p, \epsilon) \cdot n^{o(k+\sqrt{p+1/\epsilon})}$ *time algorithm which computes a* $(1 + \epsilon)$*-approximation for* 2-SCSS-$(k, 1)_{\text{PLANAR}}$.

For the proof of Theorem 2, we essentially replace the grids from the reduction of Theorem 1 with the gadgets introduced in [6,7] which are non-planar but ensure that an optimum solution is planar. Theorem 2 also implies W[1]-hardness of the 2-SCSS-$(k, 1)$ problem on general graphs parameterized by the cost p of the solution: note that this does not follow from the W[1]-hardness proof of Chitnis et al. [5] since the value of p in their reduction was a function of both k and the input size n. Since our cost function $\omega : E \to \mathbb{Z}^{\geq 1}$ has non-zero integral costs for the edges, it follows that $p \geq k$ (each $s \rightsquigarrow t$ path incurs a cost of at least one) and hence we obtain W[1]-hardness of 2-SCSS-$(k, 1)$ with respect to a larger parameter.

Finding Approximate Solutions in Polynomial Time on General Graphs. Our final two results deal with the polynomial time approximability of the 2-SCSS-$(k, 1)$ problem on general graphs. First, we show that there is a $(k + 1)$-approximation possible in polynomial time.

Theorem 3. [$\star$] *There is a polynomial time algorithm* `Repeat-k` *which provides a* $(k + 1)$*-approximation for the* 2-SCSS-$(k, 1)$ *problem on general graphs.*

As the name suggests, the `Repeat-k` algorithm computes an optimal solution for the instance (G, s, t) of 2-SCSS in $n^{O(1)}$ time [14] and then simply repeats the $s \rightsquigarrow t$ path from this solution k times. Moreover, our next result shows our analysis of the `Repeat-k` algorithm is almost tight:

Theorem 4. [$\star$] *For each* $\delta > 0$, *there is an instance* (G_δ, s, t, k) *of* 2-SCSS-$(k, 1)$ *such that the the* `Repeat-k` *algorithm outputs a solution which has cost at least* $(k - \delta)$ *times that of an optimal solution.*

1.4 Notation in the Paper

All the graphs considered in this paper are simple directed graphs, that is, they do not contain self-loops or multiple edges. We mostly use standard graph-theoretic notation [12] in the paper. A directed edge from a vertex u to a vertex v is denoted by $u \to v$; while a directed path from u to v is denoted by $u \rightsquigarrow v$. We use $[n]$ to denote the set $\{1, 2, \ldots, n\}$. For $A \subseteq V(G)$ we define $G[A]$ to be the graph induced on the vertex set A, i.e., $G[A] := (A, E_A)$ where $E_A := E(G) \cap (A \times A)$.

Due to space constraints, the results marked by [$\star$] have their proofs deferred to the full version of the paper.

2 $W[1]$-Hardness of the Planar-2-SCSS-$(k, 1)$ Problem

We start by analyzing the effect of planarity on the hardness of the 2-SCSS-$(k, 1)$ problem. Formally, the problem is defined as follows:

Planar-2-SCSS-$(k, 1)$:

Input: An edge-weighted planar directed graph $G = (V, E)$ with weight function $\omega : E \to \mathbb{R}^{\geq 0}$, two terminal vertices s, t and a positive integer k.

Question: Find a set of k paths $P_1, P_2, \ldots, P_k$ from $s \rightsquigarrow t$ and a path Q from $t \rightsquigarrow s$ such that the sum $\sum_{e \in E} \omega(e) \cdot \phi(e)$ is minimized where $\phi(e) = \max\{|\{j \in [k] : e \in P_j\}|, s(Q)\}$ with $s(Q) = 1$ if $e \in Q$ and $s(Q) = 0$ otherwise.

Chitnis et al. [5] showed that the 2-SCSS-$(k, 1)$ problem on general graphs is $W[1]$-hard parameterized by k. We extend this hardness result to show that the Planar-2-SCSS-$(k, 1)$ problem is $W[1]$-hard parameterized by k and we state it as a theorem below:

Theorem 1. *The* Planar-2-SCSS-$(k, 1)$ *problem is $W[1]$-hard parameterized by k. Moreover, under the Exponential Time Hypothesis (ETH), the* Planar-2-SCSS-$(k, 1)$ *problem on graphs with n vertices cannot be solved in $f(k) \cdot n^{o(k)}$ time for any computable function f.*

The proof follows by reduction from the k-Clique problem which is known to be $W[1]$-hard parameterized by k [13]. We start by constructing an instance of the Planar-2-SCSS-$(k, 1)$ problem.

2.1 Constructing an Instance (G', s, t, k) of Planar-2-SCSS-$(k, 1)$

Consider an instance (G, k) of the k-Clique problem on a graph $G = (V, E)$ with $V = \{v_1, v_2, \ldots, v_n\}$. We construct an instance (G', s, t, k) of the Planar-2-SCSS-$(k, 1)$ problem in two steps. First we construct a planar intermediate graph G_{int}. We then modify the (black) grid vertices of G_{int} to obtain the required graph G' on which we consider the Planar-2-SCSS-$(k, 1)$ problem.

Construction of the Intermediate Graph G_{int}: Given integers k and n, we build the directed intermediate graph G_{int} as follows:

1. **Origin**: Mark the origin on the bottom left as shown in Fig. 1. We can consider this point as the origin in a $X - Y$ coordinate system; with values increasing from left to right as well as from bottom to top. This will help in labelling the vertices of the graph.
2. **Small Grids**: For each $1 \leq i, j \leq k$ construct a $n \times n$ grid $M_{i,j}$ of vertices with row numbers increasing from 1 to n as we go from bottom to top and column numbers increasing from 1 to n as we go from left to right. For $1 \leq \ell, r \leq n$, denote the vertex at the intersection of the rth row and ℓth column as $w_{i,j}^{\ell,r}$. Construct the directed edge set (colored black; and going rightwards or downwards) of $M_{i,j}$ as follows:
   - For every $1 \leq \ell < n$, $1 \leq r \leq n$, construct the edge $w_{i,j}^{\ell,r} \to w_{i,j}^{\ell+1,r}$.
   - For every $1 \leq \ell \leq n$, $1 < r \leq n$, construct the edge $w_{i,j}^{\ell,r} \to w_{i,j}^{\ell,r-1}$.

3. **Large Grid**: Place the small grids ($M_{i,j}, 1 \leq i, j \leq k$) into a $k \times k$ grid of grids in such a way that i increases from left to right and j increases from bottom to top (based on our coordinate system).
4. **Terminal and Border vertices**: Add two vertices s, t (which we call terminal vertices) and the following $4k$ many vertices (which we call border vertices):
   - $A = \{a_1, a_2, \ldots a_k\}$
   - $B = \{b_1, b_2, \ldots b_k\}$
   - $C = \{c_1, c_2, \ldots c_k\}$
   - $D = \{d_1, d_2, \ldots d_k\}$
5. **Dummy vertices**: Add two rows and two columns of dummy vertices surrounding the large grid as shown in Fig. 1. Denote the vertices in the top and bottom rows as $w_{i,k+1}^{\ell,1}$ and $w_{i,0}^{\ell,n}$ respectively (where $1 \leq i \leq k$ and $1 \leq \ell \leq n$). Similarly denote the vertices in the left and right columns as $w_{0,j}^{n,r}$ and $w_{k+1,j}^{1,r}$ respectively (where $1 \leq j \leq k$ and $1 \leq r \leq n$).
6. **Red Edges**: Add the following sets of Vertical and Horizontal Red Edges:
   - Horizontal Red Edges: For every $i, j \in [k]$ and $r \in [n]$, construct a directed edge (colored red): $w_{i,j}^{n,r} \rightarrow w_{i+1,j}^{1,r}$. Similarly, for every $j \in [k]$ and $r \in [n]$, construct a directed edge (colored red): $w_{0,j}^{n,r} \rightarrow w_{1,j}^{1,r}$.
   - Vertical Red Edges: For every $i, j \in [k]$ and $r \in [n]$, construct a directed edge (colored red): $w_{i,j}^{\ell,1} \rightarrow w_{i,j-1}^{\ell,n}$. Similarly, for every $i \in [k]$ and $\ell \in [n]$, construct a directed edge (colored red): $w_{i,k+1}^{\ell,1} \rightarrow w_{i,k}^{\ell,n}$.
7. **Blue Border Edges**: From each border vertex, add n many directed edges (colored blue) to or from the corresponding dummy vertices as follows:
   - For each $1 \leq j \leq k$ and $1 \leq r \leq n$, construct the edge $a_j \rightarrow w_{0,j}^{n,r}$.
   - For each $1 \leq j \leq k$ and $1 \leq r \leq n$, construct the edge $w_{k+1,j}^{1,r} \rightarrow b_j$.
   - For each $1 \leq i \leq k$ and $1 \leq \ell \leq n$, construct the edge $c_i \rightarrow w_{i,k+1}^{\ell,1}$.
   - For each $1 \leq i \leq k$ and $1 \leq \ell \leq n$, construct the edge $w_{i,0}^{\ell,n} \rightarrow d_i$.
8. **Source and Sink Edges**: For each $1 \leq i \leq k$, add edges (denoted by dotted black lines) as follows:
   - Add an edge: $s \rightarrow c_i$. We call them source edges.
   - Add an edge: $d_i \rightarrow t$. We call them sink edges.
9. **Connector Edges**: Moreover, add the following edges (called connector edges and colored magenta):
   - Add the edge $t \rightarrow a_1$.
   - For $1 \leq j \leq k - 1$, add the edge $b_j \rightarrow a_{j+1}$.
   - Finally, add the edge $b_k \rightarrow s$.

Definition 1 (canonical path). For each $1 \leq j \leq k$ we define an $a_j \rightsquigarrow b_j$ *canonical path* as a directed path $a_j \rightsquigarrow b_j$ that starts with a blue border edge coming out from a_j followed by a fully horizontal path of red and black edges and ends with a blue border edge going into b_j. Similarly, for each $1 \leq i \leq k$, define a $c_i \rightsquigarrow d_i$ canonical path as a path $c_i \rightsquigarrow d_i$ which starts with a blue border

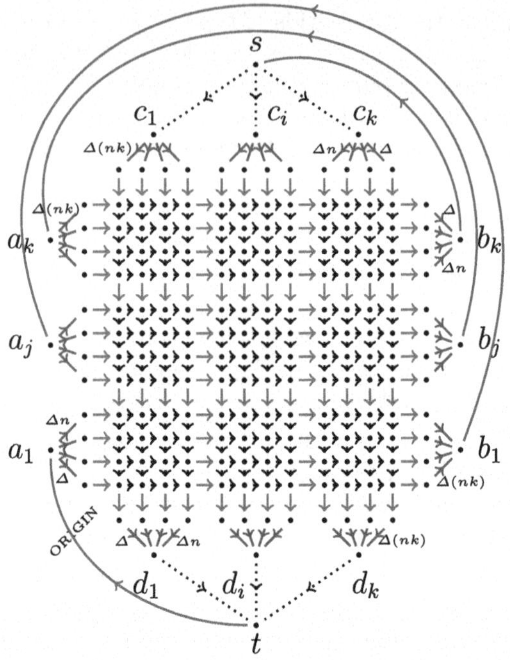

Fig. 1. The intermediate graph G_{int} constructed from an instance (G, k) of the k-CLIQUE problem.

edge coming out of c_i followed by a fully vertical path of red and black edges and ends with a blue border edge going into d_i.

Since for each $1 \leq j \leq k$ there are n many edge-disjoint $a_j \rightsquigarrow b_j$ canonical paths, we shall denote them as $Q_j^1, Q_j^2, \ldots, Q_j^n$ as viewed from bottom to top. Similarly denote the n many edge-disjoint $c_i \rightsquigarrow d_i$ canonical paths as $P_i^1, P_i^2, \ldots, P_i^n$ as viewed from left to right.

We give the following weights to the edges of G_{int}:

- **Black Edges** (in small grids): Give a weight of 4 to each of the black edges in the small grids.
- **Red Edges**: For each red edge, give a weight of $R = 38n^9$ (the idea is to choose a value of R large enough that it prevents the use of each connector edge more than once in solutions of low-cost).
- **Blue Border Edges**: For each $1 \leq j \leq k$ and $1 \leq r \leq n$, assign a weight of $\Delta(nj - n + r)$ and $\Delta(nk - nj + n + 1 - r)$ to the first and last edge respectively of Q_j^r where $\Delta = 7n^6$. Similarly, for each $1 \leq i \leq k$ and $1 \leq \ell \leq n$, assign a weight of $\Delta(nk - ni + n + 1 - \ell)$ and $\Delta(ni - n + \ell)$ to the first and last edge respectively of P_i^ℓ (the idea here is to choose a value of Δ large enough such that in any solution of low-cost, each $s \rightsquigarrow t$ path is canonical).
- **Source and Sink Edges**: Give a weight of 0 to each of the source and sink edges.

- **Connector Edges**: Give a weight of 0 to the $t \to a_1$ and $b_k \to s$ connector edges. For each $1 \leq j \leq k-1$, give a weight of 1 to the $b_j \to a_{j+1}$ connector edge.

Observe that any canonical path will use two blue border edges, $(n-1)$ black edges inside each small grid and $(k+1)$ red edges. Since each canonical path visits exactly k small grids, the total weight of the canonical path will be:

$$\alpha = \Delta(nk + 1) + 4k(n-1) + R(k+1) \tag{1}$$

We have now completed the construction of the intermediate graph G_{int}. Observe that G_{int} is planar.

Modifications in G_{int} to obtain G': Consider the intermediate graph G_{int}. We call a small grid $M_{i,j}$ a symmetric gadget if $i = j$. Otherwise call it an asymmetric gadget. Based on this notation, we perform the following gadget modifications to the small grids in G_{int}:

- **Modifications in Symmetric Gadgets**: For each $i \in [k]$ and $\ell \in [n]$, select the vertex inside the small grid $M_{i,i}$ at the intersection of the canonical paths P_i^ℓ and Q_i^ℓ and color it green. We then construct the shortcut as given in Fig. 2.
- **Modifications in Asymmetric Gadgets**: For each $i \neq j \in [k]$, if $\{v_x, v_y\} \in E(G)$, select the vertex inside the small grid $M_{i,j}$ at the intersection of the canonical paths P_i^x and Q_j^y and color it orange. We then construct the shortcut as given in Fig. 2.

Note that the graph G' is planar since the intermediate graph G_{int} is planar and the above gadget modifications do not affect the planarity.

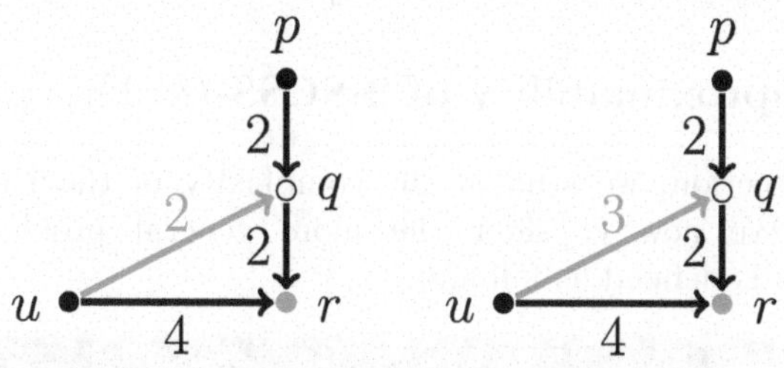

Fig. 2. Construction of shortcuts: Let p and r be two consecutive vertices on a vertical canonical path. The idea is that if both the edges (u, r) and (p, r) are being used in our solution, we can save costs by taking the edge (u, q) in the horizontal path.

We now define a threshold value β:

$$\beta = 2k \cdot \alpha + (k-1) - k(k+1) \tag{2}$$

and show in the next two subsections that G has a k-clique if and only if the instance (G', s, t, k) of Planar-2-SCSS-$(k, 1)$ has a solution of cost at most β.

2.2 G Has a k-Clique $\implies$ (G', s, t, k) Has a Solution of Weight $\leq \beta$ [$\star$]

2.3 (G', s, t, k) Has a Solution of Weight $\leq \beta \Rightarrow G$ Has a k-Clique [$\star$]

2.4 Proof of Theorem 1

We are now ready to prove Theorem 1 which is restated below:

Theorem 1. *The* PLANAR-2-SCSS-$(k, 1)$ *problem is $W[1]$-hard parameterized by k. Moreover, under the Exponential Time Hypothesis (ETH), the* PLANAR-2-SCSS-$(k, 1)$ *problem on graphs with n vertices cannot be solved in $f(k) \cdot n^{o(k)}$ time for any computable function f.*

Proof. Section 2.2 and Sect. 2.3 implies that the k-CLIQUE instance (G, k) has a solution if and only if the PLANAR-2-SCSS-$(k, 1)$ instance (G', s, t, k) has a solution with weight at most β. Moreover, the parameter k remains the same in both problems and the construction of our instance can be done in $n^{O(1)}$ time. These imply that we have obtained a parameterized reduction from the k-CLIQUE problem to the PLANAR-2-SCSS-$(k, 1)$ problem.

We know that the k-CLIQUE problem is $W[1]$-hard parameterized by k. Hence, Sect. 2.2 and Sect. 2.3 imply the $W[1]$-hardness of the PLANAR-2-SCSS-$(k, 1)$ problem as we have produced a parameterized reduction from an instance of the k-CLIQUE problem (G, k) to an instance of PLANAR-2-SCSS-$(k, 1)$ problem (G', s, t, k).

Moreover, by ETH, the k-CLIQUE problem on a graph with n vertices cannot be solved in $f(k) \cdot n^{o(k)}$ time for any computable function f. Hence, under the same reduction, since the parameter change is linear, the PLANAR-2-SCSS-$(k, 1)$ problem on a graph with n vertices cannot be solved in $f(k) \cdot n^{o(k)}$ time for any computable function f. This completes the proof.

3 FPT Inapproximability of 2-SCSS-$(k, 1)_{\text{PLANAR}}$

In the previous section, we studied the complexity of the PLANAR-2-SCSS-$(k, 1)$ problem. We now consider the more general problem of 2-SCSS-$(k, 1)_{\text{PLANAR}}$ which is defined as follows:

2-SCSS-$(k, 1)_{\text{Planar}}$:

Input: An edge-weighted directed graph $G = (V, E)$ with weight function $\omega : E \to \mathbb{Z}^{\geq 1}$, two terminal vertices s, t and a positive integer k.

Question: Find a solution to the 2-SCSS-$(k, 1)$ problem on G with cost at most p if there is a planar solution of cost at most p.

Chitnis and Feldmann [6] showed that the ℓ-SCSS$_{\text{PLANAR}}$ problem is $W[1]$-hard parameterized by $p + \ell$ (where p is the cost of the planar solution). We study the complexity of the 2-SCSS-$(k, 1)_{\text{PLANAR}}$ problem and prove the following theorem:

Theorem 2. [$\star$] *The* 2-SCSS-$(k,1)_{PLANAR}$ *problem is* $W[1]$*-hard parameterized by* $p + k$*. Moreover, under the Exponential Time Hypothesis (ETH), for any computable function* f*:*

- *there is no* $f(k,p) \cdot n^{o(k+\sqrt{p})}$ *time algorithm for* 2-SCSS-$(k,1)_{PLANAR}$
- *for every* $\epsilon > 0$*, there is no* $f(k,p,\epsilon) \cdot n^{o(k+\sqrt{p+1/\epsilon})}$ *time algorithm which computes a* $(1+\epsilon)$*-approximation for* 2-SCSS-$(k,1)_{PLANAR}$.

4 Conclusion and Open Questions

In this paper, we have studied the 2-SCSS-$(k,1)$ problem from the viewpoint of polynomial time and FPT approximation algorithms. Some natural open questions that arise are:

- Is it possible to obtain better approximation ratios for the 2-SCSS-$(k,1)$ problem in polynomial time and FPT time respectively? Would the problem be easier to approximate in planar graphs?
- Can we obtain any hardness results for the 2-SCSS-(k_1,k_2) problem when $\min(k_1,k_2) \geq 2$?

Acknowledgments. This study was supported by EPSRC New Investigator Award "New Frontiers in Fine-Grained Approximation Algorithms".

References

1. Aboulker, P., Brettell, N., Havet, F., Marx, D., Trotignon, N.: Coloring graphs with constraints on connectivity. J. Graph Theory **85**(4), 814–838 (2017). https://doi.org/10.1002/jgt.22109
2. Chakrabarty, D., Chekuri, C., Khanna, S., Korula, N.: Approximability of capacitated network design. Algorithmica **72**(2), 493–514 (2015). https://doi.org/10.1007/S00453-013-9862-4
3. Charikar, M., et al.: Approximation algorithms for directed steiner problems. J. Algorithms **33**(1), 73–91 (1999). https://doi.org/10.1006/JAGM.1999.1042
4. Chitnis, R.: A tight lower bound for edge-disjoint paths on planar dags. SIAM J. Discret. Math. **37**(2), 556–572 (2023). https://doi.org/10.1137/21M1395089
5. Chitnis, R., Esfandiari, H., Hajiaghayi, M.T., Khandekar, R., Kortsarz, G., Seddighin, S.: A tight algorithm for strongly connected steiner subgraph on two terminals with demands. Algorithmica **77**(4), 1216–1239 (2017). https://doi.org/10.1007/S00453-016-0145-8
6. Chitnis, R., Feldmann, A.E.: FPT inapproximability of directed cut and connectivity problems. In: Jansen, B.M.P., Telle, J.A. (eds.) 14th International Symposium on Parameterized and Exact Computation, IPEC 2019. LIPIcs, vol. 148, pp. 8:1–8:20. Schloss Dagstuhl - Leibniz-Zentrum für Informatik (2019). https://doi.org/10.4230/LIPICS.IPEC.2019.8

7. Chitnis, R., Feldmann, A.E., Manurangsi, P.: Parameterized approximation algorithms for bidirected steiner network problems. ACM Trans. Algorithms **17**(2), 12:1–12:68 (2021). https://doi.org/10.1145/3447584
8. Chitnis, R., Feldmann, A.E., Suchý, O.: A tight lower bound for planar steiner orientation. Algorithmica **81**(8), 3200–3216 (2019). https://doi.org/10.1007/S00453-019-00580-X
9. Chitnis, R.H., Feldmann, A.E., Hajiaghayi, M.T., Marx, D.: Tight bounds for planar strongly connected steiner subgraph with fixed number of terminals (and extensions). SIAM J. Comput. **49**(2), 318–364 (2020). https://doi.org/10.1137/18M122371X
10. Chitnis, R.H., Hajiaghayi, M., Kortsarz, G.: Fixed-parameter and approximation algorithms: a new look. In: Gutin, G.Z., Szeider, S. (eds.) Parameterized and Exact Computation - 8th International Symposium, IPEC 2013, Revised Selected Papers. LNCS, vol. 8246, pp. 110–122. Springer, Cham (2013). https://doi.org/10.1007/978-3-319-03898-8_11
11. Demaine, E.D., Fomin, F.V., Hajiaghayi, M.T., Thilikos, D.M.: Subexponential parameterized algorithms on bounded-genus graphs and H-minor-free graphs. J. ACM **52**(6), 866–893 (2005). https://doi.org/10.1145/1101821.1101823
12. Diestel, R.: Graph Theory, 5th edn. Springer, Berlin, Heidelberg (2017). https://doi.org/10.1007/978-3-662-70107-2
13. Downey, R.G., Fellows, M.R.: Parameterized Complexity. Springer, New York (2012). https://doi.org/10.1007/978-1-4612-0515-9
14. Feldman, J., Ruhl, M.: The directed steiner network problem is tractable for a constant number of terminals. SIAM J. Comput. **36**(2), 543–561 (2006). https://doi.org/10.1137/S0097539704441241
15. Feldmann, A.E., Mukherjee, A., van Leeuwen, E.J.: The parameterized complexity of the survivable network design problem. J. Comput. Syst. Sci. **148**, 103604 (2025). https://doi.org/10.1016/J.JCSS.2024.103604
16. Fomin, F., Kolay, S., Lokshtanov, D., Panolan, F., Saurabh, S.: Subexponential algorithms for rectilinear steiner tree and arborescence problems. In: 32nd International Symposium on Computational Geometry (SoCG 2016). Leibniz International Proceedings in Informatics (LIPIcs), vol. 51, pp. 39:1–39:15. Schloss Dagstuhl – Leibniz-Zentrum für Informatik, Dagstuhl, Germany (2016). https://doi.org/10.4230/LIPIcs.SoCG.2016.39
17. Fomin, F.V., Lokshtanov, D., Marx, D., Pilipczuk, M., Pilipczuk, M., Saurabh, S.: Subexponential parameterized algorithms for planar and apex-minor-free graphs via low treewidth pattern covering. In: 2016 IEEE 57th Annual Symposium on Foundations of Computer Science (FOCS), pp. 515–524 (2016). https://doi.org/10.1109/FOCS.2016.62
18. Galby, E., Kisfaludi-Bak, S., Marx, D., Sharma, R.: Subexponential parameterized directed steiner network problems on planar graphs: A complete classification. In: Bringmann, K., Grohe, M., Puppis, G., Svensson, O. (eds.) 51st International Colloquium on Automata, Languages, and Programming, ICALP 2024. LIPIcs, vol. 297, pp. 67:1–67:19. Schloss Dagstuhl - Leibniz-Zentrum für Informatik (2024). https://doi.org/10.4230/LIPICS.ICALP.2024.67
19. Goemans, M.X., Goldberg, A.V., Plotkin, S.A., Shmoys, D.B., Tardos, É., Williamson, D.P.: Improved approximation algorithms for network design problems. In: Sleator, D.D. (ed.) Proceedings of the Fifth Annual ACM-SIAM Symposium on Discrete Algorithms, pp. 223–232. ACM/SIAM (1994). http://dl.acm.org/citation.cfm?id=314464.314497

20. Halperin, E., Krauthgamer, R.: Polylogarithmic inapproximability. In: Proceedings of the 35th Annual ACM Symposium on Theory of Computing, pp. 585–594. ACM (2003). https://doi.org/10.1145/780542.780628
21. Impagliazzo, R., Paturi, R.: On the complexity of k-sat. J. Comput. Syst. Sci. **62**(2), 367–375 (2001). https://doi.org/10.1006/JCSS.2000.1727
22. Impagliazzo, R., Paturi, R., Zane, F.: Which problems have strongly exponential complexity? J. Comput. Syst. Sci. **63**(4), 512–530 (2001). https://doi.org/10.1006/JCSS.2001.1774
23. Jain, K.: A factor 2 approximation algorithm for the generalized steiner network problem. Comb. **21**(1), 39–60 (2001). https://doi.org/10.1007/S004930170004
24. Karp, R.M.: Reducibility among combinatorial problems, pp. 85–103. Springer US, Boston, MA (1972). https://doi.org/10.1007/978-1-4684-2001-2_9
25. Klein, P.N., Marx, D.: Solving planar k -terminal cut in $o(n^{c\sqrt{k}})$ time. In: Czumaj, A., Mehlhorn, K., Pitts, A.M., Wattenhofer, R. (eds.) Automata, Languages, and Programming - 39th International Colloquium, ICALP 2012, Proceedings, Part I. LNCS, vol. 7391, pp. 569–580. Springer, Cham (2012).https://doi.org/10.1007/978-3-642-31594-7_48
26. Klein, P.N., Marx, D.: A subexponential parameterized algorithm for subset tsp on planar graphs. In: Proceedings of the 2014 Annual ACM-SIAM Symposium on Discrete Algorithms (SODA), pp. 1812–1830 (2014). https://doi.org/10.1137/1.9781611973402.131
27. Ljubic, I.: Solving steiner trees: recent advances, challenges, and perspectives. Networks **77**(2), 177–204 (2021). https://doi.org/10.1002/NET.22005
28. Marx, D.: A tight lower bound for planar multiway cut with fixed number of terminals. In: Czumaj, A., Mehlhorn, K., Pitts, A.M., Wattenhofer, R. (eds.) Automata, Languages, and Programming - 39th International Colloquium, ICALP 2012, Proceedings, Part I. LNCS, vol. 7391, pp. 677–688. Springer, Cham (2012). https://doi.org/10.1007/978-3-642-31594-7_57
29. Marx, D.: The square root phenomenon in planar graphs. In: Fomin, F.V., Freivalds, R., Kwiatkowska, M.Z., Peleg, D. (eds.) Automata, Languages, and Programming - 40th International Colloquium, ICALP 2013, Riga, Latvia, July 8-12, 2013, Proceedings, Part II. LNCS, vol. 7966, p. 28. Springer, Cham (2013). https://doi.org/10.1007/978-3-642-39212-2_4
30. Marx, D., Pilipczuk, M., Pilipczuk, M.: On subexponential parameterized algorithms for steiner tree and directed subset TSP on planar graphs. In: Thorup, M. (ed.) 59th IEEE Annual Symposium on Foundations of Computer Science, FOCS 2018, pp. 474–484. IEEE Computer Society (2018). https://doi.org/10.1109/FOCS.2018.00052
31. Marx, D., Pilipczuk, M.: Optimal parameterized algorithms for planar facility location problems using voronoi diagrams. In: Bansal, N., Finocchi, I. (eds.) Algorithms - ESA 2015 - 23rd Annual European Symposium, Proceedings. LNCS, vol. 9294, pp. 865–877. Springer, Cham (2015). https://doi.org/10.1007/978-3-662-48350-3_72
32. Pilipczuk, M., Pilipczuk, M., Sankowski, P., van Leeuwen, E.J.: Subexponential-time parameterized algorithm for steiner tree on planar graphs. In: 30th International Symposium on Theoretical Aspects of Computer Science, STACS 2013. LIPIcs, vol. 20, pp. 353–364. Schloss Dagstuhl - Leibniz-Zentrum für Informatik (2013). https://doi.org/10.4230/LIPICS.STACS.2013.353

33. Steiglitz, K., Weiner, P., Kleitman, D.: The design of minimum-cost survivable networks. IEEE Trans. Circuit Theory **16**(4), 455–460 (1969). https://doi.org/10.1109/TCT.1969.1083004
34. Yang, D., Liao, W.: On multicast routing using rectilinear steiner trees for LEO satellite networks. IEEE Trans. Veh. Technol. **57**(4), 2560–2569 (2008). https://doi.org/10.1109/TVT.2007.912605

Reconfiguring Planar Perfect Matchings via Bounded Length Alternating Cycles

Robert D. Barish(✉) and Tetsuo Shibuya

Division of Medical Data Informatics, Human Genome Center, Institute of Medical Science, University of Tokyo, 4-6-1 Shirokanedai, Minato-ku, Tokyo 108-8639, Japan
rbarish@ims.u-tokyo.ac.jp, tshibuya@hgc.jp

Abstract. The k-Cycle Perfect Matching Reconfiguration Problem (k-Cycle-PM-RP) concerns the task of deciding if a pair of perfect matchings for a simple graph can be interconverted via a series of local rearrangements that allow for transitions between perfect matchings if and only if the symmetric difference of their edge sets induces a simple cycle of length $k \in 2\mathbb{N}_{>1}$ (Bonamy et al.; *Proc. 44th MFCS*; 2019). While it is known that the k-Cycle-PM-RP is PSPACE-complete in the case of bounded bandwidth and bounded degree bipartite graphs, as well as polynomial time solvable in the case of outerplanar graphs, cographs, and strongly orderable graphs, it has remained an open question whether the problem is fixed-parameter tractable in the genus of a graph. In this work, we answer this question by showing that the k-Cycle-PM-RP remains PSPACE-complete for bounded bandwidth and bounded degree planar graphs for every $k \in 2\mathbb{N}_{>1}$. On the other hand, among some other positive results, we show that it is fixed-parameter tractable in a parameter $r \in \mathbb{N}$ to add at most r edges to a planar graph G to construct a strongly orderable planar graph G', where we can observe that the $(k = 4)$-Cycle-PM-RP on strongly orderable graphs is both polynomial time tractable and guaranteed to admit a positive solution.

Keywords: combinatorial reconfiguration · reachability problem · perfect matching · PSPACE-complete · FPT · fixed-parameter tractability · edge augmentation · strongly orderable graph · cograph

1 Introduction

The notion of a perfect matching in a graph has long served as a kind of structural armature for counting and enumeration problems in computer science. Consider, for instance, that counting perfect matchings in bipartite graphs (equivalently, computing the 01-permanent) was the foundational problem for the counting complexity class $\#P$ [35,36]. Consider also the remarkable achievement of the Fisher-Kasteleyn-Temperley (FKT) algorithm [18–20,30] for exactly counting

This work was supported by MEXT/JSPS Kakenhi grants 25K21149, 23K28035, 23K18501, and 21H05052.

A. Jeż and J. Otop (Eds.): FCT 2025, LNCS 16106, pp. 45–56, 2026.
https://doi.org/10.1007/978-3-032-04700-7_4

the perfect matchings of an order n planar graph in $\mathcal{O}(n^3)$ time, where this algorithm has since been generalized to $K_{3,3}$-minor free graphs [37], as well as graphs of bounded genus [8,13,25,31]. Regarding enumeration, there have likewise been a number of successes in developing efficient enumeration algorithms for maximal and perfect matchings in bipartite [12,32,33] as well as general graphs [21,34].

In this work, we concern ourselves with the k-Cycle Perfect Matching Reconfiguration Problem (k-Cycle-PM-RP) [1] on planar graphs:

Definition 1. *k-Cycle Perfect Matching Reconfiguration Problem (k-Cycle-PM-RP)* [1]. Letting G be a simple undirected graph, letting $\mathcal{M}_\alpha$ and $\mathcal{M}_\beta$ be a pair of edge sets inducing perfect matchings for G, and letting $k \in 2\mathbb{N}_{>1}$, decide if $\mathcal{M}_\alpha$ can be converted to $\mathcal{M}_\beta$ via a series of local rearrangements that allow for transitions between perfect matchings if and only if the symmetric difference of their edge sets is a simple cycle of length k.

While the k-Cycle-PM-RP is arguably prima facie interesting on embedded graphs in the sense that it can be understood as a motion planning task – perhaps bringing to mind Sokoban-style "block-sliding" puzzles (see, e.g., [14]) – our primary motivation for looking at planar cases of this problem comes from its connection to the *perfect matching polytope*. To briefly elaborate, letting $G = (V, E)$ be a simple undirected graph, recall that the perfect matching polytope $\mathcal{P}_G$ of G is a polytope in $\mathcal{R}^{|E|}$ corresponding to the convex hull of the incidence vectors for the perfect matchings in G, and that every edge $v_x \leftrightarrow v_y$ in $\mathcal{P}_G$ corresponds to a simple cycle induced by the symmetric difference of the perfect matchings corresponding to v_x and v_y in G (see Chvátal [6]). Accordingly, we can understand the k-Cycle-PM-RP as the problem of deciding if it is possible to walk from v_x to v_y (or vice versa) along the subset of edges of $\mathcal{P}_G$ inducing simple cycles of length exactly k. For graphs of bounded genus, this then seemingly places the k-Cycle-PM-RP between the task of counting perfect matchings, made easy due to the FKT algorithm [18–20,30], and the much harder task of enumerating at most exponentially many of these matchings to directly find paths between vertices on the perfect matching polytope via methods such as Breadth First Search (BFS).

Concerning what is already known, by a result of Bonamy et al. [1], we have that the k-Cycle-PM-RP is PSPACE-complete for every $k \in 2\mathbb{N}_{>1}$ on bounded bandwidth bipartite graphs of degree at most 5. See also Ito et al. [15,16] for a related NP-hardness hardness result for finding shortest paths between vertices on the perfect matching polytope, and Bousquet et al. [3] for a related hardness result concerning reconfigurations of maximal matchings. On the other hand, Bonamy et al. [1] was also able to show that the $(k = 4)$-Cycle-PM-RP is linear time solvable on cographs, strongly orderable graphs, and outerplanar graphs. However, the complexity of the k-Cycle-PM-RP on planar graphs – or even bounded genus graphs – remains open.

Here, towards filling the knowledge gap for the complexity of the k-Cycle-PM-RP on graphs of bounded genus, we first establish for every $k \in 2\mathbb{N}_{>1}$ that the problem remains PSPACE-complete on planar graphs (see Theorem 1 in Sect. 3). Next, towards establishing *Fixed-Parameter Tractability (FPT)* results for the ($k = 4$)-Cycle-PM-RP, we concern ourselves with some relevant cases of the following graph augmentation (i.e., edge addition) problem:

Definition 2. *$\Pi(r)$ Graph Augmentation Problem ($\Pi(r)$-GAP) (see* [4] *and, e.g.,* [5,11,22,23,38]*).* Let G be a simple undirected graph, let Π be a set of graphs defined by a hereditary property where membership of a graph can be tested in time polynomial in its size, and let $r \in \mathbb{N}$ be a specified parameter. In this context, $\Pi(r)$-GAP asks us to add at most r edges to G to construct a graph $G' \in \Pi$, or to decide that this task is impossible.

Notably, we show that the $\Pi(r)$-GAP is FPT in r for the set Π of graphs defined by the hereditary property of being planar, weakly-chordal, and having no induced subgraphs isomorphic to the members of a finite set of forbidden subgraphs (see Proposition 1 in Sect. 4). This allows us to establish as a corollary that the $\Pi(r)$-GAP is FPT in r when: (1) Π is the set of planar strongly orderable graphs (see Corollary 1 in Sect. 4), where we are both guaranteed polynomial time tractability of the ($k = 4$)-Cycle-PM-RP and the existence of a positive solution per section "3.1" of Bonamy et al. [1]; or (2) when Π is the set of (planar if desired) cographs (see Corollary 2 in Sect. 4), which guarantees the polynomial time tractability of the ($k = 4$)-Cycle-PM-RP per section "3.3" of Bonamy et al. [1].

2 Preliminaries and Clarifications

2.1 Graph Theoretic Concepts and Terminology

We refer the reader to either Diestel [9] or Bondy and Murty [2] for basic graph theoretic notions and terminology. Briefly however, recall that a *perfect matching* for a simple undirected graph corresponds to a spanning (i.e., covering all vertices) 1-regular subgraph; an *orientation* for an undirected graph G is a digraph D_G generated by assigning a direction to each of the edges in G (note we may refer to G as the *underlying undirected graph* for D_G); a *hereditary property* for a graph is a property that is preserved upon taking induced subgraphs, and a *hereditary class of graphs* is a set of graphs defined by having a certain hereditary property (e.g., being planar or chordal); a graph is *chordal* if it has no induced cycles (sometimes referred to as *holes*) of length ≥ 4; and a graph is *weakly-chordal* if both it and its complement are free of induced cycles of length ≥ 5 (note that holes in a graph become *anti-holes* in the graph's complement). Concerning some less common terminology, the *bandwidth* of a graph is the minimum *matrix bandwidth* (i.e., the maximum value of $|i - j|$, where the (i, j)th matrix entry is non-zero) over all possible adjacency matrices for graphs isomorphic to G. Also, a graph $G = (V, E)$ is *strongly orderable* [10] if, for some ordering σ of V, for all vertices $v_a, v_b, v_c, v_d \in V$ where $\sigma(v_a) < \sigma(v_d)$ and $\sigma(v_b) < \sigma(v_c)$ we have that $v_a \leftrightarrow v_b, v_a \leftrightarrow v_c, v_b \leftrightarrow v_d \in E$ implies $v_c \leftrightarrow v_d \in E$. Finally, for any

$n \in \mathbb{N}_{>2}$, an *n-sun graph* is the order $2n$ and size $\frac{n}{2}(n+3)$ graph constructed by adding $\frac{n}{2}(n-1)$ edges between a maximum independent set of vertices in the cycle graph C_{2n}, resulting in a clique of order n where every vertex is adjacent to exactly two degree 2 vertices.

2.2 Gray Codes and Gray Code Graphs

A *k-bit Gray code* corresponds to a subset of k-bit binary strings ordered in such a manner that the Hamming distance between any two adjacent strings is at most 1. In this context, a *k-bit Gray code graph* is a k-dimensional hypercube graph $\mathcal{Q}_k$, where each of its 2^k vertices are assigned a distinct k-bit binary string label, and where any pair of labels for adjacent vertices must have a Hamming distance of 1 (implying that any Hamiltonian path in $\mathcal{Q}_k$ yields a Gray code). For further details, we refer the reader to a survey by Savage [26].

2.3 PSPACE Complexity

Letting n be the size of an input for a problem, and defining SPACE$(f(n))$ as the set of all decision problems solvable on a deterministic Turing machine using $\mathcal{O}(f(n))$ space for some function $f(n)$, we have that PSPACE $= \bigcup_{k \in \mathbb{N}_{>0}}$ SPACE(n^k) (see, e.g., Papadimitriou [24]). Said differently, we have that PSPACE corresponds to the union of all decision problems solvable on a deterministic Turing machine (or, equivalently, a non-deterministic Turing machine [27]) using an amount of space polynomial in the problem input size n. Recall that if a problem is PSPACE-complete, then this means that it both belongs to PSPACE and that every problem in PSPACE can be reduced to it in polynomial time.

2.4 Nondeterministic Constraint Logic (NCL)

The *Nondeterministic Constraint Logic (NCL)* model of computation provides a simple framework for establishing the PSPACE-completeness of reconfiguration and reachability problems [14]. Here, for an undirected *constraint graph* G with non-negative integral vertex and edge weights, we define an *NCL machine* with *state* $\mathcal{F}$ as an orientation of G satisfying the requirement that the weighted in-degree of each vertex (i.e., the sum over edge weights of incident edges oriented towards the vertex) is at least equal to its own vertex weight. We are then allowed to reconfigure Υ by flipping edges one at a time under the constraint that, each time we re-orient an edge, it remains the case that the weighted in-degree of each vertex is greater than or equal to its own vertex weight. In this context, we define the *state transition graph* for an NCL machine as the graph with a unique vertex for each possible *state*, and with an edge connecting a pair of vertices whenever one state can be reached from another by re-orienting exactly one edge.

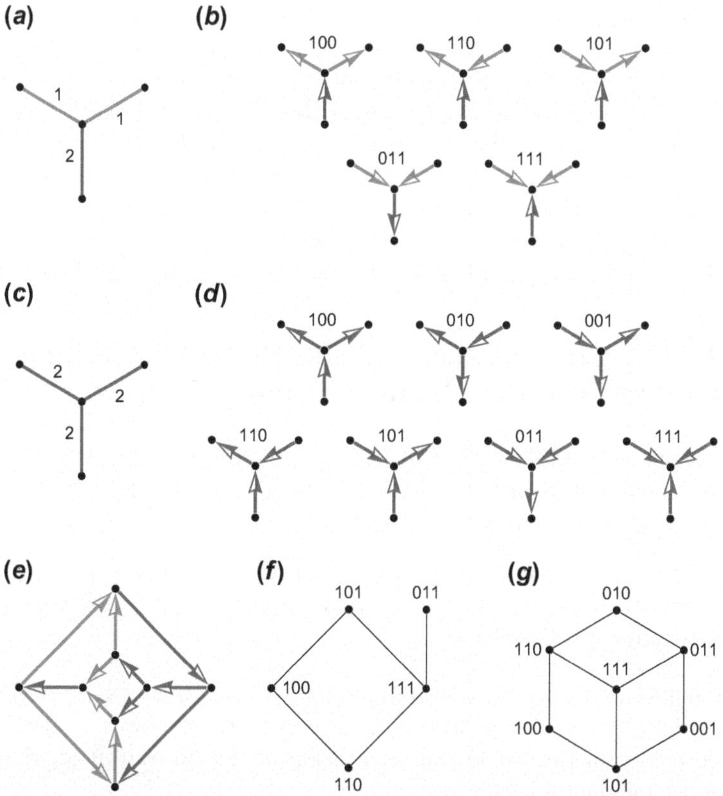

Fig. 1. Illustrations concerning NCL machines with AND/OR constraint graphs: **(a)** NCL AND vertex with two incident (red) edges of weight 1 and one incident (blue) edge of weight 2; **(b)** all 5 possible orientations of edges incident to an NCL AND vertex, each assigned a distinct 3-bit binary string; **(c)** NCL OR vertex with three incident (blue) edges of weight 2; **(d)** all 7 possible orientations of edges incident to an NCL OR vertex, each assigned a distinct 3-bit binary string; **(e)** example NCL machine using only NCL AND and NCL OR vertices; **(f)** subgraph of the 3-bit Gray code graph induced by binary strings from (b); **(g)** subgraph of the 3-bit Gray code graph induced by binary strings from (d). (Color figure online)

We remark that, for a specified NCL machine Υ with constraint graph G, and for states $\mathcal{F}_a$ and $\mathcal{F}_b$ of Υ (corresponding to allowed orientations of G), it is known to be PSPACE-complete to solve the *configuration-to-configuration* problem of deciding if there is a path between states $\mathcal{F}_a$ and $\mathcal{F}_b$ in the state transition graph for Υ [14]. Moreover, it is known that this PSPACE-completeness result holds (see [14,39]) under the constraint that all vertices are assigned a weight of 2 and that G corresponds to a bounded bandwidth planar *AND/OR constraint graph* consisting only of: (1) *NCL AND vertices* incident to exactly two edges of weight 1 and one edge of weight 2; and (2) *NCL OR vertices* incident to exactly three edges of weight 2.

For visual intuition, we refer the reader to Fig. 1(a) for an illustration of an NCL AND vertex incident to two (red) edges of weight 1 and one (blue) edge of weight 2, and Fig. 1(b) for all 5 possible orientations for these incident edges that ensure the weighted in-degree of the vertex is at least 2. Likewise, we refer the reader to Fig. 1(c) for an illustration of an NCL OR vertex incident to three (blue) edges of weight 2, and Fig. 1(d) for all 7 possible orientations for these incident edges that ensure the weighted in-degree of the vertex is at least 2. Finally, in Fig. 1(e) we show an explicit example of an orientation for an AND/OR constraint graph corresponding to a feasible NCL machine state.

3 PSPACE-Completeness of the Perfect Matching Reconfiguration Problem on Planar Graphs

In this section, we establish the PSPACE-completeness of the k-Cycle-PM-RP on bounded bandwidth planar graphs of degree at most 6 for every $k \in 2\mathbb{N}_{>1}$ (see Theorem 1). We begin with the following lemmas:

Lemma 1. *The state transition graph for an NCL AND vertex corresponds to the Fig. 1(f) subgraph of the* 3*-bit Gray code graph induced by the binary strings* $\{100, 110, 101, 011, 111\}$.

Proof. It suffices to observe the enumerated states for an NCL AND vertex shown in Fig. 1(b), where we also assign each of these states a distinct 3-bit Gray code based on (in order) the orientation of the bottom-most, right-most, and left-most incident edges. □

Lemma 2. *The state transition graph for an NCL OR vertex corresponds to the Fig. 1(g) subgraph of the* 3*-bit Gray code graph induced by the binary strings* $\{100, 010, 001, 110, 101, 011, 111\}$.

Proof. It suffices to observe the enumerated states for an NCL OR vertex shown in Fig. 1(d), where we again assign each of these states a distinct 3-bit Gray code in the same manner as in the proof argument for Lemma 1. □

We are now prepared to establish the following PSPACE-completeness results:

Theorem 1. *The* k*-Cycle-PM-RP is PSPACE-complete on bounded bandwidth planar graphs of degree at most* 6 *for every* $k \in 2\mathbb{N}_{>1}$.

Proof. Letting H be a graph hosting an instance of the k-Cycle-PM-RP, we can observe that any perfect matching for H can be described with a number of bits polynomial in the size of H. We can also observe that, for any $k \in 2\mathbb{N}_{>1}$, and for any perfect matching $\mathcal{M}$ for H, in time polynomial in the size of H it is possible to enumerate all perfect matchings for H having edge sets whose symmetric difference with the edge set for $\mathcal{M}$ induces a cycle of length k. Accordingly, and akin to the proof argument for membership of the configuration-to-configuration problem for an NCL machine in PSPACE [14], we can traverse the state space for

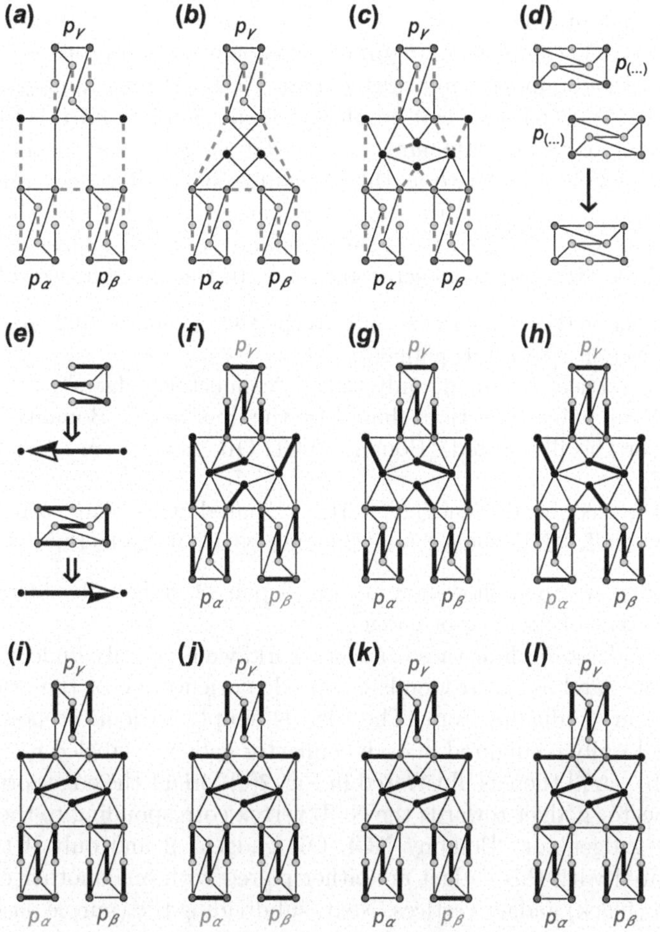

Fig. 2. Illustrations for k-Cycle-PM-RP gadgets: **(a)** planar "Bonamy NCL AND gadget" with three attached "connector gadgets"; **(b)** non-planar "Bonamy NCL OR gadget" with three attached "connector gadgets"; **(c)** novel "Planar NCL OR gadget" with three attached "connector gadgets"; **(d)** procedure for identifying "connector gadgets"; **(e)** scheme for mapping perfect matchings to edge orientations in an NCL machine; **(f–l)** representative perfect matchings for the novel "Planar NCL OR gadget", each corresponding to one of the 7 distinct incident edge orientations for an NCL OR vertex; see Theorem 1 for further details. (Color figure online)

any instance of the k-Cycle-PM-RP by nondeterministically selecting transitions between perfect matchings while storing only the current perfect matching. As Savitch's theorem [27] allows us to to make this procedure deterministic with polynomial overhead, we therefore have that the k-Cycle-PM-RP is in PSPACE.

To establish the PSPACE-hardness of the k-Cycle-PM-RP for every $k \in 2\mathbb{N}_{>1}$, we proceed by modifying the "Theorem 1" reduction in Bonamy et al. [1].

Briefly, the idea of Bonamy et al. [1] was to give a polynomial time reduction from the PSPACE-complete configuration-to-configuration problem for an NCL machine Υ with a bounded bandwidth planar AND/OR constraint graph [14,39]. To do so, they created a special graph gadget to simulate NCL AND vertices, which we refer to as the "Bonamy NCL AND gadget", and to simulate NCL OR vertices, which we refer to as the "Bonamy NCL OR gadget", as well as a "connector gadget" to join these objects together. We illustrate the "Bonamy NCL AND gadget" in Fig. 2(a) and the "Bonamy NCL OR gadget" in Fig. 2(b), each with three attached "connector gadgets". In this context, we color:

- (black) the vertices belonging only to the the "Bonamy NCL AND gadget" or the "Bonamy NCL OR gadget";
- (yellow) vertices belonging only to the "connector gadgets";
- (green) the pairs of vertices shared by the illustrated "Bonamy NCL AND gadget" or the illustrated "Bonamy NCL OR gadget" and the "connector gadgets";
- (red) the "connector gadget" vertices shared with some other adjacent "Bonamy NCL AND gadget" or "Bonamy NCL OR gadget".

In Fig. 2(d) we then show how to join a pair of gadgets by identifying the vertices in a pair of "connector gadgets".

To now understand how these gadgets work, we need only understand how a given perfect matching corresponds to an edge orientation of the original NCL machine we are reducing from. The idea is simply to look at how a perfect matching is locally configured in each connector gadget – labeled p_α, p_β, and p_γ in Fig. 2(a,b) – and then as illustrated in Fig. 2(e), orient the edge corresponding to the connector gadget towards the NCL vertex corresponding to the "Bonamy NCL AND gadget" or "Bonamy NCL OR gadget" if and only if the (green) vertices shared with this gadget are either paired with one another or are both paired with (black) gadget vertices. Next, subdividing the (purple dashed) edges in Fig. 2(a,b) exactly $k-4$ times (i.e., replacing these edges with a path having $k-3$ edges), we can observe that the only manner of "flipping" the orientations specified by the "connector gadgets" – transitioning between potentially many perfect matchings – will exactly correspond to the state transition graph shown in Fig. 1(f) for the "Bonamy NCL AND gadget" and Fig. 1(g) for the "Bonamy NCL OR gadget". Where but for the planarity constraint, this would allow us to appeal to Lemma 1 and Lemma 2 to nail the PSPACE-hardness result.

The problem in the current setting, as one can plainly see by looking at Fig. 2(b), is that the "Bonamy NCL OR gadget" is non-planar. To establish the desired PSPACE-hardness result, we replace this gadget with the novel "Planar NCL OR gadget" shown in Fig. 2(c), again with three attached instances of the "connector gadget". Here, we can explicitly check that this novel "Planar NCL OR gadget" satisfies all of the aforementioned properties of the "Bonamy NCL OR gadget". In Fig. 2(f–l) we also show representative perfect matchings corresponding to each of the 7 possible NCL OR vertex configurations from Fig. 1(d), highlighting in (blue) the $\{p_\alpha, p_\beta, p_\gamma\}$ "connector gadget" labels corresponding to edges oriented towards the NCL vertex being represented by the gadget.

It remains to observe that the configuration-to-configuration problem for an NCL machine is PSPACE-complete even if the associated AND/OR constraint graph is a bounded bandwidth planar graph [14, 39], and that our reduction using the novel "Planar NCL OR gadget" will necessarily yield a bounded bandwidth planar graph of degree at most 6. □

4 Augmenting Graphs to Make the ($k = 4$)-Cycle-PM-RP Tractable

Proposition 1. *For any finite set of forbidden subgraphs S, the $\Pi(r)$-GAP is FPT in r if Π is the set of graphs with the hereditary property of being planar (if desired), weakly-chordal (or chordal if desired), and having no induced subgraphs in S.*

Proof. From Cai [4] we have that the $\Pi(r)$-GAP problem is FPT in r whenever the set of graphs Π is defined with respect to the hereditary property of having no induced subgraphs isomorphic to graphs in a finite forbidden set S. Letting n be the order of the input graph, and letting h be the maximum order of any graph in S, the fundamental idea is that: (1) should any exist, we can find induced subgraphs isomorphic to graphs in S in $\mathcal{O}\left(|S| \cdot n^h\right)$ time; (2) unless S contains a clique (in which case the problem becomes trivial) we can "destroy" any instance of an induced subgraph isomorphic to an element of S by adding one of $\binom{h}{2}$ possible edges; (3) each time we add an edge to "destroy" a subgraph, we can determine in $\mathcal{O}\left(poly\left(n\right)\right)$ time whether the modified graph is in Π, doing so by simply checking that no induced subgraphs are isomorphic to the elements of S; and (4) if we repeat the process of finding and "destroying" subgraphs r times, we will only ever need to consider $\mathcal{O}\left(f\left(r\right) \cdot poly\left(n\right)\right)$ possibilities for edge additions, where $f\left(r\right)$ is some computable function depending only on r. Observing that the input graph for an instance of the $\Pi(r)$-GAP problem must be planar if Π is a set of planar graphs, and moreover that planarity testing can be done in linear time, these same arguments establish that $\Pi(r)$-GAP is FPT in r if Π is the set of planar graphs having no induced subgraphs isomorphic to graphs from S.

Concerning now the additional constraint that graphs in Π must be weakly-chordal (or chordal if desired), we can first observe that both chordality and weak-chordality are polynomial time testable properties (see, e.g., Spinrad & Sritharan [28]). We can next observe that, though neither weak-chordality nor chordality admit a finite forbidden subgraph characterization, it is nevertheless possible to find a hole or anti-hole of a specified length $k \geq 4$ in time polynomial in the size of a graph (see, e.g., Spinrad [29]). Finally, regarding the process of adding edges to make a graph weakly-chordal (or chordal if desired), we can once again sequentially find and "destroy" holes and anti-holes by adding edges at most r times. However, as there is no longer any fixed size bound on the order of the holes or anti-holes we might find, we modify our earlier approach by using of the "Chordalization(G, k)" procedure from section "3" from of Cai [4] to destroy holes in a graph and its complement while still ensuring that we only ever need to consider $\mathcal{O}(f(r) \cdot poly(n))$ possibilities for edge additions (see also a concurrent work of Kaplan, Shamir, & Tarjan [17] using similar arguments), where $f(r)$ is again some computable function depending only on r. Accordingly, aside from the specifics of their discovery and elimination, we can treat holes and anti-holes as any other finite forbidden induced subgraph in the set S when finding and eliminating subgraphs via edge addition. Putting everything together now yields the current proposition.

Corollary 1. *The $\Pi(r)$-GAP is fixed parameter tractable in r if Π is the set of graphs with the hereditary property of being planar and strongly orderable.*

Proof. Per "Corollary 17" of Dragan [10], we have that a graph is strongly orderable if and only if it is a "weakly-triangulated" (equivalently, weakly-chordal) graph that is free of n-sun graphs for every $n \in \mathbb{N}_{>2}$. Noting Proposition 1, and the fact that the n-sun graph is planar only for $n \in \{3, 4\}$, it now suffices to observe that Π can equivalently be defined as the set of planar and weakly-chordal graphs having no induced subgraphs isomorphic to either the 3-sun graph or 4-sun graph.

Corollary 2. *The $\Pi(r)$-GAP is fixed parameter tractable in r if Π is the set of graphs with the hereditary property of being cographs (or planar cographs if desired).*

Proof. Noting Proposition 1, it suffices to observe that cographs are equivalent to P_4-free graphs (see, e.g., Corneil et al. [7]).

5 Concluding Remarks

As we were only able to establish the PSPACE-completeness of the k-Cycle-PM-RP for every $k \in 2\mathbb{N}_{>1}$ on bounded bandwidth planar graphs of degree at most 6 (see Theorem 1 in Sect. 3), we leave open the problem of finding more stringent conditions under which this PSPACE-completeness result holds. In particular, we conjecture that the k-Cycle-PM-RP remains PSPACE-complete on bounded bandwidth bipartite planar graphs of degree at most 3 for at least some $k \in 2\mathbb{N}_{>1}$.

References

1. Bonamy, M., et al.: The perfect matching reconfiguration problem. In: Proceedings of the 44th International Symposium on Mathematical Foundations of Computer Science (MFCS), pp. 80:1–80:14 (2019)
2. Bondy, J.A., Murty, U.S.R.: Graph Theory with Applications, 1st edn. Macmillan Press, New York, NY (1976)
3. Bousquet, N., Hatanaka, T., Ito, T., Mühlenthaler, M.: Shortest reconfiguration of matchings. In: Proceedings of the 45th International Workshop on Graph-Theoretic Concepts in Computer Science (WG), pp. 162–174 (2019)

4. Cai, L.: Fixed-parameter tractability of graph modification problems for hereditary properties. Inf. Process. Lett. **58**(4), 171–176 (1996)
5. Cao, Y., Marx, D.: Chordal editing is fixed-parameter tractable. Algorithmica **75**(1), 118–137 (2016)
6. Chvátal, V.: On certain polytopes associated with graphs. J. Comb. Theory. Ser. B **18**(2), 138–154 (1975)
7. Corneil, D.G., Lerchs, H., Burlingham, L.S.: Complement reducible graphs. Discret. Appl. Math. **3**(3), 163–174 (1981)
8. Curticapean, R., Xia, M.: Parameterizing the permanent: genus, apices, minors, evaluation mod 2k. In: Proceedings of the 56th Annual Symposium on Foundations of Computer Science (FOCS), pp. 994–1009 (2015)
9. Diestel, R.: Graph Theory, 5th edn. Springer-Verlag, Heidelberg (2017).
10. Dragan, F.F.: Strongly orderable graphs A common generalization of strongly chordal and chordal bipartite graphs. Discret. Appl. Math. **99**(1–3), 427–442 (2000)
11. Eswaran, K.P., Tarjan, R.E.: Augmentation problems. SIAM J. Comput. **5**(4), 653–665 (1976)
12. Fukuda, K., Matsui, T.: Finding all the perfect matchings in bipartite graphs. Appl. Math. Lett. **7**(1), 15–18 (1994)
13. Galluccio, A., Loebl, M.: On the theory of Pfaffian orientations. I. Perfect matchings and permanents. Electron. J. Comb. **6**(R6), 1–18 (1999)
14. Hearn, R.A., Demaine, E.D.: PSPACE-completeness of sliding-block puzzles and other problems through the nondeterministic constraint logic model of computation. Theor. Comput. Sci. **343**(1–2), 72–96 (2005)
15. Ito, T., Kakimura, N., Kamiyama, N., Kobayashi, Y., Okamoto, Y.: Shortest reconfiguration of perfect matchings via alternating cycles. In: Proceedings of the 27th Annual European Symposium on Algorithms (ESA), pp. 61:1–61:15 (2019)
16. Ito, T., Kakimura, N., Kamiyama, N., Kobayashi, Y., Okamoto, Y.: Shortest reconfiguration of perfect matchings via alternating cycles. SIAM J. Discret. Math. **36**(2), 1102–1123 (2022)
17. Kaplan, H., Shamir, R., Tarjan, R.E.: Tractability of parameterized completion problems on chordal, strongly chordal, and proper interval graphs. SIAM J. Comput. **28**(5), 1906–1922 (1999)
18. Kasteleyn, P.W.: The statistics of dimers on a lattice: I. The number of dimer arrangements on a quadratic lattice. Physica **27**(12), 1209–1225 (1961)
19. Kasteleyn, P.W.: Dimer statistics and phase transitions. J. Math. Phys. **4**(2), 287–293 (1963)
20. Kasteleyn, P.W.: Graph theory and crystal physics. In: Harary, F. (ed.) Graph Theory and Theoretical Physics, pp. 43–110. Academic Press, London, UK (1967)
21. Kobayashi, Y., Kurita, K., Wasa, K.: Polynomial-delay and polynomial-space enumeration of large maximal matchings. In: Proceedings of the 48th International Workshop on Graph-Theoretic Concepts in Computer Science (WG), pp. 342–355 (2022)
22. Krishnamoorthy, M.S., Deo, N.: Node-deletion NP-complete problems. SIAM J. Comput. **8**(4), 619–625 (1979)
23. Lewis, J.M.: On the complexity of the maximum subgraph problem. In: Proceedings of the 10th Annual Symposium on Theory of Computing (STOC), pp. 265–274 (1978)
24. Papadimitriou, C.H.: Computational Complexity, 1st edn. Addison Wesley, Reading, MA (1994)
25. Regge, T., Zecchina, R.: Combinatorial and topological approach to the 3D Ising model. J. Phys. A: Math. Gen. **33**(4), 741–761 (2000)

26. Savage, C.D.: A survey of combinatorial Gray codes. SIAM Rev. **39**(4), 605–629 (1997)
27. Savitch, W.J.: Relationships between nondeterministic and deterministic tape complexities. J. Comput. Syst. Sci. **4**(2), 177–192 (1970)
28. Spinrad, J., Sritharan, R.: Algorithms for weakly triangulated graphs. Discret. Appl. Math. **59**(2), 181–191 (1995)
29. Spinrad, J.P.: Finding large holes. Inf. Process. Lett. **39**(4), 227–229 (1991)
30. Temperley, H.N.V., Fisher, M.E.: Dimer problem in statistical mechanics - an exact result. Philos. Mag. **6**(68), 1061–1063 (1961)
31. Tesler, G.: Matchings in graphs on non-orientable surfaces. J. Comb. Theory Ser. B **78**(2), 198–231 (2000)
32. Uno, T.: Algorithms for enumerating all perfect, maximum and maximal matchings in bipartite graphs. In: Leong, H.W., Imai, H., Jain, S. (eds.) ISAAC 1997. LNCS, vol. 1350, pp. 92–101. Springer, Heidelberg (1997).
33. Uno, T.: A fast algorithm for enumerating bipartite perfect matchings. In: Proceedings of the 12th International Symposium on Algorithms and Computation (ISAAC), pp. 367–379 (2001)
34. Uno, T.: A fast algorithm for enumerating non-bipartite maximal matchings. NII J. **3**, 89–97 (2001)
35. Valiant, L.G.: The complexity of computing the permanent. Theor. Comput. Sci. **8**(2), 189–201 (1979)
36. Valiant, L.G.: The complexity of enumeration and reliability problems. SIAM J. Comput. **8**(3), 410–421 (1979)
37. Vazirani, V.V.: NC algorithms for computing the number of perfect matchings in $K_{3,3}$-free graphs and related problems. Inf. Comput. **80**(2), 152–164 (1989)
38. Yannakakis, M.: Node-and edge-deletion NP-complete problems. In: Proceedings of the 10th Annual Symposium on Theory of Computing (STOC), pp. 253–264 (1978)
39. van der Zanden, T.C.: Parameterized complexity of graph constraint logic. In: Proceedings of the 10th International Symposium on Parameterized and Exact Computation (IPEC), pp. 282–293 (2015)

Inductive Tracing and the Complexity of Finding Hamiltonian Path in DAGs

Ronak Bhadra(✉) and Raghunath Tewari

Indian Institute of Technology, Kanpur, India
{ronakb,rtewari}@cse.iitk.ac.in

Abstract. The HamiltonianPath problem is a classic decision problem that is NP-complete in general graphs, but solvable in linear time and in nondeterministic logspace (NL) for directed acyclic graphs (DAGs).

We show that the HamiltonianPath problem for DAGs lies in UL∩coUL, providing an improved upper bound on its complexity. To the best of our knowledge, this is the first instance where an unambiguous space upper bound has been established for a natural problem without relying on any variant of the classic Reinhardt-Allender double inductive counting technique, which is itself based on the Immerman-Szelepcsényi inductive counting method.

Our proof introduces a novel technique, which we call *inductive tracing*. While reminiscent of inductive counting, it diverges in a key respect: it does not perform any counting across recursive stages. We also apply this routine to obtain a parameterized unambiguous space bound for the LongPath problem in DAGs.

Keywords: Unambiguous Logspace Computations · Hamiltonian Path · Directed Acyclic Graphs

1 Introduction

A Hamiltonian path in a graph G is a path in G that visits each vertex of G exactly once. Deciding whether a graph has a Hamiltonian path or not is known as the HamiltonianPath problem. A graph having a Hamiltonian path is called a *traceable* graph. The HamiltonianPath problem is a well-studied problem in the field of computational complexity, especially because of the fact that it is famously known to be NP-complete [7]. However, for certain restricted classes of graphs, the HamiltonianPath problem can be shown to be much easier. For example, the HamiltonianPath problem can be solved in linear time for rectangular grid graphs [4]. Similarly, for Directed Acyclic Graphs or DAGs, the HamiltonianPath problem can also be solved in linear time. Moreover, the HamiltonianPath problem for DAGs can be shown to be in the complexity class NL as well. This raises the question whether we can further reduce the computational complexity of the HamiltonianPath problem for DAGs.

A. Jeż and J. Otop (Eds.): FCT 2025, LNCS 16106, pp. 57–67, 2026.
https://doi.org/10.1007/978-3-032-04700-7_5

A related problem of relevance is the LongPath problem, which is deciding whether there is a path of length at least k from one vertex to another in a graph. The HamiltonianPath problem is a special case of the LongPath problem. Though the LongPath problem is NP-complete for general graphs and even for planar graphs [2], but it is known to be NL-complete for DAGs.

The complexity class UL is the unambiguous subclass of NL and coUL is the class of the complements of the languages (problems) in UL. A decision problem L is in UL if and only if there exists a non-deterministic Turing Machine M deciding L in logspace such that, for every input instance x, M has at most one accepting computation path. The first major breakthrough in the study of the class UL came with the work of Reinhardt-Allender [10], where they showed that the classes UL and NL are equivalent in non-uniform settings. They also showed that the Reachability problem in min-unique graphs is in UL $\cap$ coUL. Using their technique of double inductive counting, Reachability in several other restricted classes of graphs (for example, planar graphs [1]) as well as some other problems (for example, longest paths in planar DAGs [8], 3-connected planar graph isomorphism [12]) have been shown to be in UL$\cap$coUL. Apart from this, using the double inductive counting technique, it has also been shown that directed graph reachability is solvable in unambiguous $O(\log^2 n)$ space and polynomial time simultaneously [6], which was later further improved to unambiguous $O(\log^{1.5} n)$ space and polynomial time [9]. In fact, to the best of our knowledge, any upper bound given till now on the unambiguous space complexity of any natural problem has been proved using some variant of the double inductive counting technique of Reinhardt-Allender [10].

In this paper, we show that the HamiltonianPath problem in Directed Acyclic Graphs or DAGs is in UL $\cap$ coUL. For this purpose, we give an unambiguous and co-unambiguous non-deterministic algorithm that decides the HamiltonianPath problem in G. Our algorithm is akin to the inductive counting algorithm of Immerman-Szelepcsényi [3,11] and the double inductive counting algorithm of Reinhardt-Allender [10], which is inspired from the former. However, unlike the above mentioned algorithms, our algorithm doesn't count anything inductively. Rather, it inductively traces the Hamiltonian path if any such path exists, or else rejects the input if at any point it fails to do so. We call this novel technique *inductive tracing.* The intuition behind this technique comes from Kahn's algorithm for topological sorting of DAGs [5]. As far as we are aware, this is the first natural problem for which an upper bound on the unambiguous space complexity is proved without relying on the double inductive counting technique of Reinhardt-Allender [10]. We also give a parametrized upper bound on the unambiguous space complexity of the LongPath problem in DAGs using our UL $\cap$ coUL routine for deciding HamiltonianPath in DAGs.

This result is slightly non-intuitive (if not, surprising) because in general, the HamiltonianPath problem is a much more difficult problem than the Reachability problem. In certain restricted subclasses of graphs, the HamiltonianPath problem may be easier but it still intuitively seems to be at least as difficult as the Reachability problem for those subclasses. Any prior result regarding HamiltonianPath does not violate this intuition to the best of our knowledge. For example, in planar DAGs, though HamiltonianPath is in UL ∩ coUL but Reachability is also in UL ∩ coUL. In fact, the proof for this [8] explicitly uses Reachability to solve LongPath or HamiltonianPath for that matter. However, Reachability in DAGs is NL-complete. Our result thus implies that the HamiltonianPath problem is in some sense easier than the Reachability problem for DAGs. This of course may not be true if NL = UL, but this is a long-standing open question anyway.

Another interesting aspect that we want to highlight is that HamiltonianPath in DAGs has a natural UL algorithm because of the fact that there is at most one Hamiltonian path in a DAG. However, this automatically doesn't imply that HamiltonianPath is in coUL. In fact, proving a similar result for st-reachability in graphs where there is at most one path from s to t (for which again there is a natural UL algorithm) would prove that UL = coUL. This is unlike most other problems shown to be in UL ∩ coUL, which have no natural UL algorithm apart from some inductive counting algorithm based on the algorithm of Reinhardt-Allender [10].

The rest of the paper is organized as follows. In Sect. 2, we will be proving the following theorem, which serves as our main result.

Theorem 1. *The problem of deciding whether a given Directed Acyclic Graph has a Hamiltonian path or not is in* UL ∩ coUL.

In Sect. 3, we discuss some consequences of our main result. We give a parametrized upper bound on the unambiguous space complexity of the LongPath problem in DAGs. We also show that reachability in traceable DAGs is in UL ∩ coUL.

2 Proof of Theorem 1

In this section, we will prove Theorem 1. First we will show that a DAG has at most one Hamiltonian path and how this fact can help us devise a natural UL algorithm for HamiltonianPath problem. In order to show that HamiltonianPath is also in coUL, we will provide an inductive non-deterministic algorithm and then prove that this algorithm works unambiguously (that is, accepts or rejects exactly along one computation path) for all input instances. We will then argue for the correctness of the algorithm in deciding whether a DAG has a Hamiltonian path or not.

We will first show that there is at most one Hamiltonian path in a DAG.

Lemma 1. *There is at most one Hamiltonian path in a Directed Acyclic Graph.*

Proof. We prove this by contradiction. Let us assume there are two Hamiltonian paths in a DAG G, say P_1 and P_2. There must be at least one vertex v which is at different positions in paths P_1 and P_2, say i and j respectively. Otherwise, P_1 and P_2 are identical. Without loss of generality, let us assume $i < j$. There must be some vertex u which comes after v in path P_1 but comes before v in path P_2. Now, if we take the subpath from v to u in P_1 and then the subpath from u to v in P_2, we get a cycle $(v \ldots u \ldots v)$. This is a contradiction to our assumption that G is acyclic.

Now, our unambiguous logspace routine to decide if a DAG G has a Hamiltonian path or not is simple. Non-deterministically guess a walk of length n (number of nodes in G) in G. Accept if the guess succeeds. Reject if the guess fails. Since there is only one Hamiltonian path in a DAG and every walk in a DAG is a path, therefore we can guarantee that our algorithm will accept along exactly one computation path, if a Hamiltonian path exists in G and will reject along all computation paths if no Hamiltonian path exists in G.

The existence of a unique Hamiltonian path in a DAG is sufficient to show that the HamiltonianPath problem for a DAG is in UL. However, it doesn't automatically imply that the HamiltonianPath problem for a DAG is in coUL as well. For this purpose, we devise Algorithm 1 which uses Algorithm 2 as a subroutine. Algorithm 1 is an inductive algorithm akin to the inductive counting algorithms of ImmermanâĂŞSzelepcsényi [3,11] and Reinhardt-Allender [10], but unlike those algorithms, our algorithm doesn't actually *count* any parameter inductively. Rather, it inductively computes the next vertex in the (possible) Hamiltonian path from the previous vertex. In this way, the algorithm is able to trace the Hamiltonian path if it exists and otherwise, it is able to detect that such a path doesn't exist.

The algorithm first checks if the graph has a single source (and sink) or not. If the answer is yes, the algorithm proceeds. Otherwise, it rejects the input (that is, it declares that there is no Hamiltonian path in the graph G). The algorithm then starts by initializing a variable v as the (unique) source vertex (say s) of G. Now, the algorithm, in its i-th iteration, counts the number of vertices (say m_i) which have an incoming edge from v but no incoming edges from vertices other than those that belong to the (unique) path of length $i-1$ from s to v. This is done using the function COUNT, which non-deterministically guesses a path of length l from s to v and simultaneously counts the number of incoming edges to vertex v' from vertices on this path. The algorithm rejects if in any iteration i, m_i does not turn out to be one. The algorithm accepts only when it successfully completes n iterations. In other words, the algorithm accepts if and only if the value of m_i is one for all $i \leq n$.

We now provide the pseudocode for Algorithm 1 and the subroutine COUNT.

Algorithm 1. Determine whether a DAG G has a Hamiltonian path

```
Require: G = (V, E)
Check if the graph has exactly one source and one sink.
if no then
    return reject
else
    set s to be the source vertex of G
Initialize v = s, l = 1
while l ≠ n do
    Initialize m = 0, w = NULL
    for each v' ∈ V such that (v, v') ∈ E do
        d = COUNT(G, s, v, v', l)
        if d = "?" then
            return "?"
        if indegree(v') = d then
            m = m + 1
            w = v'
    if m ≠ 1 then
        return reject
    else
        v = w
        l = l + 1
return accept
```

Algorithm 2. Count the number of vertices, in a path of length l from s to v in G, from which there is an edge to v'

```
function COUNT(G = (V, E), s, v, v', l)
    Initialize x = s, d = 0, i = 0
    while x ≠ v and i ≠ l do
        if (x, v') ∈ E then
            d = d + 1
        non-deterministically guess a vertex x' such that (x, x') ∈ E
        if guess is correct then
            i = i + 1
            x = x'
        else
            return "?"
    if (x, v') ∈ E then
        d = d + 1
    if x = v and i = l then
        return d
    else
        return "?"
```

In order to prove that HamiltonianPath problem is in $\mathsf{UL} \cap \mathsf{coUL}$, we need to show the following things:

- Algorithm 1 accepts or rejects an input along exactly one computation path for all input instances (DAGs).
- Algorithm 1 accepts an input (a DAG G) if and only if G has a Hamiltonian path.

In Sect. 2.1, we prove that the algorithm works unambiguously for all input instances. In Sect. 2.2, we argue for the correctness of the algorithm in determining whether a DAG is Hamiltonian or not. For the sake of easier readability of the proof, we will define v_i to be the value of the variable v at the beginning of the i-th iteration of the while loop (in line 7) of Algorithm 1. Note that, by this definition, v_1 happens to be the source vertex of the input graph G, if G has a unique source vertex. We will also assume the input to Algorithm 1 is a DAG named G.

2.1 Unambiguity of Algorithm 1

In this subsection, we will show that Algorithm 1 works unambiguously, that is, accepts or rejects along exactly one computation path, for all input instances.

We first prove the following lemma.

Lemma 2. *Let v_i be the value of the variable v at the beginning of the i-th iteration of the while loop (in line 7) of Algorithm 1. There is a unique path $(v_1, v_2, \ldots, v_i)$ of length $(i-1)$ from the source v_1 to vertex v_i in G for all i. Also, there are no incoming edges to vertex v_i from any vertex other than $v_1, v_2, \ldots, v_{i-1}$.*

Proof. We will prove this by induction.

The base case is trivial. There is a unique path of length 0 from v_1 to v_1 (itself), which is the source vertex itself (v_1). There are no incoming edges to vertex v_1 since v_1 is the source vertex of G.

Let us assume that the lemma is true for all $i < j$. We will show that the lemma then must also be true for $i = j$.

Now let us assume there is another path (say P) of length $(j-1)$ from vertex v_1 to v_j, other than $(v_1, v_2, \ldots, v_j)$. There must be at least one vertex other than vertices $v_1, \ldots, v_j$ in P. Let u be the last such vertex in P. Let the vertex succeeding u in P be v_k. As per our assumption, $v_k \in \{v_1, v_2, \ldots, v_j\}$. Now, this implies that there is an edge from vertex u to vertex v_k. However, according to our induction hypothesis, if $k < j$, there is no incoming edge to v_k from any vertex other than $v_1, v_2, \ldots, v_{k-1}$, which will be a contradiction.

During the $(j-1)$-th iteration of the while loop in line 7 of Algorithm 1, the algorithm counts the total number d of incoming edges to a vertex v' (which has an edge from vertex v_{j-1}) from all vertices in the path of length $(j-2)$ from vertex v_1 to v_{j-1} (which are the vertices $v_1, v_2, \ldots, v_{j-1}$ by our induction hypothesis). The algorithm checks if d is equal to the indegree of vertex v' and increments the variable m by 1 if it is. At the end of the for loop, m indicates the number of vertices which have an incoming edge from v_{j-1} but has no incoming edge from vertices other than $v_1, v_2, \ldots, v_{j-1}$. The algorithm proceeds further only if $m = 1$ and sets the variable v to be the vertex for which the above mentioned criteria is satisfied. By definition, the value of the variable at the end of the $(j-1)$-th iteration is v_j. Thus, v_j does not have any incoming edge from vertices other than $v_1, v_2, \ldots, v_{j-1}$. Hence, having an edge from vertex u to vertex v_j will also be a contradiction. Therefore, there cannot be any path of length $(j-1)$ from vertex v_1 to vertex v_j other than $(v_1, v_2, \ldots, v_{j-1}, v_j)$.

Lemma 2 implies that whenever the function Count is called in line 10 of Algorithm 1, it returns a value of d other than "?" along exactly one computation path. Any computation path where Count returns "?" is ignored (returns "?" in line 12). Only the computation path where Count returns a non "?" value proceeds further and can either accept or reject depending on the input. Thus, for all input instances, Algorithm 1 accepts or rejects along exactly one computation path. All other computation paths are ignored (returns "?").

Remark 1. "?" is like a don't-care state. Roughly speaking, if we want an UL algorithm for our problem, we can consider *"?"* to be *reject*. If we want a coUL algorithm, we can consider *"?"* to be *accept*.

2.2 Correctness of Algorithm 1

In this subsection, we will argue for the correctness of Algorithm 1. First, we will show that if the input DAG G has a Hamiltonian path, then Algorithm 1 must accept. For this, we first state and prove the following lemma.

Lemma 3. *Let v_i be the value of the variable v at the beginning of the i-th iteration of the while loop (in line 7) of Algorithm 1. If the input DAG G has a Hamiltonian path, then v_i must be the i-th vertex of the Hamiltonian path in G.*

Proof. We will prove this by induction.

The base case is easy to see. The Hamiltonian path in a DAG must start from the source vertex of the DAG, since the source vertex cannot occur at any other position in the Hamiltonian path. As we have already mentioned, v_1 is the source vertex of G since the variable v is initialized to be the (unique) source vertex of G (in line 5 of Algorithm 1).

Let's assume that v_{i-1} is the $(i-1)$-th vertex of the Hamiltonian path in G. The value of register v at the beginning of $(i-1)$-th iteration is v_{i-1}. Let the i-th vertex of the Hamiltonian path be u. Therefore, there is an edge (v_{i-1}, u) from vertex v_{i-1} to u in G. Also, there cannot be any incoming edge to vertex u other than from vertices $v_1, v_2, \ldots, v_{i-1}$. This is because, if there is an edge to u from a vertex y that is subsequent to u in the Hamiltonian path, then it implies that there is a cycle $(u \ldots yu)$ in G, which will be a contradiction since G is promised to be acyclic. Moreover, there is only one vertex in G which has an incoming edge from v_{i-1} but has no incoming edge from any vertex other than vertices $v_1, v_2, \ldots, v_{i-1}$. This is because the vertices $v_1, v_2, \ldots, v_{i-1}$ can't have an incoming edge from vertex v_{i-1} since otherwise there will be a cycle. Any vertex which comes after u in the Hamiltonian path in G must be preceded by a vertex other than $v_1, v_2, \ldots, v_{i-1}$ and hence, must have an incoming edge from at least one vertex other than $v_1, v_2, \ldots, v_{i-1}$. Thus, u is the only vertex in G that has an incoming edge from v_{i-1} and has no incoming edge from vertices other than $v_1, v_2, \ldots, v_{i-1}$.

The COUNT routine non-deterministically guesses a path of length $(i-2)$ from v_1 to v_{i-1}. During the process, it counts the total number of incoming edges to a vertex v' from any of the vertices in the guessed path.

The COUNT routine is used for each vertex that has an incoming edge from v_{i-1}. In this way, Algorithm 1 (in lines 9–15) computes the number (m) of vertices that have an incoming edge from v_{i-1} but have no incoming edge from vertices other than $v_1, v_2, \ldots, v_{i-1}$. We have already seen that m must be exactly equal to one if G is Hamiltonian. Therefore, Algorithm 1 in this case will move to line 18, set v to be u and proceed for the next iteration. Thus, the value of the variable v at the beginning of the i-th iteration, that is v_i, is equal to u. Hence, v_i is the i-th vertex of the Hamiltonian path in G.

Lemma 3 implies that if the input G to Algorithm 1 has a Hamiltonian path, then the algorithm will run for n iterations. At the beginning of the i-th iteration, the value of the variable v will be set to the i-th vertex of the Hamiltonian path. If Algorithm 1 successfully runs for n iterations, then it must accept the input at line 21 once it comes out of the loop.

Next, we show that if Algorithm 1 accepts the input DAG G, then G must have a Hamiltonian path.

Lemma 4. *If Algorithm 1 accepts, then there must be a Hamiltonian path in G.*

Proof. We can see that for all i, v_i has an incoming edge from v_{i-1}. Therefore, for all i, $(v_1, v_2, \ldots, v_i)$ constitutes a walk and since the graph G is acyclic, this walk is also a path. Algorithm 1 accepts only when the while loop (in line 7) has successfully iterated for n times. Hence, there exists a path $(v_1, v_2, \ldots, v_n)$ in G. This must be a Hamiltonian path since it is a path consisting of n vertices. Thus, there always exists a Hamiltonian path in G if Algorithm 1 accepts.

Lemma 3 and Lemma 4 taken together prove that Algorithm 1 accepts if and only if the input DAG G has a Hamiltonian path.

Remark 2. Algorithm 1 as presented here seems to work correctly under the promise that the input graph is a DAG. Identifying whether a graph is acyclic or not is NL-complete. Hence, we don't know if the promise (of being acyclic) on the input can be independently checked in $\mathsf{UL} \cap \mathsf{coUL}$. However, if a graph which is not acyclic is provided as input, Algorithm 1 will reject it. We can prove by induction that for any k, the vertex v_k is not a part of any cycle. v_1 is the source of the graph and hence not a part of any cycle (base case). Let us first assume that v_k is not a part of any cycle for any $k < i$ (induction hypothesis). Now, Algorithm

1 ensures that v_i has incoming edges only from vertices $v_1, v_2, \ldots, v_{i-1}$. If v_i is a part of a cycle, then it must have an edge from a vertex which is also a part of that cycle. However, none of $v_1, v_2, \ldots, v_{i-1}$ are part of any cycle (according to induction hypothesis). Therefore v_i also cannot be part of any cycle. If a graph that is not acyclic is provided as input to Algorithm 1, then the vertices in the graph which are part of some cycles can never occur as v_k, for any k. Thus, the while loop at line 7 in Algorithm 1 cannot run for n times and at some point, the input will be rejected at line 16–17 in Algorithm 1. Hence, the set of traceable DAGs can be decided in $\mathsf{UL} \cap \mathsf{coUL}$ even without any promise on the input.

3 Some Consequences of Our Result

In this section, we discuss two consequences of our main result.

3.1 A Parametrized Upper Bound on the Unambiguous Space Complexity of the LongPath Problem

In this subsection, we provide a parametrized upper bound on the unambiguous non-deterministic space complexity of the LongPath problem in DAGs using our $\mathsf{UL} \cap \mathsf{coUL}$ routine for the HamiltonianPath problem in DAGs. We state this bound in the following theorem.

Theorem 2. *Given a directed acyclic graph G having n vertices and two vertices s and t in G, the problem of deciding whether there exists a path of length at least $n-k$ from s to t is solvable in unambiguous (and co-unambiguous) non-deterministic $O(k \log n)$ space.*

Proof. We assume we are given an $\mathsf{UL} \cap \mathsf{coUL}$ routine for HamiltonianPath in DAGs Theorem 1. That is, we have a non-deterministic logspace Turing machine that decides along exactly one computation path whether the given DAG has a Hamiltonian path or not. Now, consider one by one all subsets of vertices (other than s and t) of size at most k. For each such subset, eliminate the vertices in this subset from the graph and check whether the remaining subgraph has a Hamiltonian path from s to t using the $\mathsf{UL} \cap \mathsf{coUL}$ routine. If the answer is yes, then we accept since the Hamiltonian path in the subgraph is a path of length at least $n-k$ from s to t as the size of the subgraph is at least $n-k$. If the answer is no, then we move on to the next subset of vertices and continue. If the answer is no for all subsets of vertices, then we reject.

This Algorithm works because if there is a path of length at least $n-k$ from s to t, then this path is a Hamiltonian path in the subgraph that does not contain the vertices (at most k in number) not in this path. That is, there exists a path of length at least $n-k$ from s to t if and only if there is a subset of (at most k) vertices, which when removed from the graph, the remaining subgraph (of size at least $n-k$) has a Hamiltonian path from s to t.

Remark 3. Note that unlike the UL bound for the Hamiltonian path problem in DAGs, the unambiguous space bound provided in Theorem 2 is not a natural or obvious one. There can be several paths of length at least $n-k$ from s to t and hence, just non-deterministically guessing a path of length at least $n-k$ from s to t doesn't necessarily give an unambiguous algorithm. Our $\mathsf{UL} \cap \mathsf{coUL}$ routine for HamiltonianPath in DAGs is crucial for providing the unambiguous space bound in Theorem 2.

3.2 Reachability in Traceable DAGs

In this subsection, we prove the following theorem.

Theorem 3. *Reachability in traceable DAGs is decidable in* $\mathsf{UL} \cap \mathsf{coUL}$.

Proof. Given a traceable DAG G and two vertices u and v in G, we need to decide whether u is reachable from v or not. Let s and t be the source and sink of G respectively. Since G is traceable, therefore there is a unique Hamiltonian path from s to t in G (by Lemma 2). Now, our unambiguous (and co-unambiguous) routine to decide reachability is simple. Non-deterministically guess a walk (which is also a path because G is a DAG) of length n from s to t. There will be only one computation path which will correctly guess such a path since there is only one such path in G. Return "?" along the computation paths which don't make a correct guess (see Remark 1). While making the guesses of which vertex to take next in the walk, keep track of which vertex out of u and v we visit first. In the computation path that correctly guesses the Hamiltonian path from s to t, if u is visited before v, then v is reachable from u and hence accept; else, v is not reachable from u and hence reject. This routine accepts or rejects along exactly one computation path. Therefore, this routine is unambiguous as well as co-unambiguous. Thus, reachability in traceable DAGs is decidable in $\mathsf{UL} \cap \mathsf{coUL}$.

Acknowledgments. This work has been partly funded by Research-I foundation.

References

1. Bourke, C., Tewari, R., Vinodchandran, N.V.: Directed planar reachability is in unambiguous log-space. In: 22nd Annual IEEE Conference on Computational Complexity (CCC 2007), 13–16 June 2007, San Diego, California, USA, pp. 217–221 (2007). https://doi.org/10.1109/CCC.2007.9
2. Garey, M., Johnson, D., Stockmeyer, L.: Some simplified NP-complete graph problems. Theor. Comput. Sci. **1**(3), 237–267 (1976). https://doi.org/10.1016/0304-3975(76)90059-1, https://www.sciencedirect.com/science/article/pii/0304397576900591
3. Immerman, N.: Nondeterministic space is closed under complement. SIAM J. Comput. **17**, 935–938 (1988)

4. Itai, A., Papadimitriou, C.H., Szwarcfiter, J.L.: Hamilton paths in grid graphs. SIAM J. Comput. **11**(4), 676–686 (1982). https://doi.org/10.1137/0211056
5. Kahn, A.B.: Topological sorting of large networks. Commun. ACM **5**(11), 558–562 (1962). https://doi.org/10.1145/368996.369025
6. Kallampally, V.A.T., Tewari, R.: Trading determinism for time in space bounded computations. In: 41st International Symposium on Mathematical Foundations of Computer Science, MFCS 2016, 22–26 August 2016 - Kraków, Poland, pp. 10:1–10:13 (2016). https://doi.org/10.4230/LIPIcs.MFCS.2016.10
7. Karp, R.M.: Reducibility among combinatorial problems. In: Miller, R.E., Thatcher, J.W. (eds.) Proceedings of a symposium on the Complexity of Computer Computations, held March 20–22, 1972, at the IBM Thomas J. Watson Research Center, Yorktown Heights, New York, USA, pp. 85–103. Plenum Press, New York (1972). https://doi.org/10.1007/978-1-4684-2001-2_9
8. Limaye, N., Mahajan, M., Nimbhorkar, P.: Longest paths in planar dags in unambiguous logspace. In: Proceedings of the Fifteenth Australasian Symposium on Computing: The Australasian Theory - Volume 94, pp. 101–108. Australian Computer Society, Inc. (2009)
9. van Melkebeek, D., Prakriya, G.: Derandomizing isolation in space-bounded settings. SIAM J. Comput. **48**(3), 979–1021 (2019). https://doi.org/10.1137/17M1130538
10. Reinhardt, K., Allender, E.: Making nondeterminism unambiguous. SIAM J. Comput. **29**(4), 1118–1131 (2000). https://doi.org/10.1137/S0097539798339041
11. Szelepcsényi, R.: The method of forced enumeration for nondeterministic automata. Acta Informatica **26**, 279–284 (1988)
12. Thierauf, T., Wagner, F.: The isomorphism problem for planar 3-connected graphs is in unambiguous logspace. In: Albers, S., Weil, P. (eds.) STACS 2008, 25th Annual Symposium on Theoretical Aspects of Computer Science, Bordeaux, France, 21–23 February 2008, Proceedings, vol. 1, pp. 633–644. Schloss Dagstuhl - Leibniz-Zentrum für Informatik, Germany (2008). https://doi.org/10.4230/LIPICS.STACS.2008.1327

m-Eternal Domination and Variants on Some Classes of Finite and Infinite Graphs

Tiziana Calamoneri[1], Federico Corò[2], Neeldhara Misra[3], Saraswati Girish Nanoti[3], and Giacomo Paesani[1](✉)

[1] Sapienza University of Rome, Rome, Italy
paesani@di.uniroma1.it
[2] University of Padua, Padua, Italy
[3] Indian Institute of Technology, Gandhinagar, India

Abstract. We study the m-Eternal Domination problem, which is the following two-player game between a defender and an attacker on a graph: initially, the defender positions k guards on vertices of the graph; the game then proceeds in turns between the defender and the attacker, with the attacker selecting a vertex and the defender responding to the attack by moving a guard to the attacked vertex. The defender may move more than one guard on their turn, but guards can only move to neighboring vertices. The defender wins a game on a graph G with k guards if the defender has a strategy such that at every point of the game the vertices occupied by guards form a dominating set of G and the attacker wins otherwise. The *m-eternal domination number* of a graph G is the smallest value of k for which (G, k) is a defender win.

We show that m-Eternal Domination is NP-hard, as well as some of its variants, even on special classes of graphs. We also show structural results for the Domination and m-Eternal Domination problems in the context of four types of infinite regular grids: square, octagonal, hexagonal, and triangular, establishing tight bounds.

Keywords: Eternal Domination · Roman Domination · Italian Domination · NP-hardness · Bipartite Graphs · Split Graphs · Infinite Grids

1 Introduction

A subset S of vertices in a simple undirected graph G is called a *dominating set* if every vertex outside of S has a neighbor in S. Finding a smallest-sized dominating set is a fundamental computational problem, and indeed, several variations of this basic premise have been considered in the literature [19].

In this work, our focus is on the m-eternal version of the Domination problem, where we think of the vertices of the dominating set as being occupied by "guards" that can move in response to "attacks" on the vertices. Specifically,

A. Jeż and J. Otop (Eds.): FCT 2025, LNCS 16106, pp. 68–81, 2026.
https://doi.org/10.1007/978-3-032-04700-7_6

we consider the following two-player graph game. To begin with, the defender places k guards on the vertices of the graph. The game continues with alternating turns between the attacker and the defender. On each turn, the attacker chooses a vertex to attack, and the defender responds by repositioning guards, ensuring at least one guard moves onto the attacked vertex. The defender may relocate multiple guards during a turn, but each guard can only move to an adjacent vertex. The defender wins if there exists a strategy that maintains a dominating set of vertices occupied by guards throughout the entire game. Otherwise, the attacker wins. The m-eternal domination number of a finite graph G is the smallest value of k for which the defender wins.

Two variations of the domination problem are ROMAN DOMINATION and ITALIAN DOMINATION. A Roman dominating function on a graph $G = (V, E)$ is a function $f : V \to \{0, 1, 2\}$ such that every vertex v with $f(v) = 0$ has at least one neighbor u with $f(u) = 2$. The weight of a Roman dominating function is the sum $\sum_{v \in V} f(v)$. The Roman domination number $\gamma_R(G)$ is the minimum weight of a Roman dominating function on G. This concept is inspired by the defensive strategy of the Ancient Roman Empire, where guards were stationed such that every unguarded position was adjacent to a doubly guarded position. An Italian dominating function on a graph $G = (V, E)$ is a function $f : V \to \{0, 1, 2\}$ such that for every vertex v with $f(v) = 0$, the sum of $f(u)$ over all neighbors u of v is at least 2. The weight of an Italian dominating function is the sum $\sum_{v \in V} f(v)$. The Italian domination number $\gamma_I(G)$ is the minimum weight of an Italian dominating function on G. This concept generalizes Roman domination by allowing more flexible assignments of resources to vertices and their neighbors.

Given the context of warfare, it is very natural to propose studying the "eternal" variations of Roman and Italian domination: presumably, regions will be attacked more than once, and the guards will have to find ways of reconfiguring themselves so that they maintain the defense invariants that they started with.

Our Contributions. In Sect. 3, we consider the computational complexity of the m-ETERNAL DOMINATION problem and the Roman and Italian variants in the m-eternal setting. We show that m-ETERNAL DOMINATION (and its connected variant, where we require the locations of the guards to induce a connected subgraph) is NP-hard even on bipartite graphs of diameter four (Theorem 1); and the m-ETERNAL ROMAN DOMINATION and m-ETERNAL ITALIAN DOMINATION problems are NP-hard even on split graphs (Theorems 2 and 3).

It is well known [1] that for split graphs, the eternal domination number is at most its domination number plus one, and a characterization of the split graphs which achieve equality is given. The authors also show that the decision versions of DOMINATION and m-ETERNAL DOMINATION are NP-complete, even on Hamiltonian split graphs. Moreover, computing the eternal domination number can be solved in polynomial time for any subclass of split graphs for which the domination number can be computed in polynomial time, in particular for strongly chordal split graphs.

In Sect. 4, we focus on structural results for the DOMINATION and m-ETERNAL DOMINATION problems in the context of four types of infinite regular grids: square, octagonal, hexagonal, and triangular. We prove m-eternal dominating sets that are optimal according to a parameter (that will be defined later) expressing the concept of minimality in infinite graphs.

We highlight that exact results for the eternal domination number are given in [13,17,25] and, more recently, in [14], for grids with either 2 or 3 rows, and in [3] for grids with 4, 5, or 6 rows. When G is a general $n \times m$ square grid, it is clear that the eternal domination number must be at least as large as the domination number, so by the result in [18], it must be at least $\lfloor \frac{(n-2)(m-2)}{5} \rfloor - 4$; the best-known upper bound was determined in [22], and is $\frac{mn}{5} + O(n + m)$, thus showing that the eternal domination number is within $O(m+n)$ of the domination number.

2 Preliminaries

In this section, we introduce definitions and terminology that will be relevant to our discussions later. We will be dealing with simple undirected graphs denoted by $G = (V, E)$. The *degree* of a vertex v is the number of edges incident to v. An *independent set* or *clique* is a subset of vertices such that no or every possible edge is present, respectively. The *neighborhood* of a vertex v, denoted $N(v)$, is the set of all vertices adjacent to v. The *closed neighborhood* of a vertex v, denoted $N[v]$, is the set $N(v) \cup \{v\}$. A graph is *bipartite* if its vertex set V can be partitioned into two disjoint *independent sets* U and W. A *split graph* is a graph in which the vertex set V can be partitioned into a clique and an independent set. In this paper, we say that a graph $G = (V, E)$ is infinite if V is countably infinite. For more detailed graph terminology, we refer to [11], while for computational complexity terminology, we refer to [15].

2.1 Domination, m-Eternal Dominations and Variants

We first introduce the definitions of domination, Roman domination, and Italian domination. Then, we give the concept of an m-eternal domination game and define all the previous kinds of domination under this light.

Definition 1 *[2,27]. A* dominating set *for a graph $G = (V, E)$ is a subset $D \subseteq V$ such that every vertex not in D is adjacent to at least one vertex in D. The* domination number *$\gamma(G)$ is the minimum cardinality of a dominating set in G. If the subgraph induced by D in G is connected, then D is said to be a* connected dominating set.

The DOMINATION problem (which is to determine a minimum dominating set for graphs) is NP-hard even for planar graphs [15] and bipartite graphs of diameter three [24].

Definition 2 *[28]. A* Roman dominating function *on a graph $G = (V, E)$ is a function $f : V \to \{0, 1, 2\}$ such that every vertex v for which $f(v) = 0$ is adjacent to at least one vertex u for which $f(u) = 2$. The* Roman domination number *$\gamma_R(G)$ is the minimum weight $\sum_{v \in V} f(v)$ of a Roman dominating function on G. If the subgraph induced by $D = \{v \mid f(v) \neq 0\}$ in G is connected, then D is said to be a* connected Roman dominating set.

The ROMAN DOMINATION problem (which is to determine the Roman domination number of a graph) is NP-hard even for split graphs, bipartite graphs, and planar graphs [10], for chordal graphs [23], and for subgraphs of grids [26].

Definition 3 *[8, 20]. An* Italian dominating function *on a graph $G = (V, E)$ is a function $f : V \to \{0, 1, 2\}$ such that, for every vertex v for which $f(v) = 0$, the sum of the function values of the neighbors of v is at least 2. The* Italian domination number *$\gamma_I(G)$ is the minimum weight $\sum_{v \in V} f(v)$ of an Italian dominating function on G. If the subgraph induced by $D = \{v \mid f(v) \neq 0\}$ in G is connected, then D is said to be a* connected Italian dominating set.

The ITALIAN DOMINATION problem (which is to determine the Italian domination number of a graph) is NP-hard even for bipartite planar graphs, chordal bipartite graphs [12], and split graphs [9].

The *m-eternal domination game* is a two-player turn-based game on graph G: a team of guards initially occupies a dominating set on a graph G. An *attacker* then assails a vertex without a guard on it; the *defender* exploits the guards to defend against the attack: one of the guards has to move to the attacked vertex from one of its neighbors, while the defender can choose whether to move or not the remaining guards to one of its neighbor vertices. This attack-defend procedure continues eternally. The defender wins if they can eternally maintain a dominating set against any sequence of attacks; otherwise, the attacker wins.

Definition 4 *[16]. An* m-eternal domination set *for a graph $G = (V, E)$ is a subset $D \subseteq V$ where a defender can place its guard and win the m-eternal domination game. The* m-eternal domination number *of G, denoted by $\gamma^\infty(G)$, is the minimum number of guards required to defend an indefinite sequence of attacks.*

We use the term *configuration* to refer to the position of guards during the m-eternal domination game. Note that when configurations correspond to dominating functions rather than sets, the value of the function at a vertex indicates the number of guards occupying the vertex, and again multiple guards are allowed to move in one round; no vertex accommodates more than two guards in a configuration, and guards can only move to neighbor vertices. By requiring the configuration of guards at every step to correspond to a Roman dominating function or to an Italian dominating function instead of a dominating set, we obtain the *m-eternal Roman domination* and *m-eternal Italian domination games*, and the minimum numbers of guards necessary to defend forever G are the *m-eternal Roman domination* and *m-eternal Italian domination numbers* and denoted by $\gamma_R^\infty(G)$ and $\gamma_I^\infty(G)$, respectively. Similarly to what we did for m-domination, Roman and Italian domination, it is possible to define m-eternal connected domination and m-eternal Roman and Italian connected domination.

In this paper, we do not consider variations of these games where only one guard is allowed to move in one step (as, *e.g.*, in [5]). To distinguish from this variation, the $m-$ prefix highlights that multiple guards are allowed to move.

We focus on the m-Eternal Domination computational problem, consisting in taking as input a graph G and determining $\gamma^\infty(G)$. The m-Eternal Roman Domination and m-Eternal Italian Domination problems are defined analogously.

When we refer to the m-Eternal Domination problem (and variants), unless mentioned otherwise, we are referring to the optimization version of the question, as opposed to explicitly calling it the minimization problem.

2.2 Infinite Regular Grids

We work with four kinds of infinite regular grids, shown in Fig. 1. In the following definitions and the rest of the paper, we adopt the following convention. We imagine that the vertices of the infinite grids occupy integer coordinates of the Cartesian plane. Despite this, we do not mean to consider the grids as graphs embedded in a metric space. We make this assumption to describe the moving strategies in an easier way.

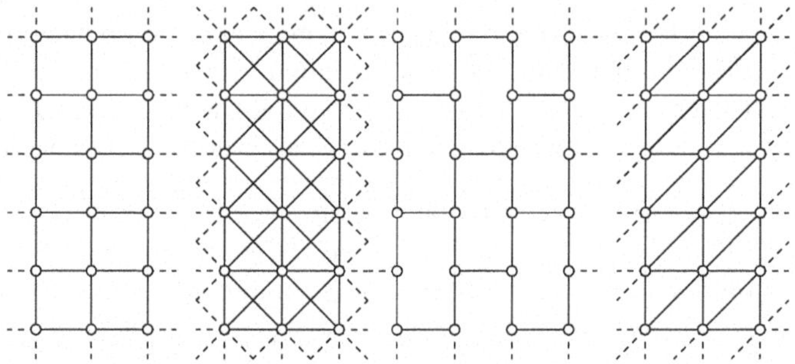

(a) T_4 (b) T_8 (c) T_3 (d) T_6

Fig. 1. The four considered infinite regular grid graphs.

We formally define the infinite regular grids as follows:

- **infinite square grid** T_4 (see Fig. 1a): every vertex (x, y) is connected with $(x+1, y)$, $(x-1, y)$, $(x, y+1)$ and $(x, y-1)$;
- **infinite octagonal grid** T_8 (see Fig. 1b): every vertex (x, y) is connected with $(x+1, y)$, $(x-1, y)$, $(x, y+1)$, $(x, y-1)$, $(x-1, y-1)$, $(x+1, y+1)$, $(x-1, y+1)$ and $(x+1, y-1)$;

- **infinite hexagonal grid** T_3 (see Fig. 1c): every vertex (x, y) is connected both with $(x, y+1)$ and with $(x, y-1)$; moreover, every vertex (x, y) is connected to $(x+1, y)$ if $x+y$ is even and to $(x-1, y)$ otherwise;
- **infinite triangular grid** T_6 (see Fig. 1d): every vertex (x, y) is connected with $(x+1, y)$, $(x-1, y)$, $(x, y+1)$, $(x, y-1)$, $(x-1, y-1)$ and $(x+1, y+1)$.

2.3 Optimal Domination

For infinite graphs, the meaning of a minimum set with respect to a set of properties is far from clear. Some related work uses notions of *density* that can be applied to infinite graphs with respect to different properties, *e.g.*, for cops configurations in Cops and Robber games [4] or for eternal vertex cover configurations on regular grids [6]. In the context of dominating sets for infinite graphs, we choose to avoid working with a density-based approach and instead adopt a local cardinality-based approach to determine when a proposed dominating set is optimal. Given two dominating sets S and T, we say that S is *at least as good as* T if there is an injective map from S to T, that is, the cardinality of S is smaller or equal to that of T. A dominating set is *optimal* if it is at least as good as every other dominating set of G. We now turn to a stronger notion of optimality.

Consider an infinite graph G with finite maximum degree. Given a subset S of $V(G)$ and a vertex $v \in V(G)$, we define the *domination index* of v with respect to S as $|S \cap N[v]|$. A dominating set S is *strongly optimal* if, for every vertex v of the graph, the domination index of v with respect to S is exactly one[1]. Note that a strongly optimal dominating set may not exist: for example, consider the infinite complete bipartite graph G with $V(G) = \{a_i \mid i \in \mathbb{N}\} \cup \{b_i \mid i \in \mathbb{N}\}$ and $E(G) = \{(a_i, b_j) \mid i, j \in \mathbb{N}\}$. However, we note that a strongly optimal dominating set, if it exists, is also optimal:

Lemma 1. *If S is a strongly optimal dominating set and T is any dominating set of an infinite graph G with finite maximum degree, then S is at least as good as T.*

Proof. Consider the following map $f : S \to T$ defined as follows: for every vertex $v \in S$, let $f(v)$ be an arbitrarily chosen vertex from $N[v] \cap T$.

We first show that f is well-defined. Since T is a dominating set, for every $v \in S$, we have $N[v] \cap T \neq \emptyset$, so there always exists a vertex to which v can be mapped. Next, we prove that f is injective. Suppose, by contradiction, that there exist two distinct vertices $u, v \in S$ such that $f(u) = f(v) = w$ for some $w \in T$. This implies that $w \in N[u] \cap N[v]$, and consequently, the domination index of w with respect to S would be at least two. However, this contradicts

[1] In the context of finite graphs, a dominating set such that every vertex is uniquely dominated is called an *exact dominating set*, and because of connections with coding theory, in some settings such dominating sets are called *perfect codes*.

the fact that S is a strongly optimal dominating set, where every vertex must have a domination index of exactly one. Therefore, f is an injective map from S to T, proving that S is at least as good as T. □

To recap, we have the following definitions:

Definition 5. *Let G be an infinite graph with finite maximum degree and $S \subseteq V(G)$. The* domination index *of a vertex $v \in V(G)$ with respect to S is defined as $|S \cap N[v]|$. A dominating set S is* optimal *if for any other dominating set T of G, there exists an injective map from S to T. A dominating set S is* strongly optimal *if for every vertex $v \in V(G)$, the domination index of v with respect to S is exactly one.*

3 Complexity Results for Variants of the m-Eternal Domination

In this section, we prove that the considered problems are all NP-hard, even for interesting subclasses of graphs. Due to space restrictions, we omitted some proofs and refer the reader to the full version [7].

We preliminarily give a property whose analogous for the m-eternal vertex cover problem has been proved in [21].

Lemma 2. *Let G be a graph and Z be a connected dominating set of G. Then $\gamma^{\infty}(G) \leq |Z| + 1$.*

Proof. Initially, place a guard on each vertex of $Z \cup \{v\}$, where v is any vertex not belonging to Z. We prove that this configuration is an m-eternal dominating set. At every step of the attacking sequence, we will maintain the following invariant: there are guards on Z and one extra guard anywhere else, which we will refer to as the *floating guard*, positioned on vertex v. Whenever any vertex v' that does not have a guard is attacked, let $v_0, \dots v_k$ be any path such that $v_0 = v$, $v_k = v'$ and $v_i \in Z$ for every $i = 1, \dots, k-1$, for some $k \geq 1$. For every $i < k$, the guard on v_i moves to v_{i+1}. After that, each vertex of $Z \cup \{v'\}$ is occupied by a guard, thus restoring both the invariant and successfully defending the attack. The size of $Z \cup \{v\}$ is $|Z| + 1$ by construction, and since G is finite, the size is preserved through sequences of dominating sets defending the attack sequence. □

Theorem 1. *The m-*ETERNAL DOMINATION *and m-*ETERNAL CONNECTED DOMINATION *problems are NP-hard, even on bipartite graphs of diameter* 4.

Proof. We describe here a reduction from the DOMINATING SET problem. Let $\langle G = (V, E); k \rangle$ be an instance of DOMINATING SET, and let $V = \{v_1, \dots, v_n\}$. Construct a bipartite graph $H = (A \cup B, F)$ as follows: let $U = \{u_1, \dots, u_n\}$ and $W = \{w_1, \dots, w_n\}$ be two copies of V; let $\{p_1, \dots, p_{n+1}\} = P$ and w be new vertices. We define $A = U \cup P$ and $B = W \cup \{w\}$. The set of edges F is defined as follows: u_i is adjacent to w_j iff $(v_i, v_j) \in E$ or $i = j$; moreover, (w, u_i) and (w, p_j) are both in F for each $i = 1, \ldots, n$ and $j = 1, \ldots, n+1$. An example of this construction is shown in Fig. 2. Now we consider the instance $\langle H; k+2 \rangle$ for the m-Eternal Domination and argue the equivalence of these instances.

Forward Direction. Let $S \subseteq V$ be a dominating set of G of size at most k. Initially, place a guard on each u_i for every $v_i \in S$. Denote this set of vertices which are occupied by guards in A as T. The set $T \cup \{w\}$ induces a connected subgraph on H because every vertex in A is adjacent to w. This set also forms a dominating set of H because every vertex in A is adjacent to w, and every vertex in B has a neighbor in T because S is a dominating set of G. By Lemma 2, $\gamma^\infty(H) \leq |T \cup \{w\}| + 1 = k + 2$.

Reverse Direction. Suppose H has a winning strategy with $k+2$ guards, and consider one of its configurations such that P and the vertex w have exactly one guard each: such configuration can be obtained from any other configuration after a vertex of P that does not have a guard is attacked. Let S be the subset of V: $v_i \in S$ if and only if u_i or w_i has a guard. Clearly, S has size at most k and is a dominating set for G. Indeed, suppose there exists a vertex $v_i \in V \setminus S$ that has no neighbor in S; thus, w_i has no guard, and has no neighbor that has a guard, and that the considered configuration is not part of a winning strategy, a contradiction.

Finally, we consider the instance $\langle H; k+2 \rangle$ for the m-Eternal Connected Domination. To prove the equivalence of these instances, we reuse the previous proof and note that all configurations in a winning strategy are connected. □

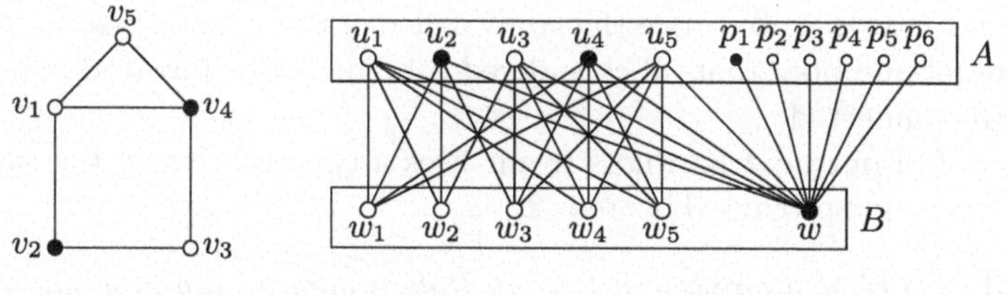

Fig. 2. The construction for the proof of Theorem 1. On the left, graph G for which the black vertices represent a (minimum size) dominating set. On the right, auxiliary bipartite graph $H = (A, B; F)$ for which the black vertices represent a configuration of an m-eternal dominating set.

Lemma 3. *Let G be a graph and Z be a connected dominating set of G. Then* $\gamma_R^\infty(G) \leq 2|Z| + 1$.

Theorem 2. *The* m-ETERNAL ROMAN DOMINATION *and* m-ETERNAL CONNECTED ROMAN DOMINATION *problems are NP-hard, even on split graphs.*

Proof. We describe here a reduction from the DOMINATING SET problem. Let $\langle G = (V, E); k\rangle$ be an instance of DOMINATING SET and let $V = \{v_1, \ldots, v_n\}$. Construct a split graph $H = (A \cup B, F)$ as follows: let $A = \{u_1, \ldots, u_n\}$ be a copy of V and let $B = \{w_i^{(j)}\}$ for $1 \leqslant i \leqslant n$ and $1 \leqslant j \leqslant 2n + 2$ contain $2n + 2$ copies of V. The edge set F is defined as follows: A induces a clique and B induces an independent set; moreover, u_i is adjacent to $w_k^{(j)}$ for all j iff $(v_i, v_k) \in E$ and u_i is adjacent to $w_i^{(j)}$ for all i, j. An example of this construction is shown in Fig. 3. Now we consider the instance $\langle H; 2k + 1\rangle$ for the m-ETERNAL ROMAN DOMINATION and argue on the equivalence of these instances.

Forward Direction. Let $S \subseteq V$ be a dominating set of size at most k. Let T be the set of vertices in A that corresponds to S, *i.e.*. $u_i \in T$ if and only if $v_i \in S$. The set T induces a connected subgraph because it is a subset of A, which induces a clique in H. By construction, T is also a dominating set of H, because S is a dominating set of G. By Lemma 3, $\gamma_R^\infty(H) \leq 2T| + 1 \leq 2k + 1$.

Reverse Direction. Suppose that H has a winning strategy with at most $2k + 1$ guards, and consider an initial configuration induced by a Roman dominating function f of H such that the weight is equal to $2k+1$. For every $1 \leqslant \mathrm{j} \leqslant 2n+2$, let $W^{(j)} := \{w_i^{(j)} \mid 1 \leqslant \mathrm{i} \leqslant n\}$. Moreover, let $S \subseteq A$ be the set of vertices in A that have value two each, *i.e.*, $s \in S$ if and only if $f(s) = 2$. By the pigeon-hole principle, there exists a $1 \leqslant j \leqslant 2n + 2$ such that $W^{(j)}$ contains only zero value vertices, *i.e.*, $f(w_i^{(j)}) = 0$ for every $i \leqslant n$. Since $W^{(j)}$ only has neighbors in A, set S must be a dominating set of G, and this means that G has a dominating set S of size at most k.

Finally, we consider the instance $\langle H; 2k+1\rangle$ for the m-ETERNAL CONNECTED ROMAN DOMINATION. To prove the equivalence of these instances, we reuse the previous proof and note that all considered configurations involved in a winning strategy are connected. □

The proof of our next hardness result uses a construction in the same style as the one used for proving Theorem 2.

Lemma 4. *Let* G *be a graph and* f *be an Italian dominating function of* G *such that the vertices with non-zero value induce a connected subgraph; moreover, let* t_f *denote the weight of* f. *Then* $\gamma_I^\infty(G) \leq t_f + 1$.

Theorem 3. *The* m-ETERNAL ITALIAN DOMINATION *and* m-ETERNAL CONNECTED ITALIAN DOMINATION *problems are NP-hard, even on split graphs.*

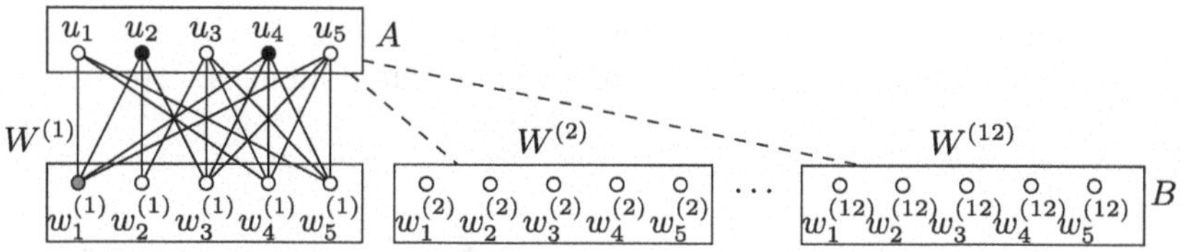

Fig. 3. The construction for the proof of Theorem 2 for graph G of Fig. 2. In the auxiliary split graph $H = (A, B; F)$, the edges between A and $W^{(1)}$ are shown in the figure, while the edges between A and every other $W^{(j)}$, $j \in [2, 12]$, are dashed lines and those within A are omitted. The color of the vertices represents a configuration of an m-eternal Roman dominating function of H: two, one, and zero guards are placed on black, grey, and white vertices, respectively.

4 m-Eternal Domination on Infinite Regular Grids

In this section, we provide a starting configuration and a strategy for the m-domination of infinite regular grids. We show that in all cases, we are able to obtain strongly optimal dominating sets that are also m-eternal dominating sets. We start with infinite square grids first.

Theorem 4 (m-Eternal Domination on the infinite Square Grid). *There is a strongly optimal dominating set for the infinite square grid T_4 that is also an m-eternal dominating set.*

Proof. We define vertex set S_4 on the infinite square grid T_4 as follows: $(0, 0) \in S_4$; moreover, if $(x, y) \in S_4$ then also $(x + 2, y + 1)$, $(x - 1, y + 2)$, $(x - 2, y - 1)$ and $(x+1, y-2)$ belong to S_4. This vertex set can be visualized in the left graph in Fig. 4.

Note that the infinite set of closed neighborhoods $N[s]$ with $s \in S_4$ forms a partition of the vertices of T_4; therefore, we claim that S_4 is a dominating set of T_4. Indeed, observe that for any guard at (x, y) there is a guard at $(x', y') = (x + 2, y + 1)$, a guard at $(x'', y'') = (x' + 1, y' - 2) = (x + 3, y - 1)$ and a guard at $(x''', y''') = (x'' + 2, y'' + 1) = (x + 5, y)$; hence, looking at the rows of T_4, for each guard at (x, y), the four vertices to its right do not contain any guard. Nevertheless, $(x + 1, y)$ is dominated by the guard at (x, y); $(x + 2, y)$ is dominated by the guard at $(x', y') = (x+2, y+1)$; $(x+3, y)$ is dominated by the guard at $(x'', y'') = (x+3, y-1)$; $(x+4, y)$ is dominated by $(x''', y''') = (x+5, y)$. The generality of the reasoning proves that S_4 is a dominating set of T_4.

Moreover, by construction, it holds $|N[v] \cap S_4| = 1$ for every $v \in V$: this means that S_4 is strongly optimal. Next, we show that S_4 is a configuration of an m-eternal dominating set of T_4. Indeed, consider an attack on a vertex $v \in V \setminus S_4$. Since S_4 is a dominating set and, by construction, there exists a unique vertex $s \in S_4$ that is a neighbor of v. It is not restrictive to consider the case $s = (i^*, j^*)$ and $v = (i^* + 1, j^*)$ (the other three cases are analogous). For every $(i, j) \in S_4$ (including (i^*, j^*)), the guard in (i, j) moves to $(i + 1, j)$. The new position of the guards is a translation of S_4 by one unit in the same direction, and thus still forms a dominating set. Therefore, with this strategy, the guards move along configurations of an m-eternal dominating set. □

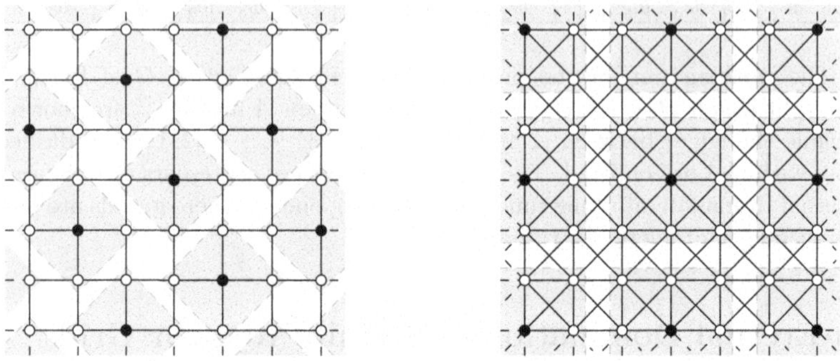

Fig. 4. On the left, graph T_4 and, on the right, graph T_8. In both figures, the black vertices represent the dominating set S described in the proof of Theorems 4 and of 5, respectively, while the shadowed zones represent how vertices of S dominate their neighbors, thus creating a partition of the vertices of the graph.

We next turn to infinite octagonal grids. The following theorem shows that we can come up with a strongly optimal dominating set that is also an eternal dominating set. The argument here is similar in spirit to square grids. The proposed solution can be visualized in the right graph in Fig. 4.

Theorem 5 (m-Eternal Domination on the infinite Octagonal Grid). *There is a strongly optimal dominating set for the infinite octagonal grid T_8 that is also an m-eternal dominating set.*

Also for the infinite hexagonal and triangular grids we obtain strongly optimal eternal dominating sets for both cases. While the proofs are deferred to the full version [7], we refer the reader to Figs. 5 and 6 for an intuition of how our dominating sets are constructed.

Theorem 6 (m-Eternal Domination on the infinite Hexagonal Grid). *There is a strongly optimal dominating set for the infinite hexagonal grid T_3 that is also an eternal dominating set.*

Theorem 7 (m-Eternal Domination on the infinite Triangular Grid). *There is a strongly optimal dominating set for the infinite triangular grid T_6 that is also an eternal dominating set.*

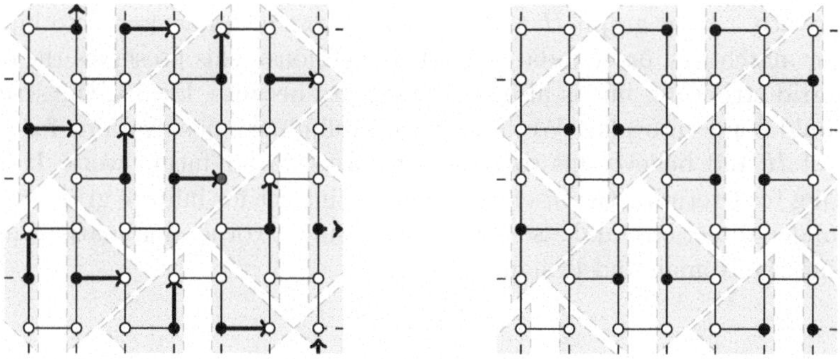

Fig. 5. The graph T_3. In both figures, the black vertices represent the dominating set S described in the proof of Theorem 6, while the shadowed zones represent how vertices of S dominate their neighbors, thus creating a partition of the vertices of the graph. On the left, the thick arrowed edges represent the movement of the guards along the edges when the red vertex is attacked.

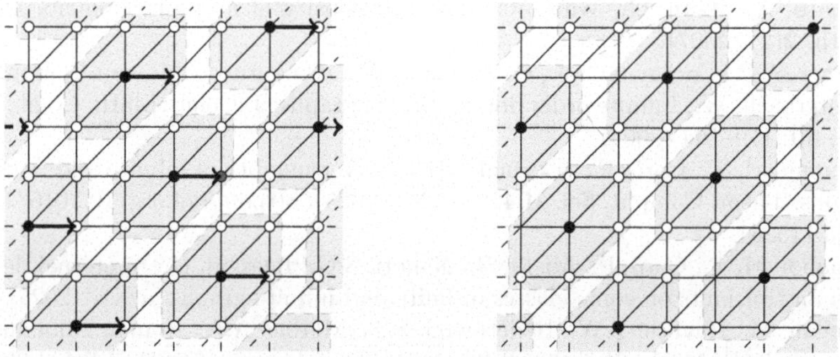

Fig. 6. The graph T_6. In both figures, the black vertices represent the dominating set S described in the proof of Theorem 7, while the shadowed zones represent how vertices of S dominate their neighbors, thus creating a partition of the vertices of the graph. On the left, the thick arrowed edges represent the movement of the guards along the edges when the red vertex is attacked.

5 Conclusions and Open Problems

We studied the m-ETERNAL DOMINATION problem and variants on classes of finite graphs and infinite regular grids. In particular, we showed that m-ETERNAL DOMINATION is NP-hard even on bipartite graphs of diameter four. Moreover, the m-ETERNAL ROMAN DOMINATION and m-ETERNAL ITALIAN DOMINATION problems are NP-hard even on split graphs. Finally, we showed optimal results for the DOMINATION and m-ETERNAL DOMINATION problems when considering four types of infinite regular grids: square, octagonal, hexagonal, and triangular.

For future work, we propose the following directions. We would like to consider infinite grids with vertex replacement, *i.e.*, each vertex in the grid is replaced with a fixed graph H and carrying across edges as complete bipartite graphs or matchings, or boundary-based connections. This preserves the overall infinite grid structure, but is also more intricate because, locally, the structure of H would come into play. We have some preliminary observations for simple choices of H and believe this to be a rich direction for future work. It is also interesting to determine the m-eternal dominating set for infinite grids with 1,2, or 3 bounded directions, and also extend our results to other domination variants (including the Roman and Italian questions).

References

1. Bard, S., Duffy, C., Edwards, M., Macgillivray, G., Yang, F.: Eternal domination in split graphs. J. Comb. Math. Comb. Comput. **101**, 121–130 (2017)
2. Berge, C.: The Theory of Graphs and Its Applications. Methuen, London (1962)
3. van Bommel, C.M., van Bommel, M.F.: Eternal domination numbers of 5× n grid graphs. J. Comb. Math. Comb. Comput. **97**(83), 102 (2016)
4. Bonato, A., Hahn, G., Wang, C.: The cop density of a graph. Contrib. Discret. Math. **2**(2) (2007)
5. Burger, A.P., Cockayne, E.J., Grundlingh, W., Mynhardt, C.M., van Vuuren, J.H., Winterbach, W.: Infinite order domination in graphs. J. Comb. Math. Comb. Comput. **50**, 179–194 (2004)
6. Calamoneri, T., Corò, F.: (Eternal) vertex cover number of infinite and finite grid graphs. Theor. Comput. Sci. **1042**, 115261 (2025). https://doi.org/10.1016/J.TCS.2025.115261
7. Calamoneri, T., Corò, F., Misra, N., Nanoti, S.G., Paesani, G.: m-eternal domination and variants on some classes of finite and infinite graphs. arXiv (2025)
8. Chellali, M., Haynes, T.W., Hedetniemi, S.T., McRae, A.A.: Roman 2-domination. Discret. Appl. Math. **204**, 22–28 (2016). https://doi.org/10.1016/J.DAM.2015.11.013
9. Chen, H., Lu, C.: A note on roman 2-domination problem in graphs. arXiv (2018)
10. Cockayne, E.J., Dreyer, P.A., Jr., Hedetniemi, S.M., Hedetniemi, S.T.: Roman domination in graphs. Discret. Math. **278**(1–3), 11–22 (2004). https://doi.org/10.1016/J.DISC.2003.06.004
11. Diestel, R.: Graph Theory, 4th Edition, Graduate texts in mathematics, vol. 173. Springer (2012)
12. Fernández, L.: Leoni, Valeria: new complexity results on roman 2-domination. RAIRO - Oper. Res. **57**(4), 1905–1912 (2023). https://doi.org/10.1051/ro/2023049
13. Finbow, S., Messinger, M., van Bommel, M.: Eternal domination in $3 \times n$ grids. Australas. J. Comb. **61**, 156–174 (2015)
14. Finbow, S., van Bommel, M.F.: The eternal domination number for $3 \times n$ grid graphs. Australas. J. Comb. **76**, 1–23 (2020)
15. Garey, M.R., Johnson, D.S.: Computers and Intractability: A Guide to the Theory of NP-Completeness. W. H. Freeman, New York (1979)
16. Goddard, W., Hedetniemi, S.M., Hedetniemi, S.T.: Eternal security in graphs. J. Comb. Math. Comb. Comput. **52**(01) (2005)
17. Goldwasser, J., Klostermeyer, W.F., Mynhardt, C.: Eternal protection in grid graphs. Utilitas Mathematica **91**, 47–64 (2013)

18. Gonçalves, D., Pinlou, A., Rao, M., Thomassé, S.: The domination number of grids. SIAM J. Discret. Math. **25**(3), 1443–1453 (2011). https://doi.org/10.1137/11082574
19. Haynes, T.W., Hedetniemi, S., Slater, P.: Fundamentals of Domination in Graphs. Pure and applied mathematics, CRC press, Boca Raton (2013)
20. Henning, M.A., Klostermeyer, W.F.: Italian domination in trees. Discret. Appl. Math. **217**, 557–564 (2017). https://doi.org/10.1016/J.DAM.2016.09.035
21. Klostermeyer, W.F., Mynhardt, C.M.: Edge protection in graphs. Australas. J. Comb. **45**, 235–250 (2009)
22. Lamprou, I., Martin, R., Schewe, S.: Eternally dominating large grids. Theor. Comput. Sci. **794**, 27–46 (2019). https://doi.org/10.1016/J.TCS.2018.09.008
23. Liu, C.H., Chang, G.J.: Roman domination on strongly chordal graphs. J. Comb. Optim. **26**(3), 608–619 (2013). https://doi.org/10.1007/S10878-012-9482-Y
24. Mahendra Kumar, R., Sadagopan, N.: A closer look at hamiltonicity and domination through the lens of diameter and convexity. Acta Informatica **61**, 357–382 (2024). https://doi.org/10.1007/s00236-024-00461-z
25. Messinger, M.E.: Closing the gap: eternal domination on $3 \times n$ grids. Contrib. Discret. Math. **12**(1) (2017)
26. Nascimento, S., Sampaio, R.: Roman domination in subgraphs of grids. Electron. Notes Discret. Math. **50**, 77–82 (2015). https://doi.org/10.1016/J.ENDM.2015.07.014
27. Ore, O.: Theory of graphs. In: Colloquium Publications. American Mathematical Society (1962)
28. ReVelle, C.: Can you protect the roman empire? Hopkins Magazine **49**(2), 40 (1997)

Parameterized Restless Temporal Path

Justine Cauvi[1,2](✉) and Laurent Viennot[1]

[1] Inria, DI ENS, Paris, France
laurent.viennot@inria.fr
[2] École Normale Supérieure de Lyon, Lyon, France
justine.cauvi@ens-lyon.fr

Abstract. Recently, Bumpus and Meeks introduced a purely temporal parameter, called vertex-interval-membership-width, which is promising for the design of fixed-parameter tractable (FPT) algorithms for vertex reachability problems in temporal graphs. We study this newly introduced parameter for the problem of restless temporal paths, in which the waiting time at each node is restricted. In this article, we prove that, in the interval model, where arcs are present for entire time intervals, finding a restless temporal path is NP-hard even if the vertex-interval-membership-width is equal to three. We exhibit FPT algorithms for the point model, where arcs are present at specific points in time, both with uniform delay one and arbitrary positive delays. In the latter case, this comes with a slight additional computational cost.

Keywords: Temporal graphs · Restless temporal paths · Parameterized algorithms · Interval-membership-width

1 Introduction

A static graph represents relationships (the edges) between entities (the nodes). However, in many real-world applications, those relationships are time-related. For instance, there is a bus at time τ from stop A to stop B in a public transit network, Alice calls Bob from time τ to τ' in a phone call network or Alice follows Bob at time τ in a social network. This kind of networks are modeled by temporal graphs. In a temporal graph, time information is associated to each edge, such as the appearance time, the delay (time it takes to traverse the edge). Depending on the application, different models can represent a temporal network [7,14]. In the simplest model called point model here, each edge appears at given points in time [2,9,19]. In the more complex model called interval model here, edges are available for entire time intervals [5,17]. From these models arise many classical problems that have been well studied [2,5,9,12,15,18,20]. Connectivity in this context is more naturally grasped by temporal walks, that is, walks where edges are traversed one after another in time. Thus, a prominent problem is to find a temporal path, which is a temporal walk in which each node is visited at most once [5,25]. An interesting variant is to find a Δ-restless temporal path (or walk) [4,8]. In a Δ-restless temporal path, the waiting time at a node cannot

A. Jeż and J. Otop (Eds.): FCT 2025, LNCS 16106, pp. 82–93, 2026.
https://doi.org/10.1007/978-3-032-04700-7_7

exceed a given value Δ. This notion of path is related to the detection of a chain of infections from an individual to another in the SIR (Susceptible-Infected-Recovered) model [3,16,21]. In this model, an individual becomes immune upon recovery. The restricted waiting time in this case is the time before an individual becomes immune after infection. Another application of this problem is in the context of public transit networks, in which an individual does not want to wait at a stop for too long.

In the point model, Casteigts, Himmel, Molter and Zschoche [8] showed that deciding whether there is a restless temporal path from a node to another is NP-complete. This motivates the design of temporal parameters that bring tractability to the problem. Bumpus and Meeks [6] introduced a novel parameter called vertex-interval-membership-width which is promising for the design of fixed parameter tractable (FPT for short) algorithms for reachability problems in temporal graphs. Intuitively, it measures the maximum number of active nodes at a given time where a node is considered active during the full interval between the first appearance of one of its incident edges and the last such appearance. Despite the recent interest in this new parameter, no FPT algorithm for the restless temporal path problem has been proposed for it so far. This article presents such an algorithm.

1.1 Related Work

Casteigts, Himmel, Molter and Zschoche [8] showed several parameterized algorithms and hardness results on restless temporal path in the point model. Among others, they showed that it is fixed-parameter tractable for three different parameters: the number of hops of the temporal path, the feedback edge number of the underlying graph and the timed feedback vertex number which is a temporal version of the feedback vertex number they introduced in the article. Thejaswi, Lauri and Gionis [24] also presented an FPT algorithm for restless temporal path with respect to the number of hops, among other results. Zschoche [26] proposed a randomized $\mathcal{O}(4^{k-d}|G|^{\mathcal{O}(1)})$ algorithm to decide whether there is a restless temporal path from s to t in G of length at most k, where d is the minimum length of a temporal st-path. The parameter we study here is not bounded by any of the parameters used so far for the design of algorithms for restless temporal path (and vice versa).

Note that computing a restless temporal walk between two nodes (rather than a temporal path) can be done in polynomial time in the point model as shown by Himmel, Nichterlein and Niedermeier [4]. However, Orda and Rom [22] proved that finding a restless temporal walk is NP-hard in the interval model.

Bumpus and Meeks [6] introduced the interval-membership-width parameter which captures the temporal aspect of the graph (unlike parameters such as the feedback edge number of the underlying graph). They worked with the point model and showed that determining if a temporal graph is temporally Eulerian (i.e. admits a temporal circuit visiting each underlying edge exactly once) is FPT when parameterized by the interval-membership-width. They obtained an FPT algorithm parameterized by the maximum number of times per edge for a problem of Akrida, Mertzios and Spirakis [1]. They also studied a problem of vertex reachability called MinReachDelete, which remains NP-hard even with bounded interval-membership-width. They thus introduced a vertex-variant of their parameter, the vertex-interval-membership-width, that brings tractability to the problem. Enright, Meeks and Molter [11] showed that counting the number of temporal paths between a given pair of nodes is FPT for the vertex-interval-membership-width. Hand, Enright and Meeks [13] presented an FPT algorithm, also with respect to the vertex-variant, for a problem called Temporal Firefighter. Christodoulou, Crescenzi, Marino, Silva and Thilikos [10] recently modified the definition of (vertex)-interval-membership-width by considering the evolution of the connected components of the temporal graph and studied several temporal problems parameterized by the new parameters. These variants are bounded by their respective counterparts, and can even be arbitrarily smaller. In particular, they translate the results of [6] for the temporal Eulerian trail and [13] for the firefighter problem to their refined parameters.

1.2 Contributions

One of the objectives in the field of temporal graphs is to find relevant parameters that bring tractability to temporal problems. Our work consists in investigating the new parameters introduced by Bumpus and Meeks [6] for the problem of restless temporal path. We show that the vertex-interval-membership-width parameter is not bounded by any of the parameters previously studied for the design of FPT algorithms for the problem of restless temporal path (those considered in [8,26]). We exhibit the NP-hardness of restless temporal path in the point model with uniform delay one even if the graph has interval-membership-width equal to three, which implies that there is no FPT algorithm parameterized by the interval-membership-width if P≠NP. We provide an FPT algorithm parameterized by the vertex-interval-membership-width in this model and adapt it in the case of arbitrary positive delays. The time complexity of these algorithms is $\mathcal{O}(M2^k k)$ and $\mathcal{O}(M2^k k(k + \log M))$, respectively, where M is the number of timed edges and k is the vertex-interval-membership-width. They do not require knowledge of the value of k. Moreover, the complexity bounds still hold if k is the variant of the parameter proposed in [10]. Finally, we prove the hardness of restless temporal path in the interval model even if the vertex-interval-membership-width is equal to three.

1.3 Structure of the Paper

Section 2 introduces the notations and provides the necessary definitions. In Sect. 3, we focus on the point model with uniform delay one (which corresponds to the classical strict setting), while Sect. 4 deals with the case of arbitrary positive delays. Section 5 focuses on the interval model. Due to space constraints, some of our proofs have been deferred to a full version of the paper.[1]

[1] The full version is available at https://inria.hal.science/hal-05144137.

2 Preliminaries

2.1 Temporal Graphs

In what follows, we introduce the definition of a temporal graph in the point model as well as useful notions and notations in this model. We call a temporal graph in this model a point temporal graph. The adapted definitions for the interval model are presented in the corresponding section for ease of reading.

A *point temporal graph* $G = (V, E)$ consists of a set of nodes V and a set of timed arcs $E \subseteq V \times V \times \mathbb{N} \times \mathbb{N}_{>0}$. Informally, a *timed arc* $(u, v, \tau, \delta) \in E$ means that if we are at node u at time exactly τ, we can arrive at v at time $\tau + \delta$ by taking the timed arc (u, v, τ, δ). Time τ is called the *appearance time* (also *departure time*) of the timed arc (u, v, τ, δ). Time $\tau + \delta$ is called the *arrival time* of (u, v, τ, δ). The value δ is called the *delay*. Note that the timed arcs are directed. We assume that both V and E are finite. The *lifetime* of a point temporal graph $G = (V, E)$ is defined as $\Lambda = \max\{\tau + \delta \mid (u, v, \tau, \delta) \in E\}$.

Let $\mathcal{T}(E) = \{\tau \in \mathbb{N} \mid \exists (u, v, \tau, \delta) \in E\}$ be the set of appearance times of G. We define $E_\tau = \{(u, v, \tau, \delta) \in E\}$, the set of timed arcs appearing at time τ. The *underlying graph* of a point temporal graph $G = (V, E)$ is the static directed graph $G_\downarrow = (V, E_\downarrow)$ with $E_\downarrow = \{(u, v) \in V^2 \mid \exists \tau \in \mathbb{N}, \delta \in \mathbb{N}_{>0}, (u, v, \tau, \delta) \in E\}$.

Let $[\![i, j]\!]$ denote the set of integers k such that $i \leq k \leq j$ for $i, j \in \mathbb{N}$. If $i > j$, the set is empty. A *temporal st-walk* of length ℓ from $s = v_0$ to $t = v_\ell$ in a point temporal graph $G = (V, E)$ is a sequence of timed arcs $W = (v_{i-1}, v_i, \tau_i, \delta_i)_{i \in [\![1, \ell]\!]}$ such that for all $i \in [\![1, \ell - 1]\!]$, $\tau_i + \delta_i \leq \tau_{i+1}$. This is a *temporal path* if $v_i \neq v_j$ for $i \neq j$, $i, j \in [\![0, \ell]\!]$. The *arrival time* of W is $\tau_\ell + \delta_\ell$. For $i \in [\![1, \ell - 1]\!]$, the *waiting time* at an intermediate node v_i is defined as $\tau_{i+1} - (\tau_i + \delta_i)$. Let $V(W)$ denote the set of vertices of W, that is $V(W) = \{v_0, \dots v_\ell\}$.

In a *Δ-restless temporal st-path*, the waiting time at any intermediate node on the path is at most Δ:

Definition 1. *A temporal path (or walk) $(v_{i-1}, v_i, \tau_i, \delta_i)_{i \in [\![1, \ell]\!]}$ in a point temporal graph $G = (V, E)$ is Δ-restless if $\tau_{i+1} - (\tau_i + \delta_i) \leq \Delta$ for all $i \in [\![1, \ell - 1]\!]$.*

Figure 1 gives an example of a 2-restless temporal path.

We assume that a point temporal graph $G = (V, E)$ is given as input as a list of timed arcs sorted by non-decreasing appearance time, which we also denote E for simplicity. We denote by $M = |E|$ the number of timed arcs. Finally, we consider the following decision problem:

Problem 1. POINT-RESTLESS-TEMPORAL-PATH: Given a point temporal graph $G = (V, E)$, a source node $s \in V$, a target node $t \in V$ and a waiting time $\Delta \in \mathbb{N}$, is there a Δ-restless temporal path from s to t in G?

A point temporal graph $G = (V, E)$ is said to have *uniform delay one* if each timed arc has delay one, that is $\forall (u, v, \tau, \delta) \in E$, $\delta = 1$. Note that a symmetric point temporal graph with uniform delay one (that is if $(u, v, \tau, 1)$ is a timed arc of G then the reverse timed arc $(v, u, \tau, 1)$ is also a timed arc of G) corresponds to an undirected temporal graph without delay in the classical strict setting

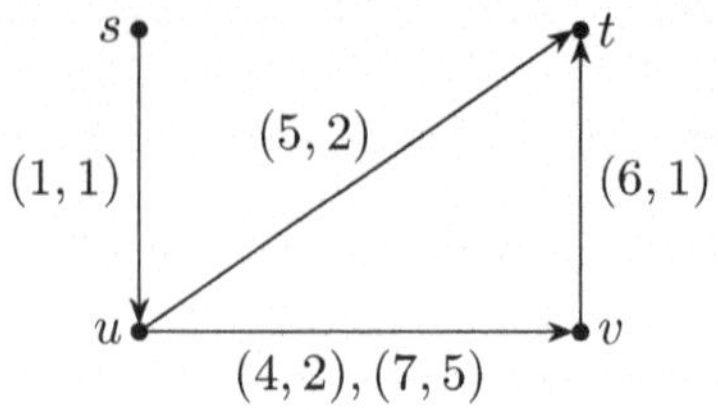

Fig. 1. Illustration of the notion of restless temporal path. A tuple (τ, δ) next to an arrow indicates the departure time τ and the delay δ of a timed arc. For instance, there is a timed arc $(s, u, 1, 1)$ and a timed arc $(u, t, 5, 2)$. The path from s to t consisting of the sequence of timed arcs $(s, u, 1, 1), (u, v, 4, 2), (v, t, 6, 1)$ is a 2-restless temporal st-path. In this point temporal graph, there is no 1-restless temporal st-path.

where appearance times are required to strictly increase along a temporal walk. In the following, when working with a point temporal graph with uniform delay one, we omit the delay for the sake of simplicity and write a timed arc (u, v, τ) instead of $(u, v, \tau, 1)$.

2.2 The Interval-Membership-Width Parameters

In this section, we adapt to our context the two parameters introduced in [6] for a simpler temporal graph model without delays, namely the interval-membership-width and the vertex-interval-membership-width.

We say that an arc $(u, v) \in E_{\downarrow}$ of the underlying graph is active over time τ if there is a timed arc $(u, v, \tau_0, \delta_0) \in E$ whose appearance time τ_0 is before or at τ and a timed arc $(u, v, \tau_1, \delta_1) \in E$ whose arrival time $\tau_1 + \delta_1$ is after or at τ, that is if $\tau^{min}(u, v) \leq \tau \leq \tau^{max}(u, v)$ with:

$$\tau^{min}(u, v) = \min\{\tau \in \mathbb{N} \mid \exists \delta \in \mathbb{N}_{>0}, (u, v, \tau, \delta) \in E\},$$
$$\tau^{max}(u, v) = \max\{\tau + \delta \in \mathbb{N} \mid (u, v, \tau, \delta) \in E\}.$$

Definition 2. *Given a point temporal graph $G = (V, E)$ with lifetime Λ and $\tau \in [\![0, \Lambda]\!]$, we define F^a_τ as the set of arcs that are active over time τ, that is:*

$$F^a_\tau = \{(u, v) \in E_{\downarrow} \mid \tau^{min}(u, v) \leq \tau \leq \tau^{max}(u, v)\}.$$

The arc-interval-membership-width $imw(G)$ is the maximum number of arcs active over any time, that is $imw(G) = max_{\tau \in [\![0, \Lambda]\!]} |F^a_\tau|$. For the sake of brevity, we will call this parameter the arc-IM-width.

Let us now define the vertex-variant of the arc-IM-width. We say that a node $u \in V$ is active over time τ if there is a timed arc $(v, w, \tau_0, \delta_0) \in E$ with $v = u$ or $w = u$ whose appearance time τ_0 is before or at τ and a timed arc $(x, y, \tau_1, \delta_1) \in E$ with $x = u$ or $y = u$ whose arrival time $\tau_1 + \delta_1$ is after or at τ, that is if $\tau^{min}(u) \leq \tau \leq \tau^{max}(u)$ with:

$$\tau^{min}(u) = \min\{\tau \in \mathbb{N} \mid \exists v \in V, \delta \in \mathbb{N}_{>0}, (u, v, \tau, \delta) \in E \text{ or } (v, u, \tau, \delta) \in E\},$$

$$\tau^{max}(u) = \max\{\tau + \delta \in \mathbb{N} \mid \exists v \in V, (u, v, \tau, \delta) \in E \text{ or } (v, u, \tau, \delta) \in E\}.$$

Definition 3. *Given a point temporal graph $G = (V, E)$ with lifetime Λ and $\tau \in [\![0, \Lambda]\!]$, we define F_τ^v, written F_τ for simplicity, as the set of nodes that are active over time τ, that is:*

$$F_\tau = \{u \in V \mid \tau^{min}(u) \leq \tau \leq \tau^{max}(u)\}.$$

The vertex-interval-membership-width $vimw(G)$ is the maximum number of nodes active over any time, that is $vimw(G) = max_{\tau\in[\![0,\Lambda]\!]}|F_\tau|$. For the sake of brevity, we will call this parameter the vertex-IM-width.

This parameter is not bounded by any of the parameters used in [8,26] for the design of algorithms for the restless temporal path problem. See the full version for details.

Note that $\tau^{min}(u)$ and $\tau^{max}(u)$ for all $u \in V$ can be computed in $\mathcal{O}(M)$ time from the list of timed arcs.

3 Point Model with Uniform Delay One (Classical Strict Setting)

We first rule out the possibility of an FPT algorithm for restless temporal path parameterized by arc-IM-width:

Proposition 1. POINT-RESTLESS-TEMPORAL-PATH *is NP-hard, even if the input point temporal graph has arc-IM-width equal to three and uniform delay one.*

This proposition is obtained by adapting the reduction in [8]. For the sake of completeness, we give it in the full version of the paper.

Our main result is an FPT algorithm parameterized by vertex-IM-width:

Theorem 1. *When the input point temporal graph $G = (V, E)$ has uniform delay one,* POINT-RESTLESS-TEMPORAL-PATH *can be solved in $\mathcal{O}(Mk2^k)$ deterministic time, where k is the vertex-IM-width of G.*

This result is the consequence of the algorithm we will now present. In the following, we want to compute the set of nodes reachable from a source s via a Δ-restless temporal path in a point temporal graph with uniform delay one, parameterized by the vertex-IM-width. Define the trace of a temporal path at time τ as the set of active nodes over τ it contains:

Definition 4. *The trace of a temporal path P at time τ is the set $V(P) \cap F_\tau$.*

For ease of informal explanation, we might refer to a trace S without specifying a time τ if the specific time is not relevant in the explanation. Note that F_τ can be viewed as a separator in the sense that any temporal uv-path where u is active before τ (that is $\tau^{max}(u) < \tau$) and v is active after τ (that is $\tau^{min}(v) > \tau$) must go through a node of F_τ.

The main idea of the algorithm is to scan timed arcs in order of appearance time τ while maintaining, for each node u, the traces of all Δ-restless su-paths. For each such trace, we store the last arrival time of a Δ-restless temporal su-path. Note that this information is sufficient to determine if a scanned timed arc (u, v, τ) can extend a Δ-restless temporal path with such a trace. More precisely, for each node u, $L[u]$ stores a list of couples (S, σ) where S is the trace of a Δ-restless su-path using timed arcs scanned so far, and σ is the last arrival time of such a path with trace S. When scanning all timed arcs (u, v, τ) with same appearance time τ, we update a similar list $L'[v]$ of traces that can be obtained with Δ-restless temporal paths ending with such timed arcs (u, v, τ). Specifically, for a couple (S, σ) in $L[u]$, we check if $v \notin S$ to respect the property of temporal path and if $\tau - \sigma \leq \Delta$ to respect the Δ-restless constraint. We then merge lists associated to each node and clean them to remove duplicates and to update traces according to which nodes are active over time τ. See Algorithm 1 for a detailed pseudo-code.

To prove the correctness of the algorithm, we introduce the following notations. Let $\mathcal{T}(E) = \{\tau_1, \dots, \tau_T\}$ with $\tau_1 < \dots < \tau_T$. We define $E^i = \{(u, v, \tau) \in E \mid \tau \leq \tau_i\}$, the set of timed arcs whose appearance time is less than or equal to τ_i, for $i \in [\![1, T]\!]$. Finally, let $G^i = (V, E^i)$. In the following $L_i[u]$ denotes $L[u]$ just after processing time τ_i. Let $L_i^S[u]$ be the set of traces in $L_i[u]$, that is $L_i^S[u] = \{S \mid \exists (S, \sigma) \in L_i[u]\}$. Given $S \in L_i^S[u]$, the time associated to S is the unique time τ_S^i such that $(S, \tau_S^i) \in L_i[u]$. Finally, for $i \in [\![1, T]\!]$ and $u \in V$, let τ_i^u be the last appearance time of a timed arc to u before or at τ_i, that is $\tau_i^u = \max\{\sigma \mid \exists (w, u, \sigma) \in E^i\}$. We can now state the following.

Lemma 1. *For all $i \in [\![1, T]\!]$, $u \in V$,*

- *$L_i^S[u]$ is the set of traces at time τ_i^u of all Δ-restless temporal su-paths in G^i;*
- *$\forall S \in L_i^S[u]$, its associated time τ_S^i is the last arrival time of a Δ-restless temporal su-path in G^i with trace S at time τ_i^u.*

Proof. The proof is by induction on i. When processing all timed arcs with appearance time τ_1, we add to $L'[v]$ the couple $(\{s, v\}, \tau_1 + 1)$ for each timed arc (s, v, τ_1) with $v \neq s$. Thus, this is true for $i = 1$.

Suppose that this is true for i with $1 \leq i \leq T - 1$ and let us prove that this is true for $i + 1$. Let $v \in V$. If there is no timed arc to v with appearance time τ_{i+1}, then $L[v]$ is not modified and we have $L_{i+1}^S[v] = L_i^S[v]$ and thus $L_{i+1}^S[v]$ is the set of traces at time $\tau_{i+1}^v = \tau_i^v$ of all Δ-restless temporal sv-paths in G^{i+1}.

Let us now suppose that there is at least one timed arc to v with appearance time τ_{i+1}. Consider any Δ-restless temporal sv-path P ending with such a timed

Algorithm 1. Δ-restless temporal path

Require: A point temporal graph $G = (V, E)$ with uniform delay one with E sorted by increasing departure time, a source $s \in V$, a maximum waiting time Δ

Ensure: A table *Reachable* such that *Reachable*$[v]$ = *True* if and only if there is a Δ-restless temporal path from s to v

1: *Reachable*$[s]$:= *True*, *Reachable*$[v]$:= *False* for all $v \in V \setminus \{s\}$
2: Compute the set $\mathcal{T}(E) = \{\tau_1, \dots, \tau_T\}$ of appearance times, with $\tau_1 < \cdots < \tau_T$
3: Compute $E_\tau = \{(u, v, \tau, \delta) \in E\}$, $V_\tau^- = \{v \in V \mid \exists(u, v, \tau) \in E_\tau\}$ for all $\tau \in \mathcal{T}(E)$
4: Compute $\tau^{max}(u)$ for all $u \in V$
5: **for all** $\tau \in \mathcal{T}(E)$ in increasing order **do**
6: $\quad L[s] := [(\{s\}, \tau)]$ ▷ We can start at s at any time
7: $\quad$ **for all** $(u, v, \tau) \in E_\tau$ **do**
8: $\quad\quad$ **for all** $(S, \sigma) \in L[u]$ **do**
9: $\quad\quad\quad$ **if** $v \notin S$ and $\tau - \sigma \leq \Delta$ **then**
10: $\quad\quad\quad\quad S' := (S \cap F_\tau) \cup \{v\}$ ▷ Note that $u \in S \cap F_\tau$
11: $\quad\quad\quad\quad$ Add $(S', \tau + 1)$ to $L'[v]$
12: $\quad\quad\quad\quad$ *Reachable*$[v]$:= *True*
13: $\quad$ **for all** $v \in V_\tau^-$ **do**
14: $\quad\quad$ Append $L'[v]$ to $L[v]$
15: $\quad\quad L'[v] := \emptyset$
16: $\quad\quad$ CleanUp(v, τ)
17: **procedure** CleanUp(v, τ) ▷ Given a node $v \in V$ and a time τ
18: $\quad$ **for all** $(S, \sigma) \in L[v]$ **do**
19: $\quad\quad$ Remove from S all u such that $\tau^{max}(u) < \tau$
20: $\quad$ Sort $L[v]$ according to the first coordinate (sets are sorted lexicographically)
21: $\quad$ For each S such that there is a $(S, \sigma) \in L[v]$, keep only the couple (S, σ) with maximal σ

arc (u, v, τ_{i+1}). Its prefix P_{pre} (i.e. the path P without the last timed arc) is in G^i since the graph has positive delays. By induction hypothesis, P_{pre} is associated to a trace S at time τ_i^u in $L_i^S[u]$ and the associated time τ_S^i is the last arrival time of a Δ-restless temporal su-path P_{max} in G^i with trace S at time τ_i^u. In Line 9 of Algorithm 1, we check that we can extend the path associated to each corresponding couple $(S, \sigma) \in L[u]$ with (u, v, τ_{i+1}): the condition $\tau - \sigma \leq \Delta$ ensures that the maximum waiting time at u is respected and the condition $v \notin S$ ensures that the extended walk is still a path. Indeed, if v was already on the path, it would be in the trace S by induction hypothesis since v is active over time τ_{i+1}. If the two conditions are satisfied, we add $((S \cap F_\tau) \cup \{v\}, \tau + 1)$ to $L'[v]$. Moreover, keeping for each $S \in L_i^S[u]$ only the last arrival time of a Δ-restless temporal su-path in G^i with trace S at time τ_i^u suffices. Indeed, consider two Δ-restless temporal su-paths P_1 and P_2 in G^i with trace S at time τ_i^u with respective arrival time σ_1 and σ_2, with $\sigma_2 > \sigma_1$. If we can extend the Δ-restless temporal path P_1 with a timed arc (u, w, τ_{i+1}), then we can also extend P_2 with the same timed arc as the waiting time at u will be smaller in the extension of P_2. Thus, after processing all the arcs with appearance time τ_{i+1} (Line 12), for each Δ-restless temporal sv-path P with arrival time $\tau_{i+1} + 1$, there is a couple

$(S', \tau_{i+1}+1)$ in $L'[v]$ where S' is the trace of P at time $\tau^v_{i+1} = \tau_{i+1}$. We add $L'[v]$ to $L[v]$ Line 14. Finally, the calls to the auxiliary function CLEANUP ensure that all sets in $L[v]$ contain only nodes in $F_{\tau_{i+1}}$ (Line 19) and that we keep for each S only the couple (and only one copy of this couple) with maximal associated σ (Line 21). Thus, the hypothesis is satisfied for $i+1$.

Theorem 1 is a consequence of the following.

Proposition 2. *Algorithm 1 computes the set of nodes reachable from a source s via a Δ-restless temporal path in a point temporal graph G with uniform delay one in $\mathcal{O}(Mk2^k)$ deterministic time, where k is the vertex-IM-width of G.*

Proof. The correctness follows immediately from Lemma 1.

Regarding the time complexity, let us first observe that, after a call to CLEANUP(v, τ) for $v \in V$ and $\tau \in \mathbb{N}$, we have that the size of $L[v]$ is at most 2^k. Indeed, the sets S in $L[v]$ after CLEANUP are all subsets of F_τ (Line 19). Thus, there are at most 2^k different subsets in $L[v]$. Moreover, we keep only one couple (S, σ) for each set S (Line 21). With this, Lines 9-12 are executed $\mathcal{O}(M2^k)$ times. Computing $S \cap F_\tau$ for a trace S (Line 10) and checking the condition $v \notin S$ (Line 9) take $\mathcal{O}(k)$ time. Indeed, $S \cap F_\tau$ can be done by scanning the set S and for each $u \in S$, checking if $\tau \leq \tau^{max}(u)$.

Computing E_τ, V^-_τ for each $\tau \in \mathcal{T}(E)$ and $\tau^{max}(u)$ for each $u \in V$ (Lines 2, 3 and 4) can be done in $\mathcal{O}(M)$ time.

Let us now analyze the cost of cleaning up. Let us denote M^v_τ the number of timed arcs to v with appearance time τ, that is $M^v_\tau = |\{u \in V \mid (u, v, \tau) \in E\}|$. When processing a node v at Line 13, the size of $L[v]$ after appending $L'[v]$ to $L[v]$ (Line 14) is at most $2^k + 2^k M^v_\tau$. Then, in CLEANUP(v, τ), the cost of removing nodes that are not active over time τ (Line 19) and sorting the list (Line 20) is $\mathcal{O}(k2^k M^v_\tau)$, using the lexicographic sort from [23] for Line 20. Thanks to this sorting, we can execute Line 21 in $\mathcal{O}(2^k M^v_\tau)$ time by scanning the list and keeping one couple with maximum σ for each S. Thus, the total cost of CLEANUP() calls is $\mathcal{O}(\sum_{\tau \in \mathcal{T}(E)} \sum_{v \in V} k2^k M^v_\tau) = \mathcal{O}(k2^k M)$.

Remark 1. The complexity still holds if k is the refined parameter of [10]. Indeed, this parameter considers the maximum number of active nodes in a connected component of the underlying graph of any G^i. In Algorithm 1, the traces we consider are indeed included in the connected component of s.

Remark 2. A Δ-restless temporal path from s to a node $v \in V$ can be retrieved in $\mathcal{O}(\ell k(k + \log M))$ time, where ℓ is the length of the path and k is the vertex-IM-width of the point temporal graph. This requires additional instructions in Algorithm 1 that are executed in $\mathcal{O}(M2^k k(k + \log M))$ deterministic time. The general idea is to store, for each trace S of a Δ-restless temporal path P arriving at v at time τ, the predecessor u of v on the path, the arrival time σ of the prefix of P and the trace of the prefix at time σ. A detailed explanation is given in the full version.

Remark 3. In the function CLEANUP(v, τ) for $v \in V \setminus \{s\}$, we can also remove from $L[v]$ the couples (S, σ) such that $\tau - \sigma > \Delta$ as we know that they will not be useful to extend a Δ-restless temporal path.

Remark 4. Note that our time complexity can be expressed in terms of the lifetime $\Lambda \geq T$ as $M \leq k^2T$, where k is the vertex-IM-width of G. Indeed, there are at most k^2 timed arcs with the same appearance time. The time complexity obtained is $\mathcal{O}(k^3 2^k \Lambda)$.

4 Point Model with Arbitrary Positive Delays

We now generalize the previous result for point temporal graphs with arbitrary (positive) delays:

Theorem 2. *POINT-RESTLESS-TEMPORAL-PATH can be solved in* $\mathcal{O}(Mk2^k(k + \log M))$ *deterministic time, where* k *is the vertex-IM-width of the graph.*

Again, we want to compute the set of nodes reachable from a source s via a Δ-restless temporal path in a point temporal graph, parameterized by the vertex-IM-width. Algorithm 1 cannot be easily adapted to arbitrary delays. The reason is that it implicitly assumes that all timed arcs with appearance time less than τ have arrival time at most τ. This does not hold anymore for arbitrary delays. Note that scanning timed arcs in order of arrival time instead of appearance time leads to a similar problem.

To deal with this, we adapt the previous algorithm by storing all possible arrival times at a node v for each possible trace S in a self-balancing binary search tree. $L[v]$ now contains couples (S, T_S) with S a trace and T_S the associated tree. When we try to extend a Δ-restless temporal su-path with trace S by adding a timed arc (u, v, τ, δ), we first look at the last arrival time $\sigma \leq \tau$ at u with trace S in the corresponding tree. If we can extend the path, then we add $\tau + \delta$ to $T^v_{S'}$ if this time is not already in it, where $T^v_{S'}$ is the tree associated to trace $S' = (S \cap F_\tau) \cup \{v\}$ in $L[v]$ (if there is no such tree, we create one that contains only $\tau + \delta$). After processing a timed arc with appearance time τ, we clean up to update traces according to which nodes are active over time τ. When performing this clean up, we might have several traces that become the same trace S' after removing the nodes that are not active over time τ. We thus need to merge the associated trees to have all the possible arrival times associated to trace S'. The detailed pseudo-code as well as the analysis and useful remarks are available in the full version.

5 Hardness Result for the Interval Model

It is straightforward to extend the definitions of arc-IM-width and vertex-IM-width to an interval model of temporal graph where arcs are present during intervals of time. We show in the full version that, in this interval model, there is no FPT algorithm parameterized by the corresponding variant of the vertex-IM-width unless $P = NP$. The idea is to reduce SUBSET-SUM to RESTLESS-TEMPORAL-PATH in an interval temporal graph of constant vertex-IM-width. This is possible because the interval model allows to encode temporal graphs that would have exponentially many edges if converted to the point model. In particular, our reduction uses intervals of exponential length.

Acknowledgments. This work was supported by the French ANR projects ANR-22-CE48-0001 (TEMPOGRAL), ANR-24-CE48-4377 (GODASse) and ANR-23-PEIA-005 (REDEEM).

Disclosure of Interests. The authors have no competing interests to declare that are relevant to the content of this article.

References

1. Akrida, E.C., Mertzios, G.B., Spirakis, P.G., Raptopoulos, C.: The temporal explorer who returns to the base. J. Comput. Syst. Sci. **120**, 179–193 (2021). https://doi.org/10.1016/j.jcss.2021.04.001
2. Akrida, E.C., Mertzios, G.B., Spirakis, P.G., Zamaraev, V.: Temporal vertex cover with a sliding time window. In: 45th International Colloquium on Automata, Languages, and Programming (ICALP 2018). Leibniz International Proceedings in Informatics (LIPIcs), vol. 107, pp. 1–14. Schloss Dagstuhl – Leibniz-Zentrum für Informatik (2018). https://doi.org/10.4230/LIPIcs.ICALP.2018.148
3. Barabási, A.: Network Science. Cambridge University Press (2016)
4. Bentert, M., Himmel, A., Nichterlein, A., Niedermeier, R.: Efficient computation of optimal temporal walks under waiting-time constraints. Appl. Netw. Sci. **5**(1), 73 (2020). https://doi.org/10.1007/S41109-020-00311-0
5. Xuan, B. B., Ferreira, A., Jarry, A.: Computing shortest, fastest, and foremost journeys in dynamic networks. Int. J. Found. Comput. Sci. **14** (2011). https://doi.org/10.1142/S0129054103001728
6. Bumpus, B.M., Meeks, K.: Edge exploration of temporal graphs. Algorithmica **85**(3), 688–716 (2022). https://doi.org/10.1007/s00453-022-01018-7
7. Casteigts, A., Flocchini, P., Quattrociocchi, W., Santoro, N.: Time-varying graphs and dynamic networks, vol. 27 (2010). https://doi.org/10.1007/978-3-642-22450-8_27
8. Casteigts, A., Himmel, A.-S., Molter, H., Zschoche, P.: Finding temporal paths under waiting time constraints. Algorithmica **83**(9), 2754–2802 (2021). https://doi.org/10.1007/s00453-021-00831-w
9. Casteigts, A., Peters, J.G., Schoeters, J.: Temporal cliques admit sparse spanners. In: 46th International Colloquium on Automata, Languages, and Programming (ICALP 2019). Leibniz International Proceedings in Informatics (LIPIcs), vol. 132, pp. 1–14. Schloss Dagstuhl – Leibniz-Zentrum für Informatik (2019). https://doi.org/10.4230/LIPIcs.ICALP.2019.134
10. Christodoulou, F., Crescenzi, P., Marino, A., Silva, A., Thilikos, D.M.: Making the interval membership width of temporal graphs connected and bidirectional. In: Rescigno, A.A., Vaccaro, U. (eds.) Combinatorial Algorithms - 35th International Workshop, IWOCA 2024, Ischia, Italy, 1–3 July 2024, Proceedings. Lecture Notes in Computer Science, vol. 14764, pp. 247–258. Springer (2024). https://doi.org/10.1007/978-3-031-63021-7_19

11. Enright, J., Meeks, K., Molter, H.: Counting temporal paths. In: 40th International Symposium on Theoretical Aspects of Computer Science (STACS 2023). Leibniz International Proceedings in Informatics (LIPIcs), vol. 254, pp. 30:1–30:19. Schloss Dagstuhl – Leibniz-Zentrum für Informatik (2023). https://doi.org/10.4230/LIPIcs.STACS.2023.30
12. Haag, R., Molter, H., Niedermeier, R., Renken, M.: Feedback edge sets in temporal graphs. Discrete Appl. Math. **307**, 65–78 (2022). https://doi.org/10.1016/j.dam.2021.09.029
13. Hand, S.D., Enright, J., Meeks, K.: Making life more confusing for firefighters. In: 11th International Conference on Fun with Algorithms (FUN 2022). Leibniz International Proceedings in Informatics (LIPIcs), vol. 226, pp. 1–15. Schloss Dagstuhl – Leibniz-Zentrum für Informatik (2022). https://doi.org/10.4230/LIPIcs.FUN.2022.15
14. Holme, P.: Modern temporal network theory: a colloquium. Eur. Phys. J. B **88**(9), 1–30 (2015). https://doi.org/10.1140/epjb/e2015-60657-4
15. Kempe, D., Kleinberg, J., Kumar, A.: Connectivity and inference problems for temporal networks. J. Comput. Syst. Sci. **64**(4), 820–842 (2002). https://doi.org/10.1006/jcss.2002.1829
16. Kermack, W.O., McKendrick, A.G.: A contribution to the mathematical theory of epidemics. Proc. Roy. Soc. Lond. Ser. A, Containing Pap. Math. Phys. Charact. **115**(772), 700–721 (1927). https://doi.org/10.1098/rspa.1927.0118
17. Latapy, M., Viard, T., Magnien, C.: Stream graphs and link streams for the modeling of interactions over time. Soc. Netw. Anal. Min. **8**(1), 1–29 (2018). https://doi.org/10.1007/s13278-018-0537-7
18. Mertzios, G.B., Molter, H., Niedermeier, R., Zamaraev, V., Zschoche, P.: Computing maximum matchings in temporal graphs. J. Comput. Syst. Sci. **137**, 1–19 (2023). https://doi.org/10.1016/j.jcss.2023.04.005
19. Michail, O.: An introduction to temporal graphs: an algorithmic perspective. Internet Math. **12** (2015). https://doi.org/10.1007/978-3-319-24024-4_18
20. Michail, O., Spirakis, P.G.: Traveling salesman problems in temporal graphs. Theoret. Comput. Sci. **634**, 1–23 (2016). https://doi.org/10.1016/j.tcs.2016.04.006
21. Newman, M.: Networks. OUP Oxford (2018)
22. Orda, A., Rom, R.: Traveling without waiting in time-dependent networks is NP-hard. Tech. rep., Technion - Israel Institute of Technology, Haifa, Israel 32000 (1989)
23. Paige, R., Tarjan, R.E.: Three partition refinement algorithms. SIAM J. Comput. **16**(6), 973–989 (1987). https://doi.org/10.1137/0216062
24. Thejaswi, S., Lauri, J., Gionis, A.: Restless reachability problems in temporal graphs (2025). https://doi.org/10.1007/s10115-025-02405-6
25. Wu, H., Cheng, J., Ke, Y., Huang, S., Huang, Y., Wu, H.: Efficient algorithms for temporal path computation. IEEE Trans. Knowl. Data Eng. **28**(11), 2927–2942 (2016). https://doi.org/10.1109/TKDE.2016.2594065
26. Zschoche, P.: Restless temporal path parameterized above lower bounds. In: 40th International Symposium on Theoretical Aspects of Computer Science (STACS 2023). Leibniz International Proceedings in Informatics (LIPIcs), vol. 254, pp. 1–16. Schloss Dagstuhl – Leibniz-Zentrum für Informatik (2023). https://doi.org/10.4230/LIPIcs.STACS.2023.55

Red-Blue Unshared Dominators

Gennaro Cordasco(✉), Luisa Gargano, and Adele A. Rescigno

Department of Computer Science, University of Salerno, Fisciano, Italy
gcordasco@unisa.it

Abstract. We introduce the Red-Blue Unshared Domination (RBUD) problem, a variant of domination problems in graphs. Given an undirected graph G with vertices colored either red or blue, the goal is to find a minimum-size subset of vertices S such that for every red-blue pair of vertices outside S, there exist two distinct vertices in S (called unshared dominators), each dominating one vertex of the pair but not the other. Specifically, for each red vertex $r \notin S$ and blue vertex $b \notin S$, there must be a vertex in S that dominates r but not b and another vertex in S that dominates b but not r.

We prove that the RBUD problem is NP-hard, even for bipartite planar subcubic graphs with large girth. We also provide an approximation algorithm for general graphs and exact efficient algorithms for trees and graphs with bounded neighborhood diversity. Additionally, we extend our algorithmic approach to graphs with bounded modular-width.

1 Introduction

Domination in graphs, used to distinguish vertices, is a well-studied problem within the broader field of identification problems. The goal is to select a dominating set D of a graph G such that the closed neighborhoods of the vertices of G have distinct intersections with D. This allows the vertices of the graph to be uniquely identified.

Such a dominating and distinguishing set D is often referred to as a code in the literature [17]. Depending on the type of dominating set and the distinguishing rule applied, several variants of the problem have been introduced and studied under various names [16,23,24].

Locating-domination is one such variant: A set D of vertices in a graph is a locating-dominating code if it is a dominating set and any two vertices not in D are dominated by distinct subsets of D.

We introduce a new problem within the area of domination-distinguishing problems: Given a graph $G = (V, E)$ with vertices colored either red or blue, compute an optimal subset of vertices S, called a Red-Blue Unshared Dominating set, that satisfies the following property: for every red-blue pair of vertices $r \in R \setminus S$ and $b \in B \setminus S$, there exist vertices $s_R \in S$ and $s_B \in S$ such that:

- s_R dominates r but does not dominate b (i.e., s_R is an unshared dominator of r with respect to b), and

A. Jeż and J. Otop (Eds.): FCT 2025, LNCS 16106, pp. 94–108, 2026.
https://doi.org/10.1007/978-3-032-04700-7_8

- s_B dominates b but does not dominate r (i.e., s_B is an unshared dominator of b with respect to r).

The idea of RBUD lies in enabling selective influence and separation in complex networks. By requiring a minimal set of nodes that can distinguish between red and blue entities, Red-Blue Unshared Dominating sets support targeted control. This has practical applications in scenarios such as network intrusion detection, where monitoring servers selectively identify compromised machines without false alarms, or social network privacy, where control nodes regulate sensitive interactions without restricting general users.

Related Work. Recently, the Red-Blue Separation problem on graphs was introduced in [12]. Given a graph with vertices colored either red or blue, the goal is to determine a (minimum-size) subset of vertices S such that for any red-blue vertex pair, there exists $s \in S$ that separates the pair, that is, the closed neighborhood of s contains exactly one vertex of the pair. The Red-Blue Separation problem is inspired by a similar problem in a geometric context [2,21].

We stress that the requirement of RBUD is different from that of Red-Blue Separation. As in the locating-domination, RBUD requires that only the vertices outside the solution are distinguished. Furthermore, the distinction between the vertices of any red-blue pair has to be such that each vertex in the pair has its own dominator (i.e., an unshared dominator) in the solution.

More generally, the Separation problem for set systems has been extensively studied from a combinatorial perspective [1,3] as well as from an algorithmic standpoint [10,11]. In the context of graphs, the Separation problem is also known as the Identifying Code [17] (see [18] for a comprehensive bibliography).

Our Results. The Red-Blue Unshared Domination problem is formally defined in Sect. 2. In Sect. 3, we establish that RBUD is NP-hard, even for bipartite planar subcubic graphs with large girth. On the positive side, we present an approximation algorithm for general graphs in Sect. 4, along with exact polynomial-time algorithms for trees and for graphs with bounded modular-width or neighborhood diversity. While the results for graphs with bounded modular-width also apply to those with bounded neighborhood diversity, the algorithm specifically designed for the latter class is more efficient.

2 The Problem

Let $G = (V, E)$ be an undirected graph. We denote by $n = |V|$ the number of vertices in G. For $v \in V$, we denote by $N_G(v) = \{u \in V \mid (u, v) \in E\}$ the open neighborhood and by $N_G[v] = N_G(v) \cup \{v\}$ the closed neighborhood of v. For $X \subseteq V$, we denote by $G[X]$ the subgraph of G induced by the vertices in X, by $N_G(X) = \bigcup_{v \in X} N_G(v) \setminus X$ and by $N_G[X] = N_G(X) \cup X$. We denote by $\overline{G}$ the complement of G that is $\overline{G} = (V, (V \times V) \setminus E)$. We omit the subscript G whenever the graph is clear from the context.

A *red-blue (R-B) coloring* of G is a partition of the vertex set V of G into two subsets R and B, i.e. $R \cup B = V$ and $R \cap B = \emptyset$. We denote the vertices in

R as the *red vertices* of G, and those in B as the *blue vertices* of G. If $R = \emptyset$ or $B = \emptyset$, we call G monochromatic; otherwise, G is called bi-chromatic.

Given a graph $G = (V, E)$ and a R-B coloring of G, we are interested in a subset $S \subseteq V$ such that for each pair of vertices $r \in R \setminus S$ and $b \in B \setminus S$, it holds that r has a neighbor in S, which is not a neighbor of b and b has a neighbor in S, which is not a neighbor of r. Formally, we use the following definition.

Definition 1. *A set $S \subseteq V$ is a* Red-Blue Unshared Dominating set *(*RBUD set*) if any $v \in V$ that is not in S has* unshared dominators *(with respect to nodes of the opposite color) in S. That is,*

1) *if $v \in R \setminus S$ then for each $b \in B \setminus S$ it holds $(N(v) \setminus N(b)) \cap S \neq \emptyset$*
2) *if $v \in B \setminus S$ then for each $r \in R \setminus S$ it holds $(N(v) \setminus N(r)) \cap S \neq \emptyset$.*

The goal is to find a Red-Blue Unshared Dominating set S of minimum size for G; we denote by MIN-RBUD such a problem. To this aim, we define

RED-BLUE UNSHARED DOMINATION (RBUD):
Input: Graph $G=(V, E)$ with a red-blue coloring R-B of G, and an integer k;
Question: Is there a Red-Blue Unshared Dominating set S for G of size k?

Since R and B are both *RBUD sets*, the answer is YES whenever $k \geq \min\{|R|, |B|\}$.

In the following, we say that a solution $S \subseteq V$ is *non-trivial* when $R \not\subseteq S$ and $B \not\subseteq S$. It is worth observing that if S is non-trivial then it also is a Dominating Set (i.e., any vertex of G is in S or has a neighbor in S). As a consequence, if the solution to the problem is non-trivial, then the size of the minimum RBUD set is at least equal to the size of the minimum dominating set. On the other hand, the size of the smallest RBUD set can be arbitrarily larger than the size of the minimum dominating set. For instance, consider a star graph with n vertices (roughly half red and half blue). In such a case, the domination number is 1, while the smallest RBUD set coincides with the smallest color class and has size $O(n)$. Moreover, the minimum RBUD set does not necessarily contain the vertices of any minimum dominating sets; an example is given in Fig. 1.

The following property will be useful in the sequel.

Lemma 1. *Given a graph $G = (V, E)$ with a red-blue coloring R-B. Let S be a RBUD set for G. For any $r \in R$ and $b \in B$ it holds that $S \cap \{r, b\} \neq \emptyset$ whenever any of the following relations holds*

$$N(r) \subseteq N(b) \text{ or } N(b) \subseteq N(r) \text{ or } N[r] \subseteq N[b] \text{ or } N[b] \subseteq N[r]. \tag{1}$$

Proof. Assume first that there exists $r \in R \setminus S$ and $b \in B \setminus S$, with $N(r) \subseteq N(b)$ (or $N(b) \subseteq N(r)$). Hence, $N(r) \setminus N(b) = \emptyset$ (or $N(b) \setminus N(r) = \emptyset$), which contradicts Definition 1.
We notice now that $N[r] \subseteq N[b]$ (or $N[b] \subseteq N[r]$) implies $N(r) \setminus N(b) = \{b\}$ (or $N(b) \setminus N(r) = \{r\}$). If we assume $b, r \notin S$ then $(N(r) \setminus N(b)) \cap S = \emptyset$ (or $(N(b) \setminus N(r)) \cap S = \emptyset$) contradicting Definition 1. $\square$

3 Hardness

In this section, we prove that RBUD is `NP-hard` even for bipartite planar subcubic graphs with girth at least 12. Moreover, we show that while the problem admits a trivial solution when $k \geq \min\{|R|, |B|\}$, it becomes hard as soon as the solution size falls below the value $\min\{|R|, |B|\}$. To this aim, we give an approximation-preserving reduction from the Dominating Set (DS) problem, which has been proved to be NP-hard when the input graph is a bipartite planar subcubic graph with girth at least 12 [25].

Dominating Set (DS): Given graph $G = (V, E)$ and an integer k, there exists $D \subseteq V$ such that $|D| \leq k$ and $D \cap N(v) \neq \emptyset$ for each $v \in V \setminus D$?

The NP-hardness of the RBUD problem can be easily derived from the DS problem. Indeed, given an instance $\langle G, k \rangle$ of DS, it is easy to see that if we consider a graph G' obtained by two disjoint copies of G with the vertices of one copy colored blue and the vertices of the other copy colored red, then the union of the dominating sets of each copy of G is a red-blue unshared dominating set for G'; vice versa, any RBUD set for G' must contain a dominating set for each copy of G to satisfy Definition 1. Hence, G has a dominating set D with $|D| \leq k$ if and only if there exists a RBUD set S for G' with $|S| \leq k' = 2k$.

However, it is possible to prove other interesting hardness results. We already observed that a trivial solution exists when $k \geq \min\{|R|, |B|\}$. In what follows, we show that the problem becomes NP-hard already when $k = \min\{|R|, |B|\} - 1$.

The proof is done by means of a polynomial reduction from the DS problem. Given an instance $\langle G, k \rangle$ of DS, we construct an instance $\langle G', R, B, k' \rangle$ of RBUD. The graph G' is built by considering a copy of G, called $G_R = (V_R, E_R)$, and a star graph $G_B = (V_B, E_B)$ with center v_B and k leaves. Then we add six new vertices $u_1, u_2, u_3, w_1, w_2, w_3$ such that u_1, u_2, u_3 form a path, w_1, w_2, w_3 form a path, and u_1 is connected to w_1. Moreover, we arbitrarily choose a vertex $v \in V(G)$, and connect the copy $v_R \in V_R$ to u_1. The center of the star v_B is connected to w_1. The colors are assigned as follows: The red vertices are $R = V_R \cup \{u_1, u_2, u_3\}$ and the blue vertices are $B = V_B \cup \{w_1, w_2, w_3\}$. Figure 2 shows the construction of G'. By the construction $|B| = k + 4$.

Lemma 2. *$\langle G, k \rangle$ is a YES-instance if and only if $\langle G', R, B, k' = k + 3 \rangle$ is a YES-instance.*

Theorem 1. *The* RBUD *problem is NP-hard for bipartite planar graphs with girth at least* 12*, even when* $k = \min\{|R|, |B|\} - 1$.

Proof. Recalling that DS is NP-hard for bipartite planar subcubic graphs with girth at least 12 [25], and that the reduction preserves such properties except for the constraint on the degrees of the vertices, by Lemma 2, the theorem holds. □

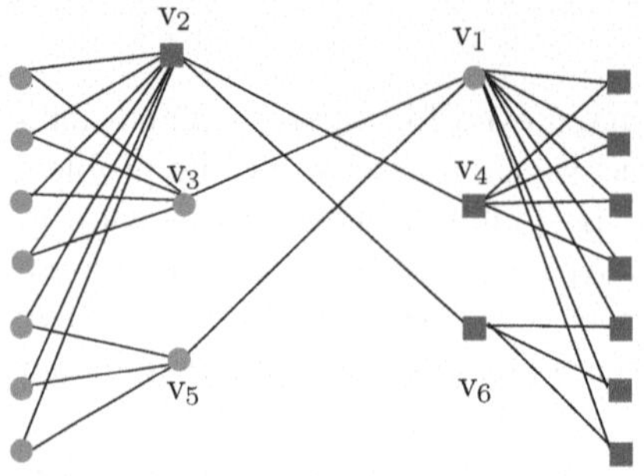

Fig. 1. The min. DS is $D = \{v_1, v_2\}$. The min RBUD set is $S = \{v_3, v_4, v_5, v_6\}$.

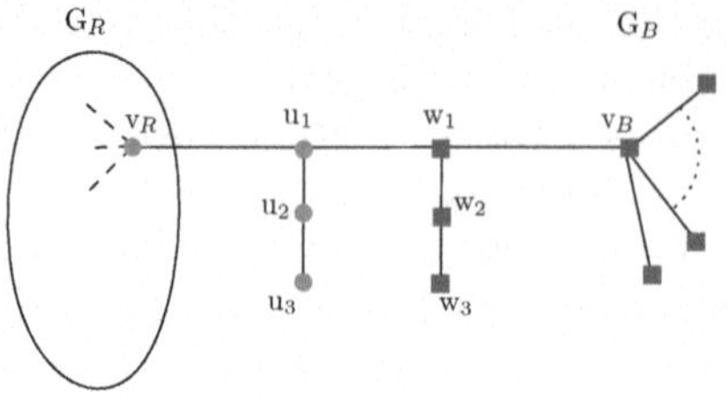

Fig. 2. Reduction from DS to RBUD.

With a slight modification to the graph G' used in the reduction of Theorem 1, one can prove that RBUD is NP-hard even when the graph is subcubic, giving up the request on k.

Theorem 2. RBUD *is NP-hard for subcubic bipartite planar graphs with girth at least* 12.

By the results for DS in [4,13], we have the following approximation hardness.

Corollary 1. *The* MIN-RBUD *problem cannot be approximated within a factor of* $(1 - \epsilon) \ln n$ *for any* $\epsilon > 0$ *unless* $NP \subseteq DTIME(n^{O(loglogn)})$.

4 Algorithms

In this section, we give a simple $O(\log n)$-approximation algorithm for general graphs and exact efficient algorithms for some graph classes.

4.1 Approximation

We design a polynomial-time approximation algorithm based on a simple reduction of Red-Blue Unshared Domination to SET COVER.

Theorem 3. MIN-RBUD *has a* $(2 \ln n + 1)$ *approximation algorithm.*

Proof. Consider the SET COVER problem:

SET COVER (SC):
Input: A universe set $U = \{1, 2, \dots, n\}$, a family $\mathcal{F} = \{F_1, F_2, \dots, F_m\}$ of subset of U, and an integer k.
Output: There exists $\mathcal{C} \subseteq \mathcal{F}$ of size $|\mathcal{C}| \leq k$ such that $\bigcup_{C \in \mathcal{C}} C = U$?

Let $\langle G = (V, E), R, B, k\rangle$ be an input instance of RBUD. We reduce it to an instance $\langle U, \mathcal{F}, k\rangle$ of SC.

The universal set U consists of all the ordered pairs (r, b) and (b, r) such that $r \in R$ and $b \in B$. The family $\mathcal{F}$ consists of all the sets F_u, for each $u \in V$, where

$$F_u = \{(v, w) \mid (v, w) \in U,\ u \in N(v) \setminus N(w)\} \cup \{(v, w) \mid (v, w) \in U,\ u \in \{v, w\}\} \quad (2)$$

The first set in equation (2) consists of all the red/blue or blue/red pairs (v, w) such that u is a neighbor of v and is not a neighbor of w while the second set includes all the blue/red or blue/red pairs containing u.

We can prove that that $\langle U, \mathcal{F}, k\rangle$ is a YES-instance of SC if and only if $\langle G, R, B, k\rangle$ is a YES-instance of RBUD.

Suppose first that $S \subseteq V$ is a RBUD set for the instance $\langle G, R, B, k\rangle$. If $\mathcal{C} = \{F_u \in \mathcal{F} \mid u \in S\}$ is not a set cover for the instance $\langle U, \mathcal{F}, k\rangle$, then there exists a pair in U not covered by $\mathcal{C}$. W.l.o.g. assume that the pair is a red/blue pair (r, b). By construction, we have that both r and b do not belong to S and does not exist $u \in S$ such that $u \in N(r) \setminus N(b)$. Hence, $(N(r) \setminus N(b)) \cap S = \emptyset$. This implies that 1) of Definition 1 does not hold, thus contradicting the assumption that S is a RBUD set.

Assume now that $\mathcal{C} \subseteq \mathcal{F}$ is a set cover for the instance $\langle U, \mathcal{F}, k\rangle$. If $S = \{u \in S \mid F_u \in \mathcal{F}\}$ is not a RBUD set, we know that Definition 1 does not hold for S. W.l.o.g., assume that there exists $v \in R \setminus S$ that does not satisfy 1) of Definition 1. This implies that there exists $b \in B \setminus S$ such that $(N(v) \setminus N(b)) \cap S = \emptyset$. Hence, there is no $F_u \in \mathcal{C}$ such that $u \in N(v) \setminus N(b)$. Moreover, since both r and b are not in S, we have that both $F_v, F_b \notin \mathcal{C}$. Consequently, the pair $(v, b) \in U$ is not covered by $\mathcal{C}$, thus contradicting the assumption that $\mathcal{C}$ is a set cover.

Applying the greedy algorithm for the minimum set cover to U, and $\mathcal{F}$, one gets an approximation factor $\ln |U| + 1$. Since, in our case $|U| \le |V|^2/2$, the resulting approximation factor for Red-Blue Unshared Domination is at most $\ln(|V|^2/2) + 1 < 2 \ln |V| + 1$ (Fig. 3). □

4.2 Trees

A dynamic programming algorithm, exploiting the structure of the input tree, enables solving the MIN-RBUD problem. We first focus on non-trivial solutions. The output will then be the smallest among the optimal non-trivial solution and the two trivial ones, R and B.

Definition 2. *Given a graph $G = (V, E)$, a red-blue coloring R-B of G, and a set $S \subseteq V$. A vertex $w \in S$ is called* red *(resp.* blue*)* Dominator *when for each $u \in N(w) \setminus S$, $u \in R$ (resp. $u \in B$). We denote by $S_R \subseteq S$ (resp. $S_B \subseteq S$) the set of* red *(resp.* blue*)* Dominators.

Lemma 3. *Let $T = (V, E)$ be a tree and let R-B be a red-blue coloring of T. Given a set $S \subseteq V$ then S is a non-trivial RBUD set if and only if for each $v \in V \setminus S$, $|N(v) \cap S| \ge 2$ or $N(v) \cap (S_R \cup S_B) \ne \emptyset$.*

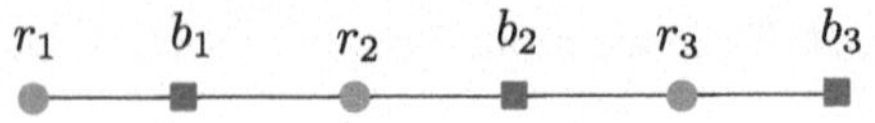

$S = \{b_1, r_3\}$

U	r_1 b_1	r_1 b_2	r_1 b_3	r_2 b_1	r_2 b_2	r_2 b_3	r_3 b_1	r_3 b_2	r_3 b_3	b_1 r_1	b_2 r_1	b_3 r_1	b_1 r_2	b_2 r_2	b_3 r_2	b_1 r_3	b_2 r_3	b_3 r_3
F_{r_1}	✓	✓	✓							✓	✓	✓	✓			✓		
F_{b_1}	✓	✓	✓	✓	✓	✓	✓			✓			✓			✓		
F_{r_2}				✓	✓	✓				✓	✓		✓	✓	✓	✓	✓	
F_{b_2}		✓		✓	✓	✓	✓	✓	✓		✓			✓			✓	
F_{r_3}							✓	✓	✓		✓	✓		✓	✓	✓	✓	✓
F_{b_3}			✓			✓	✓	✓	✓			✓			✓			✓

$\mathcal{C} = \{F_{b_1}, F_{r_3}\}$

Fig. 3. An example of reduction from an instance of RBUD to an instance of SC: (Top) A path with a red-blue coloring. The minimum RBUD-set is $S = \{b_1, r_3\}$. (Bottom) The table describing the associated SC problem. The first row depicts the set U of all the red/blue and blu/red pairs. The family of subsets $\mathcal{F} = \{F_{r_1}, F_{b_1}, F_{r_2}, F_{b_2}, F_{r_3}, F_{b_3}\}$ of U, one for each vertex in V is given in the remaining rows. Check marks indicate the set (in the column) belonging to the subset (in the row). We observe that $\mathcal{C} = \{F_{b_1}, F_{r_3}\} \subset \mathcal{F}$, associated with S, emphasized in grey in the table, is a cover of U (Color figure online).

Given a tree $T = (V, E)$, root it at arbitrary vertex z, and define $T(v)$ as the subtree rooted at vertex v, for any $v \in V$. We denote by $V(T(v))$ the vertex set of $T(v)$. Moreover, for each $v \neq z$, we denote by $p(v)$ the parent of v and by $C(v)$ the set of children of v in T.

Let R-B be a red-blue coloring of T. We assume here that the optimal RBUD set is non-trivial. The algorithm processes the vertices in a bottom-up manner, ensuring that all child vertices are visited before their parent.

Fix a vertex $v \in V$ and considering the tree $T(v)$, in order to be able to recursively reconstruct the solution, the algorithm calculates partial solutions $S \subseteq V(T(v))$, under different hypotheses, based on the following considerations.

The vertex v can either be in S (case α) or not (case β). If v belongs to S, then it is necessary to distinguish whether v is a Dominator (red or blue). This depends on the neighborhood of v, which also includes $p(v)$. Hence, when v is considered, it is still not possible to decide whether v is a Dominator or not. Therefore, two hypotheses are formulated and only one will be considered in the following steps.

In the case v does not belong to S, it is necessary to distinguish three additional cases: v has an unshared dominator in $T(v)$; v has a non-Dominator child in S (i.e., if $p(v) \in S$, then v has an unshared dominator in T); v has no neighbors in S (i.e., if $p(v)$ is a Dominator of v's color, then v has an unshared dominator in T).

Hence the following cases are considered:

case α **:** $v \in S$

case α_{RB}**:** v has at least one red and one blue child not in S (i.e., v is not a Dominator).

case α_R**:** $C(v) \cap (B \setminus S) = \emptyset$ but v is not a red Dominator in T.

case α'_R**:** $C(v) \cap (B \setminus S) = \emptyset$ and v is a red Dominator in T.

case α_B**:** $C(v) \cap (R \setminus S) = \emptyset$ but v is not a blue Dominator in T.

case α'_B**:** $C(v) \cap (R \setminus S) = \emptyset$ and v a blue Dominator in T.

case β**:** $v \notin S$

case β_d**:** $\left(\text{either } |C(v) \cap S| \geq 2 \text{ or } \begin{cases} (C(v) \cap S_R) \neq \emptyset & \text{if } v \in R \\ (C(v) \cap S_B) \neq \emptyset & \text{if } v \in B \end{cases}\right)$

case β_1**:** $|C(v) \cap S| = 1$ and $\begin{cases} (C(v) \cap S_R) = \emptyset & \text{if } v \in R \\ (C(v) \cap S_B) = \emptyset & \text{if } v \in B \end{cases}$

case β_0**:** $|C(v) \cap S| = 0$.

We notice that in all the α cases plus the case β_d we have that all the vertices in $T(v)$ have unshared dominators, and consequently, they enable us to identify a RBUD set. The last two cases are auxiliary and will be used to reconstruct the solutions related to the subtree, rooted in v, starting from the solutions of the subtrees rooted in each of its children.

Theorem 4. *Let $T = (V, E)$ be a tree of n vertices and R-B be a red-blue coloring of T. We can compute a solution for the instance $\langle T, R, B\rangle$ of* MIN-RBUD *in time $O(n\Delta^2)$, where Δ is the maximum degree of a vertex in T.*

4.3 Graphs of Bounded Modular-Width

The notion of modular decomposition of graphs was introduced by Gallai in [14] as a tool for defining hierarchical graph decompositions. The *modular-width* parameter was subsequently proposed in [15]. Consider graphs that can be constructed through a sequence of operations (in any order or number):

(O1) The creation of an isolated vertex.

(O2) The substitution of the vertices of a graph $H = (\{1, \ldots, h\}, E(H))$ by the graphs (i.e., modules) $G_1 = (V_1, E_1), \ldots, G_h = (V_h, E_h)$, denoted by $H(G_1, \ldots, G_h)$, is the graph $G = (V, E)$ with $V = \bigcup_{i=1}^{h} V_i$ and $E = \bigcup_{i=1}^{h} E_i \cup \{(u, w) \mid u \in V_i,\ w \in V_j,\ (i, j) \in E(H),\ 1 \leq i < j \leq h\}$.

The *width* of such a sequence is the maximum number of vertices in any graph H appearing in an operation (O2). The *modular-width* of G, denoted $\mathtt{mw}$, is the minimum width over all possible sequences of operations that construct G.

A *parse tree* of a graph G is a tree $T(G)$ that represents the construction of G using operations (O1) and (O2). The leaves of $T(G)$ correspond to the vertices of G (operation (O1)), while the internal nodes represent applications of operation (O2). The width of a parse tree $T(G)$ is defined as the maximum number of children of any node in $T(G)$, that is, the maximum size of a graph H used in any (O2) operation during the construction of G. A parse tree of minimum width, corresponding to the modular-width of G, can be constructed in linear time [9]. Modular-width and related parameters [6,20] have been effectively utilized to develop FPT algorithms for Domination-like problems [5–8,15,19,22].

Given $\langle G, R, B\rangle$, we classify the RBUD sets satisfying such instance into three types $\{\mu, \beta, \rho\}$. The types β and ρ correspond to solutions that leave out only blue and red vertices, respectively. Formally, given the instance $\langle G, R, B\rangle$, a solution $S \subseteq V$ is called of

$$\begin{array}{ll} \text{type } \mu & \text{if } (V \setminus S) \cap B \neq \emptyset \neq (V \setminus S) \cap R \\ \text{type } \beta & \text{if } V \setminus S \subseteq B \\ \text{type } \rho & \text{if } V \setminus S \subseteq R. \end{array} \tag{3}$$

We denote by S^x the RBUD set of minimum size among all the solutions of type x, for $x = \mu, \beta, \rho$. We notice that both S^β and S^ρ always exist and in particular $S^\beta = R$ and $S^\rho = B$. Indeed S^β and S^ρ correspond to the trivial solutions. Conversely, S^μ may not exist (e.g., in any bi-chromatic clique).

Clearly, a minimum RBUD set S for the instance $\langle G, R, B\rangle$ satisfies

$$|S| = \min\{|S^\mu|, |R|, |B|\}.$$

Algorithm 1 identifies the optimal solution S according to the above observation.

We present now Algorithm 2, which computes S^μ, if any exists. The algorithm computes S^μ leveraging the *parse tree* $T(G)$. Since the problem is trivial for single-vertex graphs (see line 1 of Algorithm 2), we focus on the case where the input graph G is constructed via operation (O2). Hence, for each module $\hat{G} = (\hat{V}, \hat{E}) = H(\hat{G}_1, \ldots, \hat{G}_h)$, represented by an internal node of $T(G)$, where the red-blue coloring is given by $\hat{B} = B \cap \hat{V}$ and $\hat{R} = R \cap \hat{V}$, the algorithm computes the optimal solution $S^\mu_{\hat{G}}$ for the instance $\langle \hat{G}, \hat{R}, \hat{B}\rangle$ of MIN-RBUD. This goal is achieved by recursively solving the problem on each module $\hat{G}_1, \ldots, \hat{G}_h$. Hence, for $i = 1, 2, \ldots, h$, the algorithm computes the solutions S^x_i for $\langle \hat{G}_i, \hat{R}_i, \hat{B}_i\rangle$ of MIN-RBUD, where $\hat{B}_i = B \cap \hat{V}_i$ and $\hat{R}_i = R \cap \hat{V}_i$. The computation depends on a set of different cases, denoted by the parameter x, which will be described below. Finally, the algorithm performs an exhaustive combination of all cases. For each combination, the algorithm computes a solution S. If the obtained solution is of type μ, represents a RBUD-set, and improves upon the best solution found so far, then the algorithm updates the stored solution (line 18 of Algorithm 2). Recalling that the input graph G is represented by the root of $T(G)$, we are able to compute the solution S^μ, if any. See Algorithm 2 for details.

In order to simplify the notation, given an instance $\langle \hat{G}, \hat{R}, \hat{B} \rangle$, we will use r to denote a red vertex in $\hat{R}$ and b to denote a blue vertex in $\hat{B}$.

Lemma 4. *Let $\hat{G} = H(\hat{G}_1, \ldots, \hat{G}_h)$, with red-blue coloring $\hat{B} = B \cap \hat{V}$ and $\hat{R} = R \cap \hat{V}$. Any RBUD set S for the instance $\langle \hat{G}, \hat{R}, \hat{B} \rangle$ must include a RBUD set S_i for the instance $\langle \hat{G}_i, R_i, B_i \rangle$, for each $i = 1, \ldots, h$.*

Proof. Consider a generic module i. If either $(\hat{V}_i \setminus S_i) \cap \hat{B}$ or $(\hat{V}_i \setminus S_i) \cap \hat{R}$ is empty then S_i is a RBUD set. Otherwise, for each $r, b \in V_i$, the vertices r and b share the same neighborhood outside G_i, that is $N(r) \cap (\hat{V} \setminus \hat{V}_i) = N(b) \cap (\hat{V} \setminus \hat{V}_i)$. Hence, any RBUD set S for the instance $\langle \hat{G}, \hat{R}, \hat{B} \rangle$ must include a RBUD set S_i for the instance $\langle \hat{G}_i, \hat{R}_i, \hat{B}_i \rangle$ so that Definition 1 is satisfied for $r, b \in \hat{V}_i$. □

Before proving the correctness of Algorithm 2, we introduce the following auxiliary problem.

Definition 3. *Consider a graph $G = (V, E)$ and $A \subseteq V$. A set $D \subseteq V$ is called subset-dominating set (*SDS*) of G with respect to A, if each vertex in $A \setminus D$ is adjacent to a vertex in D. A* SDS *D of G with respect to A is called optimal if its size is minimum among all the* SDS*s of G with respect to A.*

MINIMUM SUBSET-DOMINATING SET (MSDS): Given a graph $G = (V, E)$ and a subset $A \subseteq V$, find an optimal SDS of G with respect to A.

Using the same approach as in [22], one can establish the following result. Given a modular decomposition $H(G_1, \ldots, G_h)$ of a graph $G = (V, E)$, it can be shown that, for any $A \subseteq V$, there exists a MINIMUM SUBSET-DOMINATING SET for the instance $\langle G, A \rangle$ of MSDS in which each module contains at most one vertex from the solution. By exhaustively considering all possible combinations, we obtain:

Lemma 5. *Let $G = (V, E)$ be a graph of modular-width at most $\mathtt{mw}$ and $A \subseteq V$. A solution for the instance $\langle G, A \rangle$ of* MSDS *can be obtained in time $O(n^2\, 2^{\mathtt{mw}})$.*

Given $G = (V, E)$ and $A \subseteq V$, we denote by MSDS(G, A) the algorithm that returns a MINIMUM SUBSET-DOMINATING SET for G with respect to A.

The following Lemmas present some properties of the MINIMUM SUBSET-DOMINATING SET that will be used in the algorithm.

Lemma 6. *Let $\hat{G} = (\hat{V}, \hat{E}) = H(\hat{G}_1, \ldots, \hat{G}_h)$, with red-blue coloring $\hat{B} = B \cap \hat{V}$ and $\hat{R} = R \cap \hat{V}$. Let $A_R = \{r \in \hat{R} \mid \hat{B} \subseteq N_{\hat{G}}(r)\}$ and $A_B = \{b \in \hat{B} \mid \hat{R} \subseteq N_{\hat{G}}(b)\}$. The following properties hold:*

- *The set $C^\rho =$ MSDS$(\overline{\hat{G}[\hat{R}]}, A_R)$ is a smallest set X_C satisfying*

$$\forall r \in \hat{R} \setminus X_C \; \exists u \in (\hat{B} \cup X_C) \text{ such that } u \notin N_{\hat{G}}(r). \tag{4}$$

- *The set $C^\beta =$ MSDS$(\overline{\hat{G}[\hat{B}]}, A_B)$ is a smallest set Y_C satisfying*

$$\forall b \in \hat{B} \setminus Y_C \; \exists v \in (\hat{R} \cup Y_C) \text{ such that } v \notin N_{\hat{G}}(b). \tag{5}$$

- *The sets* C^ρ *and* C^β *can be computed in time* $O(n^2\, 2^{\mathtt{mw}})$.

Lemma 7. *Let* $\hat{G} = (\hat{V}, \hat{E}) = H(\hat{G}_1, \ldots, \hat{G}_h)$*, with red-blue coloring* $\hat{B} = B \cap \hat{V}$ *and* $\hat{R} = R \cap \hat{V}$*. The following properties hold:*

- *The set* D^ρ=MSDS$(\hat{G}[\hat{R}], \hat{V} \backslash N[\hat{B}])$ *is a smallest set* X_D *satisfying* $\hat{V} \subseteq N[\hat{B} \cup X_D]$
- *The set* D^β=MSDS$(\hat{G}[\hat{B}], \hat{V} \backslash N[\hat{R}])$ *is a smallest set* Y_D *satisfying* $\hat{V} \subseteq N[\hat{R} \cup Y_D]$
- *The sets* D^ρ *and* D^β *can be computed in time* $O(n^2\, 2^{\mathtt{mw}})$.

Given a solution S for the instance $\langle \hat{G}, \hat{R}, \hat{B} \rangle$, with $\hat{G} = (\hat{V}, \hat{E}) = H(\hat{G}_1, \ldots, \hat{G}_h)$, we denote by $S_i = S \cap \hat{V}_i$ the subset of vertices of S that belong to $\hat{G}_i$. Lemma 8 allows to characterize the sets S_i, for $i = 1, \ldots, \mathtt{mw}$.

Lemma 8. *Let* $\hat{G} = H(\hat{G}_1, \ldots, \hat{G}_h)$ *be a graph and* $\hat{R}$, $\hat{B}$ *a red-blue coloring of* $\hat{G}$*. Let* S *be a solution for the instance* $\langle \hat{G}, \hat{R}, \hat{B} \rangle$ *of the* MIN-RBUD *problem. For each* $i \in [h]$*, the set* S_i *satisfies one of the following:*

(1) $S_i = S_i^\mu$*, where* S_i^μ *is the optimal solution for* $\hat{G}_i$ *such that* $(\hat{V}_i \setminus S_i) \cap \hat{B} \neq \emptyset \neq (\hat{V}_i \setminus S_i) \cap \hat{R})$
(2) $S_i = S_i^{\rho_1} = \hat{B}_i$
(3) $S_i = S_i^{\rho_2} = \hat{B}_i \cup C_i^\rho$
(4) $S_i = S_i^{\rho_3} = \hat{B}_i \cup D_i^\rho$
(5) $S_i = S_i^{\beta_1} = \hat{R}_i$
(6) $S_i = S_i^{\beta_2} = \hat{R}_i \cup C_i^\beta$
(7) $S_i = S_i^{\beta_3} = \hat{R}_i \cup D_i^\beta$
(8) $S_i = S_i^\tau = \hat{V}_i$
(9) $S_i = S_i^\nu = \{v\}$*, where* v *is any vertex in* $\hat{V}_i$ *and* $\hat{V}_i$ *is monochromatic (i.e.,* $\hat{R}_i = \emptyset$ *or* $\hat{B}_i = \emptyset$*).*

Theorem 5. *Let* $G = (V, E)$ *be a graph of modular-width at most* $\mathtt{mw}$*, and R-B define a red-blue coloring of* G*. Algorithm* TYPE-μ-RBUD-MW(G) *decides whether a RBUD set of type* μ *exists for the instance* $\langle G, R, B \rangle$ *and, if so, returns one of minimum size in time* $O(n^3\, 8^{\mathtt{mw}})$.

Since the minimum RBUD set S for the instance $\langle G, R, B \rangle$ satisfies $|S| = \min\{|S^\mu|, |S^\beta = R|, |S^\rho = B|\}$, by Theorem 5 we obtain the following result

Theorem 6. *Let* $G = (V, E)$ *be a graph of modular-width at most* $\mathtt{mw}$*, and* R*,* B *define a red-blue coloring of* G*. A solution for the instance* $\langle G, R, B \rangle$ *of* MIN-RBUD *can be obtained in time* $O(n^3 8^{\mathtt{mw}})$.

4.4 Graphs of Bounded Neighborhood Diversity

While the results for graphs of bounded modular-width include graphs of bounded neighborhood diversity, a more efficient algorithm can be designed specifically for this latter class.

The neighborhood diversity of a graph was introduced by Lampis in [20]. In this paper, we will use the following equivalent formulation.

Algorithm 1: Algorithm RBUD-MW(G, R, B)

Input: A graph $G = H(G_1, \ldots, G_h)$ and a red-blue coloring R-B of G.
Output: S_G, a minimum RBUD set for the MIN-RBUD instance $\langle G, R, B\rangle$.

if $|R| < |B|$ **then** $S_G = R$ **else** $S_G = B$
$(Res, S^{\mu}_G) =$ TYPE-μ-RBUD-MW(G, R, B)
if $((Res = true) \wedge (|S^{\mu}_G| < |S_G|))$ **then** $S_G = S^{\mu}_G$
return S_G

Algorithm 2: Algorithm TYPE-μ-RBUD-MW $(\hat{G}, \hat{R}, \hat{B})$

Input: A graph $\hat{G}=(\hat{V}, \hat{E})=H(\hat{G}_1, \ldots, \hat{G}_h)$ and a red-blue coloring $\hat{R}, \hat{B}$ of $\hat{G}$.
Output: The couple $(Res, S^{\mu}_{\hat{G}})$. $Res =$ true if $\hat{G}$ admits RBUD set of type μ and false otherwise. When $Res =$ true, $S^{\mu}_{\hat{G}}$ is a minimum RBUD set of type μ for the MIN-RBUD instance $\langle \hat{G}, \hat{R}, \hat{B}\rangle$.

if ($\hat{G}$ *is either a clique or an ind. set*) **then** // base case: if $\hat{G}$ is a clique or an ind. set, then a red and a blue vertex cannot simultaneously lie outside a RBUD set.
    return (false, $\emptyset$)
$Res =$ false, $S^{\mu}_{\hat{G}} = \hat{V}$
for $i = 1, \cdots, h$ **do**
    $(Res^{\mu}_i, S^{\mu}_i) =$ TYPE-μ-RBUD-MW$(\hat{G}_i)$
    $Res^{\rho 1}_i =$ true, $S^{\rho 1}_i = \hat{B}_i$
    $Res^{\rho 2}_i =$ true, $S^{\rho 2}_i = \hat{B}_i \cup C^{\rho}_i$, where $C^{\rho}_i =$ MSDS$(\overline{\hat{G}_i[\hat{R}_i]}, A_R)$
    $Res^{\rho 3}_i =$ true, $S^{\rho 3}_i = \hat{B}_i \cup D^{\rho}_i$, where $D^{\rho}_i =$ MSDS$(\hat{G}_i[\hat{R}_i], \hat{V}_i \setminus N[\hat{B}_i])$
    $Res^{\beta 1}_i =$ true, $S^{\beta 1}_i = \hat{R}_i$
    $Res^{\beta 2}_i =$ true, $S^{\beta 2}_i = \hat{R}_i \cup C^{\beta}_i$, where $C^{\beta}_i =$ MSDS$(\overline{\hat{G}_i[\hat{B}_i]}, A_B)$
    $Res^{\beta 3}_i =$ true, $S^{\beta 3}_i = \hat{R}_i \cup D^{\beta}_i$, where $D^{\beta}_i =$ MSDS$(\hat{G}_i[\hat{B}_i], \hat{V}_i \setminus N[\hat{R}_i])$
    $Res^{\tau}_i =$ true, $S^{\tau}_i = V_i$
    $Res^{\nu}_i =$ true, $S^{\nu}_i = \{v\}$, where v is any vertex in V_i
for *each* $\mathbf{x} = (x_1, \ldots, x_h) \in \{\mu, \rho_1, \rho_2, \rho_3, \beta_1, \beta_2, \beta_3, \tau, \nu\}^h$ **do**
    if $\bigwedge_{i=1}^{h} Res^{x_i}_i$ **then**
        $S = \bigcup_{i=1}^{h} S^{x_i}_i$
        if $((S$ *is of type* $\mu) \wedge (S$ *is a RBUD set*$) \wedge (|S| < |S^{\mu}_{\hat{G}}|))$ **then**
        // S is of type μ: Both $(\hat{V} \setminus S) \cap B$ and $(\hat{V} \setminus S) \cap R$ are non empty.
            $Res =$ true, $S^{\mu}_{\hat{G}} = S$
return $(Res, S^{\mu}_{\hat{G}})$

Definition 4. *The neighborhood diversity of* $G = (V, E)$ *is the minimum integer* h *such that* $G = H(G_1, \ldots, G_h)$ *where* $G_i = (V_i, E_i)$ *is either a clique or an independent set, for each* $i = 1, \ldots, h$.

The neighborhood diversity of a graph G and the corresponding partition $V_1, \dots, V_h$ of V (named *type partition*), can be determined in $O(n+m)$ time [20]. In the sequel, we denote by $\mathtt{nd}$ the neighborhood diversity of the input graph G.

Given a red-blue coloring R-B of $G = H(G_1, \dots, G_{\mathtt{nd}})$, let $R_i = R \cap V_i$ and $B_i = B \cap V_i$ be the sets of red and blue vertices in G_i, respectively, for each $i \in [\mathtt{nd}]$. Moreover, given a solution S for the RBUD instance $\langle G, R, B\rangle$, we denote by $S_i = S \cap V_i$ the set of vertices of the solution S that belong to G_i. The following three Lemmas allow us to characterize the sets S_i, for $i \in [\mathtt{nd}]$.

Lemma 9. *Let $G = H(G_1, \dots, G_{\mathtt{nd}})$ and let S be a solution for the* RBUD *instance $\langle G, R, B\rangle$. For each $i \in [\mathtt{nd}]$, it holds that either $R_i \subseteq S_i$ or $B_i \subseteq S_i$.*

Proof. Clearly the Lemma holds if either $R_i = \emptyset$ or $B_i = \emptyset$. Assume then that there exist $r \in R_i \setminus S_i$ and $b \in B_i \setminus S_i$. If G_i is an independent set, then $N(r) = N(b)$, while if G_i is a clique, then $N[r] = N[b]$ and by Lemma 1, we obtain a contradiction because one between r and b must belong to S. □

Lemma 10. *Let $G = H(G_1, \dots, G_{\mathtt{nd}})$ be a graph and R, B a red-blue coloring of G. Let S be a solution for the instance $\langle G, R, B\rangle$ of the* MIN-RBUD *problem. For each $i \in [\mathtt{nd}]$, the set S_i satisfies one of the following:*

(a) $S_i = R_i$ *(i.e.,* $V_i \setminus S = B_i$*).*
(b) $S_i = B_i$ *(i.e.,* $V_i \setminus S = R_i$*).*
(c) $S_i = V_i$ *(i.e.,* $V_i \setminus S = \emptyset$*).*
(d) $S_i = \{v\}$*, where v is any vertex in V_i and V_i is monochromatic.*

Theorem 7. MIN-RBUD *can be solved in time $O(n + m + \mathtt{nd}^2\, 3^{\mathtt{nd}})$.*

Proof. First, we compute the type partition in $O(n+m)$ time. Then we note that each module corresponds to 3 possible solutions. Indeed, case (d) in Lemma 10 can only hold for monochromatic modules. Moreover, for monochromatic modules, two of the three cases (a), (b), (c) are equivalent. Hence, the number of potential solutions is $O(3^{\mathtt{nd}})$. Since vertices within the same module have the same neighborhood, we can check whether a potential solution satisfies conditions 1) and 2) in Definition 1 in $O(\mathtt{nd}^2)$ time. Overall, we can compute the best solution within $O(n + m + \mathtt{nd}^2\, 3^{\mathtt{nd}})$ time. □

5 Conclusion

In this work, we introduced the Red-Blue Unshared Domination (RBUD) problem, a variation of classical domination problems in which every pair of differently colored vertices can be distinguished by a pair of private (i.e., unshared) dominators. We proved that RBUD is NP-hard, even under strong structural constraints such as bipartite planar subcubic graphs. Despite this inherent complexity, we developed a general approximation algorithm and provided efficient exact solutions for several important graph classes, namely, trees, which represent sparse graphs, and graphs with bounded neighborhood diversity, which capture certain dense structures. Furthermore, we extended our algorithmic approach to graphs with bounded modular-width.

However, designing a fixed-parameter tractable (FPT) algorithm with respect to treewidth remains an open challenge. Indeed, our solution for trees relies on a structural property formalized in Lemma 3, which is not easily generalizable to graphs of bounded treewidth.

References

1. Bollobás, B., Scott, A.D.: On separating systems. Eur. J. Comb. **28**, 1068–1071 (2007)
2. Bonnet, É., Giannopoulos, P., Lampis, M.: On the parameterized complexity of red-blue points separation. J. Comp. Geom. **10**(1), 181–206 (2019)
3. Charbit, E., Charon, I., Cohen, G., Hudry, O., Lobstein, A.: Discriminating codes in bipartite graphs: bounds, extremal cardinalities, complexity. Adv. Math. Commun. **2**(4), 403–420 (2008)
4. Chlebík, M., Chlebíkova, J.: Approximation hardness of dominating set problems in bounded degree graphs. Inf. Comp. **206**, 1264–1275 (2008)
5. Cordasco, G., Gargano , L., Rescigno, A.A.: (t,r)-broadcast domination in networks. In: Proceedings of ICTCS 2024. https://ceur-ws.org/Vol-3811/ (2024)
6. Cordasco, G., Gargano, L., Rescigno, A.A.: Parameterized complexity for iterated type partitions and modular-width. DAM **350**, 100–122 (2024)
7. Cordasco, G., Gargano, L., Rescigno, A.A.: Distance vector domination. In: Proceedings of SOFSEM 2025. LNCS 15538 (2025)
8. Cordasco, G., Gargano, L., Rescigno, A.A.: Parameterized complexity of (d,r)-domination via modular decomposition. In: Proceedings of WALCOM 2025 (2025)
9. Corneil, D., Habib, M., Paul, C., Tedder, M.: A recursive linear time modular decomposition algorithm via LexBFS. arXiv:0710.3901 [cs.DM] (2024)
10. Crowston, R., Gutin, G., Jones, M., Muciaccia, G., Yeo, A.: Parameterizations of test cover with bounded test sizes. Algorithmica **74**(1), 367–384 (2016)
11. De Bontridder, K.M.J., Halldórsson, B.V., Halldórsson, M.M., Hurkens, C.A.J., Lenstra, J.K., Ravi, R., Stougie, L.: Approximation algorithms for the test cover problem. Math. Program. Ser. B **98**, 477–491 (2003)
12. Dev, S.R., Dey, S., Foucaud, F., Klasing, R., Lehtilä, T.: The RED-BLUE SEPARATION problem on graphs. Theor. Comput. Sci. **970** (2023)
13. Feige, U.: A threshold of ln n for approximating set cover. J. ACM 45(4), 634–652 (1998). https://doi.org/10.1145/285055.285059
14. Gallai, T.: Transitiv orientierbare Graphen. Acta Math. Acad. Scientiarum Hung. **18**, 26–66 (1967)
15. Gajarský, J., Lampis, M., Ordyniak, S.: Parameterized algorithms for modular-width. In: Gutin, G., Szeider, S. (eds.) IPEC 2013. LNCS, vol. 8246, pp. 163–176. Springer, Cham (2013). https://doi.org/10.1007/978-3-319-03898-8_15
16. Haynes, T.W., Henning, M.A., Howard, J.: Locating and total-dominating sets in trees. Discret. Appl. Math. **154**, 1293–1300 (2006)
17. Karpovsky, M.G., Chakrabarty, K., Levitin, L.B.: On a new class of codes for identifying vertices in graphs. IEEE Trans. on Inf. theory **44**(2), 599–611 (1998)

18. Lobstein, A.: Watching systems, identifying, locating-dominating and discriminating codes in graphs: a bibliography (2025). https://www.lri.fr/~lobstein/debutBIBidetlocdom.pdf
19. Lafond, M., Luo, W.: Parameterized complexity of domination problems using restricted modular partitions. MFCS **2023**, 61:1–61:14 (2023)
20. Lampis, M.: Algorithmic meta-theorems for restrictions of treewidth. Algorithmica **64**, 19–37 (2012)
21. Misra, N., Mittal, H., Sethia, A.: Red-blue point separation for points on a circle. Proc. CCCG **2020**, 266–272 (2020)
22. Romanek, M.: Parameterized algorithms for modular-width. Bachelor's Thesis, (2016)
23. Seo, S.J., Slater, P.J.: Open neighborhood locating dominating sets. Australas. J. Com. **46**, 109–119 (2010)
24. Slater, P.J.: Dominating and reference sets in a graph. J. Math. Phys. Sci. **22**, 445–455 (1988)
25. Zvervich, I.E., Zverovich, V.E.: An induced subgraph characterization of domination perfect graphs. J. Graph Theory **20**(3), 375–395 (1995)

Finding d-Cuts in Probe H-Free Graphs

Konrad K. Dabrowski[1], Tala Eagling-Vose[2(✉)], Matthew Johnson[2], Giacomo Paesani[3], and Daniël Paulusma[2]

[1] School of Computing, Newcastle University, Newcastle, UK
konrad.dabrowski@newcastle.ac.uk
[2] Department of Computer Science, Durham University, Durham, UK
{tala.j.eagling-vose,matthew.johnson2,daniel.paulusma}@durham.ac.uk
[3] Department of Computer Science, Sapienza University of Rome, Rome, Italy
paesani@di.uniroma1.it

Abstract. For an integer $d \geq 1$, the d-CUT problem is that of deciding whether a graph has an edge cut in which each vertex is adjacent to at most d vertices on the opposite side of the cut. The 1-CUT problem is the well-known MATCHING CUT problem. The d-CUT problem has been extensively studied for H-free graphs. We extend these results to the probe graph model, where we do not know all the edges of the input graph. For a graph H, a partitioned probe H-free graph (G, P, N) consists of a graph $G = (V, E)$, together with a set $P \subseteq V$ of probes and an independent set $N = V \setminus P$ of non-probes such that we can change G into an H-free graph by adding zero or more edges between vertices in N. For every graph H and every integer $d \geq 1$, we completely determine the complexity of d-CUT on partitioned probe H-free graphs.

Keywords: d-cut · probe graph · subset problem · H-free graph

1 Introduction

When studying computationally hard problems for special graph classes, it is natural to generalize polynomial-time results for certain graph classes to larger graph classes. In particular, consider a graph H' and an induced subgraph H of H'. The class of H-*free* graphs (class of graphs that do not contain H as an induced subgraph) is contained in the class of H'-free graphs. Say an NP-complete problem Π is polynomial-time solvable on H-free graphs. Is Π also polynomial-time solvable on H'-free graphs? This question leads to complexity studies for a wide range of graph problems where the goal is to obtain *complexity dichotomies* that tell us for exactly which graphs H a certain NP-complete problem is polynomial-time solvable, and for which graphs H it stays NP-complete.

We follow this line of research, but also assume that we do not know all the edges of the input graph. Before explaining the latter in more detail, we first introduce the problem that we study. Consider a connected graph $G = (V, E)$. A subset $M \subseteq E$ is an *edge cut* of G if it is possible to partition V into two non-empty sets B (*blue* vertices) and R (*red* vertices) in such a way that M is the

A. Jeż and J. Otop (Eds.): FCT 2025, LNCS 16106, pp. 109–121, 2026.
https://doi.org/10.1007/978-3-032-04700-7_9

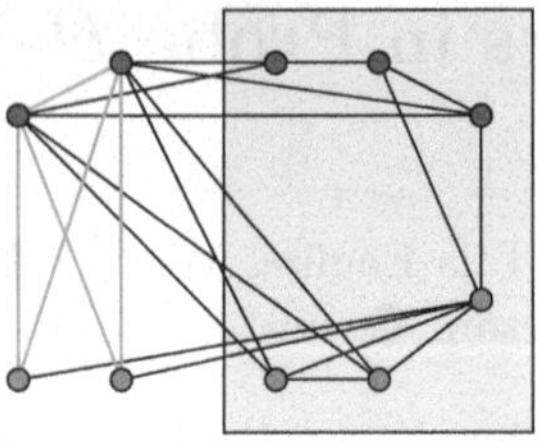

Fig. 1. A probe P_5-free graph G with a set P of probes. The Edges of G are shown in black, whereas the edges in F (which do not belong to G) are shown in purple. Note that $G + F$ is indeed P_5-free and that the red-blue colouring corresponds to a 2-cut of G. (Colour figure online)

set of all edges with one end-vertex in B and the other in R. Now, for an integer $d \geq 1$, if every blue vertex has at most d red neighbours, and every red vertex has at most d blue neighbours, then the edge cut M is said to be a *d-cut* of G. See also Fig. 1. The d-Cut problem is that of deciding whether a connected graph has a d-cut. A 1-cut is also called a *matching cut*, and the 1-Cut problem is better known as Matching Cut. For all $d \geq 1$, d-Cut is NP-complete [8,14]. Graphs with matching cuts were introduced in 1970 by Graham [15] in the context of number theory; for other applications see [2,9,11,25].

Our Focus. We consider the classical probe graph model, which was introduced by Zhang et al. [27] in 1994 to deal with partial information in genome research. In this model, the complete set of neighbours is only known for *some* vertices of the input graph G. These vertices form the set P of *probes*. The other vertices of G form the set N of *non-probes*. As we do not know the adjacencies between vertices in N, the set N is an independent set in G. However, in the probe graph model we also assume there exists a "certifying" set F of edges between (some of) the non-probes such that $G + F$ has some known global structure; again see Fig. 1. In our paper, $G + F$ is H-free. Note that $G[P]$ is already H-free.

So, a *partitioned probe H-free* graph (G, P, N) consists of a graph $G = (V, E)$, a set $P \subseteq V$ of probes and an independent set $N = V \setminus P$ of non-probes, such that $G + F$ is H-free for some edge subset $F \subseteq \binom{N}{2}$. Any H-free graph is also (partitioned) probe H-free: take $P = V$ and $N = \emptyset$. Hence, (partitioned) probe H-free graphs contain all H-free graphs, and any NP-completeness results for H-free graphs carry over to partitioned probe H-free graphs. We therefore ask:

For which H, does d-Cut stay polynomial-time solvable on probe H-free graphs?

As such, our paper belongs to a recent systematic study of graph problems on probe H-free graphs. This study was initiated by Brettell et al. [5] for Vertex Cover, whereas the previous literature on probe graphs aimed to characterize and recognize classes of probe graphs; see e.g. [3,6,7,12,13]. For example, if $H = P_4$, then probe H-free graphs can be recognized in polynomial time [7]. However, for most other graphs H, the complexity of recognizing probe H-free

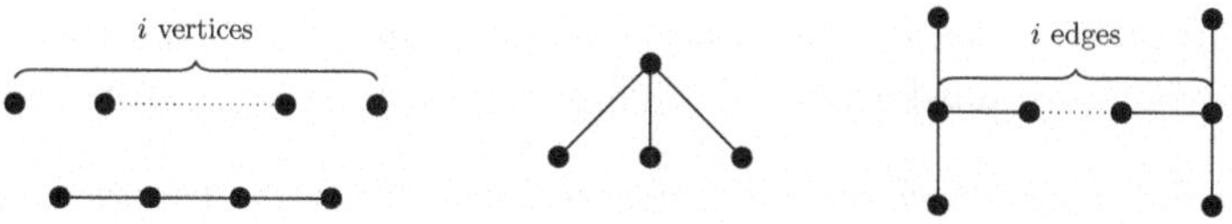

Fig. 2. The graphs $sP_1 + P_4$, $K_{1,3}$ and H_i^*, from left to right.

graphs is still unknown. Hence, for our algorithms, we assume that P and N are part of the input, that is, we will consider partitioned probe H-free graphs.

We will also consider two related problems: Maximum Matching Cut and Perfect Matching Cut on probe H-free graphs. The first is to decide if a connected graph has a matching cut of at least k edges for some integer k. The second is to decide if a connected graph has a *perfect matching cut*, that is, an edge cut that is a perfect matching. This problem is also NP-complete [16].

Known Results. For two vertex-disjoint graphs G_1 and G_2, let $G_1 + G_2 = (V(G_1) \cup V(G_2), E(G_1) \cup E(G_2))$. We let sG be the disjoint union of s copies of G. We write $G_1 \subseteq_i G_2$ if G_1 is an induced subgraph of G_2. Let C_s denote the cycle on s vertices, P_t the path on t vertices, and $K_{1,r}$ the star on $r+1$ vertices. The graph $K_{1,3}$ is known as the *claw*. Let H_1^* be the "H"-graph, which has vertices u, v, w_1, w_2, x_1, x_2 and edges $uv, uw_1, uw_2, vx_1, vx_2$. For $i \geq 2$, let H_i^* be obtained from H_1^* by subdividing uv exactly $i-1$ times. See Fig. 2.

In Theorems 1–3 we present the state-of-art for d-Cut, Perfect Matching Cut and Maximum Matching Cut for H-free graphs. Only Theorem 3 is a full dichotomy. The references in Theorem 1 are explained in [20] except for the recent result that 2-Cut is NP-complete for claw-free graphs [1]; note the jump in complexity from $d=1$ to $d=2$ for $H = 3P_2$ and $H = K_{1,3}$. For $d \geq 2$, the only three non-equivalent open cases in Theorem 1 are $H = 2P_4$, $H = P_6$ and $H = P_7$ (see also [20]). The references in Theorem 2 are explained in [23].

Theorem 1 ([1,4,8,10,17,19–22,24]). *Let H be a graph and $d \geq 1$.*

- *If $d = 1$, then d-Cut on H-free graphs is polynomial-time solvable if $H \subseteq_i sP_3 + S_{1,1,2}$, $sP_3 + P_4 + P_6$, or $sP_3 + P_7$ for some $s \geq 0$; and* NP*-complete if $H \supseteq_i K_{1,4}$, P_{14}, $2P_7$, $3P_5$, C_r for $r \geq 3$, or H_i^* for $i \geq 1$.*
- *If $d \geq 2$, then d-Cut on H-free graphs is polynomial-time solvable if $H \subseteq_i sP_1 + P_3 + P_4$ or $sP_1 + P_5$ for some $s \geq 0$; and* NP*-complete if $H \supseteq_i K_{1,3}$, $3P_2$, C_r for $r \geq 3$, or H_i^* for $i \geq 1$.*

Theorem 2 ([10,17,18,22]). Perfect Matching Cut *on H-free graphs is polynomial-time solvable if $H \subseteq_i sP_4 + S_{1,2,2}$ or $sP_4 + P_6$ for some $s \geq 0$; and* NP*-complete if $H \supseteq_i K_{1,4}$, P_{14}, $2P_7$, $3P_6$, C_r for $r \geq 3$ or H_j^* for $j \geq 1$.*

Theorem 3 ([23]). Maximum Matching Cut *on H-free graphs is polynomial-time solvable if $H \subseteq_i sP_2 + P_6$ for some $s \geq 0$; and* NP*-complete otherwise.*

Our Results. We combine NP-completeness results from Theorems 1–3 with new polynomial and hardness results (shown in Sect. 3 and 4, resp.) to prove:

Theorem 4. *For a graph H, the following four complete dichotomies hold:*

- 1-Cut, Perfect Matching Cut *and* Maximum Matching Cut *on partitioned probe H-free graphs are polynomial-time solvable if $H \subseteq_i sP_1 + P_4$ for some $s \geq 0$; and* NP*-complete otherwise;*
- *for $d \geq 2$, d-*Cut *on partitioned probe H-free graphs is polynomial-time solvable if $H \subseteq_i P_1 + P_4$; and* NP*-complete otherwise.*

From Theorems 1 and 4, it follows that d-Cut becomes harder for *probe* H-free graphs than for H-free graphs (if P $\neq$ NP) even if $H = 2P_2$ for $d \geq 1$ and $H = 4P_1$ for $d \geq 2$.

2 Preliminaries and Basic Results

Throughout the paper, we only consider finite, undirected graphs without multiple edges and self-loops. We first define some general graph terminology.

Let $G = (V, E)$ be a graph. We let $N_G(v) = \{u \in V \mid uv \in E\}$ be the *(open) neighbourhood* of v and $N_G[v] = N_G(v) \cup \{v\}$ be the *closed neighbourhood* of v. Let $S \subseteq V$. We write $G[S]$ to denote the subgraph of G induced by S. A vertex $v \notin S$ is *complete* to S if $N(v) \supseteq S$, and v is *anti-complete* to S if $N(v) \cap S = \emptyset$. Let $S' \subseteq V$ with $S' \cap S = \emptyset$. If every vertex of S is complete (anti-complete) to S', then S is *complete* (*anti-complete*) to S'.

In our paper we also define some other probe graph classes. For example, we may say that a graph G with a set P of probes and a set N of non-probes is *probe split* if there exists a set $F \subseteq \binom{N}{2}$ such that $G + F$ is a split graph (a graph whose vertex set can be partitioned into a clique and an independent set).

We now recall some colouring terminology for d-cuts from [20] that is commonly used in the context of matching cuts (see, e.g. [21]). A *red-blue colouring* of a graph G colours every vertex of G either red or blue. For $d \geq 1$, a red-blue colouring is a *red-blue d-colouring* if every blue vertex has at most d red neighbours, every red vertex has at most d blue neighbours, and G has at least one blue vertex and at least one red vertex. See Fig. 1 for an example of a graph G with a red-blue 2-colouring. For some $d \geq 1$, a red-blue d-colouring is *perfect* if and only if every red vertex has exactly d blue neighbours and vice versa. This gives us the following straightforward observation (in the case of perfectness we focus on $d = 1$: a perfect 1-cut is a perfect matching cut).

Observation 5 ([20]). *For every $d \geq 1$, a connected graph G has a (perfect) d-cut if and only if G has a (perfect) red-blue d-colouring.*

Let $d \geq 1$. Let $G = (V, E)$ be a connected graph and $X, Y \subseteq V$ be disjoint sets. A *red-blue (X, Y)-d-colouring* of G is a red-blue d-colouring of G that colours all the vertices of X red and all the vertices of Y blue. We say that

(X, Y) is a *d-precoloured pair* of G. We will usually "guess" such a pair (X, Y) as the starting point in our algorithms. On a d-precoloured pair (X, Y) we can safely apply the following two rules exhaustively.

R1. Return no (i.e. G has no red-blue (X, Y)-d-colouring) if a vertex $v \in V$ is adjacent to $d + 1$ vertices in X as well as to $d + 1$ vertices in Y.

R2. Let $v \in V \setminus (X \cup Y)$. If v is adjacent to $d + 1$ vertices in X, then put v in X, and if v is adjacent to $d + 1$ vertices in Y, then put v in Y.

Afterwards, we either returned no, or we obtained two new sets $X' \supseteq X$ and $Y' \supseteq Y$. In the latter case we say that we have *colour-processed* (X, Y) *into* (X', Y'). By construction, every vertex of $V \setminus (X' \cup Y')$ is adjacent to at most d vertices of X' and to at most d vertices of Y'. The next lemma shows that we can work safely with (X', Y') instead of (X, Y).

Lemma 6 ([20]). *Let G be a connected graph with a precoloured pair (X, Y). It is possible, in polynomial time, to either colour-process (X, Y) into a pair (X', Y') such that G has a red-blue (X, Y)-d-colouring if and only if it has a red-blue (X', Y')-d-colouring, or to find that G has no red-blue (X, Y)-d-colouring.*

3 Polynomial-Time Results

In this section, we show our polynomial-time results for Maximum Matching Cut, Perfect Matching Cut and d-Cut ($d \geq 1$). That is, we show that Maximum Matching Cut (and thus 1-Cut) and Perfect Matching Cut are polynomial-time solvable on $(sP_1 + P_4)$-free graphs, and d-Cut, for $d \geq 2$, is polynomial-time solvable on $(P_1 + P_4)$-free graphs.

Our proofs are based on combining colour-processing with the observation that we can guess the closed neighbourhood of any set of size at most some constant c: this will lead to only $\mathcal{O}(2^c n^{cd})$ branches, due to the fact that any vertex can have at most d neighbours of the opposite colour. In our algorithm we choose a constant number of constant-sized sets in such a way that afterwards the whole input graph is coloured. We show below how we do the above for our most involved result, which is our algorithm for d-Cut for $(P_1 + P_4)$-free graphs for $d \geq 2$, and we omit the proof of Theorem 7 due to page restrictions.

Theorem 7. Maximum Matching Cut *and* Perfect Matching Cut *are polynomial-time solvable on partitioned probe $(sP_1 + P_4)$-free graphs for all $s \geq 0$.*

To prove our next result (Theorem 9), we need a lemma whose proof we omit.

Lemma 8. *For any red-blue d-colouring of a connected P_4-free graph, some colour class has size at most $2d$.*

Theorem 9. *For every $d \geq 2$, d-*Cut *is polynomial-time solvable on partitioned probe $(P_1 + P_4)$-free graphs.*

Proof. Let $d \geq 2$. Let $G = (V, E)$. Let (G, P, N) be a connected partitioned probe $(P_1 + P_4)$-free graph, so there exists an edge subset $F \subseteq \binom{N}{2}$ such that $G + F$ is $(P_1 + P_4)$-free. We show how to find a d-cut of G or that none exists.

Suppose that there is a set $Q \subseteq P$ that induces a P_4 in G. Then Q dominates G, otherwise G contains an induced $P_1 + P_4$ of which at most one vertex belongs to N and so $G + F$ also contains an induced $P_1 + P_4$. For every red-blue d-colouring of G, every vertex $v \in V$ has at most d neighbours in a different colour class. That is, there are $\mathcal{O}(n^d)$ red-blue d-colourings of $N_G[v]$. As $|Q| = 4$, we can consider all $\mathcal{O}(n^d)$ colourings of $N_G[Q]$, and, as Q dominates G, this is all red-blue d-colourings of G; hence, we can solve the problem in polynomial time.

We may now assume that $G[P]$ is a P_4-free graph (cograph) and we consider three cases according to the number of connected components of $G[P]$. As G is connected and N is an independent set of G, we find that P dominates N, that is, every vertex of N has a neighbour in P. This means that G has a red-blue d-colouring that colours every vertex of P blue only if G has a red-blue d-colouring that colours exactly one vertex of N red and all other vertices of G blue. We can check this in polynomial time. Hence, from now on, we will assume that in every red-blue d-colouring of G (if such a colouring exists), there is at least one red vertex and at least one blue vertex in P.

Case 9.1: $G[P]$ has exactly one connected component.
By Lemma 8, we may assume that a set X of at most $2d$ vertices of P is coloured red. We guess X. As we may assume that P is not monochromatic, we may assume that X is a proper non-empty subset of P. We colour every vertex of $P \backslash X$ blue. Then we consider all $\mathcal{O}(n^{2d})$ possible red-blue colourings of the neighbourhood of X in N. For each of them, we consider the remaining uncoloured vertices in N. As these only have blue neighbours, we can safely colour them blue. It remains to check in polynomial time if the obtained colouring of G is indeed a red-blue d-colouring.

Case 9.2: $G[P]$ has exactly two connected components.
The proof of this case has been omitted.

Case 9.3: $G[P]$ has at least three connected components.
Let the connected components of $G[P]$ have vertex sets $C_1, \ldots, C_r$ for some $r \geq 3$. We partition N into four types of vertices. Let $v \in N$. If v is complete to P, then we say that v is of *type-A*.

Now, suppose that v is not of type-A, but that v has a neighbour in every component of $G[P]$. We say that v is of *type-B* and make the following observation. We know that v is not complete to some set C_i, so we can find adjacent vertices x_i and x'_i in C_i such that x_i but not x'_i is adjacent to v. Suppose that v is not complete to another set C_j, so there is a vertex x'_j in C_j not adjacent to v. Let x_k be a vertex adjacent to v in the vertex set of a third component. We now have that $\{x'_j\} \cup \{x'_i, x_i, v, x_k\}$ forms an induced $P_1 + P_4$ in G, and consequently in $G + F$, as it only contains one vertex of N, namely v. Hence, a type-B vertex is complete to all but one set C_i, in which it has least one neighbour.

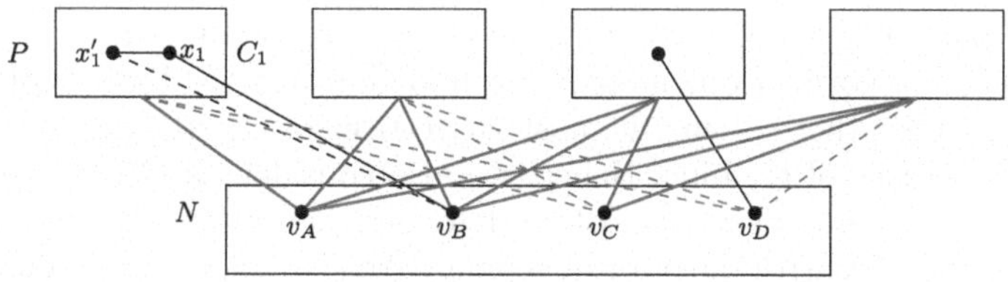

Fig. 3. Illustration of the types described in Case 9.3, specified by vertex label. A (dashed) purple line between a vertex $v \in N$ and some C_i represents that v is (anti-) complete to C_i. A black (dashed) edge between a vertex $v \in N$ and a vertex in P represents a (non)-edge. (Colour figure online)

Now, suppose that v is anti-complete to some C_h, so v is neither of type-A nor of type-B. We say that v is of *type-C* if v has a neighbour in at least two other components C_i and C_j. Suppose that v has a non-neighbour in either C_i or C_j, say in C_i. Let $x_h \in C_h$, $x_i, x_i' \in C_i$ such that $x_i x_i'$ is an edge with only x_i adjacent to v, and let $x_j \in C_j$ be adjacent to v. Now, $\{x_h\} \cup \{x_i', x_i, v, x_j\}$ induces a $P_1 + P_4$ in G, and also in $G + F$ as it has only one vertex from N, a contradiction as $G + F$ is $(P_1 + P_4)$-free. Hence, v is complete to every C_j in which it has a neighbour. So, a type-C vertex of N is complete to at least two and at most $r - 1$ sets in $\{C_1, \ldots, C_r\}$ and anti-complete to all other sets in $\{C_1, \ldots, C_r\}$.

Finally, if v is neither type-A nor type-B nor type-C, then v is of type-D. As G is connected, a type-D vertex v has at least one neighbour in one set C_i and is anti-complete to every other set in $\{C_1, \ldots, C_r\}$. See Fig. 3 for an illustration. We distinguish between the following three cases:

Case 9.3.1 N has a type-A vertex.
Let $v \in N$ be of type-A. We colour v blue and we consider all $\mathcal{O}(n^d)$ possible red-blue colourings of its neighbourhood $N(v) = P$. We then find a set $Q \subseteq P$ of at most d vertices in P that are coloured red. As we already checked whether G has a red-blue d-colouring in which P is monochromatic, we may assume that Q is a proper non-empty subset of P. We now consider all possible $\mathcal{O}(n^{d^2})$ possible red-blue colourings of the neighbourhood of Q in N. For each one of them, the uncoloured vertices in N only have blue neighbours and form an independent set, we can safely colour them blue. Note that at least one vertex is red and at least one is blue. It remains to check whether the obtained colouring of G is a red-blue d-colouring. If so, we are done, and otherwise we discard the branch. Hence, as the number of branches is polynomial, and we can process each branch in polynomial time, this case takes polynomial time.

Case 9.3.2 N has no type-A vertices, but N has a type-B vertex.
Let $v \in N$ be of type-B. We assume without loss of generality that v has a neighbour and a non-neighbour in C_1 and is complete to $C_2, \ldots, C_r$. We colour v blue, and we consider all $\mathcal{O}(n^d)$ options of choosing a set X_v of at most d red neighbours of v in P. We colour all other neighbours of v in P blue. Hence,

afterwards part of C_1 is coloured, while all vertices of every C_i with $i \in \{2, \ldots, r\}$ are coloured. Moreover, by Lemma 8, we find that any red-blue d-colouring of G (if one exists) either colours at most $2d$ vertices of C_1 red, or else it colours at most $2d$ vertices of C_1 blue. Hence, we also consider all $\mathcal{O}(n^{2d})$ options for choosing a set $X \subseteq C_1$ of size at most $2d$ that consists of these vertices. We colour the vertices of $C_1 \setminus X$ with the opposite colour. We also colour the neighbourhood of $X_v \cup X$ in N in every possible way. This yields another $\mathcal{O}(n^{3d^2})$ branches. For each branch, we colour-process. If afterwards we find an uncoloured vertex in N whose neighbourhood in P is monochromatic, then we give it the unique colour of the vertices in its neighbourhood in P.

Suppose that afterwards there still exist uncoloured vertices in N. Let b be an arbitrary uncoloured vertex in N and note that b is adjacent to both a blue vertex x and a red vertex y in P. As we coloured the neighbourhood of $X_v \cup X$, neither x nor y belongs to $X_v \cup X$. As X_v contains all red vertices in $N(v)$, it follows that y belongs to $C_1 \setminus N(v)$, and thus $y \in C_1 \setminus X$. Consequently, all vertices of X are blue. As x, which is blue, is not in X, and all vertices in $C_1 \setminus X$ are red, x belongs to some set C_i with $i \geq 2$. Hence, all uncoloured vertices have a neighbour in C_1 and a neighbour in at least one other C_i. As N has no type-A vertices, every uncoloured vertex in N is of type-B or of type-C.

First, suppose that N has an uncoloured vertex b that is of type-B. By definition, b is complete to $r-1$ sets in $\{C_1, \ldots, C_r\}$. Let Y be the union of these $r-1$ sets. We recall that b is uncoloured and that all vertices in every C_i are coloured. Hence, Y must contain at most $2d$ vertices in total, otherwise we would have given b a colour during colour-processing. We now consider all $\mathcal{O}(n^{2d^2})$ possible red-blue colourings of the neighbourhood of Y in N. Afterwards, there are no uncoloured vertices left, as every uncoloured vertex was of type-B or of type-C and thus must have a neighbour in Y. It remains to check whether the obtained colouring of G is a red-blue d-colouring. If so, we are done, and otherwise we discard the branch.

So we can now assume that all uncoloured vertices are of type-C. Let b again denote an uncoloured vertex in N. From the definition of type-C, it follows that b is either complete or anti-complete to every set C_i. Recall that all uncoloured vertices of N, and thus b, have a neighbour in C_1. Hence, b is complete to C_1. Note that b is anti-complete to X, as otherwise b would have been coloured. This means that $X = \emptyset$, and thus every vertex of C_1 is coloured red. As b is uncoloured and we have colour-processed, this means that C_1 has size at most d. Hence, we can consider all $\mathcal{O}(n^{d^2})$ possible red-blue colourings of the neighbourhood of C_1 in N. Afterwards, all uncoloured vertices in N have received a colour (as they were all complete to C_1). It remains to check whether the obtained colouring of G is a red-blue d-colouring. If so, we are done, otherwise we discard the branch.

As the number of branches is polynomial, and we can process each branch in polynomial time, our algorithm takes polynomial time if this case occurs.

Case 9.3.3 N has no type-A and no type-B vertices.

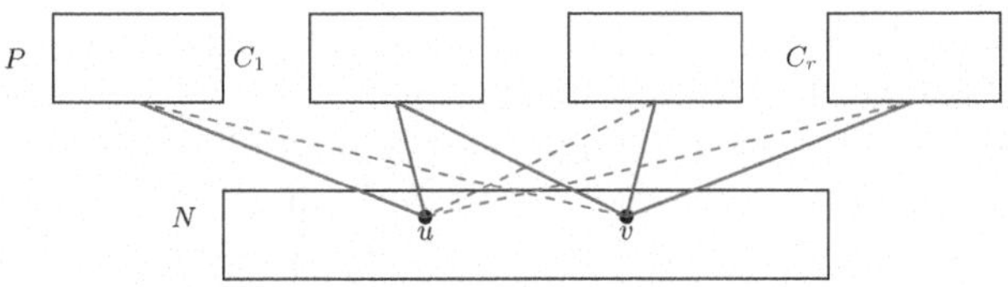

Fig. 4. A P-dominating pair $\{u, v\}$.

As G is connected and $r \geq 3$, G must have vertices that are adjacent to at least two sets C_i. As G has no vertices of type-A or type-B, all of these vertices must be of type-C.

We now make a useful observation. Let u and v be two type-C vertices, such that the following holds:

(i) u is complete to some C_h, but anti-complete to C_i;
(ii) v is anti-complete to C_h, but complete to C_i; and
(iii) u and v are both complete to some C_j.

In this case, we claim that $\{u, v\}$ is a *P-dominating pair*, that is, every C_i is complete to a least one of u, v; see Fig. 4. For a contradiction, suppose that neither u nor v is complete to some C_k (so both are anti-complete to C_k as they are of type-C). Let $z \in C_k$. If $uv \notin E(F)$, then a vertex of C_h, u, a vertex of C_j and v form, together with z, an induced $P_1 + P_4$ in $G + F$. If $uv \in E(F)$, then a vertex of C_h, u, v and a vertex of C_i, together with z, form an induced $P_1 + P_4$ in $G + F$. So, in both cases, we derive a contradiction.

We now continue with the description of our algorithm. We choose a type-C vertex $v \in N$ such that v is complete to a maximum number of sets C_i over all type-C vertices of N. We say that v is of *maximum* type-C. From the definition of type-C, it follows that v is anti-complete to at least one set in $\{C_1, \ldots, C_r\}$. Let C_h be such a set. As G is connected, G contains a path from v to the vertices in C_h. Hence, without loss of generality, there exists a type-C vertex $u \in N$ that is complete to C_h and to at least one other C_j to which v is also complete. As v is of maximum type-C and v is not complete to C_h, to which u is complete, we find that u is also anti-complete to some set C_i to which v is complete. Hence, $\{u, v\}$ satisfies (i)–(iii), so $\{u, v\}$ is a P-dominating pair.

As $\{u, v\}$ is a P-dominating pair, we can colour P as follows. First suppose u and v are coloured alike, say both are coloured blue. We proceed in exactly the same way as in Case 9.1. As u and v can each have at most d red neighbours, we may assume that a set X of at most $2d$ vertices of P is coloured red. We guess X. As we already checked whether G has a red-blue d-colouring in which P is monochromatic, we may assume that X is a proper non-empty subset of P. We colour every vertex of $P \setminus X$ red. Afterwards, we consider all $\mathcal{O}(n^{2d})$ possible red-blue colourings of the neighbourhood of X in N. For each one of them, we consider the uncoloured vertices in N. As these only have blue neighbours, we can safely colour them blue. Note that at least one vertex is red and at least one vertex is blue. It remains to check in polynomial time whether the obtained colouring of G is indeed a red-blue d-colouring.

Now suppose u is coloured red and v is coloured blue. We guess a set X_u of at most d blue neighbours of u in P. We colour all other neighbours of u in P red. Similarly, we guess a set X_v of at most d blue neighbours of v in P. We colour all other neighbours of v in P red. This gives us $\mathcal{O}(n^{2d})$ branches. We note that every vertex of P has been coloured, as $\{u, v\}$ is a P-dominating pair. We now colour the neighbourhood of $X_u \cup X_v$ in N in every possible way. This gives us a further $\mathcal{O}(n^{2d^2})$ branches.

We let $C_1^*, \ldots, C_q^*$ be those sets C_i that contain a vertex of $X_u \cup X_v$. We let $C_1^u, \ldots, C_s^u$ be those sets C_i that contain no vertex of $X_u \cup X_v$ and are coloured red. We let $C_1^v, \ldots, C_t^v$ be those sets C_i that contain no vertex of $X_u \cup X_v$ and are coloured blue. Note that every C_i belongs to one of these three families of sets, but some of these families might be empty.

As v is coloured blue, its neighbours not in X_v are blue. Hence, v is anti-complete to every C_i^u. Similarly, u is anti-complete to every C_j^v. Recall that $\{u, v\}$ is a P-dominating pair and that each of u, v, being of type-C, is either complete or anti-complete to a set C_i. Consequently, u is complete to every C_i^u and v is complete to every C_i^v.

If C_1^u exists, then we select an arbitrary vertex $x_1 \in C_1^u$, and we colour the neighbourhood of x_1 in N in every possible way. This leads to $\mathcal{O}(n^d)$ additional branches. We do the same if C_1^v exists, leading to another $\mathcal{O}(n^d)$ branches. We now colour-process. Afterwards, we colour any vertex in N with monochromatic neighbourhood in P with the unique colour of its neighbours in P.

We claim that every vertex of type-D has now been coloured. For a contradiction, suppose $z \in N$ is of type-D and has no colour yet. We recall that all C_i^u and C_j^v are monochromatic and that z, being type-D, only has neighbours in exactly one C_i. This means that z must have both a blue neighbour and a red neighbour in some C_i^*. However, one of these two neighbours of z belongs to $X_u \cup X_v$, meaning z would have been coloured.

Hence, the only vertices of G that are possibly still uncoloured are type-C vertices in N. Let $b \in N$ be an uncoloured vertex of type-C. This means that b has both a red neighbour and a blue neighbour in P. By definition, b is either complete or anti-complete to C_i for every $i \in \{1, \ldots, r\}$. This means that b is anti-complete to every C_h^*, as we coloured the neighbourhood of $C_h^* \cap (X_u \cup X_v)$, which is a non-empty set by definition. Hence, the red neighbour of b must be some $z^u \in C_i^u$, and the blue neighbour of b must be some $z^v \in C_j^v$. From the above, we find that b is anti-complete to C_1^u and C_1^v (as we coloured the neighbourhood of one of the vertices in them). Hence, we have $i \geq 2$ and $j \geq 2$. Let $y^u \in C_1^u$ and $y^v \in C_1^v$. If $bv \notin F$, then $\{y^u\} \cup \{z^u, b, z^v, v\}$ induces a $P_1 + P_4$ in $G + F$. If $bv \in F$, then $\{y^u\} \cup \{z^u, b, v, y^v\}$ induces a $P_1 + P_4$ in $G + F$. See Fig. 5 for an illustration. In both cases, we obtain a contradiction.

From the above, we conclude that there are no uncoloured vertices, and we have obtained a red-blue colouring of G. It now remains to check in polynomial time whether this is a red-blue d-colouring. If so, then we are done. Otherwise, we

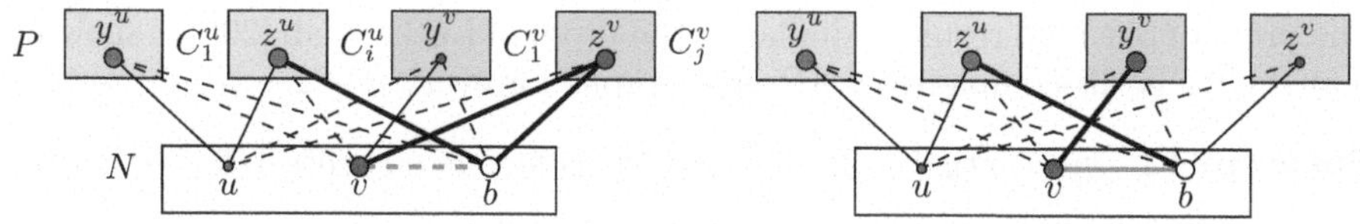

Fig. 5. Illustration of Case 9.3, where there is a P-dominating pair $\{u, v\}$ made of vertices of type-C and another type-C vertex b that has not been coloured yet. The graph $G+F$ contains an induced P_1+P_4 (highlighted by large vertices and thick edges) regardless of whether there is an edge between v and b or not.

discard this branch and move on to the next branch. As the number of branches is polynomial, and we can process each branch in polynomial time, our algorithm runs in polynomial time (its correctness follows from its description). □

4 NP-Completeness Results

To finish the proof of Theorem 4, we must show that 1-Cut and Perfect Matching Cut are NP-complete on probe $2P_2$-free graphs and probe $K_{1,3}$-free graphs and that for $d \geq 2$, d-Cut is NP-complete on probe $2P_2$-free graphs and probe $4P_1$-free graphs. All other NP-completeness results follow directly from the corresponding NP-completeness results in Theorems 2 and 3.

We start with the following result, which we recall is in contrast to the polynomial-time result of 1-Cut for $K_{1,3}$-free graphs due to Bonsma [4]. We omit its proof which is based on an observation of Moshi [24].

Theorem 10. 1-Cut *is* NP-*complete on probe* $K_{1,3}$*-free graphs.*

The diamond $\overline{2P_1 + P_2}$ is obtained from taking the complement of $2P_1 + P_2$, or equivalently, from the K_4 after removing an edge. A graph is *subcubic* if it has maximum degree at most 3. We omit the proof of our next result as well, as it is similar to a corresponding result for Vertex Cover from [5] except that we reduce from a different known NP-complete problem.

Theorem 11. Perfect Matching Cut *is* NP-*complete on the class of probe* $(K_{1,3}, \overline{2P_1 + P_2})$*-free subcubic planar graphs (and thus on probe* $K_{1,3}$*-free graphs).*

We recall that a graph is split if and only if it is $(2P_2, C_4, C_5)$-free, and we show:

Theorem 12. *For every* $d \geq 1$, d-Cut *and* Perfect Matching Cut *are* NP-*complete on probe split graphs (and thus on probe* $2P_2$*-free graphs).*

Proof. It is known that for every $d \geq 1$, d-Cut and Perfect Matching Cut are NP-complete for bipartite graphs. The former statement was shown in [24] for $d = 1$ and in [10] for $d \geq 2$. The latter statement was shown in [18].

Let G be a connected bipartite graph. Let N be one of the two bipartition classes. Note that N is an independent set. Let F consist of all the edges between vertices of N. Then $G + F$ is a split graph. Hence, the theorem follows. □

The proof of our final result is similar to the proof of Lucke et al. [20] for showing that d-Cut is NP-complete for $3P_2$-free graphs for every $d \geq 2$.

Theorem 13. *For every* $d \geq 2$, d-Cut *is* NP-*complete on probe* $4P_1$*-free graphs.*

5 Conclusions

We precisely identified those graphs H, for which polynomial-time results for the problems d-Cut ($d \geq 1$), Perfect Matching Cut and Maximum Matching Cut can be extended from H-free graphs to probe H-free graphs. This yielded complete complexity dichotomies for all three problems on probe H-free graphs. Note such a dichotomy remains unknown for H-free graphs, with Theorems 1 and 2 still containing open cases.

We conclude that, despite our new polynomial-time results, these problems are harder for *probe* H-free graphs than for H-free graphs (if $\mathsf{P} \neq \mathsf{NP}$) even for small graphs H. Moreover, we propose to study other graph problems on probe H-free graphs; so far, systematic studies have only been performed for Vertex Cover [5], Colouring [26] and the problems in this paper.

References

1. Ahn, J., Eagling-Vose, T., Lucke, F., Paulusma, D., Smith, S.: Finding d-cuts in claw-free graphs. In: Proceedings of ISAAC 2025. LIPIcs (to appear)
2. Araújo, J., Cohen, N., Giroire, F., Havet, F.: Good edge-labelling of graphs. Discret. Appl. Math. **160**, 2502–2513 (2012)
3. Berry, A., Golumbic, M.C., Lipshteyn, M.: Recognizing chordal probe graphs and cycle-bicolorable graphs. SIAM J. Discret. Math. **21**, 573–591 (2007)
4. Bonsma, P.S.: The complexity of the Matching-Cut problem for planar graphs and other graph classes. J. Graph Theory **62**, 109–126 (2009)
5. Brettell, N., Oostveen, J.J., Pandey, S., Paulusma, D., Rauch, J., van Leeuwen, E.J.: Computing subset vertex covers in H-free graphs. Theor. Comput. Sci. **1032**, 115088 (2025)
6. Chandler, D.B., Chang, M., Kloks, T., Liu, J., Peng, S.: On probe permutation graphs. Discret. Appl. Math. **157**, 2611–2619 (2009)
7. Chang, M.-S., Kloks, T., Kratsch, D., Liu, J., Peng, S.-L.: On the recognition of probe graphs of some self-complementary classes of perfect graphs. In: Wang, L. (ed.) COCOON 2005. LNCS, vol. 3595, pp. 808–817. Springer, Heidelberg (2005). https://doi.org/10.1007/11533719_82
8. Chvátal, V.: Recognizing decomposable graphs. J. Graph Theory **8**, 51–53 (1984)
9. Farley, A.M., Proskurowski, A.: Networks immune to isolated line failures. Networks **12**, 393–403 (1982)
10. Feghali, C., Lucke, F., Paulusma, D., Ries, B.: Matching cuts in graphs of high girth and H-free graphs. In: Proceedings of the ISAAC 2023. LIPIcs, vol. 283, pp. 28:1–28:16 (2023)
11. Golovach, P.A., Paulusma, D., Song, J.: Computing vertex-surjective homomorphisms to partially reflexive trees. Theoret. Comput. Sci. **457**, 86–100 (2012)
12. Golumbic, M.C., Lipshteyn, M.: Chordal probe graphs. Discret. Appl. Math. **143**, 221–237 (2004)

13. Golumbic, M.C., Maffray, F., Morel, G.: A characterization of chain probe graphs. Ann. Oper. Res. **188**, 175–183 (2011)
14. Gomes, G., Sau, I.: Finding cuts of bounded degree: complexity, FPT and exact algorithms, and kernelization. Algorithmica **83**, 1677–1706 (2021)
15. Graham, R.L.: On primitive graphs and optimal vertex assignments. Ann. N. Y. Acad. Sci. **175**, 170–186 (1970)
16. Heggernes, P., Telle, J.A.: Partitioning graphs into generalized dominating sets. Nordic J. Comput. **5**, 128–142 (1998)
17. Le, H., Le, V.B.: Complexity results for matching cut problems in graphs without long induced paths. In: Proceedings of the WG 2023. LNCS, vol. 14093, pp. 417–431 (2023)
18. Le, V.B., Telle, J.A.: The Perfect Matching Cut problem revisited. Theoret. Comput. Sci. **931**, 117–130 (2022)
19. Lucke, F., Marchand, J., Olbrich, J.: Finding minimum matching cuts in H-free graphs and graphs of bounded radius and diameter. CoRR abs/2502.18942 (2025)
20. Lucke, F., Momeni, A., Paulusma, D., Smith, S.: Finding d-cuts in graphs of bounded diameter, graphs of bounded radius and H-free graphs. In: Proceedings of the WG 2024. LNCS, vol. 14760, pp. 415–429 (2025)
21. Lucke, F., Paulusma, D., Ries, B.: On the complexity of Matching Cut for graphs of bounded radius and H-free graphs. Theoret. Comput. Sci. **936**, 33–42 (2022)
22. Lucke, F., Paulusma, D., Ries, B.: Finding matching cuts in H-free graphs. Algorithmica **85**, 3290–3322 (2023)
23. Lucke, F., Paulusma, D., Ries, B.: Dichotomies for Maximum Matching Cut: H-freeness, bounded diameter, bounded radius. Theoret. Comput. Sci. **1017**, 114795 (2024)
24. Moshi, A.M.: Matching cutsets in graphs. J. Graph Theory **13**, 527–536 (1989)
25. Patrignani, M., Pizzonia, M.: The complexity of the Matching-Cut problem. In: Brandstädt, A., Le, V.B. (eds.) WG 2001. LNCS, vol. 2204, pp. 284–295. Springer, Heidelberg (2001). https://doi.org/10.1007/3-540-45477-2_26
26. Paulusma, D., Rauch, J., van Leeuwen, E.J.: Colouring probe H-free graphs. CoRR abs/2505.20784 (2025)
27. Zhang, P., et al.: An algorithm based on graph theory for the assembly of contigs in physical mapping of DNA. Comput. Appl. Biosci. **10**, 309–317 (1994)

On Distance-d Independent Set Problems for Some Graph Classes

Sandip Das, Soura Sena Das, Sweta Das(✉), and Sk Samim Islam

Indian Statistical Institute, Kolkata, India
sweta.twin@gmail.com

Abstract. A *distance-d independent set*, for any integer $d \geq 2$, in a simple unweighted graph G is defined as a subset of vertices S such that any two vertices $u, v \in S$ are at least distance d apart in G. Given an unweighted graph G and a positive integer k, the *Distance-d Independent Set Problem* (DdIS) asks to determine whether, there exists a distance-d independent set S in G such that $|S| \geq k$. D2IS is equivalent to the well-known classical independent set problem. We denote the maximization version of DdIS as *MaxDdIS* and the parameterized version as *ParaDdIS*(k) where the parameter k is the size of a distance-d independent set. In this article, we study the computational complexity of DdIS, MaxDdIS, and ParaDdIS(k) for some popular graph classes: claw-free and bisplit graphs. We have shown that DdIS ($d \geq 3$) is NP-COMPLETE for claw-free graphs with maximum degree 12. Moreover, we prove that DdIS (for $d = 3, 4$) is NP-COMPLETE for triangle-free and C_6-free bisplit graphs. Furthermore, we have shown that, for any $\epsilon > 0$, it is NP-HARD to approximate MaxD3IS within a factor $n^{1/4-\epsilon}$ and MaxD4IS within a factor $n^{1/2-\epsilon}$ for triangle-free bisplit graphs. Further, we have shown that ParaD3IS(k) and ParaD4IS(k) are W[1]-HARD for triangle-free bisplit graphs.

Keywords: Distance-d independent set · Claw-free graphs · Bisplit graphs

1 Introduction

The INDEPENDENT SET problem [7,18,19] is a classical problem in the domain of graph theory in computer science. It has numerous applications in fields such as networks, scheduling, computer vision, pattern recognition, coding theory, map labeling, clustering, computational biology, etc. An *independent set* of a graph $G = (V, E)$ is a set $S \subseteq V$ such that each pair of vertices of S are non-adjacent in G i.e. for each u, v in S, the edge (u, v) is not in E. Given an unweighted graph G and a positive integer k, the INDEPENDENT SET (IS) problem asks if G has an independent set of size at least k. IS is one of the fundamental NP-COMPLETE problems in graph theory and complexity. The decision problem IS is NP-COMPLETE on many restricted graph classes like planar graphs [14],

A. Jeż and J. Otop (Eds.): FCT 2025, LNCS 16106, pp. 122–135, 2026.
https://doi.org/10.1007/978-3-032-04700-7_10

cubic graphs [21], triangle-free graphs [16,22], unit-disk graph [6] etc. Whereas the IS problem is polynomial-time solvable for certain graph classes like chordal graphs [15], bipartite graphs [16], bisplit graphs [16], AT-free graphs [4], fork-free graphs [2], etc. Split graphs are chordal graphs, hence IS is solvable in polynomial-time for split graphs.

For a given integer $d \geq 2$, a distance-d independent set of graph G is a vertex subset S such that the distance between any two vertices $u, v \in S$ (i.e. the number of edges on a shortest path between u and v) is at least d. This is a natural generalization of the concept of "independent set". Given an unweighted graph G and a positive integer k, the DISTANCE-d INDEPENDENT SET (DdIS) problem asks if G has a distance-d independent set of size at least k. Therefore, D2IS is nothing but IS. Let n be the number of vertices in a graph. For every integer $d \geq 2$, DdIS can be solved in linear time for interval graphs and circular-arc graphs, while for trapezoid graphs, DdIS takes $O(n(\log\log n + \log d))$ time [1]. For $d \geq 3$, DdIS is NP-COMPLETE for bipartite graphs [9]. For chordal graphs, DdIS is NP-COMPLETE when d is odd and $d \geq 3$ but polynomial-time solvable when d is even and $d \geq 2$ [9,15]. We denote the parameterized version of DdIS as ParaDdIS(k) where the parameter k is the size of a distance-d independent set. Eto et al. [9] showed that, for every fixed integer $d \geq 3$, ParaDdIS(k) is W[1]-HARD for bipartite graphs. Moreover, they showed that for every fixed odd integer $d \geq 3$, ParaDdIS(k) is W[1]-HARD for chordal graphs.

A *maximum distance-d independent set* is a largest possible distance-d independent S in G. MAXIMUM DISTANCE-d INDEPENDENT SET (MaxDdIS) problem asks a distance-d independent set of maximum size in G. The Max2IS problem, which is the maximization version of D2IS, is NP-HARD for general graphs [13]. MaxD2IS is also NP-HARD for many restricted graph classes like cubic planar graphs [12], triangle-free graphs [22], etc. Eto et al. [9] showed that MaxDdIS is NP-HARD for planar bipartite graph for any fixed $d \geq 3$. They further showed that for any $\varepsilon > 0$ and any fixed integer $d \geq 3$, it is NP-HARD to approximate MaxDdIS to within a factor of $n^{1/2-\varepsilon}$ and also ParaDdIS(k) is W[1]-HARD for bipartite graphs. They also showed that, for every fixed integers $d, r \geq 3$, MaxDdIS on r-regular graphs is APX-HARD [10]. They also proposed $O(r^{d-1})$ and $O(\frac{r^{d-2}}{d})$-factor approximation algorithms for the same and further improved their $O(\frac{r^{d-2}}{d})$-factor approximation algorithm to a 2-factor approximation result [10]. This has been improved further to 1.875-factor approximation [11]. It is known that the class of AT-free graphs is closed under taking any power, that is, if G is AT-free then G^d is AT-free for any d [5]. This implies that DdIS is polynomial-time solvable on AT-free graphs for any d as the independent set problem is polynomial-time solvable on AT-free graphs. We present Table 1 that contains the hardness results of DdIS for several graph classes and several values of d.

Our Contribution: *Claw-free graph* is a graph that does not contain a claw (or $K_{1,3}$) as an induced subgraph and *fork-free graph* is a graph that does not contain a fork (Fig. 1 depicted in the Appendix) as an induced subgraph. Since claw-free graphs are subclasses of fork-free graphs and D2IS in fork-free graphs are

Table 1. Hardness classification of DdIS based on graph classes and d-values. P denotes polynomial-time solvable problem. Here, we worked on the gray cells. ⋆ NP-COMPLETE of $DdIS$ for Claw-free graphs for $d \geq 3$ are presented here. ⋆⋆ NP-COMPLETE of $DdIS$ for bisplit graphs for $d = 3, 4$ are presented here.

Graph class	d-value	Hardness
Planar	$d \geq 2$	NP-COMPLETE [9]
Bipartite	$d = 2$	P [16]
	$d \geq 3$	NP-COMPLETE [9]
Chordal	$d \geq 2$, even	P [9,15]
	$d \geq 3$, odd	NP-COMPLETE [9]
Trapezoid	$d \geq 2$	P [1]
Circular-arc	$d \geq 2$	P [1]
Co-graph	$d \geq 2$	P [20]
AT-free	$d = 2$	P [4]
	$d \geq 3$	P [4][5]
Claw-free	$d = 2$	P [2]
	$d \geq 3$	NP-COMPLETE ⋆
Split	$d = 2$	P [15]
	$d = 3$	NP-COMPLETE [9]
	$d \geq 4$	P (Trivial)
Bisplit	$d = 2$	P [16]
	$d = 3, 4$	NP-COMPLETE ⋆⋆
	$d \geq 5$	P (Trivial)
Unit Disk	$d \geq 2$	NP-COMPLETE [6,17]

polynomial-time solvable [2], hence D2IS in claw-free graphs are also polynomial-time solvable. In this paper, we study the hardness of DdIS ($d \geq 3$) for claw-free graphs. It is known that, for $d \geq 2$, DdIS is NP-COMPLETE for cubic graphs[1] [10, 12]. We reduce D$(d-1)$IS of cubic graphs to our problem to show the following.

Theorem 1. *For every fixed integer $d \geq 3$, DdIS is NP-COMPLETE for claw-free graphs even on graphs of maximum degree* 12.

A graph $G = (V, E)$ is called a *bisplit graph* if V can be partitioned into a complete bipartite graph (or biclique) and an independent set. Bisplit graphs are comparability graphs[2] [3]. It is known that D2IS is polynomial-time solvable in comparability graphs [16]. Therefore, D2IS is polynomial-time solvable

[1] A 3-regular graph is called a *cubic graph.*

[2] The comparability graph of a partial order $(V, \leq)$ has vertex set V and edge xy whenever $x \leq y$ or $y \leq x$. G is a comparability graph if it is the comparability graph of some poset.

in bisplit graphs. In this paper, we study the hardness of DdIS ($d \geq 3$) for bisplit graphs. Moreover, we reduce a well-known NP-COMPLETE problem D3IS of general graphs to our problem to show the following.

Theorem 2. *D3IS is NP-COMPLETE for triangle-free bisplit graphs.*

Moreover, we have shown that, for any $\epsilon > 0$, it is NP-HARD to approximate MaxD3IS within a factor $n^{1/2-\epsilon}$ and ParaD3IS(k) is for triangle-free bisplit graphs.

Furthermore, we proved the following by reducing the NP-COMPLETE problem D2IS on general graphs to D4IS on triangle-free bisplit graphs.

Theorem 3. *D4IS is* NP-COMPLETE *for triangle-free bisplit graphs.*

We have shown that, for any $\epsilon > 0$, it is NP-HARD to approximate MaxD4IS within a factor $n^{1/2-\epsilon}$ and ParaD4IS(k) is W[1]-HARD for triangle-free bisplit graphs.

We reduce an NP-COMPLETE problem D2IS of triangle-free graphs [22] to D3IS and D4IS of C_6-free[3] bisplit graphs to show the following two theorems.

Theorem 4. *D3IS is* NP-COMPLETE *for C_6-free bisplit graphs.*

Theorem 5. *D4IS is* NP-COMPLETE *for C_6-free bisplit graphs.*

The results are non-surprising but tight hardness.

The proofs of the claims marked by (∗) are deferred to the appendix section.

Organisation: In Sect. 2, we present the definitions and terminologies we use throughout this article. In Sect. 3, we discuss the hardness results of claw-free graphs for the distance-d independent set problem. In Sect. 4, we discuss the hardness results of bisplit graphs for the same problem. We conclude in Sect. 5. We provide some basic definitions and notations in the Appendix.

2 Preliminaries

We further assume all graphs mentioned in this article to be simple. Let G be a graph with vertex set $V(G)$ and edge set $E(G)$. The *degree* of a vertex $u \in V(G)$ is the total number of edges incident on u, denoted by $deg(u)$. We denote $\mathrm{d}(u, v)$ as the length of a shortest path joining the vertices u and v in G. The *open neighborhood* of a vertex $u \in V(G)$, $N(u)$, is the set of all the vertices $v \in V(G)$ such that (u, v) is an edge in $E(G)$. The *closed neighborhood* of a vertex u, $N[u] = N(u) \cup \{u\}$ The *maximum degree* of a graph G, $\Delta(G) = max\{deg(v) : v \in V(G)\}$. The kth power G^k of an undirected graph G is a graph such that G^k has the same set of vertices, but in which two vertices u and v are adjacent when $d_G(u, v)$ is at most k.

Let G and H be two graphs. G is called H-free if H is not an induced subgraph of G. A complete bipartite graph $K_{1,3}$ is called a *claw*, and the vertex that is

[3] G is called C_6-free if C_6 is not an induced subgraph of G.

incident with all the edges in a claw is called the *claw center.* If a graph G does not contain a claw as an induced subgraph, it is called a *claw-free* graph. If the degree of every vertex in a graph G is d, then G is called *d-regular.* A 3-regular graph is called a *cubic graph.* A graph G is called a *bisplit graph* if $V(G)$ can partitioned into a complete bipartite graph (or biclique) and an independent set.

Fig. 1. .

3 Claw-Free Graphs

In this section, we show that finding DdIS ($d \geq 3$) of a claw-free graph with maximum degree 12 is NP-COMPLETE.

We reduce D$(d-1)$IS of cubic graphs to our problem, which is a known NP-COMPLETE problem [10]. Consider any cubic graph G_1 with vertex set $V(G_1) = \{v_1, v_2, \ldots, v_n\}$ and edge set $E(G_1)$. We construct a new graph G_2 and begin by taking a copy of G_1. Refer to Fig. 2 for a pictorial representation. We take each $u_i \in V(G_2)$ corresponds to a vertex $v_i \in V(G_1)$. For $i < j$, we subdivide each edge $(u_i, u_j) \in E(G_2)$ by adding a new vertex e_{ij}^1 i.e. if (u_i, u_j) is an edge, we replace it by two edges: (u_i, e_{ij}^1) and (e_{ij}^1, u_j). We construct a cycle C_{ij} of length $(2d-1)$ whose one vertex is e_{ij}^1. Let $C_{ij} = \langle e_{ij}^1, e_{ij}^2, \ldots, e_{ij}^{2d-1} \rangle$ (see Fig. 2(b)). Further, we introduce two edges (u_i, e_{ij}^2) and (u_j, e_{ij}^{2d-1}) in G_2 for each $(v_i, v_j) \in E(G_1)$ where $i < j$. Finally, for each $u_i \in G_2$, we form a clique whose vertex set is $N_{G_2}[u_i]$. It is clear that the reduction to G_2 can be done in polynomial-time.

Observation 1. *The graph G_2 is claw-free and $\Delta(G_2) = 12$.*

Proof. We start by showing that the maximum degree of G_2 is at most 12. Since G_1 is a cubic graph, it follows from the construction of G_2 that for every vertex $u_i \in V(G_2)$, $deg(u_i) = 6$. Now, since each e_{ij}^1 is adjacent to u_i and u_j and all the neighbors of u_i and u_j, e_{ij}^1 has exactly 12 neighbors and hence $deg(e_{ij}^1) = 12$. The vertex e_{ij}^2 (resp. e_{ij}^{2d-1}) is a part of a clique and adjacent to e_{ij}^3 (resp. e_{ij}^{2d-2}), and therefore, $deg(e_{ij}^1) = 7$, (resp. $deg(e_{ij}^{2d-1}) = 7$). Finally, observe that $e_{ij}^3, e_{ij}^4, \ldots, e_{ij}^{2d-3}, e_{ij}^{2d-2}$ have degree 2.

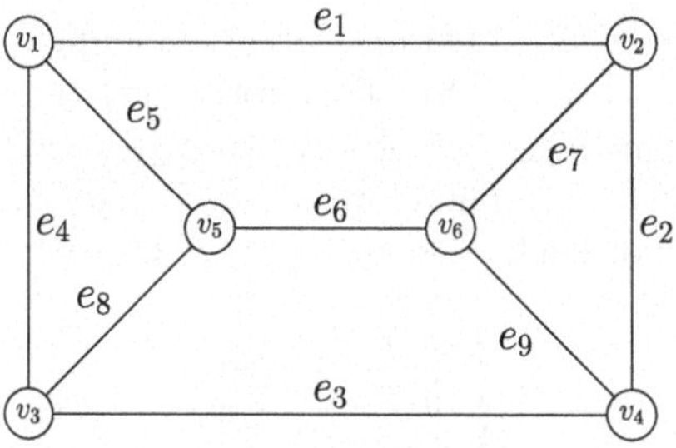

(a) Cubic graph G_1.

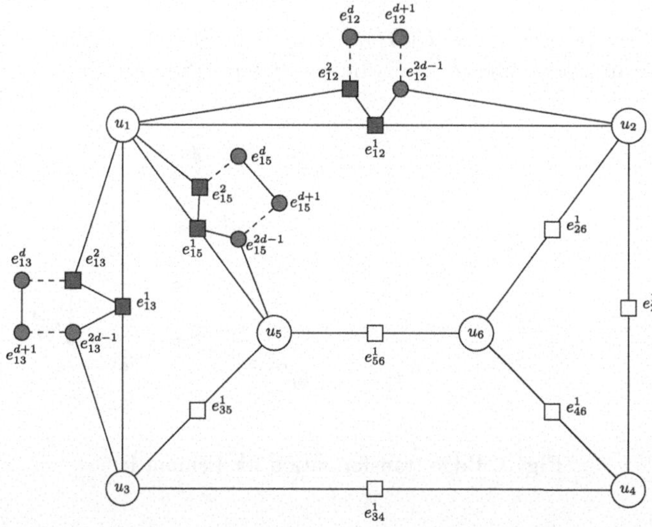

(b) Claw-free graph G_2, where $N[u_i]$ forms a clique for each u_i. In the figure, the vertices with the symbol ■ along with the vertex u_1 form a clique.

Fig. 2. Gadget construction of G_2 from cubic graph G_1 for DdIS.

Below, we show that G_2 is also claw-free. It suffices to show that no vertex in G_2 can be a claw centre. Since all the neighbours of u_i are in a clique, u_i cannot be a claw centre. Observe that e_{ij}^1 is adjacent to two cliques, thus e_{ij}^1 cannot be a claw centre. Similarly, e_{ij}^2 is adjacent to a clique and e_{ij}^3 and also e_{ij}^{2d-1} is adjacent to a clique and e_{ij}^{2d-2}. Hence e_{ij}^2 and e_{ij}^{2d-1} cannot be claw centres. Finally, vertices $e_{ij}^3, e_{ij}^4, \ldots, e_{ij}^{2d-3}, e_{ij}^{2d-2}$ have degree 2, they cannot be claw centres. Hence, no vertex in G_2 can be a claw centre. Hence the result follows. □

Lemma 1. *For $d \geq 3$, G_1 has a distance-$(d-1)$ independent set S_1 of size at least k iff G_2 has a distance-d independent set S_2 of size at least $k+m$, where $m = |E(G_1)|$.*

Proof. (If part) Suppose S_1 is a distance-$(d-1)$ independent set of G_1 of size at least k. Let $\{v_1, v_2, \ldots, v_k\} \subseteq S_1$. Therefore, $\{v_1, v_2, \ldots, v_k\}$ is a distance-$(d-1)$ independent set in G_1. Now observe that $\mathrm{d}(u_i, u_j) \geq d$ holds in G_2 for each $1 \leq i, j \leq k$ where $i \neq j$. Therefore, $S_2' = \{u_1, u_2, \ldots, u_k\}$ is a distance-d independent set in G_2. For each edge $(v_i, v_j) \in E(G_1)$ where $i < j$, we define

$$e'_{ij} = \begin{cases} e_{ij}^{d+1} & u_i \in S_2' \\ e_{ij}^{d} & u_j \in S_2' \\ e_{ij}^{d} & u_i, u_j \notin S_2' \end{cases}.$$

We denote $S_2'' = \{e'_{ij} | (v'_i, v'_j) \in E(G_1), i < j\}$. Let $S_2 = S_2' \cup S_2''$. Note that, S_2 is a distance-d independent set in G_2 of size $k + m$. (See Fig 3)

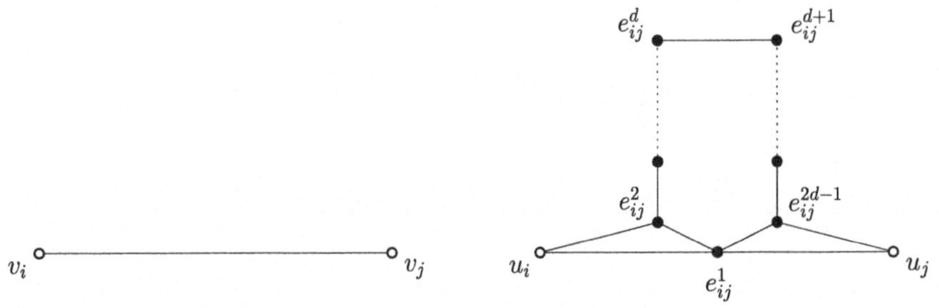

Fig. 3. Edge transformation for Lemma 1

(Only-if part) Suppose S_2 is a distance-d independent set of G_2 of size at least $k + m$. The induced cycle $C_{ij} = \langle e_{ij}^1, e_{ij}^2, \ldots, e_{ij}^{2d-1} \rangle$ has a diameter of length $\lfloor \frac{2d-1}{2} \rfloor < d$. Thus $|S_2 \cap V(C_{ij})| \leq 1$ for each C_{ij}. Therefore, $|S_2 \cap (\bigcup_{(v_i,v_j) \in E(G_1)} V(C_{ij}))| \leq m$. This leaves us $|S_2 \cap \{u_1, u_2, \ldots, u_n\}| \geq k$. Let $S_1 = \{v_i \in V(G_1) : u_i \in S_2\}$. Since S_2 is a distance-d independent set of G_2, therefore S_1 is a distance-$(d-1)$ independent set of G_1 of size at least k. □

Lemma 1 yields the following.

Theorem 1. *For every fixed integer $d \geq 3$, DdIS is NP-COMPLETE for claw-free graphs even on graphs of maximum degree* 12.

A *net* is the induced six-vertex graph G as shown in Fig. 4. We call a graph *net-free* if it does not contain a net as an induced subgraph.

Observation 2. *The graph G_2 is also net-free.*

Corollary 1. *For every fixed integer $d \geq 3$, DdIS is NP-COMPLETE even on (claw,net)-free graphs.*

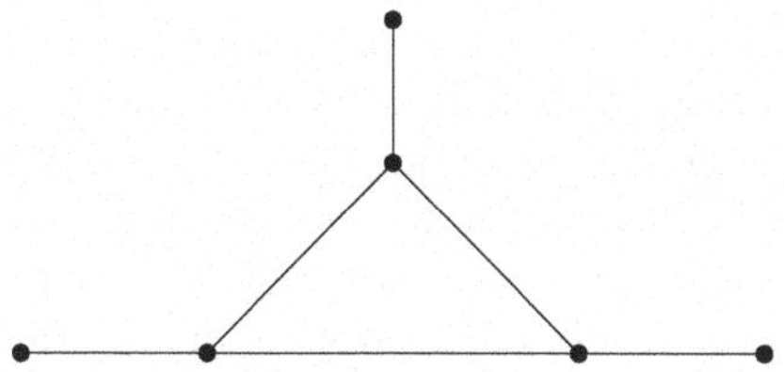

Fig. 4. The six-vertex graph known as a *net*.

4 Bisplit Graphs

In this section, we prove that D3IS is NP-COMPLETE for triangle-free bisplit graphs. Moreover, we show that, for any $\epsilon > 0$, it is NP-HARD to approximate MaxD3IS within a factor $n^{1/2-\epsilon}$ for triangle-free bisplit graphs. Further, we show that, the ParaD3IS(k) is W[1]-HARD for triangle-free bisplit graphs.

4.1 Triangle-Free Bisplit Graphs

Consider any graph G_1 with vertex set $V(G_1) = \{v_1, v_2, \ldots, v_n\}$ and edge set $E(G_1)$. We construct a new graph G_2 in the following way: (i) We add the vertex set $\{u_1, u_2, \ldots, u_n\}$ in G_2 where each u_i corresponds to the vertex $v_i \in V(G_1)$. (ii) We add the vertex set $\{w_1, w_2, \ldots, w_n\}$ in G_2 where each w_i corresponds to the vertex $u_i \in V(G_2)$ and each u_i is adjacent to w_i. (iii) We add a vertex a which is adjacent to all $u_j \in V(G_2)$. (iv) If $(v_i, v_j) \in E(G_1)$, we add an edge between w_i and u_j and an edge between u_i and w_j. (iv) We add a vertex b which is adjacent to a, (Fig. 5). Therefore G_2 is a bisplit graph since $\{w_1, w_2, \ldots, w_n\}$ forms an independent set and the sets $\{u_1, u_2, \ldots, u_n, b\}$ and $\{a\}$ forms a complete bipartite graph. Note that the graph G_2 is triangle-free.

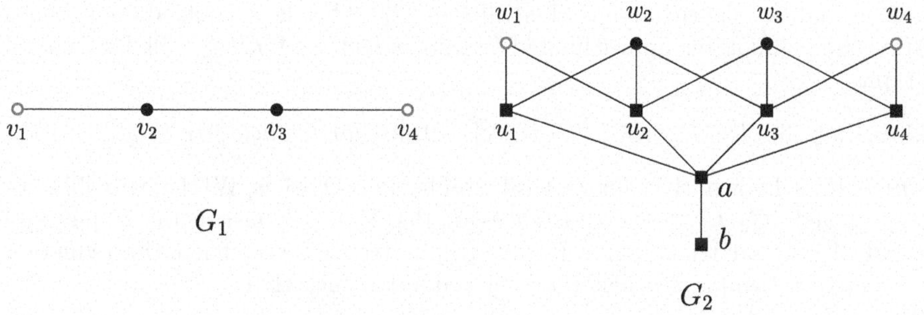

Fig. 5. Gadget construction of G_2 from G_1

Lemma 2. *The graph G_1 has a distance-3 independent set S_1 of size at least k iff G_2 has a distance-3 independent set S_2 of size at least $k+1$.*

Proof. (If part) Let the graph G_1 has a distance-3 independent set $S_1 = \{v_1, v_2, \ldots, v_k\}$. Observe that, since each pair of vertices $v_i, v_j \in S_1$ has a distance of at least 3, its corresponding vertices $w_i, w_j \in G_2$ have a distance at least 3. Moreover, the distance between b and any vertex of $\{w_i : 1 \leq i \leq n\}$ is 3. Thus, $S_2 = \{w_1, w_2, ..., w_k, b\}$ is a distance-3 independent set of G_2 of size $k + 1$.

(Only-if part) Suppose G_2 has a distance-3 independent set S_2 of size $k + 1$. Note that, the induced subgraph $G_2[u_1, u_2, \ldots, u_n, a, b]$ forms a complete bipartite graph, hence it has diameter 2. Therefore, $|S_2 \cap V(G_2[u_1, u_2, \ldots, u_n, a, b])| \leq 1$, which implies $|S_2 \cap \{w_1, w_2, \ldots w_n\}| \geq k$.

Let $S_2 \cap \{w_1, w_2, \ldots w_n\} = \{w_1, w_2, \ldots, w_k\}$. Then the corresponding vertices $\{v_1, v_2, \ldots, v_k\}$ is a distance-3 independent set of G_1 of size k. □

Lemma 2 yields the following.

Theorem 2. *D3IS is* NP-COMPLETE *for triangle-free bisplit graphs.*

The maximization version of D3IS, MaxD3IS cannot be approximated within a factor of $n^{1/2-\varepsilon}$ [9] for general graphs. We show that the reduction presented in Lemma 2 gives us the following inapproximability of MaxD3IS for triangle-free bisplit graphs.

Corollary 2. *For any $\epsilon > 0$, it is* NP-HARD *to approximate MaxD3IS within a factor $n^{1/4-\epsilon}$ for triangle-free bisplit graphs.*

Proof. We denote $OPT(G_1)$ as the number of vertices in an optimal solution of MaxD3IS in G_1, where $|V(G_1)| = n$. Let $OPT'(G_2)$ be the number of vertices in an optimal solution of MaxD3IS in G_2, where $|V(G_2)| = n'$. Note that, $n' = O(n)$. Suppose there exists a solution of MaxD3IS of size $APX'(G_2)$ in G_2 such that the ratio $\frac{OPT'(G_2)}{APX'(G_2)}$ is better than $(n)^{1/2-\epsilon}$. From Lemma 2, there exist a solution of size $APX'(G_2) - 1$ of MaxD3IS in G_1. Let $APX(G_1) = APX'(G_2) - 1$. Note that, $\frac{OPT'(G_2)}{APX'(G_2)} = \frac{OPT(G_1)+1}{APX(G_1)+1} \approx \frac{OPT(G_1)}{APX(G_1)}$. Therefore, there exist a $(n)^{1/2-\epsilon}$-approximation algorithm for MaxD3IS in G_1. This is a contradiction, since MaxD3IS cannot be approximated within a factor of $n^{1/2-\varepsilon}$ [9] for general graphs. □

Corollary 3. *The ParaD3IS(k) is* W[1]-HARD *for triangle-free bisplit graphs.*

Proof. It is known that for general graphs ParaD3IS is W[1]-HARD [9]. Let (G_1, k) and (G_2, k') be two instances of ParaD3IS(k) on general graphs and ParaD3IS(k') on triangle-free bisplit graphs respectively. Then the reduction presented in Lemma 2 is also the FPT reduction such that:
(i) $k' \leq k + 1$ and
(ii) (G_1, k) is a *yes* instance of ParaD3IS(k) on general graphs if and only if (G_2, k') is a *yes* instance of ParaD3IS(k') on triangle-free bisplit graphs. □

Now, we show that D4IS is NP-COMPLETE for triangle-free bisplit graphs. To show this, we reduce from the D2IS for general graphs, which is known to be NP-COMPLETE.

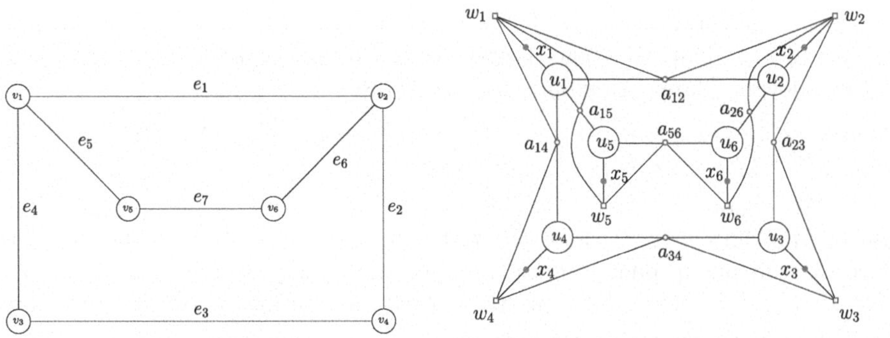

(a) Graph G_1

(b) Graph G_2. (Each x_i and a_{ij} are adjacent to every u_i.)

Fig. 6. Gadget construction from G_1 to G_2.

Given any graph G_1 with vertex set $V(G_1) = \{v_1, v_2, \ldots, v_n\}$, edge set $E(G_1)$, we construct a new graph G_2 in the following way: (i) We start by taking a copy of G_1, denoting the vertex set of G_2 as $V(G_2) = \{u_1, u_2, \ldots, u_n\}$, where each vertex $u_i \in V(G_2)$ corresponds to $v_i \in V(G_1)$. (ii) We add the length-two– path $u_i - x_i - w_i$ with each vertex u_i. (iii) We subdivide each edge $(u_i, u_j) \in E(G_2)$ with $i < j$ by adding a new vertex a_{ij}. (iv) We add edges between w_i and each a_{ij} which are adjacent to u_i. (v) Form a complete bipartite graph where the partite sets are $A = \{u_i \in V(G_2) : 1 \leq i \leq n\}$ and $B = \{x_i \in V(G_2) : 1 \leq i \leq n\} \cup \{a_{ij} \in V(G_2) : (v_i, v_j) \in E(G_1), i < j\}$. Therefore G_2 is a bisplit graph since $\{w_1, w_2, \ldots, w_n\}$ forms an independent set. Note that G_2 is triangle-free.

Lemma 3. *The graph G_1 has a distance-2 independent set of size at least k iff G_2 has a distance-4 independent set of size at least k.*

Proof. (If part) Let the graph G_1 have a distance-2 independent set $I = \{v_1, v_2, \ldots, v_k\}$. Observe that since each pair of the vertices v_i, v_j in I has a distance of at least 2, its corresponding vertices $w_i, w_j \in V(G_2)$ have a distance of at least 4 by the construction of G_2. Thus, $\{u_1, u_2, \ldots, u_k\}$ is a distance-4 independent set of size k.

(Only-if part) Let G_2 have a distance-4 independent set S of size k. Observe that if any vertex in $\{x_i, u_i, a_{ij}\} \in V(G_2)$ belongs to S, then there can not be any other vertex in S since every vertex in $V(G_2)$ is at a distance at most 3 from every vertex in $\{x_i, u_i, a_{ij}\} \in V(G_2)$. Thus S must only contain a subset of vertices from $\{w_1, w_2, \ldots, w_n\}$. Since every pair of vertices $w_i, w_j \in S$ is at a distance of at least 4, the corresponding pairs v_i, v_j must be at a distance of at least 2 in G_1.

Therefore G_1 has a distance-2 independent set of size at least k. □

Lemma 3 yields the following.

Theorem 3. *D4IS is NP-COMPLETE for triangle-free bisplit graphs.*

The maximization version of D2IS, MaxD2IS cannot be approximated within a factor of $n^{1-\varepsilon}$ [23]. We show that the reduction presented in Lemma 3 gives us the following inapproximability for MaxD4IS.

Corollary 4. *For any $\epsilon > 0$ it is* NP-HARD *to approximate MaxD4IS within a factor $n^{1/2-\epsilon}$ for triangle-free bisplit graphs.*

Proof. We denote $OPT(G_1)$ as the number of vertices in an optimal solution of MaxD2IS in G_1, where $|V(G_1)| = n$. Let $OPT'(G_2)$ be the number of vertices in an optimal solution of MaxD4IS in G_2, where $|V(G_2)| = n'$. Note that, $n' = O(n^2)$. Suppose there exists a solution of MaxD4IS of size $APX'(G_2)$ in G_2 such that the ratio $\frac{OPT'(G_2)}{APX'(G_2)}$ is better than $(n^2)^{1/2-\epsilon}$. From Lemma 3, there exist a solution of size $APX'(G_2) - 1$ of MaxD2IS in G_1. Let $APX(G_1) = APX'(G_2) - 1$. Note that, $\frac{OPT'(G_2)}{APX'(G_2)} = \frac{OPT(G_1)+1}{APX(G_1)+1} \approx \frac{OPT(G_1)}{APX(G_1)}$. Therefore, there exist a $(n^2)^{1/2-\epsilon}$-approximation algorithm for MaxD2IS in G_1. This is a contradiction, since MaxD2IS cannot be approximated within a factor of $n^{1-\varepsilon}$ [23] for general graphs. □

Corollary 5. *The ParaD4IS(k) is* W[1]-HARD *for triangle-free bisplit graphs.*

Proof. It is known that for general graphs ParaD2IS is W[1]-HARD [8]. Let (G_1, k) and (G_2, k') be two instances of ParaD2IS(k) on general graphs and ParaD4IS(k') on triangle-free bisplit graphs respectively. Then the reduction presented in Lemma 3 is also the FPT reduction such that:
(i) $k' \leq k$ and
(ii) (G_1, k) is a *yes* instance of ParaD2IS(k) on general graphs if and only if (G_2, k') is a *yes* instance of ParaD4IS(k') on triangle-free bisplit graphs. □

4.2 C_6-Free Bisplit Graphs

In this subsection, we show that deciding DdIS, $d = \{3, 4\}$ for a C_6-free bisplit graph is NP-COMPLETE.

To show that D3IS for C_6-free bisplit graphs is NP-COMPLETE, we reduce from D2IS of triangle-free graphs which is a known NP-COMPLETE problem [22].

Given a triangle-free graph G_1 with vertex set $V(G_1) = \{v_1, v_2, \ldots, v_n\}$, edge set $E(G_1)$, we construct a new graph G_2. We start by taking a copy of G_1, denoting the vertex set of G_2 as $V(G_2) = \{u_1, u_2, \ldots, u_n\}$, where each vertex $u_i \in V(G_2)$ corresponds to $v_i \in V(G_1)$. We subdivide each edge $(u_i, u_j) \in E(G_2)$ with $i < j$ by adding a new a vertex a_{ij}. Additionally, for each $u_i \in V(G_2)$, we add a vertex w_i and connect it to u_i. Then we add edges between w_i and each a_{ij} which are adjacent to u_i. Lastly, we add edges between u_i and all other a_{ij}. Refer to Fig 7 for a pictorial representation.

Observe that all the vertices $w_i \in G_2$ are independent. Furthermore, we have two sets $A = \{u_i \mid u_i \in G_2, 1 \leq i \leq n\}$ and $B = \{a_{ij} \mid a_{ij} \in G_2, 1 \leq i, j \leq n\}$, such that A and B forms a biclique. A biclique can not contain an induced C_6. Since G_1 is triangle-free, w_i and a_{ij} can not form an induced C_6. Thus G_2 can not contain an induced C_6. Therefore, G_2 is a C_6-free bisplit graph.

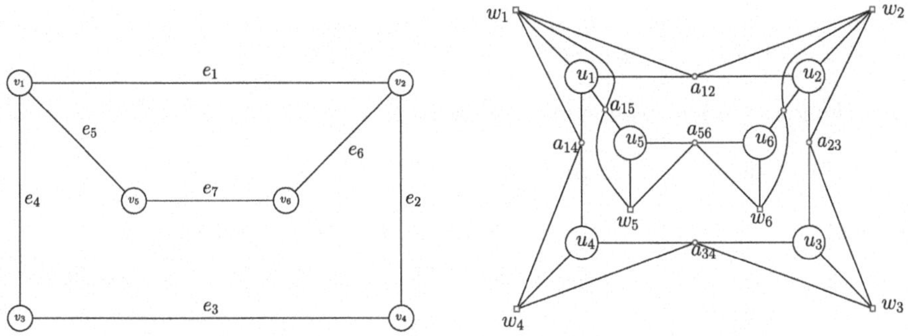

(a) triangle free graph G_1 (b) each a_{ij} adjacent to all u_i

Fig. 7. Gadget construction from G_1 to G_2.

Lemma 4. *The graph G_1 has a distance-2 independent set of size at least k iff G_2 has a distance-3 independent set of size at least k.*

Proof. (If part) Let the graph G_1 have a distance-2 independent set $I = \{v_1, v_2, \ldots, v_k\}$. Observe that since each pair of the vertices v_i, v_j in I have a distance of at least 2, its corresponding vertices $w_i, w_j \in G_2$ have a distance of at least 3 by construction. Thus, $\{w_1, w_2, \ldots, w_k\}$ is a distance-3 independent set of size k.

(Only if part) Let G_2 have a distance-3 independent set S of size at least k. Observe that if either $u_i \in V(G_2)$ or $a_{ij} \in V(G_2)$ belongs to S, then there can not be any other vertex in S since every vertex in $V(G_2)$ is at a distance at most 2 from both u_i and a_{ij} . Thus, S must contain only a subset of vertices from $\{w_1, w_2, \ldots, w_n\}$. Since every pair of vertices $w_i, w_j \in S$ is at a distance of at least 3, the corresponding pairs v_i, v_j must be at a distance of at least 2 in G_1. Therefore, G_1 has a distance-2 independent set of size at least k. □

Lemma 4 yields the following.

Theorem 4. *D3IS is NP-COMPLETE for C_6-free bisplit graphs.*

To show that D4IS for C_6-free bisplit graphs is NP-COMPLETE, we reduce from D2IS of triangle-free graphs. We follow the same construction as we did for D3IS. Additionally, for every edge $(u_i, w_i) \in E(G_2)$ further subdivided by a vertex x_i. We add edges from x_i to every $u_i \in V(G_2)$. Refer to Fig 6 for a pictorial representation.

Observe that all the vertices $w_i \in G_2$ are independent. Furthermore, we have two sets $A = \{u_i \mid u_i \in G_2, 1 \leq i \leq n\}$ and $B = \{x_i \mid x_i \in G_2, 1 \leq i \leq n\} \cup \{a_{ij} \mid a_{ij} \in G_2, 1 \leq i, j \leq n\}$, such that A and B forms a biclique. Hence, G_2 is a bisplit graph. Furthermore, vertex subsets A and B form a biclique, they do not contain an induced C_6. Since G_1 is triangle-free, w_i and a_{ij} cannot form an induced C_6. Thus G_2 cannot contain an induced C_6.

Lemma 5. *The graph G_1 has a distance-2 independent set of size at least k iff G_2 has a distance-4 independent set of size at least k.*

Proof. Follows similar arguments as Lemma 3. □

Lemma 5 yields the following.

Theorem 5. *D4IS is NP-COMPLETE for C_6-free bisplit graphs.*

5 Conclusion

We have shown that for every fixed integer $d \geq 3$, DdIS is NP-COMPLETE for claw-free graphs even for maximum degree 12 and also DdIS (for $d = 3, 4$) is NP-COMPLETE for C_6-free bisplit graphs. Some natural questions would be finding a good approximation algorithm for claw-free graphs with bounded degrees and also for C_6-free bisplit graphs.

References

1. Agnarsson, G., Damaschke, P., Halldórsson, M.M.: Powers of geometric intersection graphs and dispersion algorithms. Discret. Appl. Math. **132**(1–3), 3–16 (2003)
2. Alekseev, V.E.: Polynomial algorithm for finding the largest independent sets in graphs without forks. Discret. Appl. Math. **135**(1–3), 3–16 (2004)
3. Brandstädt, A., Hammer, P.L., Lozin, V.V., et al.: Bisplit graphs. Discret. Math. **299**(1–3), 11–32 (2005)
4. Broersma, H., Kloks, T., Kratsch, D., Müller, H.: Independent sets in asteroidal triple-free graphs. SIAM J. Discret. Math. **12**(2), 276–287 (1999)
5. Chang, J.M., Ho, C.W., Ko, M.T.: Powers of asteroidal triple-free graphs with applications. Ars Comb. **67** (2003). https://api.semanticscholar.org/CorpusID:9862925
6. Clark, B.N., Colbourn, C.J., Johnson, D.S.: Unit disk graphs. Discrete Math. **86**(1–3), 165–177 (1990)
7. Cook, S.A.: The complexity of theorem-proving procedures. In: Logic, Automata, and Computational Complexity: The Works Of Stephen A. Cook, pp. 143–152 (2023)
8. Downey, R.G., Fellows, M.R.: Fixed-parameter tractability and completeness ii: On completeness for w [1]. Theoret. Comput. Sci. **141**(1–2), 109–131 (1995)
9. Eto, H., Guo, F., Miyano, E.: Distance-d independent set problems for bipartite and chordal graphs. J. Comb. Optim. **27**(1), 88–99 (2014)
10. Eto, H., Ito, T., Liu, Z., Miyano, E.: Approximability of the distance independent set problem on regular graphs and planar graphs. In: International Conference on Combinatorial Optimization and Applications, pp. 270–284. Springer (2016)
11. Eto, H., Ito, T., Liu, Z., Miyano, E.: Approximability of the distance independent set problem on regular graphs and planar graphs. IEICE Trans. Fundam. Electron. Commun. Comput. Sci. **105**(9), 1211–1222 (2022)
12. Fleischner, H., Sabidussi, G., Sarvanov, V.I.: Maximum independent sets in 3-and 4-regular Hamiltonian graphs. Discret. Math. **310**(20), 2742–2749 (2010)

13. Garey, M.R., Johnson, D.S.: Computers and intractability, vol. 174. freeman San Francisco (1979)
14. Garey, M.R., Johnson, D.S., Stockmeyer, L.: Some simplified np-complete problems. In: Proceedings of the Sixth Annual ACM Symposium on Theory of Computing, pp. 47–63 (1974)
15. Gavril, F.: Algorithms for minimum coloring, maximum clique, minimum covering by cliques, and maximum independent set of a chordal graph. SIAM J. Comput. **1**(2), 180–187 (1972)
16. Golumbic, M.C.: Algorithmic graph theory and perfect graphs. Elsevier (2004)
17. Jena, S.K., Jallu, R.K., Das, G.K., Nandy, S.C.: The maximum distance-d independent set problem on unit disk graphs. In: Chen, J., Lu, P. (eds.) FAW 2018. LNCS, vol. 10823, pp. 68–80. Springer, Cham (2018). https://doi.org/10.1007/978-3-319-78455-7_6
18. Karp, R.M.: Reducibility among combinatorial problems. Springer (2010)
19. Karp, R.M., Wigderson, A.: A fast parallel algorithm for the maximal independent set problem. J. ACM (JACM) **32**(4), 762–773 (1985)
20. Lokshantov, D., Vatshelle, M., Villanger, Y.: Independent set in P_5-free graphs in polynomial time. In: Proceedings of The Twenty-fifth Annual ACM-SIAM Symposium on Discrete Algorithms, pp. 570–581. SIAM (2014)
21. Mohar, B.: Face covers and the genus problem for apex graphs. J. Comb. Theor. Series B **82**(1), 102–117 (2001)
22. Poljak, S.: A note on stable sets and colorings of graphs. Comment. Math. Univ. Carol. **15**(2), 307–309 (1974)
23. Zuckerman, D.: Linear degree extractors and the inapproximability of max clique and chromatic number. In: Proceedings of the Thirty-eighth Annual ACM Symposium on Theory of Computing, pp. 681–690 (2006)

On the Complexity of Problems on Graphs Defined on Groups

Bireswar Das[1(✉)], Dipan Dey[2], and Jinia Ghosh[1]

[1] IIT Gandhinagar, Gandhinagar, India
bireswar@iitgn.ac.in, jiniag@iitgn.ac.in
[2] Tata Institute of Fundamental Research, Mumbai, India
dipan.dey@tifr.res.in

Abstract. We study the complexity of graph problems on graphs defined on groups, especially power graphs. We observe that an isomorphism invariant problem, such as Hamiltonian Path, Feedback Vertex Set, Partition into Cliques, Subgraph Isomorphism, etc., cannot be NP-complete for power graphs, commuting graphs, enhanced power graphs, directed power graphs, and bounded-degree Cayley graphs, assuming the Exponential Time Hypothesis (ETH). Analogously, no isomorphism invariant group problem can be NP-complete, under ETH. We show that the Weighted Max-Cut problem is NP-complete in power graphs. We also show that, unless ETH is false, the Graph Motif problem cannot be solved in quasipolynomial time on power graphs, even for power graphs of cyclic groups. We study the recognition problem of power graphs when the adjacency matrix or list is given as input and show that for abelian groups and some classes of nilpotent groups, it is solvable in polynomial time.

Keywords: Complexity Theory · Graphs Defined on Groups · Exponential Time Hypothesis · Power Graphs · Graph Motif

1 Introduction

Several important classes of graphs are defined on groups, including power graphs, enhanced power graphs, Cayley graphs, and commuting graphs. These constructions encode algebraic relationships into combinatorial structures, creating an interplay between group theory and graph theory. In this paper, we study the computational complexity of various problems on such graphs. Since these graphs are derived from algebraic objects, a natural and compelling question is whether the underlying group structure can be leveraged to design more efficient algorithms or to better understand the inherent difficulty of the problems.

In this paper, while our primary focus is on power graphs, several of our results also extend to enhanced power graphs, directed power graphs, commuting graphs and bounded degree Cayley graphs (see [8] for a comprehensive survey). The notion of directed power graphs was introduced by Kelarev et al. [23], while

A. Jeż and J. Otop (Eds.): FCT 2025, LNCS 16106, pp. 136–149, 2026.
https://doi.org/10.1007/978-3-032-04700-7_11

the concept of power graphs was later proposed by Chakrabarty et al. [9]. As power graphs are perfect graphs [1], using results by Lovasz et al. [19] for perfect graphs, one can see that computational problems such as graph colouring, maximum clique and maximum independent set are in P.

While power graphs had been defined decades ago, to the best of our knowledge, no nontrivial NP-completeness result has been reported for this class. Our first line of investigation concerns identifying NP-complete problems on graphs defined from groups. Many classical NP-complete problems, such as Hamiltonian Cycle, Partition into Cliques, Partition into Perfect Matchings, Metric Dimension, Treewidth determination, are invariant under graph isomorphism, meaning their solutions depend only on the structure of the graph, not on the labelling of the vertices. We prove that any isomorphism-invariant graph problem is not NP-complete when restricted to power graphs, enhanced power graphs, commuting graphs, directed power graphs, or bounded-degree Cayley graphs, unless Exponential Time Hypothesis (ETH) fails (Sect. 3). This result also extends to isomorphism invariant group problems where the groups are given by their Cayley table. It is already known that isomorphism invariant problems such as the group isomorphism problem (see [28]), minimal faithful permutation degree problem [12] are not NP-complete under ETH. Our result shows that *any* isomorphism invariant group problem cannot be NP-complete, under ETH.

Therefore, to identify NP-complete problems on group-based graphs, we must focus on problems that are not invariant under graph isomorphism (Sect. 4). One such example is the Weighted Max-Cut problem. We show that Weighted Max-Cut remains NP-complete when restricted to power graphs. However, this result is not particularly surprising, as for graph classes having sufficiently large cliques, the NP-completeness of Weighted Max-Cut is expected. Motivated by this, we turn to the Graph Motif problem, which differs fundamentally in that it incorporates a connectivity constraint in addition to a colouring condition.

Originally introduced in the context of biological network analysis [24], the graph motif problem has since attracted significant interest within the computer science community. The problem remains NP-complete even when restricted to trees [24] and bipartite graphs with maximum degree 4 [14]. A variety of fixed-parameter tractable (FPT) algorithms have been developed for this problem, with parameters ranging from the number of distinct colours to the input size (see [20,29], e.g.). Ganian showed that Graph Motif can be solved in time $O^*(2^k)$ where k is the neighbourhood diversity [17]. Notably, the problem is $W[1]$-hard for trees when parameterised by the number of colours [14]. Furthermore, several lower bounds based on ETH have been established for the problem [6].

We prove that the Graph Motif problem does not admit a quasipolynomial time algorithm on power graphs, even for cyclic groups, assuming ETH (Sect. 4). This shows that even with the strong algebraic restrictions of a cyclic group, the motif problem remains hard. This hardness result is derived through a careful application of concepts from both group theory and number theory.

On the contrary, by exploiting the structure of power graphs of cyclic groups, we obtain a subexponential time algorithm to solve the Graph Motif problem for power graphs of cyclic groups. Moreover, for power graphs of p-groups, the problem admits a more efficient solution, as we present a polynomial time algorithm.

On a related note, it is worthwhile to mention that the non-hardness results for isomorphism-invariant graph (or group) problems defined on graphs (or groups), as well as the hardness results of the Graph Motif problem in power graphs, can be derived from the widely held assumption that $\mathsf{EXP} \neq \mathsf{NEXP}$, where $\mathsf{EXP} = \bigcup_{t>0} \mathsf{DTIME}(2^{n^t})$ and $\mathsf{NEXP} = \bigcup_{t>0} \mathsf{NTIME}(2^{n^t})$, by using a result by Buhrman and Homer [7].

There are two natural representations of power graphs. The first is via the Cayley table of the underlying group, where the group structure is explicitly provided. We refer to this as *the algebraic representation*. The second is through the adjacency matrix or adjacency list of the power graph, with no accompanying group information. In the latter representation, the graph is presented purely as a combinatorial object. We refer to this as the *standard graph representation*.

Within the standard graph representation, a fundamental question is the recognition problem: given a graph, determine whether it is the power graph of some group. Following the results of Arvind et al. for recognition of commuting graphs [3], one can show that this problem admits a quasipolynomial time algorithm for power graphs, enhanced power graphs, and directed power graphs. We show that it is easy to recognise power graphs of nilpotent groups of squarefree exponent. We further show that the recognition problem for power graphs is solvable in polynomial time when the graph arises from an abelian group, as well as from certain classes of nilpotent groups (Sect. 5). In particular, we consider nilpotent groups of bounded polycyclic length - a class that includes several important subclasses, such as metacyclic groups and nilpotent groups of cube-free order.

This paper presents both hardness and non-hardness results on power graphs, along with recognition algorithms for specific group classes. Except for the result on p-groups in Sect. 4, all results in Sect. 3 and 4 are representation-independent. For p-groups, however, the Graph Motif problem must be treated as a promise problem under standard representation or require algebraic input.

2 Preliminaries

For a *graph* Γ, we write $V(\Gamma)$ and $E(\Gamma)$ to denote the vertex set and edge set, respectively. We denote an *undirected edge* between two vertices u and v by $\{u, v\}$ and a *directed edge* from u to v by (u, v). Unless otherwise specified, the graphs we consider are simple undirected graphs. The *neighbourhood* of a vertex u in a graph Γ is denoted by $N_\Gamma(u)$. For a subset $L \subseteq V(\Gamma)$, $N(L) = \{v \in V(\Gamma) :$ there exists $a \in L$ with $\{a, v\} \in E(\Gamma)\}$. Two vertices u and v are *closed twins* in a *directed graph* $\mathcal{D}$ if their closed out-neighbourhoods are the same. The relation between two vertices of being closed twins is an equivalence relation, and the equivalence class under this relation is called the *closed twin-class*. The *subgraph induced on the set* $S \subseteq V(\Gamma)$ is denoted by $\Gamma[S]$. A function $f : V(\Gamma) \to V(\Gamma')$

is an *isomorphism* between two graphs Γ and Γ' if f is a bijection such that $\{u, v\} \in E(\Gamma)$ if and only if $\{f(u), f(v)\} \in E(\Gamma')$; here, Γ and Γ' are called *isomorphic* and it is denoted by $\Gamma \cong \Gamma'$. We write $col(v)$ to denote the colour of a vertex v in a vertex-coloured graph. For two vertex-coloured graphs Γ and Γ', if for an isomorphism f, $col(u) = col(f(u))$, for all $u \in V(\Gamma)$, then f is called a *colour-preserving isomorphism* (or, *colour-isomorphism*) and Γ and Γ' are said to be *colour-isomorphic*, denoted by $\Gamma \cong_c \Gamma'$.

The basic definitions and facts on group theory can be found in any standard book (e.g., [30]). The *order of a group* G and the *order of an element* g in G are denoted by $|G|$ and $ord(g)$ respectively. The *cyclic subgroup generated by* g is denoted by $\langle g \rangle$, where g is a *generator* of $\langle g \rangle$. The number of generators of $\langle g \rangle$ is $\phi(|\langle g \rangle|)$, where ϕ is the *Euler's totient function.* Any cyclic group of order n is isomorphic to $\mathbb{Z}_n$, the additive group of integers modulo n. The *exponent* of a finite group is the least common multiple of the orders of all group elements.

A group G is called a *p-group* if the order of G is a power of a prime p. It is common to refer to a group as a p-group whenever its order is a power of some prime, where the symbol p denotes an arbitrary prime. If p^m is the highest power of a prime p such that p^m divides $|G|$, then there is at least one subgroup of order p^m in G, and such a subgroup is called *Sylow p-subgroup* of G. A finite group is called a *nilpotent group* if it is a direct product of its Sylow subgroups. Moreover, each Sylow subgroup is unique in a finite nilpotent group.

Since our primary focus is on power graphs and directed power graphs, we provide their definitions only. For the definitions of other group-based graph classes, we refer the reader to [8]. The *directed power graph* of a group G, denoted by $DPow(G)$, is the directed graph with vertex set G, and edge set $\{(x, y) : y = x^m \text{ for some positive integer m}\}$. The *power graph*, denoted by $Pow(G)$, is the underlying undirected graph of $DPow(G)$, i.e., the undirected graph obtained from $DPow(G)$ after removing directions, self-loops and double arcs.

Remark 1. If $\{x, y\} \in E(Pow(G))$, then either $ord(x)|ord(y)$ or $ord(y)|ord(x)$. If G is a cyclic group, then $\{x, y\}$ is an edge in $Pow(G)$ if and only if either $ord(x)|ord(y)$ or $ord(y)|ord(x)$.

We define an equivalence relation $\sim$ on G as follows: for $x, y \in G$, $x \sim y$ if and only if $\langle x \rangle = \langle y \rangle$, i.e., x and y generate the same cyclic subgroup of G. The equivalence class containing x is denoted by $[x]$. So, all the elements of a class have the same order. Note that the size of $[x]$ is $\phi(ord(x))$.

Remark 2. In $Pow(G)$, the following two facts hold: (i) Each equivalence class $[x]$ of G is a clique of size $\phi(ord(x))$ in $Pow(G)$; (ii) For two classes $[x]$ and $[y]$, if an element $x \in [x]$ is adjacent to an element $y \in [y]$ in $Pow(G)$, then every element of $[x]$ is adjacent to every element of $[y]$.

Lemma 1 [9]. *The power graph of a finite group is complete if and only if the group is a cyclic group of prime-power order.*

The *Exponential Time Hypothesis* (ETH) is a conjecture by Impagliazzo, Paturi and Zane [22] that states that there is no $2^{o(n)}$-time algorithm for 3-SAT with n variables. A function f that takes a graph as input and outputs a graph is called a *canonization function* if a) $f(G) \cong G$ and b) $G_1 \cong G_2$ if and only if $f(G_1) = f(G_2)$. The size of a set S is denoted by $|S|$, whereas the length of an object x - that is, the number of bits in its encoding - is denoted by $||x||$.

3 Non-hardness of Isomorphism-Invariant Problems

Let $\mathcal{G}$ be a class of graphs closed under isomorphism. We consider problems whose instances are of the form $(\Gamma_1, \ldots, \Gamma_k, x)$, where $\Gamma_i \in \mathcal{G}$, k is a constant, and $||x|| = O(\log s)$, where $s = ||(\Gamma_1, \ldots, \Gamma_k, x)||$, is the size of the instance. In other words, a language representation of such problems is a subset of $\mathcal{G}^k \times \{0,1\}^*$, where k is a constant, and the last component (i.e., x) is of logarithmic length in the length of the input. We call such a language L *isomorphism invariant over* $\mathcal{G}$, if for all $\Lambda_i, \Gamma_i \in \mathcal{G}$, where $1 \leq i \leq k$ such that $\Gamma_i \cong \Lambda_i$, for all i, we have $(\Gamma_1, \ldots, \Gamma_k, s) \in L$ if and only if $(\Lambda_1, \ldots, \Lambda_k, s) \in L$. A wide range of problems are isomorphism invariant. For example, Graph Colouring: The language is $L = \{(\Gamma, r) | \Gamma$ has a vertex colouring with r colours$\}$, where r is given in binary and $||r|| = O(\log |V(\Gamma)|)$. This problem is isomorphism invariant because Γ is r-colourable if and only if an isomorphic copy of Γ is r-colourable. Similarly, the following problems are isomorphism invariant: Independent Set, Clique, Graph Colouring, Hamiltonian Path, Partition into Cliques, Oriented Diameter, etc.

The following theorem is stated for isomorphism invariant problems on power graphs. However, the theorem also holds for directed power graphs, enhanced power graphs, commuting graphs, and bounded-degree Cayley graphs.

Theorem 1. *If L is an isomorphism invariant graph problem over the class of power graphs, then L is not NP-complete under ETH.*

We prove the above theorem using a version of Mahaney's theorem [26], which can be proved easily by mimicking the proof of Mahaney's theorem (see, e.g., [18]). We say that a language L is *quasipolynomially sparse* if there exists a constant $c > 0$ such that $|L \cap \{0,1\}^n| = 2^{O(\log^c n)}$.

Theorem 2. *Under ETH, no quasipolynomially sparse language is NP-complete.*

Remark 3. An isomorphism invariant problem on power graphs need not be quasipolynomially sparse, irrespective of the representations (algebraic or standard graph representation), even though the number of non-isomorphic groups of order n is quasipolynomial [27]. This is evident from Lemma 3. The following lemma is possibly folklore and has been used previously (e.g., [13]).

Lemma 2 [$\star$][1]**.** *The number of distinct Cayley tables representing groups of order n is exponential in n.*

[1] Due to space constraints, proofs marked with a $\star$ have been deferred to the arXiv version [10] of the paper.

The number of adjacency matrices representing power graphs on n vertices can be very small: If n is a prime power, only then can the power graph be a complete graph (Lemma 1). However, this is the only case when this number is small, as shown in the next lemma.

Lemma 3 [$\star$]. *The number of distinct adjacency matrices representing power graphs on n vertices is $\Omega(\exp(\sqrt{n}))$, when n is a number with at least two distinct prime divisors. For infinitely many n, there are $\Omega(\exp(n/\log\log n))$ distinct adjacency matrices representing power graphs on n vertices.*

Proof of Theorem 1: Let $(\Gamma_1 \ldots, \Gamma_k, x)$ be an instance of problem L. We use Babai's quasipolynomial time computable graph canonization function [5], which we denote by *canon*, on each of $\Gamma_1, \ldots, \Gamma_k$ to obtain $canon(\Gamma_1), \ldots, canon(\Gamma_k)$. Since L is isomorphism invariant, $(\Gamma_1, \ldots, \Gamma_k, x) \in L$ if and only if $(canon(\Gamma_1), \ldots, canon(\Gamma_k), x) \in L$.

Let $L' = \{(canon(\Gamma_1), \ldots, canon(\Gamma_k), x) | (\Gamma_1, \ldots, \Gamma_k, x) \in L\}$. Since the number of non-isomorphic groups of order n is $2^{O(\log^3 n)}$ [27], the number of possibilities of $canon(\Gamma_i)$ is at most quasipolynomial in the size of the input. Since k is a constant, we observe that L' is a quasipolynomially sparse language. Therefore, L' is not NP-hard under ETH by Theorem 2. Since L is quasipolynomial time reducible to L', L is also not NP-hard under ETH. ◀

A similar notion of isomorphism invariant problems on groups can be defined. Let $\mathscr{G}$ be the class of finite groups closed under isomorphism. We consider problems whose instances are of the form $(\Gamma_1, \ldots, \Gamma_k, x)$, where $\Gamma_i \in \mathscr{G}$ is given by its Cayley table, k is a constant, and $||x|| = O(\log s)$, where $s = ||(\Gamma_1, \ldots, \Gamma_k, x)||$, is the size of the instance. Such a language L is called *an isomorphism invariant group problem*, if for all $\Lambda_i, \Gamma_i \in \mathscr{G}$ with $\Gamma_i \cong \Lambda_i$, where $1 \leq i \leq k$, we have $(\Gamma_1, \ldots, \Gamma_k, s) \in L$ if and only if $(\Lambda_1, \ldots, \Lambda_k, s) \in L$. The next result is similar.

Theorem 3. *If L is an isomorphism invariant group problem, then L is not NP-complete under ETH.*

Remark 4. Due to Buhrman and Homer [7], under $EXP \neq NEXP$, no quasipolynomially sparse language is NP-complete. Hence, the isomorphism invariant problems on both group-based graphs (Theorem 1) and groups (Theorem 3) are not NP-complete, under the assumption $EXP \neq NEXP$.

4 Hardness of the Weighted Max-Cut Problem and Graph Motif Problem

First, we study the Weighted Max-Cut problem: given an edge-weighted graph Γ and a number k, the goal is to find a subset $S \subset V(\Gamma)$ such that the total weight of edges across S and $V(\Gamma) \setminus S$ is at least k. We prove Theorem 4 using a polynomial time reduction from Weighted Max-Cut problem on complete graphs to its counterpart on power graphs of cyclic groups.

Theorem 4 [$\star$]. *WEIGHTED MAX-CUT is NP-complete for power graphs, even when the underlying groups are cyclic.*

We now shift our attention to the GRAPH MOTIF problem. A *motif* M is a multiset of colours; it is *colourful* if it is a set. In a vertex-coloured graph Γ, the colour multiset $col(V')$ of a set $V' \subseteq V(\Gamma)$ is the multiset of the colours of the vertices in V'. We say that M *occurs* in Γ if there exists a connected subgraph induced on $V' \subseteq V(\Gamma)$ such that $col(V') = M$, in which case V' is an *occurrence* of M. The GRAPH MOTIF problem asks, given a vertex-coloured graph Γ and a motif M, whether such an occurrence V' exists.

Theorem 5 [$\star$]. *Under ETH, GRAPH MOTIF problem on power graphs cannot be solved in time $2^{O(\log^c n)}$, for any constant $c \geq 0$, even if the underlying group is cyclic and M is colourful.*

Due to the next lemma, the proof of Theorem 5 follows easily.

Lemma 4. *There is an $n^{O(\log(\log n))}$-time many-one reduction from 3-SAT to GRAPH MOTIF in power graphs of cyclic groups.*

Proof. The proof is divided into two main steps:

(1) Given a 3-SAT formula Φ, with n variables, we construct an instance (Γ_Φ, M) of GRAPH MOTIF in a vertex-coloured graph Γ_Φ in polynomial time such that Φ is satisfiable if and only if M occurs in Γ_Φ.
(2) We embed Γ_Φ as an *induced subgraph* in $Pow(\mathbb{Z}_N)$, where $N = n^{O(\log(\log n))}$.

Step 1: Let $\Phi = C_1 \wedge C_2 \wedge \cdots \wedge C_m$ be an instance of 3-SAT with variables $x_1, x_2, \ldots, x_n$.

To construct Γ_Φ, we first create a vertex r and colour it with 0. Then for each literal $l_i \in \{x_i, \overline{x}_i\}$, $1 \leq i \leq n$, we create two vertices x_i and $\overline{x}_i$ both coloured with i. After that, we add the following edges: $\{r, x_i\}$, $\{r, \overline{x}_i\}$, for $1 \leq i \leq n$. Now for each literal l_i, we do the following: Suppose l_i appears in clauses $C_{i_1}, C_{i_2}, \ldots, C_{i_s}$; then create a clique consisting of vertices $z_{i_1}, z_{i_2}, \ldots, z_{i_s}$ that are coloured with colours $C_{i_1}, C_{i_2}, \ldots, C_{i_s}$ respectively, and add the edges: $\{l_i, z_{i_j}\}$, $1 \leq j \leq s$. Note that a new

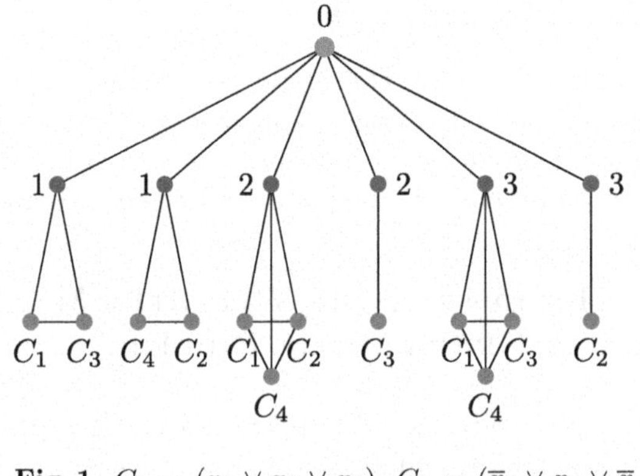

Fig. 1. $C_1 = (x_1 \vee x_2 \vee x_3)$, $C_2 = (\overline{x}_1 \vee x_2 \vee \overline{x}_3)$, $C_3 = (x_1 \vee \overline{x}_2 \vee x_3)$, $C_4 = (\overline{x}_1 \vee x_2 \vee x_3)$

set of z-vertices is added for vertices corresponding to each literal. Therefore, the total number of added z_{i_j} vertices is $3m$. The motif M is specified to be the set of colours $\{0, 1, 2, \ldots, n, C_1, C_2, \ldots, C_m\}$. Here, M is colourful. See Fig. 1 for an illustration of the construction.

Since the number of clauses in Φ is $O(n^3)$ (as Φ is a 3-SAT), the construction of Γ_Φ above is clearly polynomial in n. The correctness of this construction is stated in the next claim.

Claim 1. [$\star$] Φ is satisfiable if and only if M occurs in Γ_Φ.

Step 2: We embed Γ_Φ as an induced subgraph in $Pow(\mathbb{Z}_N)$ for some suitably chosen N. The number N is the product of the first b primes, where b is an odd number. We later show that setting $b = O(\log n)$ is sufficient. It is a well-known fact that the product of the first b primes is bounded above by $2^{(1+o(1))b \log b}$ (see, e.g., [15]). With that we have $N = 2^{O(\log n \log(\log n))}$.

First, we describe the embedding procedure. Let $N = p_1 p_2 \ldots p_b$, where p_i is the i-th natural prime number. To embed Γ_Φ in $Pow(\mathbb{Z}_N)$, we define a bijective mapping $f : V(\Gamma_\Phi) \rightarrow S$, where $S \subseteq \mathbb{Z}_N$, such that f is an isomorphism between Γ_Φ and $Pow(\mathbb{Z}_N)[S]$.

Let $\mathcal{P}_{\frac{b-1}{2}}(\{2, 3, \ldots, b\})$ denote the collection of subsets $I \subset \{2, 3, \ldots, b\}$ such that $|I| = \frac{b-1}{2}$. Let $f_\mathcal{P}$ be a one-to-one function from the set of literals of Φ to $\mathcal{P}_{\frac{b-1}{2}}(\{2, 3, \ldots, b\})$. The function f maps the vertex x_i (or $\overline{x}_i$) to an element $g \in \mathbb{Z}_N$ of order $p_1 \prod_{t \in f_\mathcal{P}(l_i)} p_t$, where $l_i = x_i$ (or $l_i = \overline{x}_i$). The element g has a neighbour h of order $\prod_{t \in f_\mathcal{P}(l_i)} p_t$ (by Remark 1). Moreover, g is adjacent to every element in the equivalence class $[h]$ (by Remark 2). Suppose $N(u) = \{r, z_{i_1}, z_{i_2}, \ldots, z_{i_k}\}$, where $u = x_i$ or $\overline{x}_i$. Then, the function f maps each z_{i_j}, $1 \leq j \leq k$, to distinct elements in $[h]$. For this mapping procedure to be one-to-one, we need $|[h]| \geq |N(u)| - 1$. The function f maps the vertex r to some element of order p_1.

Once we ensure that f is one-to-one, then taking $S = f(V(\Gamma_\Phi))$ implies that f is a bijection from $V(\Gamma_\Phi)$ to S. The next claim shows that taking $b = O(\log n)$ is sufficient for the mapping f to be one-to-one, and consequently to be an isomorphism between Γ_Φ and $Pow(\mathbb{Z}_N)[S]$.

Claim 2. If $b = O(\log n)$, then f is an isomorphism between Γ_Φ and $Pow(\mathbb{Z}_N)[S]$.

Proof. We begin by identifying two inequalities, that must be satisfied for f to be one-to-one. Subsequently, we show that $b = O(\log n)$ satisfies these inequalities.

(1) First, $f_\mathcal{P}$ has to be one-to-one. For that, we need the following:

$$\binom{b-1}{\frac{b-1}{2}} \geq 2n \tag{1}$$

(2) For each vertex $u \in \{x_i, \overline{x}_i \mid 1 \leq i \leq n\}$, we must have: $|[h]| \geq |N(u)| - 1$, where h is a neighbour of $f(u)$ in $Pow(\mathbb{Z}_N)$. Note that $|N(u)|$ is at most the

number of clauses m in Φ and $m = O(n^3)$. From the definition of f, one can see that $|[h]|$ is at least $\phi(p_2p_3\dots p_{\frac{b+1}{2}})$, as $|[h]| = \phi(ord(h))$. Hence, satisfying the following inequality is sufficient:

$$cn^3 \leq \phi(p_2p_3\dots p_{\frac{b+1}{2}}),\ \text{where } c \text{ is some constant} \tag{2}$$

We now show that for $b = 2\log(2n)+1$ and for sufficiently large n, both the above inequalities, Eq. (1) and Eq. (2), are satisfied.

Note: $\phi(p_2p_3\dots p_{\frac{b+1}{2}}) = 2.4.6.\dots.(p_{\frac{b+1}{2}} - 1) \geq 2.3.5.\dots.p_{\frac{b-1}{2}} \geq 2^{c'\frac{b-1}{2}.\log(\frac{b-1}{2})}$, for some fixed constant c'. The last inequality can be easily derived from the following two facts: (1) $lim_{n\to\infty}(p_1p_2\dots p_n)^{\frac{1}{p_n}} = e$ (see [15]); (2) prime number theorem, i.e., $\lim_{n\to\infty}\frac{p_n}{n\log n} = 1$ (see, e.g., [2]). Hence, if the following inequality holds, then Eq. (2) consequently holds:

$$cn^3 \leq 2^{c'\frac{b-1}{2}.\log(\frac{b-1}{2})} \tag{3}$$

Now, we find a suitable value of b such that Eq.n (1) and Eq. (3) simultaneously hold. Let us denote $(b-1)/2$ by k. Using Stirling's approximation, for sufficiently large k, $\binom{2k}{k} \sim \frac{2^{2k}}{\sqrt{k\pi}}$. Then Eq. (1) implies the following inequality, and solving it serves our purpose.

$$\frac{2^{2k}}{\sqrt{k\pi}} \geq 2n \tag{4}$$

For $k = \log(2n)$, we have: $2\log(2n) - \frac{1}{2}\log(\log(2n)\sqrt{\pi}) \geq \log(2n)$ and hence Eq. (4) holds. Moreover, for $b = 2\log(2n)+1$ and for sufficiently large n, Eq. (3) is also satisfied. □

We note that there exists a one-to-one mapping $f_{\mathcal{P}}$, as required, computable in time $\binom{b-1}{(b-1)/2}$, implying that f can be computed in $2^{O(\log n\log(\log n))}$ time. Following the embedding of Γ_Φ in Claim 2, an instance of Graph Motif in Pow($\mathbb{Z}_N$) can be constructed by assigning a colour not in M to all vertices in $\mathbb{Z}_N \setminus S$. This yields an $n^{O(\log\log n)}$-time reduction from 3-SAT to Graph Motif in power graphs of cyclic groups, completing the proof of the lemma. ◁

Remark 5. A similar statement to Theorem 5 can be made without using ETH. More explicitly, Graph Motif on power graphs cannot be solved in quasipolynomial time, unless $\mathsf{EXP} = \mathsf{NEXP}$. This is because, in that case Lemma4 would imply that $\mathsf{NP} \subseteq \bigcup_{t>0}\mathsf{DTIME}(2^{\log^t n})$, which in turn would imply that $\mathsf{EXP} = \mathsf{NEXP}$ [7].

We now show the existence of an algorithm with runtime $2^{o(n)}$ for power graphs of cyclic groups. Two vertices v and v' of a graph are said to have the same *type* if and only if $N(v)\setminus\{v'\} = N(v')\setminus\{v\}$. A graph $\Gamma(V,E)$ has *neighbourhood diversity* at most w if there exists a partition of $V(\Gamma)$ into at most w sets, such that all the vertices in each set have the same type [25]. Ganian [17] gave an FPT algorithm for Graph Motif parameterised by neighbourhood diversity.

Lemma 5 [17]. *Graph Motif can be solved in time* $O(2^k \cdot \sqrt{|V|}|E|)$ *on a graph* $\Gamma = (V, E)$ *of neighbourhood diversity at most* k.

Let g and g' be two elements in a group G. If $[g] = [g']$, then g and g' have the same type in $Pow(G)$. Hence, the neighbourhood diversity of $Pow(G)$ is at most the number of equivalence classes of G. If G is a cyclic group of order n, then the number of equivalence classes is the number of total divisors of n, which is bounded by $n^{O(\frac{1}{\log\log n})}$ [2]. Hence, we have the following corollary.

Corollary 1. *Graph Motif for* $Pow(\mathbb{Z}_n)$ *can be solved in* $2^{o(n)}$ *time, in particular, in* $2^{n^{c/\log\log n}}$ *time.*

Corollary 1 rules out a linear time many-one reduction from 3-SAT, under ETH. Note that when $G = (\mathbb{Z}_p)^n$, for a prime p, the number of equivalence classes is $(p^n - 1)/(p - 1)$, resulting in a $2^{O(|G|/p)}$ runtime using Ganian's algorithm (Lemma 5), which is almost exponential. Hence, his algorithm is not efficient for power graphs of p-groups. Here, we design a polynomial time algorithm for Graph Motif on power graphs of p-groups.

Theorem 6 [⋆]. *Graph Motif for power graphs of* p*-groups can be solved in time polynomial in the size of the group.*

5 Recognition of Power Graphs

In this section, we show that the recognition problem for power graphs of abelian groups and certain nilpotent groups can be solved in polynomial time (Theorem 7). The input graph is given in the standard graph representation model. The task is to decide if the graph is a power graph of some group. Due to space constraints, we outline only the main results and techniques here; full details and proofs are provided in the arXiv version [10].

Das et al. [11] gave a reconstruction algorithm that takes a power graph as input and produces a directed power graph. However, if the input is not a power graph, the algorithm may fail or produce a directed graph that is not a directed power graph. So, we may, *without loss of generality, focus on the recognition problem of directed power graphs.* Adapting the quasipolynomial time approach of Arvind, Cameron, Ma, and Maslova [3] for recognising commuting graphs (see description in the arXiv version [10]), the following lemma extends to the recognition of any group-based graph.

Lemma 6. *There are algorithms that, on the input of an undirected graph* Γ *(or directed graph* $\mathcal{D}$*) of size* n*, check whether* Γ *(or* $\mathcal{D}$*) is isomorphic to a power graph (or directed power graph), enhanced power graph, deep commuting graph or generating graph in* $2^{O(\log^c n)}$ *time.*

Two aspects of the algorithm by Arvind et al. that contribute to high runtime are: (a) the number of non-isomorphic groups of order n can be $2^{\Omega(\log^3 n)}$ [27], (b) Babai's isomorphism algorithm has quasipolynomial runtime. The polynomial time algorithm for isomorphism of power graphs for nilpotent groups by Das et al. [11] appears to be useful; but, it solves a promise problem: it is guaranteed to work correctly if the graphs are power graphs of nilpotent groups. However, if one input is not a power graph of a nilpotent group, it may give a false positive answer (for details, see the arXiv version [11, Sec. A.2]). To tackle this issue, we use the following two tools: (i) the notion of *reduced graphs* (similar to [11]); (ii) an algorithm referred to as *Gluing process*. Using these, we prove that the recognition problem of directed power graphs of nilpotent groups is reducible to the recognition problem of directed power graphs of p-groups (Lemma 8).

Reduced Graph: Given a directed graph $\mathcal{D}$, we construct a directed graph $R(\mathcal{D})$ from $\mathcal{D}$, referred to as the *reduced graph*, by the following method: (1) For each closed twin-class τ of $\mathcal{D}$, we create a vertex in $R(\mathcal{D})$; (2) The colour of each vertex τ, denoted by $col(\tau)$, of $R(\mathcal{D})$ is the out-degree of any vertex of τ in $\mathcal{D}$; (3) Put a directed edge from τ_1 to τ_2 in $R(\mathcal{D})$ where $\tau_1 \neq \tau_2$, if there is a directed edge from some vertex $x_1 \in \tau_1$ to some vertex $x_2 \in \tau_2$ in $\mathcal{D}$.

Recall that for any finite group G, the out-degree of a vertex u in $DPow(G)$ equals to $ord(u)$ in G. If (u, v) is an edge in $DPow(G)$, then $ord(v) \mid ord(u)$. Hence, if (τ_1, τ_2) is an edge in $R(DPow(G))$, then $col(\tau_2) \mid col(\tau_1)$. So, it is easy to see that $R(DPow(G))$ is a directed acyclic graph (DAG). In $DPow(G)$, two vertices u and v are closed twins if and only if $\langle u\rangle = \langle v\rangle$ in G, i.e., u and v are two generators of the same cyclic subgroup in G. So, the vertex set of $R(DPow(G))$ is $\{[x] \mid x \in G\}$. If (x, y) is an edge in $DPow(G)$, then for every $x' \in [x]$ and for every $y' \in [y]$, there is an edge $(x', y') \in DPow(G)$. Then by Remark 1, there is an edge (τ_1, τ_2) in $R(DPow(G))$. From these properties, it is not hard to see that two directed power graphs are isomorphic if and only if their reduced graphs are colour-isomorphic. Hence, for the rest of the discussion, *we assume that the input to our recognition query is a reduced directed graph.*

First, we consider the subproblem with inputs, a reduced directed graph $\mathcal{D}_1$ and a nilpotent group H; the task is to check whether $\mathcal{D}_1 \cong_c R(DPow(H))$. When the input group is a p-group, then we can directly use the linear time isomorphism algorithm of directed power graphs of p-groups by [11].

Lemma 7. *Let $\mathcal{D}_1$ be a reduced directed graph and H be a finite p-group. Then, there is a linear time algorithm that checks whether $\mathcal{D}_1 \cong_c R(DPow(H))$ and if so, returns a colour-preserving isomorphism $\psi : V(R(DPow(H))) \rightarrow V(\mathcal{D}_1)$.*

In a vertex-coloured directed acyclic graph, let vertices $u_1, u_2, \ldots, u_k$ have a common parent u. Then u is called *least common parent* of $u_1, u_2, \ldots, u_k$, if for any other existing common parent u' of $u_1, u_2, \ldots, u_k$, $col(u) \leq col(u')$.

Now consider the case when the input group is a finite nilpotent group $H = H_1 \times H_2 \times \cdots \times H_k$, where H_i is the Sylow p_i-subgroup of H. Note that any group element of H can be represented as $(x_1, x_2, \ldots, x_k)$, where $x_i \in H_i$, $1 \leq i \leq k$. We now describe the *Gluing Process*.

Gluing Process: Let $\mathcal{D}_1$ be a reduced directed graph and $H = H_1 \times \cdots \times H_k$ be a finite nilpotent group. Suppose $\mathcal{D}_{1i}$ denotes the subgraph of $\mathcal{D}_1$

induced on the set of vertices whose colours are p_i-powers, for all $1 \leq i \leq k$. Let $f_i : V(R(DPow(H_i))) \rightarrow V(\mathcal{D}_{1i})$ be a colour-isomorphim between $R(DPow(H_i))$ and $\mathcal{D}_{1i}$, for $1 \leq i \leq k$. Gluing process decides if there exists a colour-isomorphism $f : V(R(DPow(H))) \rightarrow V(\mathcal{D}_1)$ such that f restricted to $V(R(DPow(H_i)))$ is f_i. To achieve this, we define $f : V(R(DPow(H))) \rightarrow V(\mathcal{D}_1)$ as follows: f maps each vertex $[(x_1, \ldots, x_k)]$ of $R(DPow(H))$ to a least common parent, if it exists, of $f_1([x_1]), \ldots, f_k([x_k])$ in $\mathcal{D}_1$. Thereafter, we check if f is a colour-isomorphism between $R(DPow(H))$ and $\mathcal{D}_1$. For a detailed discussion of the correctness of Gluing process, we refer the reader to the arXiv version [10]. Now, we state one of the main results of this section.

Lemma 8 [$\star$]. *Let $\mathcal{X}$ be a subclass of $\mathcal{G}_{PP} = \{G$ is a p-group : p is some prime$\}$ and $Nil(\mathcal{X})$ be the class of nilpotent groups whose Sylow subgroups are in $\mathcal{X}$. Let $\mathcal{A}$ be a polynomial time algorithm that on input a reduced directed graph $\mathcal{D}'$, decides if $\mathcal{D}' \cong_c R(DPow(G))$, for some $G \in \mathcal{X}$ and if so, returns a colour-preserving isomorphism $\psi : V(R(DPow(G))) \rightarrow V(\mathcal{D}')$ along with G. Then, there is a polynomial time algorithm $\mathcal{A}_{nil}$, that, on input a reduced directed graph with n vertices, checks whether it is colour-isomorphic to $R(DPow(H))$ for some group $H \in Nil(\mathcal{X})$. Moreover, $\mathcal{A}$ is called $O(n)$ times by $\mathcal{A}_{nil}$.*

By the preceding lemma, it suffices to design a polynomial time algorithm $\mathcal{A}$ for recognising directed power graphs within certain subclasses $\mathcal{X}$ of $\mathcal{G}_{PP}$. We focus on three such subclasses: abelian groups, groups of bounded polycyclic length, and groups of prime exponent (see the arXiv version [10, Sec. 5.3]). For the first two subclasses, all candidate groups can be enumerated in polynomial time (see Lemma 9 and Lemma 10), followed by applying the isomorphism algorithm from Lemma 7. The following result summarises the key outcomes of our discussion so far.

Theorem 7. *There are polynomial time algorithms $\mathcal{A}_{ab}$, $\mathcal{A}_{polycyc}$, and $\mathcal{A}_{sqfr}$ that take a graph Γ as input and decide if Γ is isomorphic to the power graph of an abelian group, a nilpotent group of bounded polycyclic length, or a nilpotent group of squarefree exponent, respectively.*

The proof of the next lemma follows from the Fundamental Theorem of Abelian Groups and properties of the partition function. Subsequently, Theorem 7 for abelian groups follows directly.

Lemma 9 [$\star$]. *There is a polynomial time algorithm that, given a number p^m, outputs all non-isomorphic abelian groups of order p^m.*

A *polycyclic series* of a group G is a finite series $G = G_0 \geq G_1 \geq \cdots \geq G_l = \{1\}$ such that for each $0 \leq i \leq l-1$, $G_{i+1} \lhd G_i$ and the factor G_{i+1}/G_i is cyclic. A group with a polycyclic series is called a *polycyclic group*. Every finite nilpotent group is a polycyclic group. The *polycyclic length* (see [16]) of a polycyclic group is the number of non-trivial factors of a polycyclic series of shortest length.

Let $\mathcal{X}$ denote the class of groups in $\mathcal{G}_{PP}$ with bounded polycyclic length. To enumerate these groups in polynomial time (see Lemma 10), we use the concept of group *presentations*, which is derived here from their polycyclic series (see the arXiv version [10, Sec. 5.2] for a detailed definition). We first show that standard methods (e.g., [4,21]) yield logarithmic-size presentations for groups in $\mathcal{X}$. Using these, and a polynomial time *procedure* for constructing a Cayley table from a presentation (cf. [4]), we establish the next lemma.

Lemma 10 [$\star$]. *Let $c > 0$ be a constant and $\mathcal{X}$ be the set of groups G from $\mathcal{G}_{PP}$ such that the polycyclic length of G is bounded by c. Then, each $G \in \mathcal{X}$ has a presentation of length $O(\log |G|)$. Moreover, there is an algorithm that generates all non-isomorphic groups of $\mathcal{X}$ of order N in time polynomial in N.*

The above lemma, along with Lemma 8, implies Theorem 7 for nilpotent groups of bounded polycyclic length. The recognition algorithms for power graphs of nilpotent groups of d-powerfree and nilpotent groups of squarefree exponent are discussed in Sects. 5.2 and 5.3, respectively, of the arXiv version [10].

Acknowledgments. We would like to thank V. Arvind and P. J. Cameron for their valuable inputs. We also thank the three anonymous reviewers of FCT 2025 for their insightful feedback and comments.

References

1. Aalipour, G., Akbari, S., Cameron, P.J., Nikandish, R., Shaveisi, F.: On the structure of the power graph and the enhanced power graph of a group. Electron. J. Combin. **1**, P3-16 (2017)
2. Apostol, T.M.: Introduction to Analytic Number Theory. Springer, Heidelberg (2013)
3. Arvind, V., Ma, X., Cameron, P.J., Maslova, N.V.: Aspects of the commuting graph. arXiv preprint arXiv:2305.07301 (2024)
4. Arvind, V., Torán, J.: Solvable group isomorphism is (almost) in NP /spl cap/ coNP. ACM Trans. Comput. Theory (TOCT) **2**(2), 1–22 (2011)
5. Babai, L.: Graph isomorphism in quasipolynomial time. In: Proceedings of the forty-Eighth Annual ACM Symposium on Theory of Computing, pp. 684–697 (2016)
6. Bonnet, É., Sikora, F.: The graph motif problem parameterized by the structure of the input graph. Discret. Appl. Math. **231**, 78–94 (2017)
7. Buhrman, H., Homer, S.: Superpolynomial circuits, almost sparse oracles and the exponential hierarchy. In: Shyamasundar, R. (ed.) FSTTCS 1992. LNCS, vol. 652, pp. 116–127. Springer, Heidelberg (1992). https://doi.org/10.1007/3-540-56287-7_99 ISBN 978-3- 540-47507-1
8. Cameron, P.J.: Graphs defined on groups. Int. J. Group Theory **11**(2), 53–107 (2022)
9. Chakrabarty, I., Ghosh, S., Sen, M.: Undirected power graphs of semigroups. Semigroup Forum **78**, 410–426 (2009)
10. Das, B., Dey, D., Ghosh, J.: On the complexity of problems on graphs defined on groups. arXiv preprint arXiv:2507.05860 (2025)

11. Das, B., Ghosh, J., Kumar, A.: The isomorphism problem of power graphs and a question of Cameron. In: 44th IARCS Annual Conference on Foundations of Software Technology and Theoretical Computer Science (FSTTCS 2024). LIPIcs, vol. 323, pp. 20:1–20:23 (2024)
12. Das, B., Thakkar, D.: The minimal faithful permutation degree of groups without abelian normal subgroups. In: Proceedings of the 56th Annual ACM Symposium on Theory of Computing, pp. 118–129 (2024)
13. Farzan, A., Munro, J.I.: Succinct representation of finite abelian groups. In: Proceedings of the 2006 International Symposium on Symbolic and Algebraic Computation, pp. 87–92 (2006)
14. Fellows, M.R., Fertin, G., Hermelin, D., Vialette, S.: Sharp tractability borderlines for finding connected motifs in vertex-colored graphs. In: Arge, L., Cachin, C., Jurdziński, T., Tarlecki, A. (eds.) ICALP 2007. LNCS, vol. 4596, pp. 340–351. Springer, Heidelberg (2007). https://doi.org/10.1007/978-3-540-73420-8_31
15. Finch, S.R.: Mathematical constants. Cambridge University Press, Cambridge (2003)
16. Fisher, R.: The polycyclic length of linear and finite polycyclic groups. Can. J. Math. **26**(4), 1002–1009 (1974)
17. Ganian, R.: Using neighborhood diversity to solve hard problems. arXiv preprint arXiv:1201.3091 (2012)
18. Grochow, J.A.: NP-hard sets are not sparse unless P= NP: an exposition of a simple proof of Mahaney's theorem, with applications. arXiv preprint arXiv:1610.05825 (2016)
19. Grötschel, M., Lovasz, L., Schrijver, A.: Geometric algorithms and combinatorial optimization. Algorithms Combin. **2** (1988)
20. Guillemot, S., Sikora, F.: Finding and counting vertex-colored subtrees. Algorithmica **65**, 828–844 (2013)
21. Higman, G.: Enumerating P-groups. I: inequalities. Proc. London Math. Soc. **3**(1), 24–30 (1960)
22. Impagliazzo, R., Paturi, R., Zane, F.: Which problems have strongly exponential complexity? J. Comput. Syst. Sci. **63**(4), 512–530 (2001)
23. Kelarev, A.V., Quinn, S.J.: A combinatorial property and power graphs of groups. Contrib. Gener. Algebra **12**(58), 3–6 (2000)
24. Lacroix, V., Fernandes, C.G., Sagot, M.-F.: Motif search in graphs: application to metabolic networks. IEEE/ACM Trans. Comput. Biol. Bioinform. **3**(4), 360–368 (2006)
25. Lampis, M.: Algorithmic meta-theorems for restrictions of treewidth. Algorithmica **64**, 19–37 (2012)
26. Mahaney, S.R.: Sparse complete sets for NP: solution of a conjecture of Berman and Hartmanis. J. Comput. Syst. Sci. **25**(2), 130–143 (1982)
27. McIver, A., Neumann, P.M.: Enumerating finite groups. Q. J. Math. **38**(4), 473–488 (1987)
28. Miller, G.L.: On the nlog n isomorphism technique (a preliminary report). In: Proceedings of the Tenth Annual ACM Symposium on Theory of Computing, pp. 51–58 (1978)
29. Pinter, R.Y., Shachnai, H., Zehavi, M.: Deterministic parameterized algorithms for the graph motif problem. Discret. Appl. Math. **213**, 162–178 (2016)
30. Rotman, J.J.: An Introduction to the Theory of Groups. Springer, Heidelberg (2012)

On the Complexity of Locally Rainbow Path

Hiroshi Eto[1(✉)], Tesshu Hanaka[2], Eiji Miyano[1], and Shuya Yoshida[1]

[1] Kyushu Institute of Technology, Iizuka, Japan
{eto,miyano}@ai.kyutech.ac.jp, yosida.syuuya850@mail.kyutech.ac.jp
[2] Kyushu University, Fukuoka, Japan
hanaka@inf.kyushu-u.ac.jp

Abstract. This paper investigates the computational complexity of Locally Rainbow Path on undirected graphs. In this problem, given a vertex-colored graph with q colors, two vertices s, t, and two integers r, ℓ, we seek an s-t path of length at most ℓ where every subpath of length at most r contains vertices with pairwise distinct colors. Here, the length of a path is defined by the number of vertices in the path. We establish the complexity boundaries by proving that Locally Rainbow Path is NP-complete when $r \geq 4$ and $q \geq 5$, while tractable for $r \leq 2$ or when $r = 3$ and $q \leq 3$. For the case where $r = q$, we provide an $r^{O(r)}(n + m)$-time algorithm, making it fixed-parameter tractable when parameterized by r. From a structural perspective, we prove the paraNP-completeness for pathwidth on cactus graphs of maximum degree 3, while showing fixed-parameter tractability when parameterized by treewidth plus the number of colors or by vertex cover number.

Keywords: Shortest s-t path · Locally rainbow path · Parameterized algorithm

1 Introduction

Finding a path in a graph is a fundamental problem in graph theory with a wide range of applications. In practice, such path-finding problems frequently arise either directly or indirectly in various real-world domains. As a result, many variants of path-finding problems have been proposed and extensively studied from both theoretical and practical perspectives.

Among these, color-constrained path-finding problems represent one of the most well-studied variants [1,3,5,9,17,24]. In such problems, colors are typically used to model distinct attributes or categories such as communication channels, security levels, resource types, or task labels. A color-constrained path then corresponds to a valid sequence that satisfies specific requirements, for example, in

This work is partially supported by JSPS KAKENHI Grant Numbers JP22H00513, JP25K03077, JP24K02902 and JST, CRONOS, Japan Grant Number JPMJCS24K2.

A. Jeż and J. Otop (Eds.): FCT 2025, LNCS 16106, pp. 150–164, 2026.
https://doi.org/10.1007/978-3-032-04700-7_12

communication routing, secure routing, scheduling, or sequential resource allocation.

In this line of research, Fluschnik, Kellerhals, and Renken [14] introduced the LOCALLY RAINBOW PATH problem. Unlike typical color-constrained path-finding problems, LOCALLY RAINBOW PATH imposes a local color constraint. In a vertex-colored graph $G = (V, E)$, a path P is called r-locally rainbow if, in every subpath of P of length r, the vertices receive pairwise distinct colors. Formally, a path $P = \langle v_1, v_2, \ldots, v_k \rangle$ is r-locally rainbow if $v_i, v_{i+1}, \cdots, v_{i+r-1}$ receive pairwise distinct colors for every $i \in \{1, \ldots, k - r + 1\}$. The integer r is called the locality.

Figure 1 illustrates an example of a 3-locally rainbow s-t path of length 6. Given a vertex-colored undirected graph $G = (V, E)$ with q colors, two vertices $s, t \in V$, and two integers r and ℓ, the LOCALLY RAINBOW PATH problem asks whether there exists an r-locally rainbow s-t path of length at most ℓ in G. LOCALLY RAINBOW PATH is motivated by modeling scenarios that protect resources from overuse during a given time window by ensuring sufficient diversity in sequential resource utilization. In such cases, the locality r corresponds to the duration of the window, while avoiding overuse is modeled by requiring distinct colors in a subpath. LOCALLY RAINBOW PATH has practical applications in crop sequence optimization, holiday trip planning, and production process scheduling [14]. In [14], LOCALLY RAINBOW PATH is originally defined for directed graphs and shown to be NP-complete even on $r \geq 3$ whereas it is linear-time solvable if $r \leq 2$. The authors also present a fixed-parameter algorithm for LOCALLY RAINBOW PATH parameterized by $r+d$ where $d := \ell - \text{dist}(s, t)$ is the detour length between s and t.

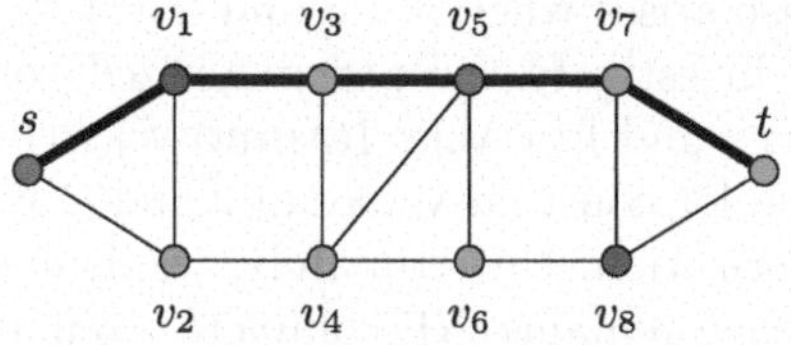

Fig. 1. The graph contains red, blue, green, and orange vertices. The path $P = \langle s, v_1, v_3, v_5, v_7, t \rangle$ is a 3-locally rainbow s-t path. Indeed, each subpath of length 3 in the path P contains vertices with different colors. On the other hand, for example, the path $\langle s, v_2, v_4, v_6, v_8, t \rangle$ is not a 3-locally rainbow path because the subpath $\langle v_2, v_4, v_6 \rangle$ contains vertices v_2, v_6 with the same color. (Color figure online)

1.1 Our Contribution

In this paper, we study the complexity of LOCALLY RAINBOW PATH on undirected graphs, focusing on the number of colors, locality, and structural parameters. We also consider reachability under the r-locally rainbow constraint. That

Table 1. Summary of our results on the complexity of LOCALLY RAINBOW REACHABILITY with respect to r and q. The symbol "-" stands for the trivial case, as LOCALLY RAINBOW REACHABILITY reduces to finding a shortest s-t path in an uncolored graph when $r > q$. The symbol "?" indicates open cases where the complexity remains unknown.

$r \setminus q$	1	2	3	4	5	≥ 6
1	P	P	P	P	P	P
2	-	P	P	P	P	P
3	-	-	P	?	?	?
4	-	-	-	P	NP-c	NP-c
≥ 5	-	-	-	-	P	NP-c

is, given a vertex-colored undirected graph $G = (V, E)$ with q colors, two vertices $s, t \in V$, and an integer r, LOCALLY RAINBOW REACHABILITY asks whether G contains an r-locally rainbow sâĂŞt path. A summary of our results on the complexity of LOCALLY RAINBOW PATH is presented in Table 1. Throughout this paper, we assume that $r \leq q$; otherwise, LOCALLY RAINBOW PATH is solvable in polynomial time, as it reduces to finding a shortest sâĂŞt path in an uncolored graph.

In Sect. 2, focusing on the number q of colors and the locality parameter r, we first establish that LOCALLY RAINBOW REACHABILITY is NP-complete for $r \geq 4$ and $q \geq 5$ on undirected graphs. In contrast, we develop an $r^{O(r)}(m+n)$-time algorithm for the case where $r = q$. Consequently, when $r (= q)$ is constant, LOCALLY RAINBOW REACHABILITY can be solved in linear time. The only remaining open case arises when $r = 3$ and $q \geq 4$.

In Sect. 3, we then investigate the parameterized complexity of LOCALLY RAINBOW REACHABILITY and LOCALLY RAINBOW PATH. Given that the problem is NP-complete even for constant values of r and q, it is natural to examine the structural parameterization. Unfortunately, we demonstrate that LOCALLY RAINBOW REACHABILITY remains NP-complete even on cacti of maximum degree 3 and pathwidth at most 2, even when $r = q$. To complement these negative results, we show that LOCALLY RAINBOW PATH is fixed-parameter tractable when parameterized by (i) the combined parameter of treewidth and q, and (ii) vertex cover number. Furthermore, we prove that LOCALLY RAINBOW PATH is in XP when parameterized by treedepth. Note that for treewidth tw, pathwidth pw, treedepth td, and vertex cover number vc, $tw \leq pw \leq td \leq vc + 1$ holds on any graph [23].

Since LOCALLY RAINBOW PATH generalizes LOCALLY RAINBOW REACHABILITY, any hardness result for LOCALLY RAINBOW REACHABILITY also applies to LOCALLY RAINBOW PATH. Moreover, all of our results could be expanded to directed graphs.

1.2 Related Work

From both graph-theoretic and algorithmic perspectives, color-constrained path-finding problems are of considerable interest, and various types of such problems have been proposed to date.

Rainbow Path, which is also known as Rainbow Vertex s-t-Connectivity, is one of the typical color-constrained path-finding problems. In the problem, we are given a vertex-colored graph $G = (V, E)$ and two vertices $s, t \in V$. Then Rainbow Path asks whether there exists an s-t path consisting of vertices with distinct colors, called a rainbow path. By the definition, if $r = n - 1$, where n is the number of vertices in G, Locally Rainbow Path is equivalent to Rainbow Path. The concept of Rainbow Path is introduced by Chartrand, Johns, McKeon, and Zhang [7] from the graph-theoretic view. Initially, it is defined for edge-colored graphs, and later Krivelevich and Yuster [18] defined it for vertex-colored graphs. A rainbow path can be interpreted as a secure path in a communication network [20]. From the algorithmic view, Rainbow Path is shown to be NP-complete [8]. It remains NP-complete even on graphs of bounded treewidth or diameter 2, whereas a polynomial-time algorithm for outerplanar graphs and a fixed-parameter algorithm parameterized by the number of colors are proposed [24]. For the edge-variant of Rainbow Path, it is shown to be NP-complete even on outerplanar graphs [24] and block graphs [19].

Rainbow Path on a graph with k colors is the Colorful Path problem, which is used for a subroutine of a color-coding-based algorithm for Longest Path [1]. In this problem, given a vertex-colored graph, an integer k, and $s, t \in V$, the goal is to find an s-t path of length k consisting of vertices with distinct colors. This problem is used for a subroutine of a color-coding-based algorithm for Longest Path [1]. Colorful Path can be solved in time $2^{O(k)}n^{O(1)}$ by dynamic programming.

Given a vertex-colored graph $G = (V, E)$, two vertices $s, t \in V$, and an integer k, Maximum Tropical Path is the problem of finding a path containing at least k different colors. The problem is NP-complete even on cactus graphs and interval graphs [9]. On the other hand, Fomin et al. [15] present a randomized algorithm that runs in time $2^k n^{O(1)}$. The minimization version of Maximum Tropical Path is called Colored Path, which is the problem of determining whether there is an (s, t)-path in G containing at most k different colors. Interestingly, the problem is W[SAT]-hard and is in W[P] when parameterized by k [13] in contrast to Maximum Tropical Path.

Balance-Fair Shortest Path is the problem of finding a shortest fair path in a vertex-colored graph that contains an equal number of vertices of each color. This problem is NP-complete even on restricted graph classes [4]. The parameterized complexity of the problem is studied in [4,5].

1.3 Preliminaries

We use standard graph-theoretic terminology and refer the reader to the standard textbook [12] on parameterized complexity for definitions of basic concepts such as treewidth, pathwidth, and fixed-parameter tractability. We also refer the reader to [23] on other graph parameters.

Let $G = (V, E)$ be an undirected vertex-colored graph with a coloring function $c : V \to \{1, \dots, q\}$ where q is the number of colors. Let $C_i = \{v \in V \mid c(v) = i\}$ for each $i \in \{1, \dots, q\}$. A path P of k vertices from a vertex v_1 to a vertex v_k is represented by a sequence $P = \langle v_1, v_2, \dots, v_k \rangle$ of distinct k vertices such that $\{v_i, v_{i+1}\}$ is an edge for all $i \in \{1, \dots k-1\}$, and we also simply write $P = v_1, v_2, \dots, v_k$. Throughout this paper, all paths are assumed to be simple paths, meaning that no vertex appears more than once in the path. The length of the path P is defined by the number of vertices in P.

2 Complexity for Colors and Locality

This section establishes complexity boundaries for Locally Rainbow Path based on parameters r and q. We prove the problem is NP-complete when $r \geq 4$ and $q \geq 5$, while showing it can be efficiently solved when (i) $r \leq 2$ and (ii) $r = q$.

2.1 NP-Completeness

In this subsection, we show the NP-completeness of Locally Rainbow Path for constant q and r.

Theorem 1. *Locally Rainbow Reachability is NP-complete if* $r = 4$ *and* $q = 5$.

Proof. We give a polynomial-time reduction from 3-SAT to Locally Rainbow Reachability. Let ϕ be a 3CNF formula given as an instance of 3-SAT. Let $U = \{u_1, u_2, \dots, u_n\}$ be the set of n variables and $W = \{w_1, w_2, \dots, w_m\}$ be the set of m clauses in ϕ. We construct a graph $G = (V, E)$ with two vertices s and t such that G has a 4-locally rainbow s-t path if and only if ϕ is satisfiable.

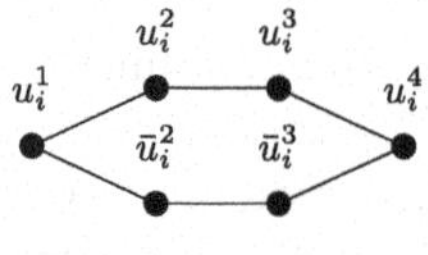

Fig. 2. Variable gadget G_{u_i}.

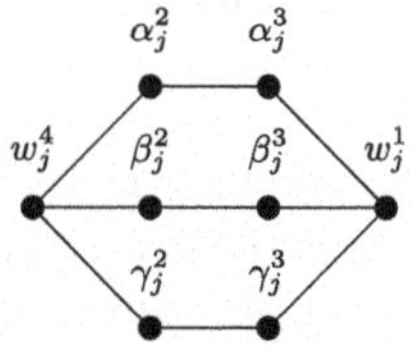

Fig. 3. Clause gadget G_{w_j}.

The graph G consists of two gadgets: (1) the variable gadget G_{u_i} associated with each variable u_i and (2) the clause gadget G_{w_j} associated with each clause w_j.

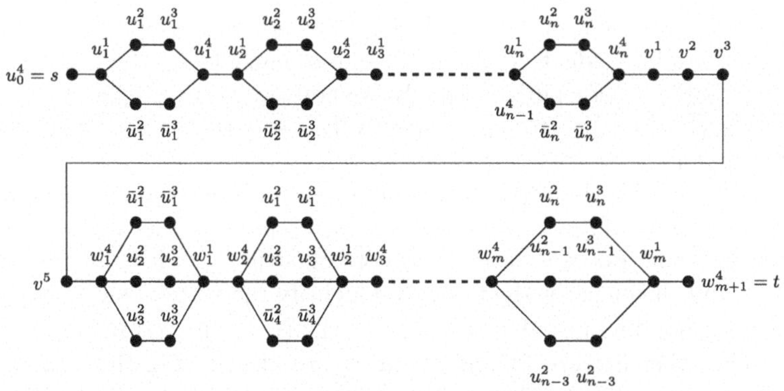

Fig. 4. The connection between variable gadgets, clause gadgets, s, t, and the connecting path Q_{uw}.

(1) Variable Gadget: Figure 2 shows the variable gadget G_{u_i} for a variable u_i. It consists of two parallel paths of length 4 between vertices u_i^1 and u_i^4 as follows (see Fig. 2):

$$V(G_{u_i}) = \{u_i^1, u_i^2, u_i^3, \bar{u}_i^2, \bar{u}_i^3, u_i^4\},$$
$$E(G_{u_i}) = \{\{u_i^1, u_i^2\}, \{u_i^1, \bar{u}_i^2\}, \{u_i^2, u_i^3\}, \{\bar{u}_i^2, \bar{u}_i^3\}, \{u_i^3, u_i^4\}, \{\bar{u}_i^3, u_i^4\}\}.$$

The edge $\{u_i^2, u_i^3\}$ corresponds to the positive literal, and the edge $\{\bar{u}_i^2, \bar{u}_i^3\}$ corresponds to the negative literal. We refer to these as the positive edge and the negative edge, respectively.

In this construction, we color a vertex based on its superscript; for example, a vertex v^k is assigned color k, i.e., $c(v^k) = k$.

(2) Clause Gadget: Figure 3 illustrates the clause gadget G_{w_j} for a clause w_j. The gadget consists of three parallel paths of length 4 between the vertices w_j^4 and w_j^1. The specific paths are $\langle w_j^4, \alpha_j^2, \alpha_j^3, w_j^1\rangle$, $\langle w_j^4, \beta_j^2, \beta_j^3, w_j^1\rangle$, and $\langle w_j^4, \gamma_j^2, \gamma_j^3, w_j^1\rangle$. The gadget is formally defined as follows:

$$V(G_{w_j}) = \{w_j^4, \alpha_j^2, \alpha_j^3, \beta_j^2, \beta_j^3, \gamma_j^2, \gamma_j^3, w_j^1\},$$
$$E(G_{w_j}) = \{\{w_j^4, \alpha_j^2\}, \{\alpha_j^2, \alpha_j^3\}, \{\alpha_j^3, w_j^1\},$$
$$\{w_j^4, \beta_j^2\}, \{\beta_j^2, \beta_j^3\}, \{\beta_j^3, w_j^1\},$$
$$\{w_j^4, \gamma_j^2\}, \{\gamma_j^2, \gamma_j^3\}, \{\gamma_j^3, w_j^1\}\}.$$

The three literals α_j, β_j, and γ_j in the clause w_j correspond to the placeholder vertex pairs (α_j^2, α_j^3), (β_j^2, β_j^3), and (γ_j^2, γ_j^3), respectively. Each of these placeholder pairs is later identified with the corresponding vertex pair in the appropriate variable gadget.

Next we link each gadget as shown in Fig. 4. First, for each $i \in \{1, \ldots, n-1\}$, we connect G_{u_i} and $G_{u_{i+1}}$ by an edge $\{u_i^4, u_{i+1}^1\}$. Similarly, for each $j \in$

$\{1,\ldots,m-1\}$, we connect G_{w_j} and $G_{W_{j+1}}$ by an edge $\{w_j^1, w_{j+1}^4\}$. We next connect the variable gadget G_{u_n} and the clause gadget G_{W_1} by a path $Q_{uw} = \langle u_n^4, v^1, v^2, v^3, v^5, w_1^4 \rangle$ of length 6. Finally, we add two vertices u_0^4 and w_{m+1}^4, and attach u_0^4 to u_1^1 and w_{m+1}^4 to w_m^1, respectively. Let s be the vertex u_0^4 and t be the vertex w_{m+1}^4.

We finally identify the vertices corresponding to literals in the variable gadgets and the clause gadgets as follows. For each clause w_j, let α_j be a positive (resp., negative) literal of a variable u_i contained in w_j. Then, in the variable gadget G_{u_i} and the clause gadget G_{w_j}, we identify the vertex pair α_j^2, α_j^3 in G_{w_j} with the corresponding literal vertex pair: u_i^2, u_i^3 (resp., $\bar{u}_i^2, \bar{u}_i^3$) in G_{u_i}, as shown in Fig. 5. The other literals, β_j and γ_j, in w_j are identified in the same manner with their corresponding vertices in the appropriate variable gadgets. This completes the construction of the graph G. It is easy to see how the construction can be accomplished in polynomial time.

In the following, we show that G has a 4-locally rainbow s-t path if and only if ϕ is satisfiable. First, suppose that ϕ is satisfiable, and let there be a satisfying truth assignment. We construct a 4-locally rainbow s-t path P as follows. For each variable u_i, we select a subpath through the gadget G_{u_i} based on its truth value:

- If u_i is true, the path P traverses $\langle u_i^1, \bar{u}_i^2, \bar{u}_i^3, u_i^4 \rangle$. This leaves the vertices corresponding to the positive literal, $\{u_i^2, u_i^3\}$, available for later use.
- If u_i is false, the path P traverses $\langle u_i^1, u_i^2, u_i^3, u_i^4 \rangle$. This leaves the vertices corresponding to the negative literal, $\{\bar{u}_i^2, \bar{u}_i^3\}$, available.

Concatenating these subpaths and the connecting path Q_{uw} creates a continuous 4-locally rainbow path from the start vertex s to the entrance of the clause gadgets, w_1^4.

This path is then extended through the sequence of clause gadgets. Since the given truth assignment is satisfying ϕ, each clause w_j contains at least one true literal. Our path-selection rule for the variable gadgets ensures that the vertices corresponding to any true literal are left available. Therefore, for each clause gadget G_{w_j}, there is at least one available path between w_j^4 and w_j^1 for P to traverse. By selecting one such path for each clause from $j = 1, \ldots, m$, we can extend P sequentially to the terminal vertex t. The resulting path P from s to t is continuous and, by the nature of the construction, is 4-locally rainbow. Thus, if ϕ is satisfiable, such a path exists.

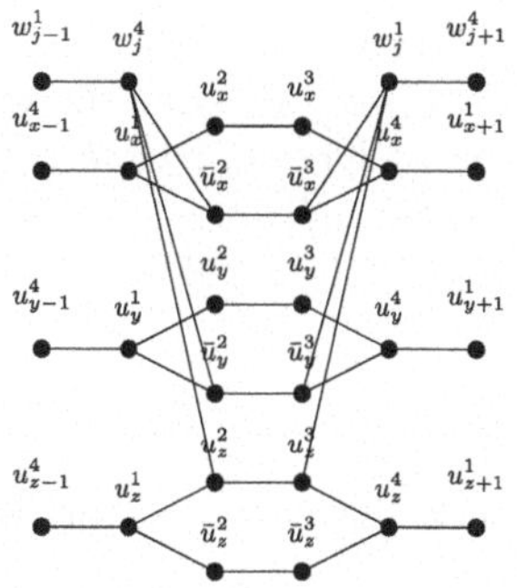

Fig. 5. An example of vertex identification for variable gadgets G_{u_x}, G_{u_y}, G_{u_z}, and Clause gadget G_{w_j} where the clause w_j is $w_j = (\bar{u}_x, \bar{u}_y, u_z)$. The vertices $\bar{u}_x^2, \bar{u}_x^3$ are identified with α_j^2, α_j^3 in G_{w_j}; vertices $\bar{u}_y^2, \bar{u}_y^3$ are identified with β_j^2, β_j^3 in G_{w_j}; and vertices u_z^2, u_z^3 are identified with γ_j^2, γ_j^3 in G_{w_j}.

Conversely, suppose that the graph G has a 4-locally rainbow s-t path P. By construction, to maintain the 4-local rainbow property, the path P is constrained to follow specific routes within each gadget. In any variable gadget G_{u_i}, P must traverse one of two primary paths: either $\langle u_i^1, \bar{u}_i^2, \bar{u}_i^3, u_i^4 \rangle$ or $\langle u_i^1, u_i^2, u_i^3, u_i^4 \rangle$. Similarly, in any clause gadget G_{w_j}, P must traverse one of its three paths corresponding to its literals.

Furthermore, any path that deviates from this sequential traversal violates the coloring constraints. For example, a "shortcut" path like $\langle u_{i-1}^4, u_i^1, u_i^2, w_j^4 \rangle$ is not 4-locally rainbow because its vertex colors would be $\langle 4, 1, 2, 4 \rangle$. Similarly, a shortcut path $\langle u_i^1, u_i^2, u_i^3, w_j^1 \rangle$ is not 4-locally rainbow because its vertex colors would be $\langle 1, 2, 3, 1 \rangle$. This implies that the s-t path P must sequentially traverse all variable gadgets, the connecting path Q_{uw}, and then all clause gadgets to reach t.

This rigid structure allows us to define a truth assignment for ϕ from the path P. For each variable u_i, we define its truth value as follows:

- We assign $u_i = \mathsf{true}$ if P traverses the path $\langle u_i^1, \bar{u}_i^2, \bar{u}_i^3, u_i^4 \rangle$.
- We assign $u_i = \mathsf{false}$ if P traverses the path $\langle u_i^1, u_i^2, u_i^3, u_i^4 \rangle$.

For this assignment to be valid, we must show it satisfies ϕ. The existence of the path P means it must successfully pass through every clause gadget G_{w_j}. To do so, P must use one of the three internal paths, corresponding to some literal ℓ. This is only possible if the vertices for ℓ were left unused by the path segment in its variable gadget. By our assignment rule, this implies that the literal ℓ is true. Since P traverses all clause gadgets, every clause must contain at least one true literal under this assignment. Therefore, the formula ϕ is satisfiable. □

Theorem 2. *Locally Rainbow Reachability is NP-complete if* $r \geq 4$ *and* $q = r + 1$.

Theorem 2 can be easily shown by making minor changes to each variable gadget and each clause gadget. First, we define the vertex s as u_0^r and the vertex t as u_{m+1}^r. Also, in each variable gadget G_{u_i}, we replace the positive edge $\{u_i^2, u_i^3\}$ with a positive path of length $r-1$: $\langle u_i^2, u_i^3, \dots, u_i^r \rangle$.Similarly, we replace the edge $\{\bar{u}_i^2, \bar{u}_i^3\}$ with a negative path $\langle \bar{u}_i^2, \bar{u}_i^3, \dots, \bar{u}_i^r \rangle$.

2.2 Algorithms

In this subsection, we present a linear-time algorithm for Locally Rainbow Path if (i) $r \leq 2$ and (ii) $r = q$.

Theorem 3. *Locally Rainbow Path can be solved in linear time if* $r \leq 2$.

Proof. When $r = 1$, Locally Rainbow Path is equivalent to s-t path that asks whether there exists an s-t path in G. Thus, it can be solved in linear time.

Similarly, when $r = 2$, Locally Rainbow Path is finding an s-t path P such that any two adjacent vertices have distinct color in P. Thus, it is sufficient to find an s-t path in the graph obtained from G by deleting all the edges whose endpoints have the same color. Thus, we can solve it in linear time. □

In Theorem 1, we showed that Locally Rainbow Path is NP-complete even if $q \geq r + 1$. On the other hand, if $q < r$, Locally Rainbow Path is trivially solvable because it is equivalent to s-t path. The remaining case is $q = r$. We show that this case is tractable.

Theorem 4. *There exists an* $r^{O(r)}(n+m)$*-time algorithm for Locally Rainbow Path when* $q = r$.

Proof. We first guess the order of the appearance of colors in a locally rainbow s-t path. The number of orderings is at most $r^{O(r)}$. Fix such an order $c_1, \dots, c_r$ where $c(s) = c_1$. Then we construct a (uncolored) directed graph $G' = (V, A)$ with the same vertex set V as G such that an arc $(u, v) \in A$ exists if and only if there exists an edge $\{u, v\} \in E$ so that $c(u) = c_i$ and $c(v) = c_{(i+1) \bmod r}$ for some i. Note that c_i and c_{i+1} are distinct colors. Then it is easily seen that there exists an s-t path such that the colors appear in the order $c_1, \dots, c_r$ in G if and only if there exists an s-t path in G'. Therefore, by applying BFS on G' for each order of the appearance of colors, we can solve Locally Rainbow Path in $r^{O(r)}(n+m)$ time. □

Theorem 5 shows that Locally Rainbow Path is fixed-parameter tractable when parameterized by r if $r = q$. In the next section, we show that Locally Rainbow Path is NP-complete when r and q are given as inputs even if $r = q$.

3 Hardness and Algorithms Using Structural Parameters

3.1 Hardness on Cacti

In this subsection, we show that Locally Rainbow Reachability is NP-complete on cacti of maximum degree 3. Since the treewidth of a cactus is at most 2, Locally Rainbow Reachability is paraNP-complete for treewidth. Note that Locally Rainbow Reachability and Locally Rainbow Path are linear-time solvable on trees because the s-t path on a tree is unique. Thus, this result provides a sharp contrast between trees and cacti, highlighting that even a slight generalization in the graph structure leads to a significant difficulty of Locally Rainbow Reachability.

Theorem 5. *Locally Rainbow Reachability is NP-complete on cacti of maximum degree 3 and pathwidth 2, even if* $r = q$.

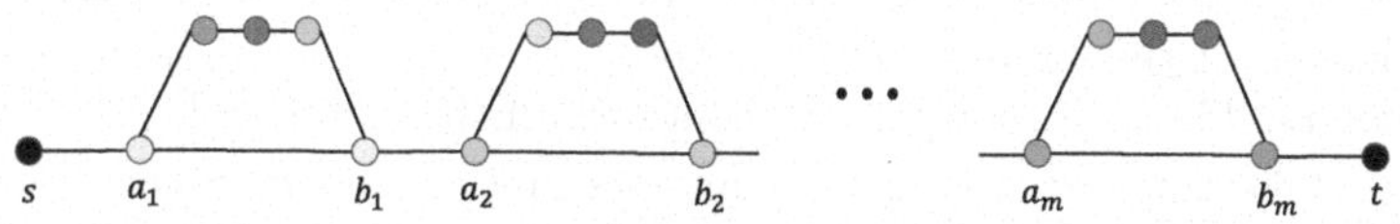

Fig. 6. The reduction from Exact Cover by 3-Sets to Locally Rainbow Path.

Proof. We give a reduction from Exact Cover by 3-Sets defined as follows: Given a set of elements $E = \{e_1, \ldots, e_{3n}\}$ and a family of subsets of elements $\mathcal{S} = \{S_1, \ldots, S_m\}$ where $|S_j| = 3$ for any j, the task is to determine whether there exists a subfamily $\mathcal{S}'$ in $\mathcal{S}$ of size n such that $\bigcup_{S \in \mathcal{S}'} S = E$, called an exact cover for E. This problem is known to be NP-complete [16].

We first create a path P of length $2m+2$. We designate the two endpoints of P as s and t. Moreover, we label internal vertices of P by $a_1, b_1, \ldots, a_m, b_m$. We color s and t in black and the vertices of P in $2m$ different colors, each per color. We abuse a_j, b_j as the color of a_j, b_j, respectively. For $j \in \{1, \ldots, m\}$, we prepare a path P_j of length 5 and identify its endpoints as a_j and b_j, respectively. The three elements in $V(P_j) \setminus \{a_j, b_j\}$ correspond to elements in the set $S_j \in \mathcal{S}$. Next, we associate each element e_i with a color. We also abuse e_i as a color. Then, if S_j contains an element e_i, we color one of the three vertices of P_j in color e_i. Let G be the constructed graph. Through this construction, every vertex is colored, and the number q of colors is $3n+2m+1$. Furthermore, the constructed graph G is a cactus of maximum degree 3. Indeed, a_j, b_j are cut vertices and of degree 3 for each j. Finally, the pathwidth is at most 2 by making bags as follows: $\{s, a_1\}$, $\{b_m, t\}$, and $\{a_j, e_{j_1}, b_j\}$, $\{e_{j_1}, e_{j_2}, b_i\}$, $\{e_{j_2}, e_{j_3}, b_j\}$ for each $j \in \{1, \ldots, m\}$ where e_{j_k} for $k \in \{1, 2, 3\}$ is an element contained in S_j. Fig. 6 shows the constructed graph.

In the following, we show that there exists an exact cover $\mathcal{S}' \subseteq \mathcal{S}$ for E if and only if there exists a $(3n+2m+1)$-locally rainbow s-t path in G. Note that $r = q = 3n + 2m + 1$.

Given an exact cover $\mathcal{S}' \subseteq \mathcal{S}$, we construct an s-t path P in G passing through a_j, b_j for all $j \in [m]$ and P_j's for $S_j \in \mathcal{S}'$. The length of P is $3n + 2m + 2$. Since $\mathcal{S}'$ is an exact cover for E, the vertices except for s, t have different colors from each other. Moreover, s and t have color black and distance between s and t in P is $3n + 2m + 2$, P is an $(3n + 2m + 1)$-locally rainbow s-t path.

Conversely, let P be a $(3n + 2m + 1)$-locally rainbow s-t path in G. Since s and t have the same color, the length of P must be at least $3n + 2m + 2$.

Then we claim that the length of P is exactly $3n + 2m + 2$.

Claim. The length of P is exactly $3n + 2m + 2$.

Proof of the Claim. It is sufficient to show that the length of P is at most $3n+2m+2$. Towards a contradiction, suppose that the length of P is more than $3n+2m+2$. In this case, P contains at least $n+1$ P_j's. Let $P_{j_1}, \dots, P_{j_{n+1}}$ be the first $n + 1$ subpaths traversed by P from s. Then consider the subpath P' of P from a_{j_1} to b_{j_n} containing $P_{j_1}, \dots, P_{j_n}$. Since P is $(3n+2m+1)$-locally rainbow and the length of the subpath P' is at most $3n + 2m$, the vertices in P' have distinct colors. Thus, the subpath P' contains all the colors of $\{e_1, \dots, e_{3n}\}$.

Let $v_1^{j_{n+1}}$ be the vertex that first appears in $P_{j_{n+1}}$ on P. Since P' contains all the colors of $\{e_1, \dots, e_{3n}\}$, $v_1^{j_{n+1}}$ has the same color as some vertex u in $\bigcup_{k=1}^{n} P_{j_k}$. However, the length of the u-$v_1^{j_{n+1}}$ subpath of P is at most $3n + 2m$. This contradicts that P is $(3n + 2m + 1)$-locally rainbow. ■

By this claim, P contains exactly n P_j's, and each internal vertex has a distinct color. Therefore, $\{S_j \mid P_j \subseteq P\}$ is an exact cover for E. □

3.2 Treewidth Plus the Number of Colors

In this subsection, we show that Locally Rainbow Path is fixed-parameter tractable when parameterized by $tw + q$. We express the property of an r-locally rainbow s-t path as monadic second-order logic (MSO_2).

Theorem 6. *Locally Rainbow Path is fixed-parameter tractable when parameterized by $tw + q$.*

Proof. We give an MSO_2 formula $\mathsf{LRP}(E')$ expressing the property of an r-locally rainbow s-t path.

We define an MSO formula $\mathsf{LR}(E')$ that verifies E' is r-locally rainbow, which implies that there is no subpath of length at most r containing two vertices with the same color in the subgraph induced by E'.

$$\mathsf{LR}(E') := \neg \bigvee_{r'=2}^{r} \Big[\exists e_1, \dots, e_{r'-1} \in E'. \exists v_1, \dots, v_{r'} \in V.$$

$$\bigwedge_{i=1}^{r'-1} \mathsf{inc}(v_i, e_i) \wedge \mathsf{inc}(v_{i+1}, e_i) \wedge \bigvee_{i \neq j \in [r']} \bigvee_{k=1}^{q} (v_i \in C_k \wedge v_j \in C_k)\Big].$$

It is well-known that for a subset $E' \subseteq E$, verifying whether E' is an s-t path in G can be formulated as an MSO formula of constant length [2,10,21], denoted by $\mathsf{stpath}(E')$. Consequently, $\mathsf{LRP}(E')$ that verifies E' is an r-locally rainbow s-t path in G as follows.

$$\mathsf{LRP}(E') := \mathsf{stpath}(E') \wedge \mathsf{LR}(E').$$

It is easily seen that the length of $\mathsf{LRP}(E')$ is bounded by a polynomial in $r+q$. Here, given a graph G and an MSO_2 formula $\varphi(X)$ with a free set variable X, the problem finds a minimum set S such that G models $\varphi(S)$ is known to be fixed-parameter tractable when parameterized by $tw+||\varphi||$ [2,6,11,21]. Since $r \leq q$, Locally Rainbow Path is fixed-parameter tractable when parameterized $tw+q$. □

3.3 Treedepth

From Theorem 6, we also show that Locally Rainbow Path is in XP when parameterized by treedepth. To show this, we use the following proposition.

Proposition 1. *([22]) The length of a longest path of a graph with treedepth td is at most 2^{td}.*

Due to Proposition 1, any s-t path in G with q colors contains at most 2^{td} colors. Then, we guess the color set of cardinality at most 2^{td} contained in a r-locally rainbow s-t path if it exists. By applying Theorem 6 to the graph obtained from G by deleting vertices not contained in the guessed color set, we can determine whether G has a r-locally rainbow s-t path consisting of vertices having colors in the guessed color set. Since the cardinality of the colored set is at most 2^{td} and the number of possible colored sets is at most $q^{2^{td}} = n^{2^{td}}$, Locally Rainbow Path is in XP when parameterized by treedepth.

Theorem 7. *Locally Rainbow Path is in XP when parameterized by treedepth.*

3.4 Vertex Cover Number

Theorem 8. *Locally Rainbow Path is FPT when parameterized by vertex cover number.*

Proof. Let S be a vertex cover of size vc. We first partition the vertices in $V \setminus S$ into the set T_i of twin vertices, called type i, such that $N(u) = N(v)$ holds for a pair of vertices u, v in T_i. We note that the number of sets of twin vertices is at most 2^{vc}.

We then observe that any s-t path contains at most vc vertices in S and at most $vc+1$ vertices in $V \setminus S$ because $V \setminus S$ is an independent set. Thus, the length of any s-t path is at most $2vc+1$. From the observation, if the number of vertices with the same color in T_i is more than $vc+1$, we can safely delete them by $vc+1$ vertices. Moreover, if T_i contains at least $2vc+1$ colors, we retain only $2vc+1$ colors and delete vertices with other colors. This reduction is safe. Indeed, suppose that G has an r-rainbow s-t path P containing a deleted vertex v. Then in the resulting graph, there exists a vertex v that is in the same type of u and has a different color from any vertex in P. Thus, the resulting graph has an alternative r-rainbow s-t path P' containing v instead of u. Consequently, it contains at most $2^{vc}(vc+1)(2vc+1)+vc = O(2^{vc}vc^2)$ vertices, which implies that LOCALLY RAINBOW PATH has an exponential size kernel parameterized by vertex cover number. Thus, LOCALLY RAINBOW PATH is fixed-parameter tractable for vertex cover number. □

4 Conclusion

In this paper, we studied the complexity of LOCALLY RAINBOW PATH on undirected graphs. We established that LOCALLY RAINBOW REACHABILITY is computationally intractable when $r \geq 4$ and $q \geq 5$, while remaining efficiently solvable for $r \leq 2$ or when $r = 3$ and $q \leq 3$. For $r = q$, we demonstrated an $r^{O(r)}(n+m)$-time algorithm, creating an interesting contrast with the hardness results when $q \geq r+1$.

From a structural parameterization perspective, LOCALLY RAINBOW REACHABILITY is paraNP-complete when parameterized by pathwidth even on cacti of maximum degree 3. This highlights a significant complexity gap between trees, which allow linear-time solutions due to path uniqueness, and cacti. To complement this hardness result, we proved that LOCALLY RAINBOW PATH is fixed-parameter tractable when parameterized by treewidth plus number of colors, and also with respect to vertex cover number. Furthermore, we prove that LOCALLY RAINBOW PATH is in XP when parameterized by treedepth.

Finally, we observe that the notion of r-locally rainbowness admits a natural generalization to arbitrary subgraphs. Formally, a vertex-colored graph $G = (V, E)$ is said to be r-locally rainbow if every path of length at most r in G consists of vertices with pairwise distinct colors. This generalized framework gives rise to a variety of problems subject to the r-locally rainbow constraint. In this context, Theorem 5 yields the following corollary.

Corollary 1. *Let Π be a graph property expressible by an MSO_2 formula ϕ. Given a vertex-colored graph $G = (V, E)$ with q colors, finding a minimum edge subgraph satisfying Π is fixed-parameter tractable when parameterized by $tw + q + |\phi|$.*

For future work, determining the complexity of LRP when $r = 3$ and $q \geq 4$, which remains open, along with exploring locally rainbow variants of other fundamental graph problems, would be valuable directions. Moreover, the parameterized complexity of LOCALLY RAINBOW PATH for other graph parameters such as modular-width, cluster deletion number, and feedback vertex set number, would be interesting.

References

1. Alon, N., Yuster, R., Zwick, U.: Color-coding. J. ACM **42**(4), 844–856 (1995). https://doi.org/10.1145/210332.210337
2. Arnborg, S., Lagergren, J., Seese, D.: Easy problems for tree-decomposable graphs. J. Algorithms **12**(2), 308–340 (1991). https://doi.org/10.1016/0196-6774(91)90006-K
3. Bendele, O., Rautenbach, D.: Optimal colorings with rainbow paths. Graphs Comb. **33**(4), 729–734 (2017). https://doi.org/10.1007/S00373-017-1808-1
4. Bentert, M., Kellerhals, L., Niedermeier, R.: The structural complexity landscape of finding balance-fair shortest paths. Theor. Comput. Sci. **933**, 149–162 (2022). https://doi.org/10.1016/J.TCS.2022.08.023
5. Bentert, M., Kellerhals, L., Niedermeier, R.: Fair short paths in vertex-colored graphs. In: Williams, B., Chen, Y., Neville, J. (eds.) Thirty-Seventh AAAI Conference on Artificial Intelligence, AAAI 2023, Thirty-Fifth Conference on Innovative Applications of Artificial Intelligence, IAAI 2023, Thirteenth Symposium on Educational Advances in Artificial Intelligence, EAAI 2023, Washington, DC, USA, February 7-14, 2023, pp. 12346–12354. AAAI Press (2023). https://doi.org/10.1609/AAAI.V37I10.26455
6. Borie, R.B., Parker, R.G., Tovey, C.A.: Automatic generation of linear-time algorithms from predicate calculus descriptions of problems on recursively constructed graph families. Algorithmica **7**(5&6), 555–581 (1992). https://doi.org/10.1007/BF01758777
7. Chartrand, G., Johns, G.L., McKeon, K.A., Zhang, P.: The rainbow connectivity of a graph. Networks **54**(2), 75–81 (2009). https://doi.org/10.1002/NET.20296
8. Chen, L., Li, X., Shi, Y.: The complexity of determining the rainbow vertex-connection of a graph. Theor. Comput. Sci. **412**(35), 4531–4535 (2011). https://doi.org/10.1016/J.TCS.2011.04.032
9. Cohen, J., Italiano, G.F., Manoussakis, Y., Nguyen, K.T., Pham, H.P.: Tropical paths in vertex-colored graphs. J. Comb. Optim. **42**(3), 476–498 (2021). https://doi.org/10.1007/S10878-019-00416-Y
10. Courcelle, B.: The monadic second-order logic of graphs. i. recognizable sets of finite graphs. Inf. Comput. **85**(1), 12–75 (1990). https://doi.org/10.1016/0890-5401(90)90043-H, https://doi.org/10.1016/0890-5401(90)90043-H
11. Courcelle, B.: The monadic second-order logic of graphs. i. recognizable sets of finite graphs. Inf. Comput. **85**(1), 12–75 (1990). https://doi.org/10.1016/0890-5401(90)90043-H
12. Parameterized Algorithms. Springer, Cham (2015). https://doi.org/10.1007/978-3-319-21275-3
13. Eiben, E., Kanj, I.: A colored path problem and its applications. ACM Trans. Algorithms **16**(4), 47:1–47:48 (2020). https://doi.org/10.1145/3396573
14. Fluschnik, T., Kellerhals, L., Renken, M.: Locally rainbow paths. In: Wooldridge, M.J., Dy, J.G., Natarajan, S. (eds.) Thirty-Eighth AAAI Conference on Artificial Intelligence, AAAI 2024, Thirty-Sixth Conference on Innovative Applications of Artificial Intelligence, IAAI 2024, Fourteenth Symposium on Educational Advances in Artificial Intelligence, EAAI 2014, February 20-27, 2024, Vancouver, Canada, pp. 8013–8021. AAAI Press (2024). https://doi.org/10.1609/AAAI.V38I8.28639
15. Fomin, F.V., Golovach, P.A., Korhonen, T., Simonov, K., Stamoulis, G.: Fixed-parameter tractability of maximum colored path and beyond. ACM Trans. Algorithms **20**(4), 36:1–36:48 (2024). https://doi.org/10.1145/3674835

16. Garey, M.R., Johnson, D.S.: Computers and Intractability: A Guide to the Theory of NP-Completeness. Freeman, W. H (1979)
17. Kowalik, L., Lauri, J.: On finding rainbow and colorful paths. Theor. Comput. Sci. **628**, 110–114 (2016). https://doi.org/10.1016/J.TCS.2016.03.017
18. Krivelevich, M., Yuster, R.: The rainbow connection of a graph is (at most) reciprocal to its minimum degree. J. Graph Theory **63**(3), 185–191 (2010). https://doi.org/10.1002/JGT.20418
19. Lauri, J.: Further hardness results on rainbow and strong rainbow connectivity. Discret. Appl. Math. **201**, 191–200 (2016). https://doi.org/10.1016/J.DAM.2015.07.041
20. Li, X., Shi, Y., Sun, Y.: Rainbow connections of graphs: a survey. Graphs Comb. **29**(1), 1–38 (2013). https://doi.org/10.1007/S00373-012-1243-2
21. Libkin, L.: Elements of finite model theory. texts in theoretical computer science. An EATCS Series, Springer (2004). https://doi.org/10.1007/978-3-662-07003-1, http://www.cs.toronto.edu/%7Elibkin/fmt
22. Sparsity. AC, vol. 28. Springer, Heidelberg (2012). https://doi.org/10.1007/978-3-642-27875-4
23. Sorge1, M., Weller, M.: The graph parameter hierarchy (2019). https://manyu.pro/assets/parameter-hierarchy.pdf
24. Uchizawa, K., Aoki, T., Ito, T., Suzuki, A., Zhou, X.: On the rainbow connectivity of graphs: complexity and FPT algorithms. Algorithmica **67**(2), 161–179 (2013). https://doi.org/10.1007/S00453-012-9689-4

Parameterized Algorithms for Editing to Uniform Cluster Graph

Ajinkya Gaikwad(✉), Hitendra Kumar, and Soumen Maity

Indian Institute of Science Education and Research, Pune, Pune, India
{ajinkya.gaikwad,sahu.hitendrakumar}@students.iiserpune.ac.in,
soumen@iiserpune.ac.in

Abstract. Graph modification problems, which involve transforming graphs through vertex or edge operations, are pivotal in theoretical computer science and parameterized complexity. Given a graph $G = (V, E)$ and an integer $k \in \mathbb{N}$, we study Uniform Cluster Vertex Deletion (resp. Uniform Cluster Edge Deletion), where the goal is to remove at most k vertices (resp. edges) such that the connected components of the resulting graph are equal-sized cliques. Graphs satisfying this property are referred to as *uniform cluster graphs*. We present a kernelization result with a vertex kernel of size $\mathcal{O}(k^3)$ for Uniform Cluster Vertex Deletion (UCVD) and an FPT algorithm running in $\mathcal{O}^*(2^k)$ time, improving upon the best-known results in the literature. We also provide a linear vertex kernel for Uniform Cluster Edge Deletion (UCED) of size $6k$. Through this work, we resolve several open questions posed by Misra, Mittal, Saurabh & Thakkar [ISAAC 2023] regarding the parameterized complexity of these problems, thus offering a comprehensive view of the landscape surrounding uniform cluster graphs.

1 Introduction

In clustering, many real-world challenges involve grouping objects or data points into well-defined clusters. By representing these challenges as graphs, where nodes represent objects and edges represent relationships, we can utilize graph modification techniques to enforce structural properties, such as equal-sized clusters or clusters containing highly similar elements. The ability to edit graphs by deleting or adding edges, or by removing vertices facilitates more accurate or computationally feasible clustering, particularly when the initial data does not naturally conform to ideal clusters.

A classic problem in this domain is the Cluster Vertex Deletion problem, where given a graph $G = (V, E)$ and an integer $k \in \mathbb{N}$, the goal is to remove at most k vertices so that each connected component of the resulting graph is a clique, called the *cluster graph*.

The parameterized version of Cluster Vertex Deletion problem has been the subject of extensive study, leading to the development of the best-known FPT algorithm with a running time of $\mathcal{O}^*(1.811^k)$[1] [16], and a kernel of

[1] The $\mathcal{O}^*(f(k))$ notation hides polynomial factors depending only on the input size.

A. Jeż and J. Otop (Eds.): FCT 2025, LNCS 16106, pp. 165–179, 2026.
https://doi.org/10.1007/978-3-032-04700-7_13

size $\mathcal{O}(k^2)$ [1]. Earlier contributions to this line of work have explored various facets of the problem, including complexity analysis and algorithmic strategies [3,4,12]. Another notable problem in clustering is the CLUSTER EDGE DELETION problem. In this problem, the goal is to decide whether a graph can be transformed into a cluster graph by deleting at most k edges. The parameterized version of CLUSTER EDGE DELETION has been intensively studied in the literature [2,5–7,10,11,17]. The problem is solvable in time $\mathcal{O}^*(1.404^k)$ [17], and it admits a kernel with at most $2k$ vertices [5,6]. In this paper, we study another variant of the clustering problem. We focus on computational problems where the objective is to modify a given graph either by vertex deletions or edge deletions to transform it into a collection of equal-sized cliques, which we refer to as *uniform cluster graphs*. We now formally define the problems considered in this paper:

UNIFORM CLUSTER VERTEX DELETION (UCVD)
Input: An undirected graph $G = (V, E)$ and a positive integer k.
Question: Is there a subset $S \subseteq V$ with $|S| \leq k$ such that $G - S$ is a disjoint union of equal-sized cliques?

UNIFORM CLUSTER EDGE DELETION (UCED)
Input: An undirected graph $G = (V, E)$ and a positive integer k.
Question: Is there a subset $F \subseteq E$ with $|F| \leq k$ such that $G - F = (V, E \setminus F)$ is a disjoint union of equal-sized cliques?

1.1 Known Results

Recently, Misra, Mittal, Saurabh & Thakkar [14] examined these problems in a different context, investigating the complexity of reducing the number of distinct eigenvalues of an input graph by removing vertices or editing edges. A special case where the number of eigenvalues is reduced to two is equivalent to transforming the graph into a uniform cluster graph, as shown in previous works [8,9]. It has been proved that the adjacency matrix of a graph has at most two distinct eigenvalues if and only if the graph is a disjoint union of equal-sized cliques. Misra et al. [14] provided several results for this special case, which we discuss later.

Furthermore, Madathil and Meeks [13] recently studied a generalization of this problem termed as BALANCED CLUSTER EDGE MODIFICATION. In this problem, given a graph G, and two integers $0 \leq \eta \leq n$ and k, the goal is to perform at most k edge modifications so that the resulting graph is a cluster graph in which the size difference between any two connected components does not exceed η. Notably, when $\eta = 0$, this problem reduces to the UNIFORM CLUSTER EDGE MODIFICATION problem. In their work, the authors presented polynomial kernels for the BALANCED CLUSTER COMPLETION, BALANCED CLUSTER DELETION, and BALANCED CLUSTER EDITING. While they provided polynomial kernels for the general cases, we have obtained improved bounds for the special case when $\eta = 0$. Our results were obtained simultaneously and independently with those of Madathil and Meeks.

Before the work of Madathil and Meeks [13], a closely related variant of BALANCED CLUSTER EDGE MODIFICATION, termed DIFFERENCE-δ BALANCED CLUSTERING was studied in the master thesis of Steinvik [15]. In this variant, δ plays the same role as η in BALANCED CLUSTER EDGE MODIFICATION. In addition to edge modification, vertex deletion was also considered in the thesis. Steinvik also introduced a generalization of the UNIFORM CLUSTER GRAPH MODIFICATION problem, called FACTOR-α BALANCED CLUSTERING. In this problem, given a graph G and integers α and k, the goal is to perform at most k edit operations to obtain a cluster graph in which, for every pair of components C_i and C_j, the inequality $\alpha \cdot |V(C_i)| \geq |V(C_j)|$ holds. When $\alpha = 1$, this problem coincides with UNIFORM CLUSTER GRAPH MODIFICATION. In his thesis, Steinvik provided a vertex kernel of size $\mathcal{O}(k^4)$ for both DIFFERENCE-δ BALANCED CLUSTER VERTEX DELETION and FACTOR-α BALANCED CLUSTER VERTEX DELETION. We obtain improved kernel for the special case when $\delta = 0$ or $\alpha = 1$.

The classical complexity of UCVD and UCED has been studied in [14,15], where both problems were shown to be NP-complete. In addition to classical complexity, numerous results in parameterized complexity, including FPT algorithms and kernelization algorithms for the UCVD and UCED problems, are available in the literature [13–15]. The best-known FPT algorithms [14,15] for UCVD run in time $\mathcal{O}^*(3^k)$. FPT algorithms for UCED can be found in [13–15] Moreover, the classical and parameterized complexity of UNIFORM CLUSTER EDGE ADDITION and UNIFORM CLUSTER EDGE EDITING have also been studied [13–15].

1.2 Our Results

Our main contributions include kernelization and an FPT algorithm for UCVD. By using simple yet powerful structural observations along with non-trivial ideas, we develop a combinatorial algorithm for the problem. We prove that UCVD admits a kernel with $\mathcal{O}(k^3)$ vertices, improving upon the $\mathcal{O}(k^4)$-vertex kernel presented in [15]. We note that our kernel for UCVD uses arguments quite similar to those in [15]. Additionally, we present an FPT algorithm running in time $\mathcal{O}^*(2^k)$, which improves upon the previously known $\mathcal{O}^*(3^k)$ algorithms given in [14,15]. Furthermore, we present a linear vertex kernel of size $6k$ for UCED, improving on the quadratic kernel previously given in [15].

1.3 Preliminaries

Throughout this article, $G = (V, E)$ denotes a finite, simple and undirected graph with $|V(G)| = n$ and $|E(G)| = m$. For $u \in V$, we define $N(u) = \{v \in V : (u, v) \in E\}$ and $N[u] = N(u) \cup \{u\}$. The *degree* of $u \in V$ is $|N(u)|$ and denoted by $d_G(u)$. A *clique* C in an undirected graph $G = (V, E)$ is a subset of the vertices $C \subseteq V$ such that every two distinct vertices are adjacent. The *size* of a clique refers to the number of vertices it contains, not the number of edges.

A *cluster graph* is a graph where every component is a clique. Observe that a graph is a cluster graph if and only if it does not have an induced P_3, that is, an induced path on three vertices. Let $U \subseteq V$ be a subset of vertices of G and $F \subseteq \binom{V}{2}$ be a subset of pairs of vertices of G. The subgraph induced by $U \subseteq V$ is denoted by $G[U]$. We define $G - U = G[V \setminus U]$ and $G - F = (V, E \setminus F)$. If $U = \{u\}$ or $F = \{e\}$ then we simply write $G - u$ and $G - e$ for $G - U$ and $G - F$, respectively.

2 Kernelization and FPT Algorithm for UCVD Parameterized by Solution Size

In this section, we provide kernelization and fixed-parameter tractable (FPT) algorithm for the Uniform Cluster Vertex Deletion (UCVD) problem.

Theorem 1. *The* UCVD *problem parameterized by solution size admits a kernel of size $\mathcal{O}(k^3)$ vertices.*

Proof of Theorem 1: Note that a graph G is a cluster graph if and only if it does not have an induced path on three vertices. The first step of the kernelization process is to compute a maximal set $\mathcal{P}_3$ of vertex-disjoint induced P_3s in G. If $|\mathcal{P}_3| > k$, we have a no-instance. So we assume that $|\mathcal{P}_3| \leq k$, and let S be the vertices of $\mathcal{P}_3$. We have $|S| \leq 3k$. Let us denote the set of cliques of $G - S$ by $\mathcal{C}$. We begin with the following reduction rule that provides two sufficient conditions for a vertex $s \in S$ to be in a solution. Note that the two conditions are either inspired or taken from Reduction Rules given in [15].

Reduction Rule UCVD 1. *Let $s \in S$ be a vertex such that one of the following conditions holds:*

- *s has neighbors in at least $k + 2$ distinct cliques in $\mathcal{C}$, or*
- *s has at least $k + 1$ neighbors in a clique $C \in \mathcal{C}$ and more than k neighbors in cliques other than C.*

Then, delete s from G and decrement the parameter k by 1. The new instance becomes $(G - s, k - 1)$.

This is correct because in both conditions, deleting any k vertices excluding s is insufficient to destroy all induced P_3s in the graph. Therefore, if (G, k) is a yes-instance, s must be included in the solution.

Next, we partition the set of cliques in $\mathcal{C}$ based on whether they have neighbours in S or not. We define: $\mathcal{C}_0 = \{C \in \mathcal{C} : \text{no vertex in } C \text{ has a neighbour in } S\}$ and $\mathcal{C}_1 = \{C \in \mathcal{C} : \text{some vertex in } C \text{ has a neighbour in } S\}$. An r-clique is a clique of size r. Now, we have the following reduction rule.

Reduction Rule UCVD 2. *If for some $r \in \{1, 2, \ldots, n\}$ there are more than $k + 1$ r-cliques in $\mathcal{C}_0$, then remove all but $k + 1$ r-cliques from G.*

Lemma 1. *Reduction Rule UCVD 2 is safe.*

Proof. Suppose (G, k) is a yes-instance and X is a uniform cluster vertex deletion set of G of size at most k. Since the removal of X can alter the sizes of at most k cliques, all cliques in $G - X$ have size r. Since the solution size is at most k, this allows us to remove all but $k + 1$ r-cliques from G. □

Reduction Rule UCVD 3. *If $\mathcal{C}_0$ contains cliques of more than $k+1$ distinct sizes then return a no instance.*

This is correct because by deleting at most k vertices we can alter the sizes of at most k cliques. Therefore, in this case, there cannot exist a uniform cluster vertex deletion set of size at most k.

Lemma 2. *After applying Reduction Rules 1, 2 and 3, the set $\mathcal{C} = \mathcal{C}_0 \cup \mathcal{C}_1$ contains at most $\mathcal{O}(k^2)$ cliques.*

Proof. Due to Reduction Rules 2 and 3, we know that $|\mathcal{C}_0| \leq (k+1)(k+1)$. After the exhaustive application of Reduction Rule UCVD 1, each vertex of S can have neighbours in at most $k+1$ cliques of $\mathcal{C}_1$. Since $|S| \leq 3k$ and each vertex of S can have neighbours in at most $k+1$ cliques of $\mathcal{C}_1$, there are at most $3k(k+1)$ cliques in $\mathcal{C}_1$. Therefore, we have $|\mathcal{C}_1| \leq 3k(k+1)$. Combining the two, we get $|\mathcal{C}| = |\mathcal{C}_0| + |\mathcal{C}_1| \leq (k+1)(k+1) + 3k(k+1) = 4k^2 + 5k + 1$. □

Now, we consider two cases based on the maximum size of a clique in $\mathcal{C}$. Let $\omega(\mathcal{C}) = \max_{C \in \mathcal{C}}\{|C|\}$.

Case 1: $\omega(\mathcal{C}) < 8k$. In this case, we obtain a vertex kernel of size: $|S| + \sum_{C \in \mathcal{C}} |C| < 3k + 8k(4k^2 + 5k + 1) = \mathcal{O}(k^3)$ and we are done.

Case 2: $\omega(\mathcal{C}) \geq 8k$. Let C^* be a clique in $\mathcal{C}$ such that $\omega(\mathcal{C}) = |C^*|$. Then, C^* becomes a clique of size at least $|C^*| - k \geq 7k$ in $G - X$, where X is a uniform cluster vertex deletion set of G of size at most k. In this case, the resulting graph $G - X$ must be a collection of r-cliques, where $r \geq |C_0| - k \geq 7k$.

Reduction Rule UCVD 4. *If there exist $C_1, C_2 \in \mathcal{C}$ such that $|C_1| - |C_2| > 4k$, then delete all the vertices in C_2 and decrement the parameter k by $|C_2|$. The new instance is $(G - C_2, k - |C_2|)$.*

Lemma 3. *Reduction Rule UCVD 4 is safe,*

Proof. For the sake of contradiction, let us assume that $v \in C_2$ does not belong to some solution X of size at most k. First, since C_2 is a clique, v has $|C_2| - 1$ neighbours in C_2. Second, v can have at most $3k$ neighbours in S. Thus, v can be part of a clique of size at most $(|C_2| - 1) + 3k + 1 = |C_2| + 3k$.

On the other hand, there always exists a vertex $u \in C_1$ such that u is not contained in X. Clearly, u will have at least $|C_1| - k - 1$ neighbours in $G - X$. Therefore, u must be contained in a clique of size at least $|C_1| - k > |C_2| + 4k - k = |C_2| + 3k$. Therefore, u and v can never be contained in equal-sized cliques. This contradicts the fact that $G - X$ is a collection of equal sized-cliques. Therefore, we must include all vertices of C_2 in every uniform cluster vertex deletion set X of G of size at most k. □

Observation 1. *After applying Reduction Rule UCVD 4,* $\min_{C \in \mathcal{C}} |C| \geq 4k$.

The above observation is true because we have assumed that $\omega(\mathcal{C}) \geq 8k$. We now introduce the concept of a *heavy neighbour* of a vertex $s \in S$ in $\mathcal{C}$, which is a key element for explaining the forthcoming reduction rules.

Definition 1. *A clique $C \in \mathcal{C}$ is called a heavy neighbour of $s \in S$ if s is adjacent to at least* $\max\{|C| - 4k, k + 1\}$ *vertices in C.*

The next two reduction rules show that each vertex s has a unique heavy neighbour in $\mathcal{C}$.

Reduction Rule UCVD 5. *If $s \in S$ has no heavy neighbour, remove s from G, and decrement the parameter k by 1. The new instance is* $(G - s, k - 1)$.

Lemma 4. *Reduction Rule UCVD 5 is safe.*

Proof. For the sake of contradiction, let us assume that $s \in S$ has no heavy neighbour and does not belong to some solution X of size at most k. It means s must belong to some clique in $G - X$. Since s has no heavy neighbour in $\mathcal{C}$, s is adjacent to at most $\max\{|C| - 4k, k+1\} - 1$ vertices in some $C \in \mathcal{C}$. Additionally, s can have at most $3k - 1$ neighbours in S. Now consider two cases:

1. If $\max\{|C| - 4k, k + 1\} - 1 = |C| - 4k - 1$, then s can be part of a clique of size at most $(|C| - 4k - 1) + (3k - 1) + 1 = |C| - k - 1$. This contradicts the fact that $G - X$ is a collection of r-cliques, where $r \geq |C| - k$.
2. If $\max\{|C| - 4k, k + 1\} - 1 = k$, then s can be part of a clique of size at most $k + (3k - 1) + 1 = 4k$. This contradicts the fact that $G - X$ is a collection of r-cliques, where $r \geq 7k$.

Therefore, s must be part of every solution of size at most k. □

Lemma 5. *After applying the Reduction Rules 1 and 5 exhaustively, each s has exactly one heavy neighbour.*

Proof. The condition 2 in the Reduction Rule UCVD 1 makes sure that s has at most one heavy neighbour. On the other hand, the Reduction Rule UCVD 5 makes sure that s has at least one heavy neighbour. Therefore, it shows that each s has exactly one neighbour. □

For each $C \in \mathcal{C}$, let $S_C \subseteq S$ denotes the set of vertices in S whose heavy neighbour is C. We define

$$\Gamma(C) := \bigcup_{s \in S_C} \Big(C \setminus N(s)\Big) \cup \bigcup_{s \in S \setminus S_C} \Big(N(s) \cap C\Big) \quad \text{and} \quad \Gamma(\mathcal{C}) = \bigcup_{C \in \mathcal{C}} \Gamma(C).$$

Next, we show that the vertices of $C \setminus \Gamma(C)$ are true twins.

Lemma 6. *For each $C \in \mathcal{C}$, every pair of adjacent vertices x and y in $C \setminus \Gamma(C)$ are true twins, that is,* $N[x] = N[y]$.

Proof. By the definition of $\Gamma(C)$, note that for any vertex $v \in C \setminus \Gamma(C)$, we have $N(v) \cap S = S_C$. Recall that for the vertices in S_C, C is their heavy neighbour. Since $v \in C$ and C is a clique, v is adjacent to all other vertices of C. Therefore, for every $v \in C \setminus \Gamma(C)$, we have $N[v] = C \cup S_C$. □

Reduction Rule UCVD 6. *If* $\min_{C \in \mathcal{C}} \{|C \setminus \Gamma(C)|\} > k+1$, *then arbitrarily delete exactly one vertex from each* $C \setminus \Gamma(C)$ *for every* $C \in \mathcal{C}$. *The new instance is* $(G - W, k)$, *where* W *is the set of vertices deleted from* G.

Lemma 7. *Reduction Rule UCVD 6 is safe.*

Proof. Let $I' = (G', k)$ be an instance of UCVD obtained from $I = (G, k)$ by exhaustively applying Reduction Rule UCVD 6. We will show that I is a yes-instance if and only if I' is a yes-instance.

Let $X \subseteq V(G)$ be a solution of size at most k for the instance I. We will construct a solution X' of the same size as X for the instance I'. We know $G - X$ is a collection of equal sized cliques, say $Q_1, Q_2, \ldots, Q_r$. We observe that there is no clique Q_i such that $Q_i \subseteq S$. This is true because otherwise $|Q_i| \leq 3k$ as $|S| \leq 3k$, and we have already observed that the size of cliques must be at least $7k$. We construct X' from X as follows: if $u \in X \cap (C \setminus \Gamma(C))$ for some $C \in \mathcal{C}$ and u is deleted during the execution of Reduction Rule UCVD 6, then remove u from X and include u's true twin $v \notin X$. We have to show that $G' - X'$ is a collection of equal-sized cliques. It is easy to see that $G' - X'$ is a collection of cliques $Q'_i = Q_i \setminus \{u\}$ where u is a vertex deleted by Reduction Rule UCVD 6 from the clique Q_i. As $|Q_i| = |Q_j|$, we get $|Q'_i| = |Q'_j|$ for all $i \neq j$. This shows that I' is a yes-instance.

In the other direction, let us assume that X' is a solution of size at most k for the instance I'. Let us denote the equal sized cliques in $G' - X'$ by $Q'_1, Q'_2, \ldots, Q'_r$. We claim that $X := X'$ is a solution for the instance I. Note that $G - X$ is again a collection of cliques $Q_1, Q_2, \ldots, Q_r$ where $Q_i = Q'_i \cup \{u\}$ where u is a vertex deleted by Reduction Rule UCVD 6 from Q_i. Clearly, we have $|Q_i| = |Q_j|$ as we know that $|Q'_i| = |Q'_j|$ for all $i \neq j$. □

Lemma 8. *After the exhaustive application of the above reduction rules, we have* $|\Gamma(\mathcal{C})| = \mathcal{O}(k^2)$.

Proof. Let us assume that C_s denotes the unique heavy neighbour of s. We have

$$\Gamma(\mathcal{C}) = \bigcup_{C \in \mathcal{C}} \left[\bigcup_{s \in S_C} \left(C \setminus N(s)\right) \cup \bigcup_{s \in S \setminus S_C} \left(N(s) \cap C\right) \right]$$

$$= \bigcup_{s \in S} \left[\left(C_s \setminus N(s)\right) \cup \bigcup_{C \in \mathcal{C} \setminus C_s} \left(N(s) \cap C\right) \right]$$

Note that by the Reduction Rule UCVD 1, for any fixed $s \in S$, we have $\left|\bigcup_{C \in \mathcal{C} \setminus C_s} \left(N(s) \cap C\right)\right| < (k+1)$. Also, for each $s \in S$, we have

$$\begin{aligned}|C_s \setminus N(s)| &\leq |C| - (|C| - 4k) = 4k \text{ if } \max\{|C| - 4k, k+1\} = |C| - 4k \\ &\leq |C| - (k+1) \leq 4k \text{ if } \max\{|C| - 4k, k+1\} = k+1\end{aligned}$$

As there are at most $3k$ vertices in S, we get that $|\Gamma(\mathcal{C})| \leq 3k(k+4k) = 15k^2 = \mathcal{O}(k^2)$.

Lemma 9. *After the exhaustive application of the above reduction rules, we have* $\max_{C \in \mathcal{C}} |C| \leq \mathcal{O}(k^2)$.

Proof. After exhaustively applying Reduction Rule UCVD 6, we know that there exists a clique, say $C_q \in \mathcal{C}$, such that $|C_q| - |\Gamma(C_q)| \leq k+1$. Note that $\max_{C \in \mathcal{C}} |C| - |C_q| \leq 4k$, as ensured by Reduction Rule UCVD 4. Therefore, we get $\max_{C \in \mathcal{C}} |C| - |\Gamma(C_q)| \leq 5k+1$. Due to Lemma 8, we have $|\Gamma(C_q)| = \mathcal{O}(k^2)$. Hence, we get $\max_{C \in \mathcal{C}} |C| \leq \mathcal{O}(k^2)$. □

Note that, we have

$$V(G) = S \cup \bigcup_{C \in \mathcal{C}} C = S \cup \bigcup_{C \in \mathcal{C}} (\Gamma(C) \cup (C \setminus \Gamma(C))) = S \cup \Gamma(\mathcal{C}) \cup \bigcup_{C \in \mathcal{C}} (C \setminus \Gamma(C))$$

To obtain a cubic kernel, it suffices to show that the size of the set $\bigcup_{C \in \mathcal{C}} (C \setminus \Gamma(C))$ is bounded by a cubic function of the parameter. To see this, we partition the set of cliques in $\mathcal{C}$ into two parts: $\mathcal{H} = \{C \in \mathcal{C} \mid C \text{ is a heavy neighbour of some } s \in S\}$ and $\mathcal{L} = \mathcal{C} \setminus \mathcal{H}$. As every vertex $s \in S$ has exactly one heavy neighbour, we have $|\mathcal{H}| \leq 3k$. Therefore, we obtain $\bigcup_{C \in \mathcal{H}} |(C \setminus \Gamma(C))| = \mathcal{O}(k^3)$, since $\max_{C \in \mathcal{C}} |C| \leq \mathcal{O}(k^2)$. Now, we focus on the set $\mathcal{L}$. Define $\mathcal{L}' = \{C \in \mathcal{L} \mid |C \setminus \Gamma(C)| \geq k+1\}$ and $\mathcal{L}'' = \mathcal{L} \setminus \mathcal{L}'$. It is clear that $\bigcup_{C \in \mathcal{L}''} |(C \setminus \Gamma(C))| \leq \mathcal{O}(k^3)$, since $|\mathcal{L}''| \leq |\mathcal{C}| \leq \mathcal{O}(k^2)$. We can assume that $|\mathcal{L}'| \geq 2k+1$; otherwise, we already obtain the desired cubic kernel.

Lemma 10. *If* $|\mathcal{L}'| \geq 2k+1$ *then in polynomial time we can determine the size* c *of equal sized cliques obtained after deleting the vertices in a solution, assuming that the input is a yes-instance.*

Proof. Let us denote $\mathcal{L}' = \{C'_1, C'_2, \ldots, C'_l\}$, where $l \geq 2k+1$. Let X be a solution of size at most k of the input instance. As $|C'_i \setminus \Gamma(C'_i)| \geq k+1$, at least one vertex must survive from each set $C'_i \setminus \Gamma(C'_i)$ in $G - X$. Since the vertices in $C'_i \setminus \Gamma(C'_i)$ do not have neighbours in S, the deletion of at most k vertices can change the degree of vertices in at most k sets $C'_i \setminus \Gamma(C'_i)$. This implies that vertices in all but at most k distinct sets $C'_i \setminus \Gamma(C'_i)$ must have the same degree. □

It is important to note that having $|\mathcal{L}'| \geq 2k + 1$ is necessary; otherwise, identifying c is not possible. If it is the case, then let us denote the degree by d. One can easily see that $c = d + 1$. The following set of reduction rules are only applicable when $|\mathcal{L}'| \geq 2k + 1$.

Reduction Rule UCVD 7. *If there is a vertex v such that $d(v) < d$, then add it to the solution. The new instance is $(G - v, k - 1)$.*

Due to Reduction Rule UCVD 7, we can assume that all the cliques in $\mathcal{L}'$ have size at least c.

Reduction Rule UCVD 8. *If there is a clique $C' \in \mathcal{L}'$ such that $|C'| = c$, then delete all the vertices in $N(C')$ from the graph G. The new instance is $(G - N(C'), k - |N(C')|)$.*

Lemma 11. *Reduction Rule UCVD 8 is safe.*

Proof. Note that at least one vertex $u \in C' \setminus \Gamma(C')$ must survive after deleting the vertices in a solution. This vertex must be part of a clique of size c. Therefore, the clique C' must appear unchanged in the resulting graph. Hence, we must delete all the neighbours of C'. □

Let us denote by $\mathcal{L}'_{iso,c}$ the set of isolated cliques of size c in $\mathcal{L}'$ obtained after applying Reduction Rule UCVD 8. Similarly, denote by $\mathcal{L}'_{>c}$ the set of cliques of size more than c in $\mathcal{L}'$. Note that, we have $\mathcal{L}' = \mathcal{L}'_{iso,c} \cup \mathcal{L}'_{>c}$.

Reduction Rule UCVD 9. *If $|\mathcal{L}'_{>c}| > k$, then conclude that we are dealing with a no-instance.*

This is correct because we need to delete more than k vertices to ensure that each clique in $\mathcal{L}'_{>c}$ becomes a clique of size c in the resulting graph.
Due to Reduction Rule UCVD 2, we get $|\mathcal{L}'_{iso,c}| \leq k+1$. After applying Reduction Rules 1 to 9 exhaustively, we have $|\mathcal{L}'| = |\mathcal{L}'_{iso,c}| + |\mathcal{L}'_{>c}| \leq (k+1) + k \leq 2k+1$. Hence, we have $\bigcup_{C \in \mathcal{L}'} |C \setminus \Gamma(C)| \leq \mathcal{O}(k^3)$. Therefore, we get a kernel of size $\mathcal{O}(k^3)$.
This completes the proof of Theorem 1. □

Theorem 2. *The* UCVD *problem parameterized by solution size k can be solved in $2^k \cdot n^{\mathcal{O}(1)}$ time.*

Proof. We utilize an algorithm from [16] to find a cluster vertex deletion set X of size at most k in $\mathcal{O}^*(1.811^k)$ time. If this algorithm determines that no such set exists, conclude that we are dealing with a no-instance. Otherwise, this algorithm returns a cluster vertex deletion set X of size at most k. To decide whether G contains a uniform cluster vertex deletion set S of size at most k, we proceed as follows:

- We guess the intersection of S with X, that is, we guess the set $X_{in} = X \cap S$, delete X_{in} from G and reduce parameter k by $|X_{in}|$.

- For each guess of X_{in}, we set $X_{out} = X \setminus X_{in}$ and solve DISJOINT UCVD on the instance $(G - X_{in}, X_{out}, k - |X_{in}|)$.

For each guess X_{in}, we try to find a uniform cluster vertex deletion set S' in $G - X_{in}$ of size at most $k - |X_{in}|$ that is disjoint from X_{out}. This problem is called DISJOINT UCVD. If for some guess X_{in}, we find a uniform cluster vertex deletion set S' in $G - X_{in}$ of size at most $k - |X_{in}|$ that is disjoint from X_{out}, then we output $S = X_{in} \cup S'$. Otherwise, we conclude that the given instance of the DISJOINT UCVD problem is a no-instance. The number of all guesses is bounded by 2^k. Therefore, to obtain an FPT algorithm for UCVD, it is sufficient to solve DISJOINT UCVD in polynomial time.

An algorithm for DISJOINT UCVD: Let $(G - X_{in}, X_{out}, k)$ be an instance of DISJOINT UCVD, and let $H = G - X$, where $X = X_{in} \cup X_{out}$. We denote $G' = G - X_{in}$.

Reduction Rule Disjoint-UCVD 1. *If $G[X_{out}]$ is not a disjoint union of cliques or $G - X$ is not a disjoint union of cliques, then return that (G', X_{out}, k) is a no-instance.*

Reduction Rule Disjoint-UCVD 2. *If there is a vertex v in $G - X$ which has neighbours in two distinct cliques in $G[X_{out}]$, then delete v from G and decrement the parameter k by 1. The new instance is $(G' - \{v\}, X_{out}, k - 1)$.*

The previous reduction rule is correct because we will get an induced P_3 where the only vertex that we are allowed to delete is v. Therefore, we must add v to the solution.

So from now onwards, we assume that $G[X_{out}]$ is a disjoint union of cliques and $H = G - X$ is a disjoint union of cliques. We also guess the exact size of the solution as k'. The possible guesses for k' range from 1 to k. This will be helpful in the final part of the algorithm. We will solve this problem by guessing the size c of cliques in the resulting graph. At the end, we will compare the solution sizes for each c and return the minimum one. The possible guesses for c ranges from 1 to n.

Reduction Rule Disjoint-UCVD 3. *If there is a clique of size exactly c in $G[X_{out}]$ then delete $N(C)$ and add all the vertices of the set $N(C)$ to the solution. The new instance is $(G' - N[C], X_{out}, k - |N(C)|)$.*

Observation 2. *If there is a clique C of size larger than c in $G[X_{out}]$, then we can discard such a guess for c.*

Observation 3. *Every clique in $G[X_{out}]$ of size less than c must utilize vertices from exactly one clique in $G - X$ to form a clique of size c.*

Due to Observation 3, for every clique C of size less than c in $G[X_{out}]$, there is a unique clique in $G - X$ that helps to form a clique of size c containing vertices from C.

We now construct a bipartite graph with bipartition (A, B) in the following way. For a given guess (k', c), observe that the number of equal-sized cliques in the resulting graph is $p = \frac{|V(G')|-k'}{c}$. We need a total of p vertices in A. Suppose the number of cliques in $G[X_{out}]$ is p_1. We add a vertex $a \in A$ corresponding to each clique in $G[X_{out}]$. If $p_1 < p$, we add $p - p_1$ dummy vertices in A to ensure the number of vertices in A equals p. Similarly, we add a vertex $b \in B$ corresponding to each clique in $G - X$. Due to Reduction Rule Disjoint-UCVD 3 and Observation 2, it follows that every clique in $G[X_{out}]$ has size less than c. We add an edge between $a \in A$ and $b \in B$ if the clique corresponding to a can be transformed into a clique of size c using some vertices from the clique corresponding to b. Additionally, some cliques in $G - X$ can be transformed into cliques of size c by deleting the necessary number of vertices from the respective cliques. We make $b \in B$ adjacent to all dummy vertices in A if the clique corresponding to b can be transformed into a clique of size c by deleting the necessary number of vertices.

Lemma 12. *There exists an A-saturated matching if and only if there is a set of p disjoint cliques, each of size exactly c, obtained by deleting exactly k' vertices from $G - X$, such that the graph induced by these cliques forms a uniform cluster graph.*

Proof. In the forward direction, suppose there exists an A-saturated matching. We can identify p cliques in the resulting graph as follows:

- If $(a, b) \in M$ and a is not a dummy vertex, the clique corresponding to $a \in A$ will be transformed into a clique of size c using some vertices from the clique corresponding to $b \in B$.
- If $(a, b) \in M$ and a is a dummy vertex, the clique corresponding to $b \in B$ will be adjusted to a clique of size c by deleting the necessary vertices from that clique.

Thus, we can identify the exact cliques in the final graph. Since $p \cdot c = |V(G')| - k'$, we know exactly which k' vertices need to be deleted to form these cliques of size c.

In the backward direction, assume there is a set of p disjoint cliques, each of size exactly c, obtained by deleting exactly k' vertices from $G - X$, and that the induced subgraph on these cliques forms a uniform cluster graph. By our construction, we can obtain an edge in the bipartite graph (A, B) corresponding to each clique, forming a matching. Since there are p cliques, this matching will be A-saturated. □

We now try to find an A-saturated matching in this bipartite graph. This can be done in polynomial time. Due to the previous claim, if an A-saturated matching does not exist, we discard such a guess (k', c). Otherwise, we find an A-saturated matching M of size p and return a yes-instance. This completes the proof of Theorem 2. □

3 Kernelization for UCED Parameterized by Solution Size

In this section, we present a kernelization algorithm for UCED.

Theorem 3. *The* UCED *problem parameterized by solution size admits a kernel with at most* $6k$ *vertices.*

We begin by giving a preprocessing step which we will refer to as the *Preparation step.*

Preparation Step: We compute a maximal set $\mathcal{P}_3$ of vertex-disjoint induced P_3s in G. If $|\mathcal{P}_3| > k$, then we have a no-instance. Therefore, we assume that $|\mathcal{P}_3| \leq k$, and let S be the vertices of $\mathcal{P}_3$. We have $|S| \leq 3k$. Let $\mathcal{C}$ denotes the set of cliques of $G - S$. Let F be a uniform cluster edge deletion set of size at most k in G, and let V_F be the set of vertices adjacent to edges in F. Then we have $|V_F| \leq 2k$.

We observe that if there indeed exists a uniform cluster edge deletion set F of size at most k, then the vertices in $V \setminus V_F$ have equal degree in both G and $G - F$. In other words, for any two distinct vertices $x, y \in V \setminus V_F$, we have $d_G(x) = d_G(y) = d_{G-F}(x) = d_{G-F}(y)$. We assume that G has at least $4k + 1$ vertices; otherwise, we would have a kernel of size $4k$. We know that if the input is a yes-instance, then all but $2k$ vertices have the same degree d. Therefore, we first check if the input instance (G, k) satisfies this condition. If the input instance does not satisfy this condition, then conclude that we are dealing with a no-instance. If the input instance satisfies this condition, we can calculate the exact value of d in polynomial time. Assuming that the input is a yes-instance, let c be the size of equal sized cliques in $G - F$. It follows that c must be equal to $d+1$ in $G-F$. The maximum degree of G is at most $d+k$, because executing k edge deletion operations can decreases the degree of a vertex by at most k. The preparation step ends here.

Proof of Theorem 3. We start with the same preparation step as designed above. The correctness of the following two reduction rules follows from the preparation step.

Reduction Rule UCED 1. *If* $\delta(G) < d$ *or* $\Delta(G) > d + k$*, then conclude that we are dealing with a no-instance.*

Reduction Rule UCED 2. *If the number of vertices in* $V(G)$ *with degrees not equal to* d *exceeds* $2k$*, then conclude that we are dealing with a no-instance.*

Reduction Rule UCED 3. *If* $k < 0$*, then* (G, k) *is a no-instance.*

Reduction Rule UCED 4. *If there exists a vertex* u *of degree* d *in* G *such that the subgraph induced by* $N[u]$ *is not a clique, then conclude that we are dealing with a no-instance.*

This is correct because u can be part of a clique of size $d+1$ only if $N[u]$ forms a clique.

Reduction Rule UCED 5. *Let $u \in V(G)$ be a vertex with degree d. If u has a neighbour v with a degree strictly greater than d, then delete all the edges $(v, w) \in E(G)$ such that w is neither u nor a neighbour of u. The set of edges deleted is denoted as W. The resulting instance is $G = (V(G), k - |W|)$. Also if $|W| \geq k+1$ then conclude that we are dealing with a no-instance.*

Lemma 13. *Reduction Rule UCED 5 is safe.*

Proof. As we have already noted that a vertex of degree d can be part of a clique of size $d+1$ only if $N[v]$ forms a clique. We also observe that the connected component containing v must be $G[N[v]]$ after deleting the solution. Therefore, we must delete the edges $(v, w) \in E(G)$ such that $v \in N[u]$ but $w \in V(G) \setminus N[u]$. □

Reduction Rule UCED 6. *If there are isolated cliques of size $d+1$, then we retain only x cliques of size $d+1$ where $x(d+1) \geq 2k+1$ and remove the remaining cliques of size $d+1$.*

Lemma 14. *Reduction Rule UCED 6 is safe.*

Proof. The condition $x(d+1) \geq 2k+1$ in Reduction Rule UCED 6 ensures that at least $2k+1$ vertices have degree exactly d, which in turn ensures that the size of cliques, after deleting at most k edges, remains exactly $d+1$. Therefore deleting the extra cliques does not change the size of a solution. □

We consider two cases based on the size of d.

Case 1: $d \geq 2k$

Reduction Rule UCED 7. *If $k \geq 0$, then we return that (G, k) is a yes-instance.*

Lemma 15. *Reduction Rule UCED 7 is safe.*

Proof. First we will show that every vertex in the reduced graph has degree d. For the sake of contradiction assume there exists a vertex v with a degree of at least $d+1$. If $N(v)$ contains a vertex with degree d, then Reduction Rule UCED 5 could have been applied. Therefore, we assume that all vertices in $N(v)$ have a degree of at least $d+1$. As a result, the number of vertices with degrees not equal to d is at least $d+2 \geq 2k+2$ as $d \geq 2k$. However, in this scenario, Reduction Rule UCED 2 could have been applied because the number of vertices with degrees not equal to d exceeds $2k$. Consequently, in a reduced graph, every vertex has a degree exactly d. After exhaustively applying Reduction Rules UCED 4, 5 and 6, we obtain a clique of size $d+1$. This is a uniform cluster graph. Therefore, we can safely return that (G, k) is a yes-instance. □

Therefore, in Case 1, we obtain a kernel of size $\mathcal{O}(1)$.

Case 2: $d < 2k$
After exhaustively applying Reduction Rules UCED 4 and UCED 5, the components containing a vertex of degree d must form a $(d+1)$-clique. After applying Reduction Rule UCED 6, the cliques of size $d+1$ collectively contain at most $4k$ vertices. According to Reduction Rule UCED 2, we have at most $2k$ vertices that do not have degree d. Therefore, we have $|V(G)| \leq 6k$.

This completes the proof of Theorem 3. □

4 Conclusion

We presented a kernel with $\mathcal{O}(k^3)$ vertices for UCVD and a $6k$-vertex kernel for UCED, along with an FPT algorithm for UCVD running in time $\mathcal{O}^*(2^k)$. Obtaining a kernel with $\mathcal{O}(k^2)$ vertices and further improving the FPT running time for UCVD remain open challenges.

References

1. Abu-Khzam, F.N.: A kernelization algorithm for d-hitting set. J. Comput. Syst. Sci. **76**(7), 524–531 (2010)
2. Böcker, S., Damaschke, P.: Even faster parameterized cluster deletion and cluster editing. Inf. Process. Lett. **111**(14), 717–721 (2011)
3. Boral, A., Cygan, M., Kociumaka, T., Pilipczuk, M.: A fast branching algorithm for cluster vertex deletion. Theory Comput. Syst. **58**, 357–376 (2013)
4. Cai, L.: Fixed-parameter tractability of graph modification problems for hereditary properties. Inf. Process. Lett. **58**(4), 171–176 (1996)
5. Cao, Y., Chen, J.: Cluster editing: kernelization based on edge cuts. Algorithmica **64**, 152–169 (2012)
6. Cao, Y., Ke, Y.: Improved kernels for edge modification problems. arXiv preprint arXiv:2104.14510 (2021)
7. Damaschke, P.: Bounded-degree techniques accelerate some parameterized graph algorithms. In: Chen, J., Fomin, F.V. (eds.) IWPEC 2009. LNCS, vol. 5917, pp. 98–109. Springer, Heidelberg (2009). https://doi.org/10.1007/978-3-642-11269-0_8
8. Doob, M.: On characterizing certain graphs with four eigenvalues by their spectra. Linear Algebra Appl. **3**, 461–482 (1970)
9. Goldberg, F., Kirkland, S., Varghese, A., Vijayakumar, A.: On split graphs with four distinct eigenvalues. Discret. Appl. Math. **277**, 163–171 (2020)
10. Gramm, J., Guo, J., Hüffner, F., Niedermeier, R.: Graph-modeled data clustering: exact algorithms for clique generation. Theory Comput. Syst. **38**, 373–392 (2005)
11. Guo, J.: A more effective linear kernelization for cluster editing. Theoret. Comput. Sci. **410**(8–10), 718–726 (2009)
12. Hüffner, F., Komusiewicz, C., Moser, H., Niedermeier, R.: Fixed-parameter algorithms for cluster vertex deletion. In: Laber, E.S., Bornstein, C., Nogueira, L.T., Faria, L. (eds.) LATIN 2008. LNCS, vol. 4957, pp. 711–722. Springer, Heidelberg (2008). https://doi.org/10.1007/978-3-540-78773-0_61
13. Madathil, J., Meeks, K.: Parameterized algorithms for balanced cluster edge modification problems (2024)

14. Misra, N., Mittal, H., Saurabh, S., Thakkar, D.: On the complexity of the eigenvalue deletion problem. In: Iwata, S., Kakimura, N., eds, 34th International Symposium on Algorithms and Computation (ISAAC 2023). Leibniz International Proceedings in Informatics (LIPIcs), vol. 283, pp. 53:1–53:17. Dagstuhl, Germany. Schloss Dagstuhl – Leibniz-Zentrum für Informatik (2023)
15. Steinvik, A.: Kernelization for balanced graph clustering. Master's thesis, The University of Bergen (2020)
16. Tsur, D.: Faster parameterized algorithm for cluster vertex deletion. Theory Comput. Syst. **65**, 1–21 (2021)
17. Tsur, D.: Cluster deletion revisited. Inf. Process. Lett. **173**, 106171 (2022)

Computing Diverse and Nice Triangulations

Waldo Gálvez[1], Mayank Goswami[2], Arturo Merino[3], GiBeom Park[4(✉)], and Meng-Tsung Tsai[5]

[1] Universidad de Concepción, Concepción, Chile
[2] Queens College of CUNY, New York, USA
[3] Universidad de O'Higgins, Rancagua, Chile
[4] CUNY Graduate Center, New York, USA
gpark1@gradcenter.cuny.edu
[5] Academia Sinica, Taipei, Taiwan

Abstract. We initiate the study of computing diverse triangulations of a given polygon. Given a simple n-gon P, an integer $k \geq 2$, a quality measure σ on the set of triangulations of P and a factor $\alpha \geq 1$, we formulate the Diverse and Nice Triangulations (DNT) problem that asks to compute k *distinct* triangulations $T_1, \ldots, T_k$ of P such that a) their diversity, $\sum_{i<j} d(T_i, T_j)$, is as large as possible *and* b) they are nice, i.e., $\sigma(T_i) \leq \alpha\sigma^*$ for all $1 \leq i \leq k$. Here, d denotes the symmetric difference of edge sets of two triangulations, and σ^* denotes the best quality of triangulations of P, e.g., the minimum Euclidean length.

As our main result, we provide a poly(n, k)-time approximation algorithm for the DNT problem that returns a collection of k distinct triangulations whose diversity is at least $1 - \Theta(1/k)$ of the optimal, and each triangulation satisfies the quality constraint. This is accomplished by studying *bi-criteria triangulations* (BCT), which are triangulations that simultaneously optimize two criteria, a topic of independent interest. We complement our approximation algorithms by showing that the DNT problem and the BCT problem are NP-hard.

Finally, for the version where diversity is defined as $\min_{i<j} d(T_i, T_j)$, we show a reduction from the problem of computing optimal Hamming codes, and provide an $n^{O(k)}$-time $\frac{1}{2}$-approximation algorithm. This improves over the naive $\binom{C_{n-2}}{k} \approx 2^{O(nk)}$ time bound for enumerating all k-tuples among the triangulations of a simple n-gon, where C_n denotes the n-th Catalan number.

Keywords: Triangulation · Diversity · Delaunay · Bi-Criteria Triangulation

A preliminary version of this paper was presented at the 31st Annual Fall Workshop on Computational Geometry 2024.

A. Jeż and J. Otop (Eds.): FCT 2025, LNCS 16106, pp. 180–193, 2026.
https://doi.org/10.1007/978-3-032-04700-7_14

1 Introduction

There has been considerable recent interest in computing diverse solutions to optimization problems, also called the diverseX paradigm. Instead of computing a single (approximately) optimal solution to a (perhaps hard) optimization problem, this paradigm asks to compute a set of $k \geq 2$ many good solutions that are as different from each other as possible. There are several motivations for this, such as providing more choices to the user [9], fairness [4,5], robustness and security [31], and portfolio optimization [16].

Computing diverse solutions to optimization problems has been studied for problems such as vertex covers and hitting sets [8,9], independent sets and vertex covers in planar graphs [29], matchings [24], shortest paths [27,33], spanning trees [34], approximate minimum spanning trees, global min cuts, matchings, short paths [31], s-t min cuts [11], and SAT [6,39].

In this work, we focus on computing diverse solutions in computational geometry (CG). One of the first topics in a CG course is the problem of computing triangulations of a given polygon or a point set. What if we want to compute k very different triangulations? When $k = 2$, we want to compute the diametral pair of triangulations according to some metric on the space of triangulations. This problem has been studied qualitatively when the metric is the flip distance between triangulations: even the simple case of convex polygons turns out to be fascinating, as computing the triangulations of a convex polygon with maximal flip distance is related to the rotation distance between binary search trees and the diameter of associahedra [43]. However, for arbitrary polygons even computing the flip distance between two given triangulations is NP-complete [3]. In this work we focus on another natural metric, the symmetric difference of the edge sets of two triangulations. We will consider non-convex polygons as well as obtaining $k > 2$ solutions. Moreover, since many applications require triangulations that are "nice" with respect to some measure, we will study the *diverse and nice triangulations* problem where we want to return a maximally diverse set of k nice triangulations.

The diverse and nice triangulations problem is not only of theoretical interest but also of potential practical significance. Decomposing a complex shape into simple pieces is required for many applications ranging from computer graphics and vision [2,10] to numerical analysis [37]. Presenting the user with many *different*-looking triangulations not only increases available choices but may also increase the robustness of these procedures. As another example, consider an application in numerical analysis - the finite element method used to solve differential equations [7]. A result achieved after averaging the computation over diverse triangulations may provide a more robust, more domain-dependent and less triangulation-dependent answer, as opposed to performing the computation over just one triangulation, or a few similar triangulations.

Moreover, in many applications certain triangulations are preferred over others. Well-studied examples include the *minimum (Euclidean) length triangulation, the Delaunay triangulation that maximizes the minimum angle*, and others. In this setting, we want triangulations that are not only diverse, but also "nice" with respect to such quality measures. This leads us to formulate the *diverse and nice triangulations* problem, which we formally define next.

1.1 Problem Statement

Unless otherwise stated, we assume that P is a simple n-gon, i.e., a polygon without holes. Define a triangulation of P as a maximal set of non-crossing diagonals of P. We first define our diversity measure for the triangulations.

Definition 1 (Diversity Measure). *Given triangulations $T_1, \ldots, T_k$ of P, we consider two diversity measures:*

$$\sum_{\mathrm{SD}}(T_1, \ldots, T_k) \stackrel{def}{=} \sum_{1 \le i < j \le k} |T_i \Delta T_j|, \quad and \quad \min_{\mathrm{SD}}(T_1, \ldots, T_k) \stackrel{def}{=} \min_{1 \le i < j \le k} |T_i \Delta T_j|$$

Our goal is to develop algorithms that maximize these two diversity measures. Note that maximizing $\sum_{\mathrm{SD}}$ is equivalent to maximizing the *average* pairwise distance between the triangulations. In addition, we will not only care about the diversity among a collection of triangulations, but with different applications in mind, also care about the *quality of each triangulation*. Let $\mathcal{T}$ denote the set of all triangulations of P. Given a quality measure $\sigma : \mathcal{T} \to \mathbb{R}_{\ge 0}$ and an approximation factor $\alpha \ge 1$, we say a triangulation T is *nice* if $\sigma(T) \le \alpha\sigma^*$, where by convention, we assume that $\sigma^* = \min_{T \in \mathcal{T}} \sigma(T)$ is the quality of a best triangulation (w.r.t. σ) of P.

Definition 2 (Sum-DNT and Min-DNT). *Input: A simple n-gon P without holes, $k \ge 2$, a quality measure σ and an $\alpha \ge 1$. Output: A collection of k distinct triangulations $T_1, \ldots, T_k \in \mathcal{T}$, if they exist, such that the following holds:*

1. *(α-optimal) For every $i \in \{1, \ldots, k\}$, $\sigma(T_i) \le \alpha \cdot \sigma^*$. These α-optimal triangulations are called "nice" triangulations.*
2. *(Maximally diverse) For any set of k distinct α-optimal triangulations $T_1^{'}$, ..., $T_k^{'} \in \mathcal{T}$ (i.e., satisfying 1 above), $\sum_{\mathrm{SD}}(T_1, \ldots, T_k) \ge \sum_{\mathrm{SD}}(T_1', \ldots, T_k')$ (Sum-DNT) or $\min_{\mathrm{SD}}(T_1, \ldots, T_k) \ge \min_{\mathrm{SD}}(T_1', \ldots, T_k')$ (Min-DNT).*

Note that the DNT problem comes with two measures - the diversity measure $\sum_{\mathrm{SD}}$ or $\min_{\mathrm{SD}}$ on a collection of k triangulations, and the quality measure σ for each triangulation. If the quality measure σ is not considered (we just set it to the constant function), omit the "N" and call these variants of problem the Diverse Triangulations (Sum-DT or Min-DT) problems.

1.2 Related Work

As far as we know, we are the first to study the problem of finding diverse triangulations. In the field of computational geometry, the only existing work on finding diverse solutions is the recent interesting work by Klute and van Kreveld [35] that considers diverse sets of geometric objects, such as polygons and point sets, with several diversity metrics. The authors investigate the maximum size of a fully diverse set, defined as a set of k objects such that $\min_{i \neq j} d(x_i, x_j)$ is at least a constant fraction of the diameter of the space. The focus is on quantitative upper and lower bounds on the sizes of such sets, and not algorithmic. The authors mention a simple randomized algorithm that samples an object and adds it to the collection if it is sufficiently far away from all current objects in the collection, but the runtime is not analyzed. Apart from bounding the number of samples needed, an additional issue in implementing this algorithm for the nice triangulation problem is that while sampling a triangulation at random can be done in polynomial time [21], it is not clear how to sample uniformly a nice triangulation (e.g., one with Euclidean weight at most twice that of the minimum) in polynomial time. For a survey on quality measures of triangulation, see [15].

A related but fundamentally different problem is that of *diversity optimization*, or *dispersion*, that has been studied extensively [1,13,22]. In the latter problem, one wants to obtain a set of k maximally dispersed points from a metric space of N points. Note that for our problem, the metric space is the space of all triangulations, and so poly(N, k) algorithms for dispersion translate directly to exp(n)poly(k) algorithms for the diverse solutions.

The concept of diversity has also been explored *within a solution.* Recent examples include work on packing [30], diverse convex and Voronoi partitions [36], and the dispersive art gallery problem [46].

2 Our Results

We state our results for the Sum-DNT problem first, followed by the results for the Min-DNT problem. Due to space constraints, complete proofs and additional technical details are provided in the full version of this paper [28].

2.1 Sum-DNT Results

Hardness: Observe that a successful output of the DNT problem consists of k triangulations and hence takes space $\Omega(kn)$, which is not polynomial in the size of input already. However, we show that even the decision version of the DNT problem (DDNT) is NP-hard (see the full version for the proof). For this hardness, our quality measure $\sigma(T)$ on a triangulation T is the sum of the Euclidean lengths of all the $(n-3)$ diagonals in the triangulation. We call this quantity $\sigma(T)$ the Euclidean length of the triangulation.

Theorem 1 (NP-hardness of DDNT). *Given a polygon P with n vertices, an integer $k \geq 2$, a threshold $D \geq 1$ and a quality factor $\alpha' \in (1, \infty)$, it is NP-hard to decide whether P admits k distinct triangulations $T_1, \ldots, T_k$ such that $\sum_{\mathrm{SD}}[T_1, \ldots, T_k] \geq D$ and $\sigma(T_i) \leq \alpha' \sigma^*$ for all $i \in [k]$, where σ^* is the Euclidean length of a minimum Euclidean length triangulation of P.*

Decomposable Quality Measure: In light of the hardness above, our first contribution to the DNT problem is identifying a class of quality measures that we call *decomposable*. It will turn out that not only does this class contain most of the quality measures on triangulations studied in the computational geometry community, but we can also develop approximation algorithms for these quality measures! Assume first that a quality measure σ' for a single edge, or a single triangle is given; e.g., σ' measures the length of an edge, or the minimum angle of a triangle. A *decomposable* quality measure σ on a triangulation T is one that is either a min, max or sum, of the measure σ' over the edges or triangles in T.

Definition 3 (Decomposable Quality Measures). *Let* $\odot \in \{\sum, \min, \max\}$*, and let* σ' *be a nonnegative function defined on the set of diagonals or triangles of the polygon. A quality measure* $\sigma : \mathcal{T} \rightarrow \mathbb{R}_{\geq 0}$ *is said to be edge-decomposable if* $\sigma(T) = \odot_{e \in T} \sigma'(e)$ *for any triangulation* T*, and triangle-decomposable if* $\sigma(T) = \odot_{t \in tr(T)} \sigma'(t)$ *for any triangulation* $T \in \mathcal{T}$*.*

By a decomposable measure we will mean a measure that is either edge-decomposable or triangle-decomposable. Examples of decomposable quality measures that have been considered extensively in the literature are the Euclidean length [40,45], maximum length (of the $(n-3)$ diagonals) [18], minimum length (of the $(n-3)$ diagonals) [23], maximum angle of the triangulation [19], and the minimum angle of the triangulation, a measure that is maximized by the popular Delaunay triangulation. There is also research on triangulations that are close-to-Delaunay [12]; see Fig. 1.

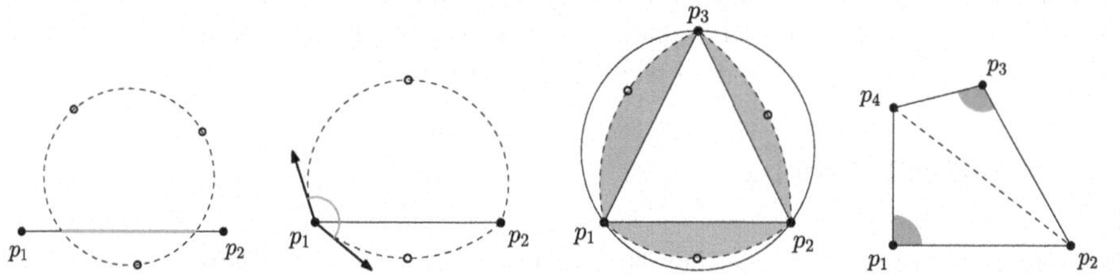

(a) Shrunk Circle (b) Lens (c) Triangular Lens (d) Opposing Angles

Fig. 1. The following measures, called near-Delaunay measures, were proposed by [12, 41]. (a) Shrunk-circle measure (D), equals $\sum_{e \in T} \sigma(e)$, where $\sigma(e)$ is the maximum fraction of the diagonal e overlapped with the largest empty circle. (b) Lens-based measure (D), equals $\sum_{e \in T} \sigma(e)$, where $\sigma(e) = \min\{1, \theta(e)/\pi\}$ and $\theta(e)$ is the angle formed by the tangent vectors of the two largest empty arcs on both sides. (c) Triangular-lens measure (D), equals $\sum_{t \in tr(T)} \sigma(t)$, where $\sigma(t)$ is the fraction of the area in the circle but outside the triangle that is covered by the shown lens. (d) Opposing angles measure (D), equals $\sum_{e \in T} \sigma(e)$, where $\sigma(e) = \max\{0, -1 + \theta(e)/\pi\}$ and $\theta(e)$ is the sum of opposing angles in the quadrilateral that has e as a diagonal. Note that this measure is an example of a near-Delaunay measure that is not decomposable, since the opposing angles depend *not* only on the diagonal e, but also on the triangles in T adjacent to e.

Approximation Algorithms: We now present our main algorithmic results for the DNT problem with respect to the $\sum_{\text{SD}}$ measure. All of our algorithms will have an approximation factor of $\beta := \max\left\{\frac{1}{2}, 1 - \frac{2}{k+1}\right\}$ for the diversity. β equals 1/2 when $k \leq 3$, and approaches 1 as $k \to \infty$.

Theorem 2 (Algorithms for Sum-DNT). *Let σ be any decomposable measure. Let P, k and α be given as in Definition 2 of the DNT problem, and let $T_1, \ldots, T_k$ and $T'_1, \ldots, T'_k$ in the following denote distinct triangulations. Then,*

1. *For $\alpha \in (1, \infty)$, there exists an $O(n^5 k^5 \log k)$-time algorithm that returns k distinct α-optimal triangulations $T_1, \ldots, T_k$ such that $\sum_{\text{SD}}(T_1, \ldots, T_k)$ is at least $\beta \cdot \sum_{\text{SD}}(T'_1, \ldots, T'_k)$ for any α-optimal triangulations $T'_1, \ldots, T'_k$.*
2. *For $\alpha \in (1, \infty)$, there exists an algorithm that runs in time $O\left(\varepsilon^{-2} \cdot n^5 k^3 \log k\right)$ and returns k distinct $\alpha(1+\varepsilon)$-optimal triangulations $T_1, \ldots, T_k$ such that $\sum_{\text{SD}}(T_1, \ldots, T_k)$ is at least $\beta \cdot \sum_{\text{SD}}(T'_1, \ldots, T'_k)$ for any α-optimal triangulations $T'_1, \ldots, T'_k$.*

For some special cases, the Sum-DNT problem has faster algorithms or even admits PTAS. See the full version for the proofs of these special cases.

Theorem 3 (Special Cases for Sum-DNT).

1. *When σ is decomposable and $\alpha = 1$, there exists an $O(n^3 k^3 \log k)$-time algorithm for Sum-DNT that returns k distinct triangulations $T_1, \ldots, T_k$ with diversity at least β of optimal.*
2. *When P admits at least k Delaunay triangulations, there is an algorithm that runs in time $k^3 n^{O(1/\varepsilon)}$ and returns k distinct Delaunay triangulations of P whose diversity is at least $(1-\varepsilon)$ of the optimal.*
3. *When P is a convex polygon, there is an algorithm for Sum-DT that runs in time $2^{O(1/\varepsilon^2)} n^5 k^5$ and returns k distinct triangulations whose diversity is at least $(1-\varepsilon)$ of the optimal.*

2.2 Min-DNT Results

We now state our results on triangulations that maximize the minimum distance between two triangulations, denoted by $\min_{\text{SD}}$. In general, the max-min version of dispersion is considered harder than the max-sum version, and some hardness results for matroids are given by Fomin et al. [25]. For this version, we will ignore the niceness constraint, and only focus on the min-DT problem.

For simple polygons, we relate the min-DT problem to the problem of computing $A_q(n, d)$, which is the maximum number of q-ary codewords of length n with pairwise Hamming distance at least d. If we let $D_q(n, m, d)$ denote the decision version $A_q(n, d) \geq m$, then we can show the following interesting result, whose proof can be found in the full version.

Theorem 4 (Reduction to Hamming Codes).*cctrihardness Assume there is an algorithm that, given a polygon with n vertices and an integer $k = O(n)$,* *runs in time in* poly(n) *and outputs* k *diverse triangulations maximizing* $\min_{\text{SD}}$. *Then there is an algorithm for computing* $A_2(n, d)$ *for any* $d > n/2$ *in time* poly(n).

As far as we know, computing $A_2(n, d)$ is still open, and only a limited number of instances are currently known, e.g., see [42]. We now present an algorithm for the min-DT problem; its proof is also provided in the full version of this paper.

Theorem 5 (Algorithm for Min-DT). *Let* P *and* k *be given as in Definition 2 of the DNT problem. Define* $r = 2(n-3) - \frac{d_{\text{OPT}}}{2}$, *where* d_{OPT} *denotes optimal diversity in* $\min_{\text{SD}}$ *measure. Then, there exists an* $r^{O(k)}$*-time algorithm that outputs* k *triangulations such that*

$$\min_{\text{SD}}(T_1, \ldots, T_k) \geq \frac{1}{2} \cdot \min_{\text{SD}}(T'_1, \ldots, T'_k)$$

for any triangulations $T'_1, \ldots, T'_k$ *of* P.

Note that the algorithm in Theorem 5 runs fast if d_{OPT} is large, i.e., if P has very diverse triangulations, our algorithm can find those faster.

3 Technical Overview for Sum-DNT: Enter Bi-criteria Triangulations

In the remainder of the main body of this abstract, we will focus on the Sum-DNT version, since we consider the polynomial time approximation algorithms for this version as our main result. Proofs of the hardness of Sum-DNT (Theorem 1) can be found in the full version.

In the Sum-DNT problem (Definition 2), the diversity is the sum, or the average, of the pairwise distances between the k triangulations. Let us recall the classical problem of dispersion, where one wants to obtain a set of k maximally dispersed (w.r.t. either the sum or the minimum of the pairwise distances) points $x_1, \cdots, x_k$ from a metric space $\mathcal{M}$ of N points. While NP-complete, an application of the farthest insertion method gives an approximation algorithm for both the sum and the minimum versions of this problem. Set x_1 to be an arbitrary point in $\mathcal{M}$, and iteratively select $x_i = \text{argmax}_{p \in \mathcal{M}} \sum_{1 \leq j \leq i-1} d(p, x_j)$ for the sum version, and $x_i = \text{argmax}_{p \in \mathcal{M}} \min_{1 \leq j \leq i-1} d(p, x_j)$ for the minimum version. Both algorithms give a (1/2)-approximation guarantee to their respective objective functions [13,44], and clearly run in time poly(N, k). To see why farthest insertion gives a 1/2-approximation, consider the problem of finding a diametral pair—that is, two points at maximum mutual distance. Let (p, p') be such a diametral pair, and let x be any point in the set with x' its farthest neighbor. Then by the triangle inequality, $d(p, p') \leq d(p, x) + d(x, p') \leq d(x, x') + d(x, x') = 2\, d(x, x')$. Hence $d(x, x') \geq \frac{1}{2} \cdot d(p, p')$, giving the desired 1/2-factor. This argument extends to the selection of k dispersed points; see [13,44] for the full proof.

For now let us aim for a (1/2)-approximation factor for the Sum-DNT problem; later we will show how to boost it to $\max\{\frac{1}{2}, 1-\frac{2}{k+1}\}$. Let $\mathcal{T}$ be the space of all triangulations of a polygon with n vertices, and $\mathcal{T}_\alpha$ the space of α-optimal, or nice triangulations as in Definition 2. Applying the dispersion algorithms (for the sum of pairwise distances) from [13] translates to the following: having found i triangulations $T_1, \cdots, T_i$, we set T_{i+1} as the triangulation $T \in \mathcal{T}_\alpha$ that maximizes $\sum_{j=1}^{i} |T_j \Delta T|$. This would give a $\frac{1}{2}$-approximation to the $\sum_{\text{SD}}$.

3.1 Bi-criteria Triangulations

It turns out that finding T_{i+1} is a special case of the following problem.

Definition 4 (BCT$(\cdot,\cdot)$). *Given a polygon P, two measures* weight *and* quality *$w, \sigma : \mathcal{T} \to \mathbb{R}_{\geq 0}$ and a bound $B \geq 0$, the Bi-Criteria Triangulation is a solution to the following program:*

$$\operatorname{argmin}\{w(T) : T \in \mathcal{T} \text{ and } \sigma(T) \leq B\}.$$

When $B = \alpha\sigma^*$, the obtained solution is called α-optimal BCT. *Additionally, k distinct triangulations $T_1, \ldots, T_k$ of P are said to be the k-best enumeration for BCT provided that 1) $w(T_i) \leq B$ for each $i \in [k]$ and 2) $w(T_1) \leq \ldots \leq w(T_k) \leq w(T')$ for any $T' \in \mathcal{T} \setminus \{T_1, \ldots, T_k\}$ such that $\sigma(T') \leq B$. The notion of k-best α-optimal* BCT*s are defined analogously.*

Remark. Although the BCT is defined as a minimization problem with "$\leq$", min and "$\leq$", respectively, can be replaced with max and $\geq$ independently.

When $k = 1$, the BCT problem asks for a *triangulation that is simultaneously good w.r.t. two measures.* Surprisingly, this problem has not appeared in literature, even though historically very related questions have been asked, such as whether the Delaunay triangulation has low Euclidean length [38].

Reduction of Farthest Insertion to BCT. Assuming one has an algorithm for BCT, we show how to compute T_{i+1} (which is $\operatorname{argmax}_{T \in \mathcal{T}_\alpha} \sum_{j=1}^{i} |T_j \Delta T|$). This follows from a general framework by [31,32], which we translate to our setting in the following proposition.

Proposition 1. *Let $T_1, \ldots, T_i$ be α-optimal triangulations of P. For any $T \in \mathcal{T}_\alpha$, define $w_i(T) := \sum_{e \in T} \sum_{j=1}^{i} \mathbb{1}(e \in T_j)$. Then,*

$$\operatorname*{argmax}_{T \in \mathcal{T}_\alpha} \sum_{j=1}^{i} |T_j \Delta T| = \operatorname*{argmin}_{T \in \mathcal{T}_\alpha} w_i(T). \tag{1}$$

Proof. Let $T_1, \ldots, T_i$ be α-optimal triangulations. Then,

$$\max_{T \in \mathcal{T}_\alpha} \sum_{j=1}^{i} |T \Delta T_j| = \max_{T \in \mathcal{T}_\alpha} \sum_{j=1}^{i} \Big(2(n-3) - 2 T \cap T_j|\Big)$$

$$= 2(n-3)i - 2 \min_{T \in \mathcal{T}_\alpha} \sum_{j=1}^{i} |T \cap T_j|$$

$$= 2(n-3) - 2 \min_{T \in \mathcal{T}_\alpha} \sum_{e \in T} \sum_{j=1}^{i} \mathbb{1}(e \in T_j),$$

and this completes the proof.

Note that the right hand side of Equation (1) is exactly the α-optimal BCT with measures w_i and σ. Hence, computing a farthest α-optimal triangulation is equivalent to solving an α-optimal BCT.

We now illustrate the algorithm. For the time being, we denote $f(w_i, \sigma)$ and $f_k(w_i, \sigma)$, respectively, the running times for solving the α-optimal BCT(w_i, σ) and the k-best α-optimal BCT(w_i, σ).

Initially, set the weight w_0 of each allowed diagonal of P to 0, σ as in the input to the Sum-DNT problem, and B as $\alpha\sigma^*$. Then, solve the associated BCT problem, i.e., BCT(w_0, σ), obtaining T_1. Next, for every $e \in T_1$, increase the weight of e by 1, solve BCT(w_1, σ), and call the output triangulation T_2. Increase the weight of each $e \in T_2$ by 1, solve BCT(w_2, σ), and so on. Note that during the process a copy of one of the preceding triangulations might be obtained again, e.g., when all allowed edges of P have appeared exactly the same number of times. To avoid duplicates, we utilize the k-best enumeration procedure: given i triangulations, instead of finding a single BCT, find $(i+1)$ distinct BCTs, $T'_1, \ldots, T'_{i+1}$ such that $w_i(T'_1) \leq \ldots \leq w_i(T'_{i+1}) \leq w_i(T')$ for all $T' \in \mathcal{T}_\alpha \setminus \{T'_1, \ldots, T'_i\}$. Then, one of them must be distinct from all previous triangulations, and we take this triangulation as our T_{i+1}. Therefore, to obtain $\frac{1}{2}$-approximation in diversity, it is enough to run the k-best α-optimal BCT program k times, which incurs a total running time of $O(k f_k(w_k, \sigma))$.

Improving $1/2$ **to** $\max\{\frac{1}{2}, 1 - \frac{2}{k+1}\}$**.** We now illustrate how to obtain the improved $\max\{\frac{1}{2}, 1 - \frac{2}{k+1}\}$ factor for large k. It turns out that the symmetric difference of two sets is a *negative type* metric [17], and for such metrics it is possible to obtain the improved approximation factor by using a local-search based swapping algorithm [14,32]. This swapping algorithm begins with any set of k solutions, say S_0. At step i, then, the algorithm finds two triangulations T^* from outside S_i and T_j inside S_i that maximize the diversity of the current solution set when they are swapped. Note that such T^* can be found by comparing the farthest triangulations of $S_i \setminus \{T_j\}$ for all $j \in [k]$. Therefore, each step of the swapping algorithm can be done by performing farthest insertions k times. Cevallos et al. [14] guarantee that one needs $O(k \log k)$ iterations of each step to obtain the desired diversity. Consequently, the overall running time for finding k triangulations with diversity of $\max\{\frac{1}{2}, 1 - \frac{2}{k+1}\}$ is $O(f_k(w_k, \sigma) k^2 \log k)$.

3.2 Results on BCT

With the general reduction above, we now state our algorithms for the BCT problem that will deliver the promised approximation algorithms for the Sum-DNT problem in Theorem 2 by showing $f_k(w_k, \sigma) \in \text{poly}(n, k)$. Before we state the approximation algorithms for the BCT problem, and since the BCT problem is interesting on its own, we mention a potentially important but tangential (for our purposes) observation, where proof can be found in the full version.

Theorem 6 (Hardness of BCT). *Let the weight w on a diagonal equal its Euclidean length. Let σ be any of the near-Delaunay measures in [12], and $B \geq 0$ be given. Then solving the* BCT *problem w.r.t. w and σ is* NP-hard.

Despite the hardness result above, we now state the promised positive algorithms for BCT. Fortunately, it turns out that the BCT problem admits pseudo-polynomial time algorithms if both weight and quality measures w and σ are decomposable. Furthermore, we also prove that the BCT problem admits an FPTAS. These two results will deliver the two results stated in Theorem 2.

Theorem 7 (Algorithms for BCT). *Consider the problem* $\text{BCT}(w, \sigma)$ *problem with bound $B \geq 0$, where both w and σ are decomposable. Then for any integer $k \geq 1$ the following hold:*

1. ***(Integer-valued measure)*** *The k-best enumeration for* $\text{BCT}(w, \sigma)$ *can be solved in time $O\big((M+1)^2\, k\, n^3\big)$, where $M = W$ in the case $w(\cdot)$ is integer-valued and $w(T) \in [0, W]$ for some integer $W \geq 0$, and $M = B$ in the case $\sigma(\cdot)$ is integer-valued.*
2. ***(FPTAS)*** *For any $\varepsilon > 0$, let $T^* = \text{BCT}(w, \sigma)$. Then, there is an $O(\varepsilon^{-2} n^5)$-time algorithm that returns a triangulation $\tilde{T}$ such that $w(\tilde{T}) \leq w(T^*)$ and $\sigma(\tilde{T}) \leq (1+\varepsilon)B$. In particular, its k-best version runs in time $O\left(\varepsilon^{-2} n^5 k\right)$.*

Proof. For brevity we prove only a special case of 1; the complete proof appears in the full version. Assume $w(\cdot)$ is integer-valued, additive, triangle-decomposable, and satisfies $w(T) \in [0, W]$ for some integer $W \geq 0$. We further assume $\sigma(\cdot)$ is additive and triangle-decomposable; extending the algorithm to edge-decomposable measures is straightforward.

Let $P[i:j]$ denote the closed chain of vertices of P, from i to j, then returning to i in a counterclockwise direction, and let $\mathcal{T}[i:j]$ denote the collection of all triangulations of $P[i:j]$. Let $OPT_k(W', i, j)$ denote the sorted multiset of the k smallest $\sigma(T)$ among $T \in \mathcal{T}[i:j]$ with $w(T) = W'$. If fewer than k such triangulations exist, we pad with ∞. Whenever $\overline{ij}$ is not an edge or diagonal of P, we set $OPT_k(W', i, j) = \{\infty, \dots, \infty\}$; if $j = i+2$ and $\overline{ij}$ *is* a diagonal of P, we set $OPT_k(W', i, i+2) = \{\sigma(\Delta(i, i+1, i+2)), \infty, \dots, \infty\}$.

For $j > i+2$, since every triangulation in $\mathcal{T}[i:j]$ chooses some vertex $m \in [i+2, j-2]$ to form the triangle Δimj, splitting at m yields the following:

$$OPT_k(W', i, j) = \theta_k\Big(\Big(\bigcup OPT_k(W_1', i, m) + OPT_k(W_2', m, j) + \sigma(\Delta imj)\Big)\Big),$$

where $\theta_k(A)$ denotes the smallest k elements of A, the union is taken over all $m \in [i+2, j-2]$ and all $W_1', W_2' \in [0, W' - w(\triangle imj)]$ such that $W_1' + W_2' = W' - w(\triangle imj)$, and $A + B + c$ denotes $\{a + b + c \mid a \in A, b \in B\}$.

Thus, dynamic programming on triples (W', i, j) with $W' \in [0, W]$ and $1 \leq i < j \leq n$ fills a table of size $(W+1)n^2$. For fixed m, W_1' and W_2', we can select the k smallest elements of $OPT_k(W_1', i, m) + OPT_k(W_2', m, j) + \sigma(\triangle imj)$ in time $O(k)$ by using the selection algorithm in [26], giving $O(Wkn)$ time per table entry and $O(W^2kn^3)$ time overall. Finally, starting from $W' = 0$, scan the cells $OPT_k(W', 1, n)$ for values $\leq B$, and return the desired set of up to k smallest weights. If fewer than k such values exist, we report that the given polygon has less than k nice triangulations. As usual, we can compute the actual triangulations with standard bookkeeping, without affecting the overall running time. We omit the details here.

3.3 Putting Everything Together: Proof of Theorem 2

Putting Everything Together: Proof of Theorem 2

Recall that at the end of Sect. 3.1, we proved that we can obtain k distinct α-optimal triangulations with diversity at least $\max\{\frac{1}{2}, 1 - \frac{2}{k+1}\}$ of optimal, in time $O(f_k(w_k, \sigma)k^2 \log k)$, where $f_k(w_k, \sigma)$ denotes the running time for computing the k-best α-optimal BCT(w_k, σ).

(1) Note that any allowed diagonal of P can appear at most k times in any set of k triangulations of P. Therefore, $w_k(T)$ is at most nk. Hence, by Theorem 7(1), $f_k(w_k, \sigma) = O((nk)^2kn^3) = O(k^3n^5)$. Therefore, the running time of the DNT algorithm for $\alpha > 1$ is $O(n^5k^5 \log k)$.

(2) Follow the similar lines in the preceding proof except to use Theorem 7(2).

4 Conclusion

In this paper, we introduced and studied the Diverse and Nice Triangulations (DNT) problem, which seeks to find k triangulations of a given simple polygon that maximize their diversity measured by symmetric differences, while ensuring each triangulation meets a predefined quality criterion. For the sum-DNT problem, we provided a polynomial-time approximation algorithm for the general case and presented a PTAS for certain special cases. For the min-DT problem, we highlighted potential computational hardness through a reduction from the Hamming code problem and showed a $\frac{1}{2}$-approximation algorithm.

A natural next step is to handle arbitrary point sets and polygons with holes. Our techniques are unlikely to generalize because MWT is NP-hard on these inputs [40].

An intriguing direction for future research is to explore diversity measures based on *flip distance*. While Aichholzer et al. [3] showed that computing the flip distance between two given triangulations is NP-complete, it is not clear if this result applies to computing $k = 2$ most diverse solutions, a.k.a. the *diameter* of the flip graph. Another interesting direction would be to consider specific inputs, such as point sets without empty pentagons, for which the flip distance between triangulations *can* be computed in polynomial time by a result of Eppstein [20]. In fact, Eppstein's methods can be used to define an *earth mover's distance* between two triangulations, which can be computed in polynomial time and is a lower bound on the flip distance. Finding diverse triangulations w.r.t. earth mover's distance would be a natural—and perhaps tractable—way to guarantee triangulations that are also diverse w.r.t. flip distance.

Acknowledgments. This work was supported by the following grants:

– Waldo Gálvez: Supported by ANID through grants SIA85220118 and Fondecyt de Iniciación 11230663.

– Mayank Goswami and GiBeom Park: Supported by NSF grant CCF-2503086.

– Arturo Merino: Supported by ANID through Fondecyt de Iniciación 11251528.

Disclosure of Interests. The authors have no competing interests to declare that are relevant to the content of this article

References

1. Abboud, A., Cohen-Addad, V., Lee, E., Manurangsi, P.: Improved approximation algorithms and lower bounds for search-diversification problems. In: 49th International Colloquium on Automata, Languages, and Programming (ICALP 2022). Schloss-Dagstuhl-Leibniz Zentrum für Informatik (2022)
2. Abrahamsen, M., Bille Meyling, W., Nusser, A.: Constructing concise convex covers via clique covers (cg challenge). In: 39th International Symposium on Computational Geometry (SoCG 2023), pp. 66–1. Schloss Dagstuhl–Leibniz-Zentrum für Informatik (2023)
3. Aichholzer, O., Mulzer, W., Pilz, A.: Flip distance between triangulations of a simple polygon is NP-complete. Discrete Comput. Geom. **54**, 368–389 (2015)
4. Aumüller, M., Har-Peled, S., Mahabadi, S., Pagh, R., Silvestri, F.: Fair near neighbor search via sampling. ACM SIGMOD Rec. **50**(1), 42–49 (2021)
5. Aumüller, M., Pagh, R., Silvestri, F.: Fair near neighbor search: independent range sampling in high dimensions. In: Proceedings of the 39th ACM SIGMOD-SIGACT-SIGAI Symposium on Principles of Database Systems, pp. 191–204 (2020)
6. Austrin, P., Bercea, I.O., Goswami, M., Limaye, N., Srinivasan, A.: Algorithms for the diverse-k-SAT problem: the geometry of satisfying assignments. In: ICALP (2025)
7. Babuška, I., Aziz, A.K.: On the angle condition in the finite element method. SIAM J. Numer. Anal. **13**(2) (1976)
8. Baste, J., et al.: Diversity of solutions: an exploration through the lens of fixed-parameter tractability theory. Artif. Intell. **303**, 103644 (2022)
9. Baste, J., Jaffke, L., Masařík, T., Philip, G., Rote, G.: FPT algorithms for diverse collections of hitting sets. Algorithms **12**(12), 254 (2019)
10. de Berg, M., Cheong, O., van Kreveld, M., Overmars, M.: Computational Geometry: Algorithms and Applications, 3rd edn. Springer, Cham (2008)
11. de Berg, M., Martínez, A.L., Spieksma, F.: Finding diverse minimum S-T cuts. In: 34th International Symposium on Algorithms and Computation (2023)

12. van Beusekom, N., Buchin, K.A., Koerts, H.O., Meulemans, W., Rodatz, B., Speckmann, B.: Near-delaunay metrics. In: 33rd Canadian Conference on Computational Geometry (CCCG 2021) (2021)
13. Borodin, A., Jain, A., Lee, H.C., Ye, Y.: Max-sum diversification, monotone submodular functions, and dynamic updates. ACM Trans. Algorithms **13**(3), 41:1–41:25 (2017)
14. Cevallos, A., Eisenbrand, F., Zenklusen, R.: An improved analysis of local search for MAX-SUM diversification. Math. Oper. Res. **44**(4), 1494–1509 (2019)
15. De Loera, J.A., Rambau, J., Santos, F.: Triangulations: Structures for Algorithms and Applications, Algorithms and Computation in Mathematics, vol. 25. Springer, Cham (2010)
16. Drygala, M., Lattanzi, S., Maggiori, A., Stouras, M., Svensson, O., Vassilvitskii, S.: Data-driven solution portfolios. arXiv preprint arXiv:2412.00717 (2024)
17. Duin, R., Pękalska, E.: The dissimilarity representation for pattern recognition: a tutorial. Technical report (2009)
18. Edelsbrunner, H., Tan, T.S.: A quadratic time algorithm for the MinMax length triangulation. SIAM J. Comput. **22**(3), 527–551 (1993)
19. Edelsbrunner, H., Tan, T.S., Waupotitsch, R.: An $O(n^2 \log n)$ time algorithm for the minmax angle triangulation. In: Proceedings of the Sixth Annual Symposium on Computational Geometry, pp. 44–52 (1990)
20. Eppstein, D.: Happy endings for flip graphs. In: Proceedings of the Twenty-Third Annual Symposium on Computational Geometry, pp. 92–101 (2007)
21. Epstein, P., Sack, J.R.: Generating triangulations at random. ACM Trans. Model. Comput. Simul. (TOMACS) **4**(3), 267–278 (1994)
22. Erkut, E.: The discrete p-dispersion problem. Eur. J. Oper. Res. **46**(1), 48–60 (1990)
23. Fekete, S.P., Hellmann, W., Hemmer, M., Schmidt, A., Troegel, J.: Computing MaxMin edge length triangulations. In: 2015 Proceedings of the Seventeenth Workshop on Algorithm Engineering and Experiments (ALENEX), pp. 55–69. SIAM (2014)
24. Fomin, F.V., Golovach, P.A., Jaffke, L., Philip, G., Sagunov, D.: Diverse pairs of matchings. In: 31st International Symposium on Algorithms and Computation (ISAAC 2020). Schloss Dagstuhl-Leibniz-Zentrum für Informatik (2020)
25. Fomin, F.V., Golovach, P.A., Panolan, F., Philip, G., Saurabh, S.: Diverse collections in matroids and graphs. Math. Program. 1–33 (2023)
26. Frederickson, G.N., Johnson, D.B.: The complexity of selection and ranking in x+y and matrices with sorted columns. J. Comput. Syst. Sci. **24**(2), 197–208 (1982)
27. Funayama, R., Kobayashi, Y., Uno, T.: Parameterized complexity of finding dissimilar shortest paths. arXiv preprint arXiv:2402.14376 (2024)
28. Gálvez, W., Goswami, M., Merino, A., Park, G., Tsai, M.T.: Computing diverse and nice triangulations. arXiv preprint arXiv:2506.01323 (2025)
29. Gálvez, W., Goswami, M., Merino, A., Park, G., Tsai, M.T., Verdugo, V.: A framework for the design of efficient diversification algorithms to np-hard problems. arXiv preprint arXiv:2501.12261 (2025)
30. Gálvez, W., Verdugo, V.: Approximation schemes for packing problems with ℓ_p-norm diversity constraints. In: Castañeda, A., Rodríguez-Henríquez, F. (eds) LATIN 2022. LNCS, vol. 13568, pp. 204–221. Springer, Cham (2022). https://doi.org/10.1007/978-3-031-20624-5_13
31. Gao, J., Goswami, M., Karthik, C., Tsai, M.T., Tsai, S.Y., Yang, H.T.: Obtaining approximately optimal and diverse solutions via dispersion. In: Castañeda,

A., Rodríguez-Henríquez, F. (eds) LATIN 2022. LNCS, vol. 13568, pp. 222–239. Springer, Cham (2022). https://doi.org/10.1007/978-3-031-20624-5_14
32. Hanaka, T., Kiyomi, M., Kobayashi, Y., Kobayashi, Y., Kurita, K., Otachi, Y.: A framework to design approximation algorithms for finding diverse solutions in combinatorial problems. In: Williams, B., Chen, Y., Neville, J. (eds.) Thirty-Seventh AAAI Conference on Artificial Intelligence, AAAI, pp. 3968–3976. AAAI Press (2023)
33. Hanaka, T., Kobayashi, Y., Kurita, K., Lee, S.W., Otachi, Y.: Computing diverse shortest paths efficiently: a theoretical and experimental study. In: Thirty-Sixth AAAI Conference on Artificial Intelligence, AAAI, pp. 3758–3766. AAAI Press (2022)
34. Hanaka, T., Kobayashi, Y., Kurita, K., Otachi, Y.: Finding diverse trees, paths, and more. In: Thirty-Fifth AAAI Conference on Artificial Intelligence (AAAI), pp. 3778–3786. AAAI Press (2021)
35. Klute, F., van Kreveld, M.: On fully diverse sets of geometric objects and graphs. In: Bekos, M.A., Kaufmann, M. (eds.) WG 2022. LNCS, vol. 13453, pp. 328–341. Springer, Cham (2022). https://doi.org/10.1007/978-3-031-15914-5_24
36. van Kreveld, M., Speckmann, B., Urhausen, J.: Diverse Partitions of Colored Points. In: Lubiw, A., Salavatipour, M. (eds.) WADS 2021. LNCS, vol. 12808, pp. 641–654. Springer, Cham (2021). https://doi.org/10.1007/978-3-030-83508-8_46
37. Lawson, C.L.: Software for C1 surface interpolation. Math. Softw. (1977)
38. Lloyd, E.L.: On triangulations of a set of points in the plane. In: 18th Annual Symposium on Foundations of Computer Science (SFCS 1977), pp. 228–240. IEEE (1977)
39. Misra, N., Mittal, H., Rai, A.: On the parameterized complexity of diverse SAT. In: 35th International Symposium on Algorithms and Computation (ISAAC 2024), pp. 50–1. Schloss Dagstuhl–Leibniz-Zentrum für Informatik (2024)
40. Mulzer, W., Rote, G.: Minimum-weight triangulation is NP-hard. J. ACM (JACM) **55**(2), 1–29 (2008)
41. O'Rourke, J.: Open problems from CCCG 2017. In: 29th Canadian Conference on Computational Geometry (2017)
42. Ostergard, P.R.: On the size of optimal three-error-correcting binary codes of length 16. IEEE Trans. Inf. Theory **57**(10), 6824–6826 (2011)
43. Pournin, L.: The diameter of associahedra. Adv. Math. **259**, 13–42 (2014)
44. Ravi, S.S., Rosenkrantz, D.J., Tayi, G.K.: Heuristic and special case algorithms for dispersion problems. Oper. Res. **42**(2), 299–310 (1994)
45. Remy, J., Steger, A.: A quasi-polynomial time approximation scheme for minimum weight triangulation. J. ACM (JACM) **56**(3), 1–47 (2009)
46. Rieck, C., Scheffer, C.: The dispersive art gallery problem. Comput. Geom. **117**, 102054 (2024)

Determining the Metric Dimension of $K_a \times K_b \times K_c$ by Static Black-Peg Mastermind

Valentin Gledel[1] and Gerold Jäger[2(✉)]

[1] CNRS UMR5127, LAMA, Université Savoie Mont Blanc, 73000 Chambéry, France
valentin-gledel@univ-smb.fr

[2] Department of Mathematics and Mathematical Statistics, University of Umeå, 901-87 Umeå, Sweden
gerold.jager@umu.se

Abstract. In this paper we determine the metric dimension of $K_a \times K_b \times K_c$ for all $a, b, c \in \mathbb{N}$ with $a \leq b \leq c$ as follows. For $3a < b + c$ and $2b \leq c$, this value is $c - 1$, for $3a < b + c$ and $2b > c$, it is $\lfloor \frac{2}{3}(b + c - 1) \rfloor$, and for $3a = b + c$, it is $\lfloor \frac{a+b+c}{2} \rfloor - 1$. The only open case is $3a > b + c$, where two values are possible, namely $\lfloor \frac{a+b+c}{2} \rfloor - 1$ and $\lfloor \frac{a+b+c}{2} \rfloor$. This result extends previous results of [4], who computed the metric dimension of $K_a \times K_b$, and of [14], who computed the metric dimension of $K_a \times K_a \times K_a$.

We prove our result by introducing and analyzing a new variant of Static Black-Peg Mastermind, in which each peg has its own permitted set of colors. For all cases, we present strategies which we prove to be both feasible and optimal. Our main result follows, as the number of questions of these strategies is equal to the metric dimension of $K_a \times K_b \times K_c$.

1 Introduction

The *metric dimension* is an important property of an undirected and unweighted graph $G = (V, E)$ and is defined as follows. A set $U \subseteq V$ so that every vertex $v \in V$ is uniquely determined by the vector of distances between v and the vertices in U is called *resolving set* of G, and the metric dimension is defined as the cardinalilty of a minimum resolving set. The concept of metric dimension has been mentioned first in [1] and was then independently re-introduced in [12,19]. The metric dimension has been studied for several graph classes, see e.g. [2,7, 18,20]. Determining whether the metric dimension of a graph is smaller than or equal to a given value is NP-complete [8,17].

In this paper, we aim to determine the metric dimension of the cartesian product of complete graphs K_a for $a \geq 1$, where in comparison to [14] we use the notation K_a instead of Z_a, as this notation of complete graphs is more common in graph theory. The following results are known for this problem. The metric dimension of $K_p^c = K_p \times K_p \times \cdots \times K_p$ is in $\Omega(p/\log p)$, if c is constant [3], the metric dimension of $K_b \times K_c$ is $\lfloor \frac{2}{3}(b + c - 1) \rfloor$ if $b \leq c \leq 2b$ and $c - 1$ if

A. Jeż and J. Otop (Eds.): FCT 2025, LNCS 16106, pp. 194–207, 2026.
https://doi.org/10.1007/978-3-032-04700-7_15

$c > 2b$ [4], and as a special case it is $\lceil(4a-1)/3-1\rceil$ for $K_a \times K_a$ [13], and finally, the metric dimension of $K_a \times K_a \times K_a$ is $\lfloor \frac{3}{2}a \rfloor$ for $a \geq 2$.

We make our investigation into the metric dimension of $K_a \times K_b \times K_c$ by introducing a variant of Mastermind, which is a famous board game invented by Meirowitz in 1970. The original game is played by two players, the *codemaker* and the *codebreaker*, where the codemaker chooses a secret code consisting of 4 pegs and 6 possible colors for each peg and where the codebreaker must discover this code by making a sequence of guesses (or questions) from the set of possible secrets until the correct secret has been found. In each answer, the codemaker gives black and white pegs, one black peg for each peg of the question which is correct in both position and color, and one white peg for each peg which is correct only in color. The problem we consider extends Mastermind in the following ways.

- We generalize the number of pegs from 4 to p.
- Each peg i has its own set of colors c_i for given positive integers $\{c_i | i \in [p]\}$.
- The codemaker gives only the number of black pegs as answers (and thus no longer white pegs, i.e., no information regarding correct colors in incorrect positions).
- The game is static, i.e., the codebreaker gives all questions already at the beginning of the game, waits for all answers and has a final question to give the (correct) question.

Almost all such variations of Mastermind have already been investigated (see for example [5,6,15] for Mastermind with p pegs and c colors, [11,16] for Black-Peg Mastermind, [5,10,13,14] for Static Mastermind).

The only novel variation is the second one, i.e., the extension of each peg to an own color set. More concretely, we generalize Mastermind to Static Black-Peg (a, b, c)-Mastermind, where there are only three pegs, each having its own set of colors.

We call it (a, b, c)-Mastermind for short, and define it as follows. We have two players, the *codemaker* and the *codebreaker*. The codemaker chooses a secret code by assigning a color to each peg, chosen from a set of a colors for the first peg, b colors for the second peg and c colors for the third peg. The codebreaker must discover this code by making a sequence of questions of possible secrets at the beginning of the game, where the set of secrets is the same as the set of questions. Each answer of the codemaker consists of black pegs, one black peg for each peg of the question which is correct in both position and color. The codebreaker then receives all answers and finally has to reply with the (not any more) secret code. The goal is to find the a sequence of questions with minimum cardinality so that the codebreaker knows the secret from the received answers. We call this an optimal strategy. Note that it can be assumed w.l.o.g that $a \leq b \leq c$. To simplify the study, we will suppose it to be the case through this paper, unless stated otherwise.

As explained in [14] for the particular case of (a, a, a)-Mastermind, the cardinality of an optimal strategy for (a, b, c)-Mastermind is equal to the metric dimension of $K_a \times K_b \times K_c$. Indeed, each question would correspond to a vertex of $K_a \times K_b \times K_c$ in the resolving set, 3 minus the number of black pegs that a question gives to a secret corresponds to the distance between the associated vertices in the graph, and being able to distinguish every secret corresponds to being able to distinguish every vertex with the resolving set. To obtain our result, we extend methods from [14]. We find that optimal strategies mostly consist of questions of the following kind: on one peg the corresponding color occurs only once over all questions, and on the other two pegs the colors occur exactly twice. As result, we obtain different values for the four main cases $3a > b+c$, $3a = b+c$, $3a < b+c$ with $2b \leq c$ and $3a < b+c$ with $2b > c$. Our results are summarized in Table 1.

Table 1. Summary of the results appearing in this paper, where $f(a,b,c)$ is the metric dimension of $K_a \times K_b \times K_c$.

$3a > b+c$	$\lfloor \frac{a+b+c}{2} \rfloor - 1 \leq f(a,b,c) \leq \lfloor \frac{a+b+c}{2} \rfloor$
	$f(a,a,2a-1) = 2a-1 = \lfloor \frac{a+a+2a-1}{2} \rfloor$
	$f(a,a,a) = \lfloor \frac{3a}{2} \rfloor$ (result from [14])
$3a = b+c$	$(a,b,c) \neq (4,6,6) \rightarrow f(a,b,c) = \lfloor \frac{a+b+c}{2} \rfloor - 1 = 2a-1$
	$f(4,6,6) = 8 = \lfloor \frac{4+6+6}{2} \rfloor$
$3a < b+c$	$2b \leq c \rightarrow f(a,b,c) = c-1$
	$2b > c \rightarrow f(a,b,c) = \lfloor \frac{2}{3}(b+c-1) \rfloor$

The outline of this paper is as follows. We start with some notations and preliminaries in Sect. 2. We then provide three general lower bounds for the metric dimension of $K_a \times K_b \times K_c$ in Sect. 3. In Sect. 4, we show upper bounds for this metric dimension, often reaching the lower bound. The results of these two sections are summarized in Table 1. Finally, in Sect. 5, we discuss our results and the remaining open case, and as possible future work we suggest to prove a conjecture. Note that for some proofs of Sec. 4 we refer to the full version of this paper.

2 Notations and Preliminaries

Let $\mathbb{N}$ be the set of positive integers, and for $n \in \mathbb{N}$ let $[n]$ be the set $\{1, 2, \ldots, n\}$. We aim to determine the metric dimension of $K_a \times K_b \times K_c$, which we denote by $f(a,b,c)$ for $a, b, c \in \mathbb{N}$.

We define a *strategy* for (a,b,c)-Mastermind to be a set of questions. For a secret $s = (s_1, s_2, s_3)$ and a question $q = (q_1, q_2, q_3)$, let $g(q,s) = i$ for $i \in \{0,1,2,3\}$, if the question q receives i black pegs for the secret s. A strategy Q is *feasible* if for all secrets s, s' with $s \neq s'$ there is a question $q \in Q$ such that $g(q,s) \neq g(q,s')$. Thus, a *feasible strategy* is one which lets the codebreaker win no matter which secret the codemaker chooses, An *optimal* strategy is one which is minimal in size.

We make use of the following terminology from [14].

Definition 1. **(a)** *Two questions* (d_1, d_2, d_3) *and* (e_1, e_2, e_3) *are* neighboring *if there is exactly one index* $i \in [3]$ *such that* $d_i = e_i$.

(b) *Two questions* (d_1, d_2, d_3) *and* (e_1, e_2, e_3) *are* double-neighboring, *if there are exactly two indices* $i \in [3]$ *such that* $d_i = e_i$.

(c) *For a given strategy and* $f_1, f_2, f_3 \in \mathbb{N}$, *a question* q *is a* (f_1, f_2, f_3)-question, *if for* $i \in [3]$, *the* i*-th color of* q *occurs* f_i *times on the* i*-th peg (throughout the entire strategy). We extend this by allowing* f_i *to be* $\star$, *meaning that it is not relevant how often the* i*-th color occurs on the* i*-th peg.*

Definition 2. *Consider a strategy for* (a, b, c)*-Mastermind. The* strategy graph $G = (V, E)$ *of this strategy is defined as the undirected multi-graph whose vertex set* V *is the set of all questions of the strategy, and in which two vertices are linked by one edge if and only if they are neighboring questions and two questions are linked by two edges if and only if they are double-neighboring.*

In the following, we often consider strategies consisting of the following types of questions, each of which is named by a letter:

$(1,2,2)$-questions	$\rightarrow$ type A	$(2,1,1)$-questions	$\rightarrow$ type D
$(2,1,2)$-questions	$\rightarrow$ type B	$(1,2,1)$-questions	$\rightarrow$ type E
$(2,2,1)$-questions	$\rightarrow$ type C	$(1,1,2)$-questions	$\rightarrow$ type F

In the corresponding strategy graph each vertex is of degree 1 or 2. Thus, the graph consists of connected components each of which is either a path or a cycle.

In the following, we list the questions in a given strategy as follows. We order the blocks in an arbitrary way. We start with the questions of one block, continue with the questions of another block until the last block. Each path starts with one of its extreme vertices while each cycle starts with an arbitrary vertex. Thus, in one block, the next question is one such that the questions are neighboring, i.e., that the corresponding vertices are adjacent.

3 Lower Bounds

In this section, we prove three general lower bounds on $f(a, b, c)$. Note that for different triples (a, b, c), different ones are best.

For the first one, we make use of the following lemma [14, Lemma 3, Lemma 4(a),(b)]. In this paper the proof was in the case $a = b = c$, but it works just as well without this assumption.

Lemma 1. *Let* $a, b, c \in \mathbb{N}$. *Every feasible strategy for* (a, b, c)*-Mastermind has the following properties:*

(a) There exists at most one color which does not occur on the first peg of any question.
Analogous statements hold for the second peg and for the third peg.

(b) There is at most one $(1,1,\star)$*-question.*
Analogous statements hold for the cases of $(1,\star,1)$*-questions and* $(\star,1,1)$*-questions.*
In particular, setting $\star = 1$*, there can be at most one* $(1,1,1)$*-question.*

(c) The strategy does not contain three distinct questions one of which is a $(1,1,\star)$*-question, one of which is a* $(1,\star,1)$*-question, and the third one of which is a* $(\star,1,1)$*-question.*

(d) Assume that there is a color which does not occur on the first peg of any question and a color which does not occur on the second peg of any question. Then the strategy cannot contain a $(1,1,\star)$*-question.*
Analogous statements hold for the first and the third peg, and for the second and the third peg.

(e) Assume that there is a color which does not occur on the first peg of any question and a color which does not occur on the second peg of any question. Then the strategy cannot contain both a $(1,\star,1)$*-question and a* $(\star,1,1)$*-question.*
Analogous statements hold for the first and the third peg (where the strategy cannot contain both a $(1,1,\star)$*-question and a* $(\star,1,1)$*-question), and for the second and the third peg (where the strategy cannot contain both a* $(1,1,\star)$*-question and a* $(1,\star,1)$*-question).*

(f) If the strategy contains two double-neighboring $(2,2,1)$*-questions, then it contains no* $(1,1,\star)$*-question, and it holds that on at least one of the first two pegs all colors occur.*
Analogous questions hold for two double-neighboring $(2,1,2)$*-questions and for two double-neighboring* $(1,2,2)$*-questions.*

(g) Assume that the strategy contains a $(2,2,1)$*-question, a* $(2,\star,1)$*-question and a* $(\star,2,1)$*-question, where the first and the second question are neighboring on the first peg, and the first and the third question are neighboring on the second peg. Then the strategy contains no* $(1,1,\star)$*-question, and it holds that on at least one of the first two pegs all colors occur in the questions.*
Analogous statements hold for a $(2,1,2)$*-question, a* $(\star,1,2)$*-question and a* $(2,1,\star)$*-question, and for a* $(1,2,2)$*-question, a* $(1,\star,2)$*-question and a* $(1,2,\star)$*-question.*

We can now give the first general lower bound with the following theorem.

Theorem 1. *Let* $a,b,c \in \mathbb{N}$*. Then* $f(a,b,c) \geq \lfloor \frac{a+b+c}{2} \rfloor - 1$ *holds.*

Proof. Assume that Q is a feasible strategy for (a,b,c)-Mastermind, with x colors that occur only once on the first peg, y colors that occur only once on the second peg and z colors that occur only once on the third peg. Assume also that $|Q| = \lfloor \frac{a+b+c}{2} \rfloor - k$, and thus

$$2(a+b+c) \geq 4|Q| + 4k. \tag{1}$$

By Lemma 1(a), on each peg at most one color is missing.

Case 1: On each peg one color is missing

Recall that x is the number of colors that occur exactly once on the first peg. As all other colors except for the missing one occur at least twice, $|Q| \geq x + 2(a - x - 1)$, and thus $x \geq 2a - |Q| - 2$. We obtain in the same way that $y \geq 2b - |Q| - 2$ and $z \geq 2c - |Q| - 2$.

Thus, by (1),

$x + y + z \;\geq\; 2(a + b + c) - 3|Q| - 6 \;\geq\; |Q| + 4k - 6.$

If $x+y+z \geq |Q|+1$, there must be a $(1, 1, \star)$, a $(1, \star, 1)$ or a $(\star, 1, 1)$-question, but by Lemma 1(d) this is impossible, since there is a missing color on each peg. So $x + y + z \leq |Q|$ and thus $4k \leq 6$, which shows that $k \leq 1$.

Case 2: There are two pegs with one color missing on each peg.

W.l.o.g., the missing colors occur on the first and the second peg and the third peg has no missing color. Analogously to Case 1, for the first two pegs we have $x \geq 2a - |Q| - 2$ and $y \geq 2b - |Q| - 2$. On the third peg, since there is no missing color, $|Q| \geq z + 2(c - z)$ holds and $z \geq 2c - |Q|$ follows. Thus, by (1),

$x + y + z \;\geq\; 2(a + b + c) - 3|Q| - 4 \;\geq\; |Q| + 4k - 4.$

Since there is a missing color on the first and on the second peg, by Lemma 1(d), no $(1, 1, \star)$-question can occur, and by Lemma 1(e), it is not possible that there is both a $(1, \star, 1)$-question and a $(\star, 1, 1)$-question. Furthermore, by Lemma 1(b), if one of these questions occurs, there is no other question of the same type. If $x+y+z \geq |Q|+2$, one of these situations would have to be the case, which shows that $x + y + z \leq |Q| + 1$. Thus, we must have $4k - 4 \leq 1$ and so $k \leq 1$.

Case 3: There is at most one peg having a missing color

W.l.o.g., there are no missing colors occurring on the second and the third peg. As above, this means $y \geq 2b - |Q|$ and $z \geq 2c - |Q|$. As there there might be a missing color on the first peg, we have $x \geq 2a - |Q| - 2$. Thus, by (1),

$x + y + z \;\geq\; 2(a + b + c) - 3|Q| - 2 \;\geq\; |Q| + 4k - 2.$

By Lemma 1(b), at most one $(1, 1, \star)$-question, at most one $(1, \star.1)$-question, and at most one $(\star.1, 1)$-question occurs. By Lemma 1(c), it is not possible that all of them occur at the same time. If $x + y + z \geq |Q| + 3$, this would be the case, which shows that $x + y + z \leq |Q| + 2$, $4k - 2 \leq 2$ and so $k \leq 1$.

Thus, the feasibility of Q implies that $k \leq 1$, which proves that $|Q| \geq \lfloor \frac{a+b+c}{2} \rfloor - 1$ for all feasible strategies. □

The second general lower bound is an easy corollary of Lemma 1.

Corollary 1. *Let $a, b, c \in \mathbb{N}$ with $a \leq b \leq c$. Then it holds that $f(a, b, c) \geq c - 1$.*

Proof. This follows directly from Lemma 1(a), as having less questions would lead to at least two missing colors in the last peg. □

The third general lower bound is based on results from Cáceres *et al.* [4].

Theorem 2. *Let $a, b, c \in \mathbb{N}$ with $a \leq b \leq c$. Then it holds that $f(a, b, c) \geq \lfloor \frac{2}{3}(b + c - 1) \rfloor$.*

Proof. First, let $c < 2b$. Then by [4, Theorem 6.1], the metric dimension of $K_b \times K_c$ is $\lfloor \frac{2}{3}(b+c-1) \rfloor$, therefore $f(1,b,c) = \lfloor \frac{2}{3}(b+c-1) \rfloor$. Moreover, by [4, Corollary 3.2], the metric dimension of $K_a \times K_b \times K_c$ is greater than or equal to $K_b \times K_c$ and therefore $f(1,b,c) \leq f(a,b,c)$ holds. Second, let $c \geq 2b$. Then by Corollary 1, $\lfloor \frac{2}{3}(b+c-1) \rfloor \leq \lfloor c - \frac{2}{3} \rfloor = c - 1 \leq f(a,b,c)$. The assertion follows in both cases. □

4 Upper Bounds

In this section, we will give upper bounds for $f(a,b,c)$. As the methods are similar, we will give all the details for the case $3a \geq b+c$, and the details for the other cases can be found in the full version of this paper.

4.1 Basic Lemma

In the remainder of this paper, we will introduce several strategies using questions of type A, B, C. We will use the following lemma to prove the feasibility of these strategies. To do so, we provide a list of patterns such that strategies avoiding all these patterns are feasible.

Let $T \in \{A, B, C\}$ be a type of question. The *single peg* of T is the peg that contains the color occurring only in the question of type T, and the two other pegs are the *double pegs* of T. For example, the single peg of a question of type A is the first peg and the other two pegs are its double pegs.

We say that a set of k questions is *a sequence of type* $T_1T_2 \ldots T_k$ if these questions can be arranged in a sequence $q_1, q_2, \ldots, q_k$ such that

(a) q_i and q_{i+1} are neighboring to each other for all $i \in [k-1]$, and
(b) q_i is of type $T_i \in \{A, B, C\}$ for all $i \in [k]$.

Lemma 2. *Let Q be a set of questions containing only questions of type A, B and C and avoiding the following patterns:*

(1) *two colors missing on the same peg,*
(2) *a missing color on each peg,*
(3) *a sequence of type TTT and missing colors on both of the double pegs of T.*
(4) *a sequence of type TTT and a pair of double-neighboring TT,*
(5) *a sequence of type $T_1T_1T_2T_2$ ($T_1 \neq T_2$) with a missing color on the peg that is double for both T_1 and T_2,*
(6) *a sequence of type $T_1T_1T_2T_3T_3$, where all three types distinct,*
(7) *two different sequences of the same type T_1T_2 ($T_1 \neq T_2$),*
(8) *a sequence of type T_1T_2 ($T_1 \neq T_2$), a sequence of type T_2T_2 of double-neighboring questions, and a missing color on the single peg of T_2,*
(9) *a sequence of type TT of double-neighboring questions, with missing colors on both the double pegs of T,*
(10) *two different sequences of type TT of double-neighboring questions,*

(11) *two different sequences of types T_1T_1 and T_2T_2, both of these consisting of double-neighboring questions, and missing colors on the single pegs of both T_1 and T_2, and*
(12) *sequences of types AA, BB, and CC of double-neighboring questions.*

Then Q is a feasible strategy.

The proof of Lemma 2 can be found in the full version. It shows that every pair of secrets is indeed distinguished by such strategies by doing a case analysis on the differences/similarities of the colors in the secrets.

Remark 1. Let a strategy have only questions of type A, B or C. Then by Lemma 2, avoiding the patterns 1 to 12 is sufficient for the feasibility of a strategy. On the other hand, it is easy to see that if a strategy contains one of the patterns (except possibly pattern 2), it is not feasible.

4.2 The Case of $(\mathbf{a}, \mathbf{b}, \mathbf{c})$-Mastermind with $3\mathbf{a} \geq \mathbf{b} + \mathbf{c}$

In this section, we prove the following theorem.

Theorem 3. *Let $a, b, c \in \mathbb{N}$ with $a \leq b \leq c$ and $3a \geq b + c$. Then it holds that $f(a, b, c) \leq \lfloor \frac{a+b+c}{2} \rfloor$.*

To establish this theorem, we define suitable feasible strategies. To do so, let $c' = c$ if $a + b + c$ is even and $c' = c - 1$ if $a + b + c$ is odd, so that $a + b + c'$ is an even number. We can now define x, y and z, such that the strategies to be constructed below contain x questions of type A, y questions of type B, and z questions of type C:

$$x := 2a - \frac{a + b + c'}{2}, \quad y := 2b - \frac{a + b + c'}{2}, \quad z := 2c' - \frac{a + b + c'}{2}.$$

Note that x, y and z are integers, because $a + b + c' \equiv 0 \pmod 2$, and that $x + y + z = \lfloor \frac{a+b+c}{2} \rfloor$. Moreover, since $3a \geq b + c$ and $b \geq a$, x and y are non-negative. If $c' = c$ then z is also non-negative.

In the remaining case, where $c' = c - 1$, z is non-negative if $3(c - 1) \geq a + b$. Assume now that $3(c - 1) < a + b$. It follows that $c - 1 < a$, because otherwise we would have $3(c-1) \geq 3a \geq b+c \geq a+b$. Together, the inequalities $a \leq b \leq c$ and $c - 1 < a$ imply $a = b = c$. So, $a + b + c - 3 = 3(c - 1) < a + b$ and thus $c < 3$. This leaves two cases: $a = b = c = 1$, where $f(a, b, c) = 0$ and the statement of the theorem holds, or $a = b = c = 2$, which does not fall into the case $c' = c - 1$ since $a + b + c$ is even. Therefore, disregarding the uninteresting case $a = b = c = 1$, z is always non-negative, too.

Strategy for the case $a + b + c \equiv 0, 1 \pmod 4$:
Observe that in the case $3a \geq b + c$ and $a + b + c \equiv 0, 1 \pmod 4$, x, y and z are even.

As mentioned earlier, the strategy consists of x questions of type A, y questions of type B, and z questions of type C. Each of these question types will form a block (recall that a block is a connected component of the strategy graph).

Strategy 1 *($\lfloor\frac{a+b+c}{2}\rfloor$-strategy for (a,b,c)-Mastermind with $3a \geq b+c$, $a+b+c \equiv 0, 1 \pmod 4$ and $(a,b,c) \notin \{(4,4,4),(4,4,5)\}$)*

First peg:
- First block: colors 1, 2, …, x.
- Second block: colors $x+1$, $x+1$, $x+2$, $x+2$, …, $x+y/2$, $x+y/2$.
- Third block: colors $x+y/2+1$, $x+y/2+1$, $x+y/2+2$, $x+y/2+2$, …, a, a.

Second peg:
- First block: colors 1, 1, 2, 2, …, $x/2$, $x/2$.
- Second block: colors $x/2+1$, $x/2+2$, …, $x/2+y-1$, $x/2+y$.
- Third block: colors $x/2+y+1$, $x/2+y+2$, $x/2+y+2$, …, b, b, $x/2+y+1$.

Third peg:
- First block: colors 1, 2, 2, …, $x/2$, $x/2$, 1.
- Second block: colors $x/2+1$, $x/2+2$, $x/2+2$, …, $x/2+y/2$, $x/2+y/2$, $x/2+1$.
- Third block: colors $x/2+y/2+1$, $x/2+y/2+2$, …, $c'-1$, c'.

Remark 2. Strategy 1 has the following characteristics:

(a) It contains three (possibly empty) blocks, each forming a cycle of questions: one cycle consisting of x questions of type A, one consisting of y questions of type B, and one consisting of z questions of type C.
(b) Each block can contain a double-neighboring pair of questions (if $x = 2$, $y = 2$ or $z = 2$). However, because of $(a,b,c) \notin \{(4,4,4),(4,4,5)\}$, there can be at most two double-neighboring pairs of questions in total.
(c) There are no missing colors on the first two pegs, and there is a single missing color on the third peg only in the case $a+b+c \equiv 1 \pmod 4$.

Two examples for different triples (a,b,c) are given in Table 2(b), (c).

Lemma 3. *Strategy 1 is feasible for (a,b,c)-Mastermind where $a \leq b \leq c$, $3a \geq b+c$, $a+b+c \equiv 0, 1 \pmod 4$ and $(a,b,c) \notin \{(4,4,4),(4,4,5)\}$.*

Proof. We use Lemma 2 to prove that Strategy 1 is feasible. By Remark 2(a), the condition of having only questions of type A, B or C is fulfilled, and by Remark 2(c) the patterns 1, 2, 3, 9 and 11 do not occur.

Since no questions of different types are neighboring to each other, the patterns 5, 6, 7 and 8 do not occur.

If two questions are double-neighboring, then the strategy contains no other questions of the same type. Therefore, the patterns 4 and 10 do not occur.

Lastly, note that since $(a,b,c) \notin \{(4,4,4),(4,4,5)\}$ the strategy does not consist of three double-neighboring pairs of questions. To show this, let $x = y = z = 2$ hold. Indeed, since $x = 2a - (a+b+c')/2 = 2a - (x+y+z)$, it follows that $a = (2x+y+z)/2 = 4$. Similarly, it follows that $b = 4$ and $c' = 4$. Since we excluded this case, there are at most two double-neighboring questions and pattern 12 does not occur.

Therefore, none of the forbidden patterns of Lemma 2 occurs, and Q is feasible. □

Table 2. Three examples of strategies for (a,b,c)-Mastermind with $3a \geq b+c$ and $a+b+c \equiv 0,1 \pmod 4$.

Peg	1	2	3
q_1	1	1	1
q_2	1	1	2
q_3	1	2	3
q_4	2	3	3
q_5	3	4	4
q_6	4	4	4

(a) $(a,b,c) = (4,4,c)$ for $c \in \{4,5\}$ and 6 questions.

Peg	1	2	3	Type
q_1	1	1	1	B
q_2	1	2	2	B
q_3	2	3	2	B
q_4	2	4	1	B
q_5	3	5	3	C
q_6	3	6	4	C
q_7	4	6	5	C
q_8	4	5	6	C

(b) $(a,b,c) = (4,6,6)$, yielding $x = 0$ and $y = z = 4$, and thus 8 questions. (Note that for $(4,6,7)$ the condition $3a \geq b+c$ does not hold.)

Peg	1	2	3	Type
q_1	1	1	1	A
q_2	2	1	1	A
q_3	3	2	2	B
q_4	3	3	3	B
q_5	4	4	3	B
q_6	4	5	2	B
q_7	5	6	4	C
q_8	5	7	5	C
q_9	6	7	6	C
q_{10}	6	8	7	C
q_{11}	7	8	8	C
q_{12}	7	6	9	C

(c) $(a,b,c) = (7,8,c)$ for $c \in \{9,10\}$, yielding $x = 2$, $y = 4$, $z = 6$ and thus 12 questions.

Remark 3. As stated in Lemma 3, Strategy 1 cannot be applied to $(4,4,4)$ and $(4,4,5)$. Note, however, that these cases can nevertheless be solved with 6 questions. Indeed, the strategy described in Table 2(a) is feasible for both of these cases, too. For the feasibility we refer to a computer program which checks the feasibility of strategies by exhaustive search [9].

Note that there is a gap of 1 between the lower bound of Theorem 1 and the upper bound given by Theorem 3. In fact, in Theorems 4 and 6, we will show that it is possible to obtain better strategies in some cases.

Strategy for the case $a+b+c \equiv 2,3 \pmod 4$: We now provide a strategy for the case $3a \geq b+c$ and $a+b+c \equiv 2,3 \pmod 4$. Observe that in this case, $x = 2a - \frac{a+b+c'}{2}$, $y = 2b - \frac{a+b+c'}{2}$ and $z = 2c' - \frac{a+b+c'}{2}$ are odd.

As before, the strategy has x questions of type A, y questions of type B, and z questions of type C. The first two blocks consist of $x-1$ and $y-1$ questions of type A and B, respectively. The third block starts with one question of type A and one of type B, and continues with z questions of type C.

Strategy 2 *($\lfloor \frac{a+b+c}{2} \rfloor$-strategy for (a, b, c)-Mastermind with $3a \geq b+c$, and $a+b+c \equiv 2, 3 \pmod 4$))*

First peg:
- *First block: colors* $1,\ 2,\ \ldots,\ x-1$.
- *Second block: colors* $x,\ x,\ x+1,\ x+1,\ \ldots,\ x-1+(y-1)/2,\ x-1+(y-1)/2$.
- *Third block: colors* $x+(y-1)/2,\ x+(y-1)/2+1,\ x+(y-1)/2+1,\ x+(y-1)/2+2,\ x+(y-1)/2+2,\ \ldots,\ a,\ a$.

Second peg:
- *First block: colors* $1,\ 1,\ 2,\ 2,\ \ldots,\ (x-1)/2,\ (x-1)/2$.
- *Second block: colors* $(x-1)/2+1,\ (x-1)/2+2,\ \ldots,\ (x-1)/2+y-2,\ (x-1)/2+y-1$.
- *Third block: colors* $(x-1)/2+y,\ (x-1)/2+y+1,\ (x-1)/2+y+2,\ (x-1)/2+y+2,\ \ldots,\ b,\ b,\ (x-1)/2+y$.

Third peg:
- *First block: colors* $1,\ 2,\ 2,\ \ldots,\ (x-1)/2,\ (x-1)/2,\ 1$.
- *Second block: colors* $(x-1)/2+1,\ (x-1)/2+2,\ (x-1)/2+2,\ \ldots,\ (x-1)/2+(y-1)/2,\ (x-1)/2+(y-1)/2,\ (x-1)/2+1$.
- *Third block: colors* $(x-1)/2+(y-1)/2+1,\ (x-1)/2+(y-1)/2+1,\ (x-1)/2+(y-1)/2+2,\ (x-1)/2+(y-1)/2+3,\ \ldots,\ c'-1,\ c'$.

Remark 4. Strategy 2 has the following characteristics:

(a) It contains three (possibly empty) blocks, each forming a cycle of questions. The first consists of $x-1$ questions of type A and the second of $y-1$ questions of type B. The third block contains one question of type A, one of type B, and z of type C.
(b) Both the first and the second block can contain a double-neighboring pair of questions, but there cannot be any other double-neighboring pairs.
(c) There is no missing color on the first two pegs, and only one missing color on the third peg for the case $a+b+c \equiv 3 \pmod 4$.

Two examples are given in Table 3.

Lemma 4. *Strategy 2 is feasible for (a, b, c)-Mastermind where $a \leq b \leq c$, $3a \geq b+c$ and $a+b+c \equiv 2, 3 \pmod 4$.*

Proof. We will use Lemma 2 to prove that Strategy 2 is feasible. By Remark 4(a), the condition of having only questions of types A, B or C is fulfilled. By Remark 4(c), none of the patterns 1, 2, 3, 9 and 11 can occur, and Remark 4(b) excludes pattern 12. Moreover, since there are no double neighboring questions of type C, and the only possible missing color is on the third peg, which is the single peg of type C, pattern 8 does not occur.

Since all the questions of type A except one form a cycle in the first block, and all those of type B except one form a cycle in the second block, the patterns 4 and 10 do not occur.

Lastly, the only question of type A that is not neighboring to another one of type A is in the cycle of the third block, which contains only one question of

Table 3. Two examples of strategies for (a, b, c)-Mastermind with $3a \geq b + c$ and $a + b + c \equiv 2, 3 \pmod 4$.

Peg	1	2	3	Type
q_1	1	1	1	A
q_2	2	2	1	B
q_3	2	3	2	C
q_4	3	3	3	C
q_5	3	1	4	C

(a) $(a, b, c) = (3, 3, c)$ for $c \in \{4, 5\}$, yielding $x = y = 1$ and $z = 3$, and thus 5 questions.

Peg	1	2	3	Type
q_1	1	1	1	A
q_2	2	1	1	A
q_3	3	2	2	B
q_4	3	3	3	B
q_5	4	4	3	B
q_6	4	5	2	B
q_7	5	6	4	A
q_8	6	7	4	B
q_9	6	8	5	C
q_{10}	7	8	6	C
q_{11}	7	9	7	C
q_{12}	8	9	8	C
q_{13}	8	6	9	C

(b) $(a, b, c) = (8, 9, c)$ for $c \in \{9, 10\}$, yielding $x = 3$ and $y = z = 5$, and thus 13 questions.

type A. Similarly, the only question of type B that is not neighboring to another one is in the cycle of the third block, which contains only one question of type B. Therefore, none of the patterns 5, 6 and 7 occurs.

In summary, we have excluded all of the forbidden patterns of Lemma 2, which proves that the strategy is feasible. □

We can now prove Theorem 3:

Proof (of Theorem 3). Our computer program [9] shows that the strategy of Table 2(a), which has $\lfloor \frac{a+b+c}{2} \rfloor = 6$ questions, is feasible for the cases $(4, 4, 4)$ and $(4, 4, 5)$. Lemmas 3 and 4 prove for all other cases that Strategy 1 is feasible for $a+b+c \equiv 0, 1 \pmod 4$ and Strategy 2 is feasible for $a+b+c \equiv 2, 3 \pmod 4$, both of which consist of $\lfloor \frac{a+b+c}{2} \rfloor$ questions. This covers all cases, showing that $f(a, b, c) \leq \lfloor \frac{a+b+c}{2} \rfloor$. □

4.3 Other Cases

The other results given in this paper follow a similar proof structure as the one in Sect. 4.2. The details can be found in the full version.

The other results achieved in this paper are the following.

Theorem 4. *Let* $a \in \mathbb{N}$. *Then* $f(a, a, 2a) = 2a - 1$ *holds.*

Theorem 5. *Let $a \in \mathbb{N}$. Then $f(a, a, 2a-1) = 2a-1$ holds.*

Theorem 6. *Let $a, b, c \in \mathbb{N}$ with $a \leq b \leq c$, $3a = b + c$, $(a, b, c) \neq (a, a, 2a)$ and $(a, b, c) \neq (4, 6, 6)$. Then it holds that $f(a, b, c) = \lfloor \frac{a+b+c}{2} \rfloor - 1$.*

Note that Theorem 4 is included in Theorem 6 but the proof of Theorem 6 uses Theorem 4, so it is still needed.

Further note that we have shown by a computer program based on exhaustive search [9] that $f(4, 6, 6) = 8$ holds.

Theorem 7. *Let $a, b, c \in \mathbb{N}$ with $a \leq b \leq c$, $3a < b + c$ and $2b \leq c$. Then $f(a, b, c) = c - 1$ holds.*

Theorem 8. *Let $a, b, c \in \mathbb{N}$ with $a \leq b \leq c$ and $3a < b + c$ and $2b > c$. Then $f(a, b, c) = \lfloor \frac{2}{3}(b + c - 1) \rfloor$ holds.*

Note that the lower bounds of Theorems 4 and 6 directly follow from Theorem 1, the one of Theorem 7 from Corollary 1, and the one of Theorem 8 from Theorem 2

The proofs of Theorems 4–8 can be found in the full version.

5 Discussion and Future Work

In this paper we have determined the metric dimension of $K_a \times K_b \times K_c$ for $a, b, c \in \mathbb{N}$. The only not completely solved case is $3a > b + c$, where two values $\lfloor \frac{a+b+c}{2} \rfloor - 1$ and $\lfloor \frac{a+b+c}{2} \rfloor$ are possible, see Table 1.

Thus, we performed an exhaustive search by our computer program [9] on all of those triples (a, b, c) with $a + b + c \leq 21$, for which $f(a, b, c)$ has not been obtained by the theoretical results of this paper, namely $3a > b + c$ and $(a, b, c) \notin \{(a, a, a), (a, a, 2a-1)\}$. More concretely, these are the following triples:

(3,3,4), (3,4,4), (4,4,5), (4,4,6), (4,5,5), (4,5,6),
(5,5,6), (5, 5,7), (5,5,8), (5,6,6), (5,6,7), (5,6,8),
(5,7,7), (6,6,7), (6,6,8), (6,6,9), (6,7,7), (6,7,8).

The computer runs showed that for all these cases, the upper bound is attained. Thus, we make the following conjecture:

Conjecture 1. Let $a, b, c \in \mathbb{N}$ with $a \leq b \leq c$. If $3a > b + c$ then it holds that $f(a, b, c) = \lfloor \frac{a+b+c}{2} \rfloor$.

Naturally, one aim of our future work is to prove or disprove this conjecture and close the remaining gap of this paper.

Acknowledgement. We would like to thank Frank Drewes for many helpful discussions and proofreading of this paper.

This research was supported by the Kempe Foundation Grant No. JCK-2022.1 (Sweden).

References

1. Blumenthal, L.M.: Theory and Applications of Distance Geometry. Clarendon Press, Oxford (1953)
2. Chartrand, G., Eroh, L., Johnson, M.A., Oellermann, O.R.: Resolvability in graphs and the metric dimension of a graph. Discret. Appl. Math. **105**(1–3), 99–113 (2000)
3. Chvátal, V.: Mastermind. Combinatorica **3**, 325–329 (1983)
4. Cáceres, J., et al.: On the metric dimension of cartesian products of graphs. SIAM J. Discret. Math. **21**(2), 423–441 (2007)
5. Doerr, B., Doerr, C., Spöhel, R., Thomas, H.: Playing Mastermind with many colors. J. ACM **63**(5), 42:1–42:23 (2016)
6. Erdös, P., Rényi, A.: On two problems of information theory. Publ. Math. Inst. Hung. Acad. Sci. **8**, 241–254 (1963)
7. Fernau, H., Heggernes, P., van't Hof, P., Meister, D., Saei, R.: Computing the metric dimension for chain graphs. Inform. Process. Lett. **115**(9), 671–676 (2015)
8. Garey, M.R., Johnson, D.S.: Computers and Intractability: A Guide to the Theory of NP-Completeness. W.H. Freeman & Co., New York (1979)
9. Gledel, V., Jäger, G.: Source code of the computer program of this article (2025). https://github.com/vgledel/metric-dimension-abc
10. Goddard, W.: Static mastermind. J. Combin. Math. Combin. Comput. **47**, 225–236 (2003)
11. Goodrich, M.T.: On the algorithmic complexity of the Mastermind game with black-peg results. Inform. Process. Lett. **109**(13), 675–678 (2009)
12. Harary, F., Melter, R.A.: On the metric dimension of a graph. Ars Combin. **2**, 191–195 (1976)
13. Jäger, G.: An optimal strategy for static black-peg mastermind with two pegs. In: Chan, T.-H.H., Li, M., Wang, L. (eds.) COCOA 2016. LNCS, vol. 10043, pp. 670–682. Springer, Cham (2016). https://doi.org/10.1007/978-3-319-48749-6_48
14. Jäger, G., Drewes, F.: The metric dimension of $\mathbb{Z}_n \times \mathbb{Z}_n \times \mathbb{Z}_n$ is $\lfloor 3n/2 \rfloor$. Theoret. Comput. Sci. **806**, 344–362 (2020)
15. Jäger, G., Peczarski, M.: The number of pessimistic guesses in Generalized Mastermind. Inform. Process. Lett. **109**(12), 635–641 (2009)
16. Jäger, G., Peczarski, M.: The number of pessimistic guesses in Generalized Black-Peg Mastermind. Inform. Process. Lett. **111**(19), 933–940 (2011)
17. Khuller, S., Raghavachari, B., Rosenfeld, A.: Landmarks in graphs. Discret. Appl. Math. **70**(3), 217–229 (1996)
18. Rodríguez-Velázquez, J.A., Kuziak, D., Yero, I.G., Sigarreta, J.M.: The metric dimension of strong product graphs. Carpathian J. Math. **31**(2), 261–268 (2015)
19. Slater, P.J.: Leaves of trees. Congre. Numer. **14**, 549–559 (1975)
20. Yero, I.G., Kuziak, D., Rodríguez-Velázquez, J.A.: On the metric dimension of corona product graphs. Comput. Math. Appl. **61**(9), 2793–2798 (2011)

Complexity of Identifying Fitting-Free Groups

Joshua A. Grochow[1], Dan Johnson[2], and Michael Levet[2(✉)]

[1] University of Colorado Boulder, Boulder, CO 80309, USA
jgrochow@colorado.edu
[2] College of Charleston, Charleston, SC 29492, USA
johnsondd2@g.cofc.edu, levetm@cofc.edu

Abstract. We exhibit an AC^3 isomorphism test for Fitting-free groups given by their Cayley tables, a class for which isomorphism testing was previously known to be in P (Babai, Codenotti, & Qiao; ICALP '12). In sharp contrast, we show that for permutation groups, isomorphism testing of Fitting-free groups is at least as hard as Graph Isomorphism and Linear Code Equivalence (the latter being GI-hard and having no known subexponential-time algorithm). Lastly, we show that any Fitting-free group of order n is identified by FO formulas (without counting) using only $O(\log \log n)$ variables. This is in contrast to the fact that there are infinite families of Abelian groups that are not identified by FO formulas with $o(\log n)$ variables (Grochow & Levet, FCT '23).

Keywords: Group Isomorphism · Circuit Complexity

1 Introduction

The Group Isomorphism problem (GpI) takes as input two finite groups G and H, and asks if $G \cong H$. When the groups are given by their Cayley tables, it is known that GpI belongs to $\mathsf{NP} \cap \mathsf{coAM}$. The generator-enumeration strategy has time complexity $n^{\log(n)+O(1)}$, where n is the order of the group [22,35]. The parallel complexity of generator-enumeration has been gradually improved– see [18] for a survey. The best known runtime for GpI is $n^{(1/4)\log(n)+O(1)}$ (see [31, Sec. 2.2]). Even the impressive body of work on practical algorithms in more succinct input models (see e.g., [9,10,15,21]) still results in an $n^{\Theta(\log n)}$-time algorithm in the general case (see [40, Page 2]).

The Cayley model is impractical when working with computer algebra software. Instead, the groups are typically given by generators as permutations or matrices, or more general succinct models that can be modeled as black boxes. For permutation groups, GpI belongs to NP [33]. For matrix groups and black-box groups, GpI belongs to $\mathsf{Promise}\Sigma_2^p$ [7]; it is open as to whether GpI belongs to NP or coNP in such succinct models. There have been significant advances on algorithms with worst-case guarantees on the serial runtime for special cases of this problem—see [20,25] for a survey.

A. Jeż and J. Otop (Eds.): FCT 2025, LNCS 16106, pp. 208–220, 2026.
https://doi.org/10.1007/978-3-032-04700-7_16

Key motivation for GPI comes from its relationship to the GRAPH ISOMORPHISM problem (GI). When the groups are given verbosely by their Cayley tables, GPI $\leq_m^{\mathsf{AC}^0}$ GI [42]. However, there is no AC^0-reduction from GI to GPI [16]. In light of Babai's result that GI $\in$ QuasiP [4], verbose GPI is a key barrier to improving the complexity of GI. There is considerable evidence suggesting that GI, and hence GPI, is not NP-complete [1,4,13,29,30,37].

There is a natural divide-and-conquer strategy for GPI which proceeds as follows. Given groups G_1, G_2, we first decompose the groups in terms of normal subgroups $N_i \trianglelefteq G_i$ ($i \in [2]$) and the corresponding quotients G_i/N_i. It is then necessary (but not sufficient) to decide if $N_1 \cong N_2$ and $G_1/N_1 \cong G_2/N_2$. The solvable radical of a finite group G is the (unique) largest normal solvable subgroup of G. Thus every finite group can be decomposed into its solvable radical $N = \text{Rad}(G)$ and the quotient $G/\text{Rad}(G)$. The latter quotient has no Abelian normal subgroups; we will refer to the class of such groups as *Fitting-free*. Fitting-free groups are thus a natural class of groups in a variety of settings, and serve not only as a necessary special case for GPI, but also as a potentially useful stepping stone towards the general case (e.g., [15,26,31]).

Main Results. In this paper, we investigate the complexity of identifying Fitting-free groups. Our first main result is the following.

Theorem 1. *Isomorphism of Fitting-free groups given by their multiplication tables can be decided in* AC^3.

There are two key motivations for Theorem 1. The first comes from GI, whose best known lower-bound is DET [39], which is contained in $\mathsf{NC}^2 \subseteq \mathsf{AC}^2$. There is a large body of work spanning almost 40 years on NC algorithms for GI—see [32] for a survey. In contrast, the work on NC algorithms for GPI is very new. As GPI is strictly easier than GI under AC^0-reductions [16], it is surprising that we know more about parallelizing subclasses of GI than subclasses of GPI.

The second motivation comes from the fact that complexity-theoretic lower bounds for GPI remain wide open, even against depth-2 AC circuits. In contrast, GPI is solvable using depth-4 $\mathsf{quasiAC}^0$ circuits [18]. Isomorphism testing is in P for a number of families of groups, and closing the gap between P and AC^0 for special cases is a key step in trying to characterize the complexity of the general GPI problem.

In sharp contrast, in the model of permutation groups we establish the following lower bound.

Theorem 2. *GI—and even LINEAR CODE EQUIVALENCE (LINCODEEQ)—is polynomial-time reducible to isomorphism testing of Fitting-free groups, when the groups are specified by generating sequences of permutations.*

While the hardness afforded by GI-hardness is tempered by Babai's result [4] that GI is in QuasiP, for LINCODEEQ, no algorithm with running time $2^{o(n)}$ is currently known (that it is solvable in time $2^{O(n)}$ was established by Babai in [5]; recently extended to MONOMIAL CODEEQ and with improved randomized and quantum algorithms by Bennett *et al.* [8]). GI is known to reduce to LINCODEEQ

[33,36]. Furthermore, although PERMUTATION-GPI was known to be GI-hard [33], Fitting-free groups were *not* previously known to be GI-hard for *any* succinct input model [41].

Remark 1. Cannon and Holt [15] exhibited an algorithm for isomorphism testing of Fitting-free groups specified by generating sequences of permutations. While their algorithm performs well in practice, they do not formally analyze the run-time complexity. Theorem 2 provides formal evidence that their algorithm is unlikely to run in polynomial-time.

We now return to the Cayley table model. Determining hard instances of GPI remains a long-standing open problem. Class-2 p-groups of exponent p have long been considered bottleneck instances of GPI (see, e.g., [5,27]). Although a formal reduction from general GPI remains elusive, existing evidence [5,6,18,20] indicates that groups whose orders are *not* large prime powers admit efficient isomorphism tests.

In the absence of formal reductions, one could ask for a complexity measure $\mathcal{C}$ and classes of groups $\mathcal{X}, \mathcal{Y}$ such that $\mathcal{C}(\mathcal{X}) < \mathcal{C}(\mathcal{Y})$. Take our complexity measure $\mathcal{C}$ to be the minimum number of variables needed in a first-order (FO) formula (without counting) that identifies a group uniquely up to isomorphism. Previous work [25] established that there are infinite families of Abelian groups that are not identified by FO formulas with $o(\log|G|)$ variables. In contrast, we show:

Theorem 3. *Every Fitting-free group is identified by a first-order formula using* $O(\log\log|G|)$ *variables.*

Despite isomorphism testing of Abelian groups in the Cayley table model being vastly simpler for serial or parallel algorithms than isomorphism testing of Fitting-free groups, in more succinct input models the situation is less clear. However, the fact that Fitting-free groups are more easily identified by first-order logic *without counting* than Abelian groups accords well with the idea that when dealing with isomorphism of groups containing Abelian normal subgroups, one must work with automorphism groups that are subgroups of matrix groups [31], which clearly "require counting" and are notoriously harder to work with than the more combinatorial permutation groups that arise in testing isomorphism of Fitting-free groups [5,6] (see also [34, Sec. 4, "Non-Abelian linear algebra"]).

It is interesting to note that isomorphism testing of Abelian groups can be done using only a single counting gate [16,18] or a single counting quantifier (see e.g., [12,25]), and the same was later established for other classes of groups [19], raising the question of a possible complexity hierarchy in the setting of first-order logic based on the number of counting quantifiers used (fixing the total number of variables and/or quantifiers).

Further Related Work. The class of Fitting-free groups has received significant attention. In the Cayley table model, it took a series of two papers [5,6] to establish a polynomial-time isomorphism test for this family. For permutation groups, Cannon & Holt [15] exhibited an efficient practical isomorphism test. Fitting-free groups have also been investigated in the context of the Weisfeiler–Leman algorithm [11,24].

2 Preliminaries

Codes. Let $\mathbb{F}$ be a field. $\mathrm{GL}_m(\mathbb{F})$ denotes the set of $m \times m$ invertible matrices over $\mathbb{F}$. A *linear code* of length m is a subspace $U \leq \mathbb{F}^m$. A $d \times m$ matrix A over $\mathbb{F}$ *generates* the code U if the rows of A span U. Let U, W be d–dimensional codes of length m over $\mathbb{F}$, generated by $d \times m$ matrices A, B, respectively. Then U and V are *equivalent* if and only there exists a permutation matrix $P \in \mathrm{GL}_m(\mathbb{F})$ and a matrix $T \in \mathrm{GL}_d(\mathbb{F})$ such that $B = TAP$.

Logics. We recall the central aspects of first-order logic (FO). We have a countable set of variables $\{x_1, x_2, \ldots, \}$. Formulas are defined inductively. As our basis, $x_i = x_j$ is a formula for all pairs of variables. If φ, ψ are formulas, then so are $\varphi \wedge \psi, \varphi \vee \psi, \neg\varphi, \exists x_i\, \varphi$, and $\forall x_i\, \varphi$. In order to define logics on groups, it is necessary to define a relation that captures the group multiplication. We recall the Version I logic introduced in [12]. Here, we add a ternary relation R where $R(x_i, x_j, x_\ell) = 1$ iff $x_i x_j = x_\ell$ in the group. We refer to the first-order logic with relation R as $\mathcal{L}^I$ and its k-variable fragment as $\mathcal{L}^I_k$. If furthermore we restrict the formulas to have quantifier depth at most r, we denote this fragment as $\mathcal{L}^I_{k,r}$. Given a formula φ with k free variables, and a k-tuple $\bar{g} \in G^k$, we say that $(G, \bar{g})$ *satisfies* φ, denoted $(G, \bar{g}) \models \varphi$, if, after plugging in $g_1, \ldots, g_k$ into the unquantified variables, the resulting formula is true in G.

Pebbling Game. Ehrenfeucht–Fraïssé pebble games are often used to show inexpressibility results in the corresponding logics. We recall the pebble game corresponding to $\mathcal{L}^I$ for groups [14,25,28]. The count-free $(k+1)$-pebble game consists of two players: Spoiler and Duplicator, as well as $(k+1)$ pebble pairs (p, p'). Intuitively, Spoiler wishes to show that the two groups G, H are not isomorphic, while Duplicator wishes to show that they are. Each round of the game proceeds as follows.

1. Spoiler picks up a pebble pair (p_i, p'_i).
2. Spoiler places one of the pebbles on some group element (either p_i on some element of G or p'_i on some element of H).
3. Duplicator places the other pebble on some element of the other group.
4. The winning condition is checked. This will be formalized later.

Let $v_1, \ldots, v_m$ be the pebbled elements of G at the end of step 4, and let $v'_1, \ldots, v'_m$ be the corresponding pebbled vertices of H. Spoiler wins precisely if the map $v_\ell \mapsto v'_\ell$ does not satisfy both of the following: (a) $v_i v_j = v_k \iff v'_i v'_j = v'_k$ for all $i, j, k \in [m]$, and (b) $v_i = v_j \iff v'_i = v'_j$ for all $i, j \in [m]$. Duplicator wins otherwise. Spoiler wins, by definition, at round 0 if $|G| \neq |H|$. Let $\bar{u} \in G^k, \bar{v} \in H^k$. Provided $|G| = |H|$, the following are equivalent: (i) Duplicator wins the first r rounds of the Version I count-free $(k+1)$-pebble game;

(ii) there exists a formula $\varphi \in \mathcal{L}^I_{k+1,r}$ such that $(G, \bar{g}) \models \varphi$ and $(H, \bar{h}) \not\models \varphi$; (iii) the first r rounds of count-free k-dimensional Weisfeiler–Leman (WL) Version I[1] fail to distinguish $\bar{u}$ and $\bar{v}$ (see e.g., [14,25,28]).

3 Proof Strategy for Theorem 1

3.1 Additional Preliminaries

In [5,6], the authors consider isomorphisms of the socle Soc(G) (=the subgroup generated by the minimal normal subgroups of G) respecting *diagonal* products.

Definition 1 ([5, Def. 2.2]). *Let $V_1, \ldots, V_r$ be isomorphic groups, where each $V_i \cong T$. A diagonal of $(V_1, \ldots, V_r)$ is an embedding $\phi : T \to \prod_{i=1}^r V_i$ such that Im(ϕ) is a subdirect product of the V_i. More generally, suppose we have a system of groups $(V_{1,1}, \ldots, V_{1,k_1}), \ldots, (V_{r,1}, \ldots, V_{r,k_r})$, where for every $i \leq r$ and every $j \leq k_i$, $V_{i,j} \cong T_i$. A diagonal product of the system is an embedding: $\phi_1 \times \cdots \times \phi_r : T_1 \times \cdots \times T_r \to \prod_{j=1}^{k_1} V_{1,j} \times \cdots \times \prod_{j=1}^{k_r} V_{r,j}$, where each ϕ_i is a diagonal of $(V_{i,j})_{j=1}^{k_i}$. The standard diagonal of T^k is the map $\Delta : t \mapsto (t, \ldots, t)$. Similarly, the standard diagonal product of $\prod_{i=1}^r T_i^{k_i}$ is the map $\Delta = \Delta_1 \times \cdots \times \Delta_r$, where each Δ_i is the standard diagonal of $T_i^{k_i}$.*

We will be particularly interested in diagonals that respect the (unique) decomposition of Soc(G) into a direct product of non-Abelian simple groups.

Definition 2 ([5, Def. 2.3]). *Let X, Y be groups, where $X = \prod_{i=1}^r \prod_{j=1}^{k_i} V_{i,j}$, and $Y = \prod_{i=1}^r \prod_{j=1}^{k_i} U_{i,j}$, and for all $i \in [r], j \in [k_i]$, $U_{i,j} \cong V_{i,j} \cong T_i$. We say that an isomorphism $\chi : G \to H$ respects the decompositions $\mathcal{V} = (V_{i,j})$ and $\mathcal{U} = (U_{i,j})$ if for every $i \in [r]$ and every $j \in [k_i]$, there exists $j' \in [k_i]$ such that $\chi(V_{i,j}) = U_{i,j'}$. We denote the set of isomorphisms that respect the decompositions $\mathcal{V}$ and $\mathcal{U}$ by ISOp$((X, \mathcal{V}), (Y, \mathcal{U}))$, where p stands for product decomposition. If the decompositions are understood from context, we write ISOp(X, Y).*

Definition 3 ([5, Def. 2.4]). *Let X, Y be as in Definition 2. Let φ be a diagonal product of $\mathcal{V} = (V_{i,j})$, and let ψ be a diagonal product of $\mathcal{U} = (U_{i,j})$. We say that an isomorphism $\chi \in$ ISOp$((X, \mathcal{V}), (Y, \mathcal{U}))$ respects the diagonal products φ, ψ if $\varphi \circ \chi = \psi$. We denote the set of diagonal product respecting automorphisms by ISOd$((X, \mathcal{V}), (Y, \mathcal{U}); \varphi, \psi)$.*

Lemma 1 ([5, Lem. 2.2]). *Let X, Y be as in Definition 2. Let φ be a diagonal product of $\mathcal{V} = (V_{i,j})$, and let $\mathcal{D}$ be the set of all diagonal products of $\mathcal{U}$. Then: ISOp$((X, \mathcal{V}), (Y, \mathcal{U})) = \bigcup_{\psi \in \mathcal{D}}$ ISOd$((X, \mathcal{V}), (Y, \mathcal{U}); \varphi, \psi)$.*

For Fitting-free groups, we can reduce to considering diagonal-respecting isomorphisms of the socle. Before we can make this precise, we must first recall the following lemma.

[1] To save space, details about the standard count-free WL coloring procedure are in the full version. We will work exclusively in the pebble game characterization.

Lemma 2 ([5, Lem. 3.1], cf.[15, §3]). *Let G and H be groups, with $R \lhd G$ and $S \lhd H$ groups with trivial centralizers. Let $\alpha : G \to G^* \leq \text{Aut}(R)$ and $\beta : H \to H^* \leq \text{Aut}(S)$ be faithful permutation representations of G and H via the conjugation action on R and S, respectively. Let $f : R \to S$ be an isomorphism. Then f extends to an isomorphism $\hat{f} : G \to H$ if and only if f is a permutational isomorphism between G^* and H^*; and if so, $\hat{f} = \alpha f^* \beta^{-1}$, where $f^* : G^* \to H^*$ is the isomorphism induced by f.*

Let G, H be Fitting-free groups. Let $R = \text{Soc}(G)$ and $S = \text{Soc}(H)$. For diagonal products ϕ, ψ of R, S respectively, we denote the set of isomorphisms respecting the diagonal products of the socle as: $\text{ISOds}(G, H; \phi, \psi) = \{\chi \in \text{ISO}(G, H) : \chi_{\text{Soc}(G)} \in \text{ISOd}(R, S; \varphi, \psi)\}$. In particular, we have the following:

Corollary 1 ([5, Cor. 4.1]). *Let G, H be Fitting-free, φ a diagonal product of Soc(G), and $\mathcal{D}$ the set of diagonal products of Soc(H). Then ISO(G, H) = $\bigcup_{\psi \in \mathcal{D}}$ ISOds($G, H; \varphi, \psi$).*

Lemma 3 ([5, Lem. 4.1]). *Let $H, \mathcal{D}$ be defined as in Corollary 1. Then $|\mathcal{D}| \leq |H|^2$.*

Definition 4 ([6] Twisted Equivalence). *Let $\Gamma_1, \ldots, \Gamma_r$ be disjoint finite alphabets. A code of length $(k_1, \ldots, k_r)$ with domain $(A_1, \ldots, A_r)$ is a subset $\mathcal{A} \subseteq \prod_{i=1}^{r} \Gamma_i^{A_i}$. For each $i \in [r]$, let $G_i \leq$ Sym(Γ_i) be a group acting on Γ_i given by a faithful permutation representation, and let $\mathcal{G} := (G_1, \ldots, G_r)$. Given bijections $\pi_i : A_i \to B_i$ and functions $g : B_i \to G_i$ ($i \in [r]$), let $\psi_i := (\pi_i, g_i)$ and $\psi := (\psi_1, \ldots, \psi_r)$. For $a = (a_1, \ldots, a_r) \in \prod_{i=1}^{r} \Gamma_i^{A_i}$, define $a^\psi := (a_1^{\psi_1}, \ldots, a_r^{\psi_r})$. A $\mathcal{G}$-twisted equivalence of two codes $\mathcal{A} \subseteq \prod_{i=1}^{r} \Gamma_i^{A_i}$ and $\mathcal{B} \subseteq \prod_{i=1}^{r} \Gamma_i^{B_i}$ is a ψ such that $\mathcal{A}^\psi = \mathcal{B}$. The set of all $\mathcal{G}$-twisted equivalences of $\mathcal{A} \to \mathcal{B}$ is denoted $EQ_{\mathcal{G}}(\mathcal{A}, \mathcal{B})$. The* Twisted Code Equivalence *problem takes as input $\mathcal{A}, \mathcal{B}, \mathcal{G}$ and decides whether $EQ_{\mathcal{G}}(\mathcal{A}, \mathcal{B}) \neq \emptyset$.*

3.2 Proof Strategy

We outline the proof strategy for Theorem 1. PKer(G) is the subgroup of elements that do not permute the simple direct factors of Soc(G). Babai, Codenotti, and Qiao [6] crucially utilize that $G/\text{PKer}(G)$ forms a permutation group of degree $O(\log |G|)$, in tandem with *small* (size polylog($|G|$)) instances of GI-hard problems such as Coset Intersection and Twisted Code Equivalence. In order to obtain our AC^3 bound, we also take advantage of the fact that $G/\text{PKer}(G)$ can be viewed as a permutation group of domain size $O(\log |G|)$. We then bring to bear an extensive library of NC algorithms for permutation groups [3]. As our permutation domain is *small* ($O(\log |G|)$), each of these parallel permutation group algorithms can be implemented with a uniform AC circuit of depth poly($\log \log n$) and poly(n) size. We refer to $\mathsf{FOLL}^{O(1)}$ as the class of languages decidable by uniform circuits of depth poly($\log \log n$) and polynomial size. Still, we require several additional ingredients, which we now outline.

Step 1. The Coset Intersection (CI) problem takes as input $G, H \leq S_m$ and $x, y \in S_m$, and asks for $Gx \cap Hy$. We show that on a permutation domain of $O(\log |G|)$ points, CI can be implemented in $\mathsf{FOLL}^{O(1)}$. Our main contribution is to precisely determine the circuit depth. We do so by taking advantage of the fact that we have a *small* permutation group, in tandem with a suite of existing parallel algorithms for permutation groups [3]. Our analysis is specific to the algorithm at hand; if Luks's algorithm used polynomial space and simply-exponential time, just scaling down to permutation groups on domain size $O(\log n)$ and using the standard simulation of space by depth would only yield $\text{poly}(\log n)$-depth circuits of quasi-polynomial size. In contrast, by carefully analyzing Luks's algorithm we are able to get $\text{poly}(\log \log n)$ depth and $\text{poly}(n)$ size. For full details see the full version [23, §3].

Step 2. We next consider the TwistCodeEq problem. Precisely, we establish the following.

Theorem 4 (cf. [6,17]). *Let $m \in O(\log n)$. Let $\mathcal{A} \subseteq \prod_{i=1}^{r} \Gamma_i^{A_i}$ and $\mathcal{B} \subseteq \prod_{i=1}^{r} \Gamma_i^{B_i}$ be codes of total length m. For each $i \in [r]$, let G_i be a group acting on Γ_i given by its Cayley table. Let $G_{\max}$ and $\Gamma_{\max}$ such that $|G_{\max}| = \max_{i \in [r]} |G_i|$, and $|\Gamma_{\max}| = \max_{i \in [r]} |\Gamma_i|$. Then $EQ_{(G_1,\ldots,G_r)}(\mathcal{A}, \mathcal{B})$ can be found using an AC circuit of depth $O((\log n) \cdot (\log |\Gamma_{\max}|) \cdot \text{poly}(\log \log n))$ and size $\text{poly}(n)$.*

The strategy follows [6,17]. We build a dynamic programming table, which has $O(\log^2 n)$ cells. We use $O(\log n)$ iterations to fill in this table. Each iteration is computable using an AC circuit of depth $O(\log |\Gamma_{\max}| \, \text{poly}(\log \log n))$ and $\text{poly}(n)$-size. Each iteration utilizes CI (Step 1), which is $\mathsf{FOLL}^{O(1)}$-computable. For full details, see [23, Thm. 4.1].

Step 3. Let $m \in O(\log n)$, and let $G, H \leq S_m$ be transitive. Babai, Codenotti, & Qiao [6] previously established that the permutational automorphism group satisfies $|\text{PAut}(G)| \leq |G| \cdot \text{poly}(n)$ and the permutational isomorphisms $\text{PISO}(G, H)$ can be listed in polynomial-time. We strengthen their result to show that PISO can be listed in AC^3. Here, we carefully analyze the combinatorics of the associated structure trees, and show that the polynomial-time subroutines can be parallelized. The technical details are nontrivial and require care to execute (see [23, §6] for full details).

Step 4. Consider the decompositions of $\text{Soc}(G)$ and $\text{Soc}(H)$ into direct products of minimal normal subgroups: $\text{Soc}(G) = \prod_{i=1}^{d} \prod_{j=1}^{z_i} N_{i,j}$ and $\text{Soc}(H) = \prod_{i=1}^{d} \prod_{i=1}^{z_i} M_{i,j}$. Suppose that for all i, j, $N_{i,j} \cong M_{i,j} \cong K_i$, where K_i is characteristically simple. Write $\mathcal{K} := \prod_{i=1}^{d} K_i^{z_i}$. Let γ, ξ be diagonal products of $(N_{i,j}), (M_{i,j})$ respectively. Let $\alpha_\gamma \in \text{ISOd}(\text{Soc}(G), \mathcal{K}; \gamma, \Delta)$ and $\beta_\xi \in \text{ISOd}(\text{Soc}(H), \mathcal{K}; \xi, \Delta)$, where Δ is the standard diagonal product. By Lemma 2, the conjugation action of G and H on their socles yields the respective embeddings $\alpha_\gamma^* : G \hookrightarrow \text{Aut}(\mathcal{K})$ and $\beta_\xi^* : H \hookrightarrow \text{Aut}(\mathcal{K})$. By Lemma 2, we have that: $\text{Iso}(G, H) = \alpha_\gamma^* \text{ISO}(G^*, H^*)(\beta_\xi^*)^{-1}$. By Corollary 1, we will need to consider diagonal products of $\text{Soc}(G)$ (resp. $\text{Soc}(H)$) that respect the decomposition of

Soc(G) (resp. Soc(H)) into a direct product of non-Abelian simple groups. Now let φ be such a diagonal product of Soc(G). Define ψ analogously for Soc(H). Let Δ be the standard diagonal product of the socles into non-Abelian simple factors. We have the following: $\text{ISOds}(G, H; \varphi, \psi) = \alpha_\gamma^* \text{ISOds}^*(G^*, H^*; \Delta, \Delta)(\beta_\xi^*)^{-1}$ [5].

The problem thus reduces to finding $\text{ISOds}^*(G^*, H^*, \Delta, \Delta)$. We will recall from [6] how to reduce $\text{ISOds}^*(G^*, H^*, \Delta, \Delta)$ to TwistCodeEq. Let G^*_{ij} be the restriction of G^* to the jth copy of K_i, and H^*_{ij} the restriction of H^* to the jth copy of K_i. For every i, j, i', j', compute $\text{ISOds}^*(G^*_{ij}, H^*_{i'j'}; \Delta, \Delta)$. As $G^*_{ij}, H^*_{i'j'}$ each have a unique minimal normal subgroup, we can by list the isomorphisms between them that preserve the standard diagonals in AC^3 (Step 3).

Let $\Gamma_1, \ldots, \Gamma_r$ be representatives of these isomorphism types, which will be our alphabets. For each pair (i, j), choose an arbitrary isomorphism χ_{ij} that preserve isomorphism between G^*_{ij} and the corresponding representative Γ_h, and let δ_{ij} be the corresponding isomorphism for H^*_{ij}. Create codes $\overline{G}, \overline{H}$ over $\Gamma_1, \ldots, \Gamma_r$ as follows. Let $\sigma \in G^*$, and let $\sigma_{ij} \in G^*_{ij}$ denote the restriction of σ to N_{ij}. The string associated with σ is $\overline{\sigma} = (\chi_{ij}(\sigma_{ij}))$. Now $\overline{G} := \{\overline{\sigma} : \sigma \in G^*\}$. Define $\overline{H}$ analogously by setting $\overline{\pi} := (\delta_{ij}(\pi_{ij}))_{ij}$, for $\pi \in H$. For each $\ell \in [r]$, let $W_\ell = \text{Autds}^*(\Gamma_\ell; \Delta, \Delta) := \text{ISOds}^*(\Gamma_\ell, \Gamma_\ell; \Delta, \Delta)$ be the group of automorphisms that preserve the standard diagonal. Let m_ℓ denote the number of the G^*_{ij} that are isomorphic to Γ_ℓ. Without loss of generality, suppose that the number of H^*_{ij} isomorphic to Γ_ℓ is also m_ℓ (otherwise, G^* and H^* are not isomorphic). Then $\overline{G}, \overline{H} \leq \prod_{\ell=1}^r \Gamma_\ell^{m_\ell}$ are subgroups, and so we view them as group codes as well. We have that $\text{ISOds}^*(\overline{G}, \overline{H}; \Delta, \Delta) = \widehat{\text{EQ}}_{(W_1, \ldots, W_r)}(\overline{G}, \overline{H})$. In particular, this reduction is AC^3-computable (see [23, §7.4]).

Step 5. In L, do the following: identify Soc(G) (resp. Soc(H)) decompose it into a direct product of minimal normal subgroups in G (resp. H) [25]; fix a diagonal product φ for Soc(G) and in parallel, consider the $\leq n^2$ diagonal products of Soc(H) [38]. Now reduce to Step 4.

4 Proof of Theorem 2

Proof (Sketch). Petrank & Roth give a reduction from GI to LinCodeEq over $\mathbb{F}_2$. Graphs with n vertices and m edges yield codes that are m-dimensional subspaces of $\mathbb{F}_2^{3m+n}$. The key to our reduction is thus a reduction from LinCodeEq over $\mathbb{F}_2$ to (abstract) isomorphism of Fitting-free permutation groups.

Let $\pi \colon S_5^N \to (\mathbb{Z}/2\mathbb{Z})^N$ be the coordinate-wise sign homomorphism $(\sigma_1, \cdots, \sigma_N) \overset{\pi}{\mapsto} (\text{sgn}(\sigma_1), \cdots, \text{sgn}(\sigma_N))$. Given a code $C \leq \mathbb{F}_2^N$, define a group $G_C := \pi^{-1}(C)$, the preimage of $C \leq \mathbb{F}_2^N$ under π. We first show how to construct a small generating set of permutations for G_C. Since $G_C \leq S_5^N$, we will build our permutations as elements of S_5^N, thought of as a subgroup of S_{5N} in the natural way. For each of the N normal copies of A_5, we include a size-2 generating set of A_5 as part of our generating set of G_C. Let $v_1, \cdots, v_k \in \mathbb{F}_2^N$ be a basis for C over $\mathbb{F}_2$; for the purposes of the reduction, we are given C by a generator matrix, and we can take $v_1, \cdots, v_k$ to be the rows of that matrix. For each such vector v_i, let $\sigma_i = ((\sigma_i)_1, \cdots, (\sigma_i)_N) \in S_5^N$ be the permutation in S_5^N whose j-th coordinate is $(\sigma_i)_j = (12)$ if $(v_i)_j = 1$, and $(\sigma_i)_j = \text{id}$ if $(v_i)_j = 0$. It is readily verified that the generating set just described consists of $2N + k$ elements, and that they generate G_C. It remains to show that if $C' \leq \mathbb{F}_2^N$ is another code, then $G_C \cong G_{C'}$ iff C and C' are equivalent linear codes. We leave these details for the full version.

5 Proof Strategy for Theorem 3

Brachter [11, Lem. 5.4.4] established the analogous result for the counting variant of the more powerful Version II logic, and without controlling for the quantifier depth. We essentially show that Brachter's pebbling strategy from the counting game can be implemented with the count-free game. In order to obtain $O(\log \log n)$ rounds in the game, we will take advantage of the following result.

Theorem 5 ([2]). *There exists a constant C, such that for every finite simple group S, there exists a generating set $g_1, \ldots, g_7 \in S$ such that every element of S can be realized as a word over $\{g_1, \ldots, g_7\}$ of length at most $C \log |S|$.*

We will now recall some preliminaries from [19,24].

Definition 5 (cf. [24, Def. 4.9]). *Let G be a Fitting-free group. Let $Soc(G) = S_1 \times \cdots \times S_k$, where each S_i is a non-Abelian simple subgroup of $Soc(G)$. For any $s \in Soc(G)$, write $s = s_1 \cdots s_k$, where each $s_i \in S_i$. Define the weight of s, $wt(s)$, as the number of i's such that $s_i \neq 1$. For $s \notin Soc(G)$, we define $wt(s) = \infty$.*

Grochow and Levet [25] generalized weight to the notion of rank.

Definition 6 ([25, Def. 4.1]). *Let $C \subseteq G$ be a subset of a group G that is closed under taking inverses. We define the C-rank of $g \in G$, denoted $\text{rk}_C(g)$, as the minimum m such that g can be written as a word of length m in the elements of C. If g cannot be so written, we define $\text{rk}_C(g) = \infty$.*

Lemma 4 ([25, Rank Lemma 4.3]). *Let $C \subseteq G$ be a subset of G that is closed under taking inverses. Suppose that if $x \in C$ and $y \notin C$, then Spoiler can win in the count-free Version I pebble game from the initial configuration $x \mapsto y$, using k pebbles and r rounds. If $\text{rk}_C(g) \neq \text{rk}_C(h)$, then Spoiler can win with $k+1$ pebbles and $r + \log(d) + O(1)$ rounds, where $d = diam(Cay(\langle C \rangle, C)) \leq |\langle C \rangle| \leq |G|$.*

We now turn to proving Theorem 3.

Definition 7 (Pebbling Strategy). *Let $G = \prod_{i=1}^k S_i$ be a group of order n, where each S_i is non-Abelian simple. For each $i \in [k]$, let $(x_{i,1}, \ldots, x_{i,7})$ be the generating sequence prescribed by Theorem 5. For each $j \in [7]$, Spoiler pebbles $x_j := \prod_{i=1}^k x_{i,j}$. Let $\ell := \lceil \log_2(k) \rceil$. For each $h \in [\ell]$, Spoiler pebbles: $p_h := \prod_{i=1}^h \left(\prod_{m=1+(i-1)\cdot k/2^{i-1}}^{k/2^i+(i-1)k/2^h} x_m \right)$.*

Proof (Sketch of Theorem 3). By [19, Lem. 6.6], if G is Fitting-free and H is not Fitting-free, then Spoiler can win with $O(1)$ pebbles and $O(1)$ rounds. So suppose that H is Fitting-free. Let $\text{Fac}(\text{Soc}(G)) = \{S_1, \ldots, S_k\}$ be the indecomposable direct factors of $\text{Soc}(G)$ and $\text{Fac}(\text{Soc}(H)) = \{S'_1, \ldots, S'_{k'}\}$. By Lemma 4, if $k \neq k'$ Spoiler wins with $O(1)$ additional pebbles and $O(\log \log n)$ additional rounds. Let $\ell := \lceil \log_2(k) \rceil$. Spoiler now applies the pebbling strategy of Definition 7 to $\text{Soc}(G)$, to pebble $x_1, \ldots, x_7, p_1, \ldots, p_\ell$. Let $x'_1, \ldots, x'_7, p'_1, \ldots, p'_\ell$ be Duplicator's response. In particular, our pebbling strategy fixes an isomorphism $\varphi : \text{Soc}(G) \to \text{Soc}(H)$ (or Spoiler wins with $O(1)$ pebbles and $O(\log \log n)$ rounds).

By [5, Lem. 3.1], φ extends in at most one way to an isomorphism $\widehat{\varphi}$ of G and H. In particular, as $G \ncong H$, there exists $g \in G$, such that for all $h \in H$, there exists $s_{gh} \in \text{Soc}(G)$ such that $\varphi(g s_{gh} g^{-1}) \neq h\varphi(s_{gh})h^{-1}$. There exists such an s_{gh} satisfying $\text{wt}(s_{gh}) = 1$ [24, Thm. 4.17]. Spoiler begins by pebbling $g \in G$. Let h be Duplicator's response. Let $s_{gh} \in \text{Soc}(G)$ such that $\varphi(g s_{gh} g^{-1}) \neq h\varphi(s_{gh})h^{-1}$. Without loss of generality, suppose that $s_{gh} \in S_1$. Spoiler pebbles s_{gh}. Let s' be Duplicator's response. Spoiler now wins with $O(1)$ additional pebbles and $O(\log \log n)$ additional rounds.

Acknowledgments. ML thanks P. Brooksbank, E.K. Gnang, T. Miyazaki, and J.B. Wilson for helpful discussions; and E. Robeva, C. Voll, and J.B. Wilson for organizing TAGA 2024, where parts of this work began. ML was partially supported by CRC 358 Integral Structures in Geometry and Number Theory at Bielefeld and Paderborn, Germany; the Department of Mathematics at Colorado State University; J.B. Wilson's NSF grant DMS-2319370; travel support from the Department of Computer Science at the College of Charleston; and SURF Grant SU2024-06 from the College of Charleston. DJ was partially supported by the Department of Computer Science at the College of Charleston, SURF Grant SU2024-06 from the College of Charleston, and ML startup funds. JAG was partially supported by NSF CAREER award CCF-2047756.

References

1. Arvind, V., Kurur, P.P.: Graph isomorphism is in SPP. Inf. Comput. **204**(5), 835–852 (2006). https://doi.org/10.1016/j.ic.2006.02.002
2. Babai, L., Kantor, W., Lubotsky, A.: Small-diameter Cayley graphs for finite simple groups. Eur. J. Combin. **10**(6), 507–522 (1989). https://doi.org/10.1016/S0195-6698(89)80067-8
3. Babai, L., Luks, E., Seress, A.: Permutation groups in NC. In: STOC 1987, pp. 409–420. Association for Computing Machinery (1987). https://doi.org/10.1145/28395.28439
4. Babai, L.: Graph isomorphism in quasipolynomial time [extended abstract]. In: STOC 2016, pp. 684–697. ACM (2016). https://doi.org/10.1145/2897518.2897542, preprint of full version at arXiv:1512.03547v2 [cs.DS]
5. Babai, L., Codenotti, P., Grochow, J.A., Qiao, Y.: Code equivalence and group isomorphism. In: SODA 2011, pp. 1395–1408. SIAM (2011). https://doi.org/10.1137/1.9781611973082.107

6. Babai, L., Codenotti, P., Qiao, Y.: Polynomial-time isomorphism test for groups with no abelian normal subgroups. In: Czumaj, A., Mehlhorn, K., Pitts, A., Wattenhofer, R. (eds.) ICALP 2012. LNCS, vol. 7391, pp. 51–62. Springer, Heidelberg (2012). https://doi.org/10.1007/978-3-642-31594-7_5
7. Babai, L., Szemerédi, E.: On the complexity of matrix group problems I. In: FOCS 1984, pp. 229–240. IEEE Computer Society (1984). https://doi.org/10.1109/SFCS.1984.715919
8. Bennett, H., et al.: Asymptotic improvements to provable algorithms for the code equivalence problem. Cryptology ePrint Archive, Paper 2025/187 (2025). https://eprint.iacr.org/2025/187, to appear, ISIT 2025
9. Besche, H.U., Eick, B.: Construction of finite groups. J. Symb. Comput. **27**(4), 387–404 (1999). https://doi.org/10.1006/jsco.1998.0258
10. Besche, H.U., Eick, B., O'Brien, E.: A millennium project: constructing small groups. Intern. J. Alg. and Comput **12**, 623–644 (2002). https://doi.org/10.1142/S0218196702001115
11. Brachter, J.: Combinatorial approaches to the group isomorphism problem. Ph.D. thesis, TU Darmstadt (2023). https://tuprints.ulb.tu-darmstadt.de/26387/1/thesis_brachter.pdf
12. Brachter, J., Schweitzer, P.: On the Weisfeiler–Leman dimension of finite groups. In: Hermanns, H., Zhang, L., Kobayashi, N., Miller, D. (eds.) LICS 2020, pp. 287–300. ACM (2020). https://doi.org/10.1145/3373718.3394786
13. Buhrman, H., Homer, S.: Superpolynomial circuits, almost sparse oracles and the exponential hierarchy. In: Shyamasundar, R. (ed.) FSTTCS 1992. LNCS, vol. 652, pp. 116–127. Springer, Heidelberg (1992). https://doi.org/10.1007/3-540-56287-7_99
14. Cai, J.Y., Fürer, M., Immerman, N.: An optimal lower bound on the number of variables for graph identification. Combinatorica **12**(4), 389–410 (1992). https://doi.org/10.1007/BF01305232, originally appeared in SFCS 1989
15. Cannon, J.J., Holt, D.F.: Automorphism group computation and isomorphism testing in finite groups. J. Symb. Comput. **35**, 241–267 (2003). https://doi.org/10.1016/S0747-7171(02)00133-5
16. Chattopadhyay, A., Torán, J., Wagner, F.: Graph isomorphism is not AC^0-reducible to group isomorphism. ACM Trans. Comput. Theory **5**(4), Article no. 13 (2013). https://doi.org/10.1145/2540088, preliminary version appeared in FSTTCS 2010; ECCC Tech. Report TR10-117
17. Codenotti, P.: Testing isomorphism of combinatorial and algebraic structures. Ph.D. thesis, University of Chicago (2011). https://people.cs.uchicago.edu/~laci/students/codenotti.pdf
18. Collins, N.A., Grochow, J.A., Levet, M., Weiß, A.: Constant depth circuit complexity for generating quasigroups. In: Hauenstein, J.D., Lee, W., Chen, S. (eds.) Proceedings of the 2024 International Symposium on Symbolic and Algebraic Computation, ISSAC 2024, Raleigh, NC, USA, 16–19 July 2024, pp. 198–207. ACM (2024). https://doi.org/10.1145/3666000.3669691
19. Collins, N.A., Levet, M.: Count-free Weisfeiler-Leman and group isomorphism. Int. J. Algebra Comput. **34**(03), 283–330 (2024). https://doi.org/10.1142/S0218196724500103
20. Dietrich, H., Wilson, J.B.: Group isomorphism is nearly-linear time for most orders. In: 2021 IEEE 62nd Annual Symposium on Foundations of Computer Science (FOCS), pp. 457–467 (2022). https://doi.org/10.1109/FOCS52979.2021.00053

21. Eick, B., Leedham-Green, C.R., O'Brien, E.A.: Constructing automorphism groups of p-groups. Comm. Algebra **30**(5), 2271–2295 (2002). https://doi.org/10.1081/AGB-120003468
22. Felsch, V., Neubüser, J.: On a programme for the determination of the automorphism group of a finite group. In: Computational Problems in Abstract Algebra (Proceedings of the Conference, Oxford, 1967), pp. 59–60 (1970)
23. Grochow, J.A., Johnson, D., Levet, M.: On the complexity of identifying groups without abelian normal subgroups: parallel, first order, and GI-hardness. arXiv:2504.19777 [cs.CC] (2025)
24. Grochow, J.A., Levet, M.: On the descriptive complexity of groups without abelian normal subgroups (extended abstract). In: Achilleos, A., Monica, D.D. (eds.) Proceedings of the Fourteenth International Symposium on Games, Automata, Logics, and Formal Verification, GandALF 2023, Udine, Italy, 18-20th September 2023. EPTCS, vol. 390, pp. 185–202 (2023). https://doi.org/10.4204/EPTCS.390.12
25. Grochow, J.A., Levet, M.: On the parallel complexity of group isomorphism via Weisfeiler-Leman. In: Fernau, H., Jansen, K. (eds.) FCT 2023. LNCS, vol. 14292, pp. 234–247. Springer, Cham (2023). https://doi.org/10.1007/978-3-031-43587-4_17
26. Grochow, J.A., Qiao, Y.: Algorithms for group isomorphism via group extensions and cohomology. SIAM J. Comput. **46**(4), 1153–1216 (2017). https://doi.org/10.1137/15M1009767
27. Grochow, J.A., Qiao, Y.: On p-group isomorphism: search-to-decision, counting-to-decision, and nilpotency class reductions via tensors. ACM Trans. Comput. Theory **16**(1), 2:1–2:39 (2024). https://doi.org/10.1145/3625308, extended abstract at CCC 2021; preliminary version was part of arXiv:1907.00309
28. Immerman, N., Lander, E.: Describing graphs: a first-order approach to graph canonization. In: Selman, A.L. (ed.) Complexity Theory Retrospective, pp. 59–81. Springer, New York (1990)
29. Impagliazzo, R., Paturi, R., Zane, F.: Which problems have strongly exponential complexity? J. Comput. Syst. Sci. **63**(4), 512–530 (2001). https://doi.org/10.1006/jcss.2001.1774
30. Köbler, J., Schöning, U., Torán, J.: Graph isomorphism is low for pp. Comput. Complex. **2**, 301–330 (1992). https://doi.org/10.1007/BF01200427
31. Le Gall, F., Rosenbaum, D.J.: On the group and color isomorphism problems. arXiv:1609.08253 [cs.CC] (2016)
32. Levet, M., Rombach, P., Sieger, N.: Canonizing graphs of bounded rank-width in parallel via Weisfeiler-Leman. In: Bodlaender, H.L. (ed.) 19th Scandinavian Symposium and Workshops on Algorithm Theory (SWAT 2024). Leibniz International Proceedings in Informatics (LIPIcs), vol. 294, pp. 32:1–32:18. Schloss Dagstuhl – Leibniz-Zentrum für Informatik, Dagstuhl, Germany (2024). https://doi.org/10.4230/LIPIcs.SWAT.2024.32
33. Luks, E.: Permutation groups and polynomial-time computation. DIMACS Ser. Discret. Math. Theor. Comput. Sci. **11** (1993). https://doi.org/10.1090/dimacs/011/11
34. Luks, E.M.: Parallel algorithms for permutation groups and graph isomorphism. In: 27th Annual Symposium on Foundations of Computer Science, Toronto, Canada, 27–29 October 1986, pp. 292–302. IEEE Computer Society (1986). https://doi.org/10.1109/SFCS.1986.39
35. Miller, G.L.: On the $n^{\log n}$ isomorphism technique (a preliminary report). In: Proceedings of the Tenth Annual ACM Symposium on Theory of Computing, STOC

1978, pp. 51–58. Association for Computing Machinery, New York (1978). https://doi.org/10.1145/800133.804331
36. Petrank, E., Roth, R.: Is code equivalence easy to decide? IEEE Trans. Info. Theory **43**(5), 1602–1604 (1997). https://doi.org/10.1109/18.623157
37. Schöning, U.: Graph isomorphism is in the low hierarchy. J. Comput. Syst. Sci. **37**(3), 312–323 (1988). https://doi.org/10.1016/0022-0000(88)90010-4
38. Tang, B.: Towards understanding satisfiability, group isomorphism and their connections. Ph.D. thesis, Tsinghua University (2013)
39. Torán, J.: On the hardness of graph isomorphism. SIAM J. Comput. **33**(5), 1093–1108 (2004). https://doi.org/10.1137/S009753970241096X
40. Wilson, J.B.: The threshold for subgroup profiles to agree is logarithmic. Theory Comput. **15**(19), 1–25 (2019). https://doi.org/10.4086/toc.2019.v015a019
41. Wilson, J.B.: Isomorphism testing of permutation groups. Pers. Commun. (2024)
42. Zemlyachenko, V.N., Korneenko, N.M., Tyshkevich, R.I.: Graph isomorphism problem. J. Soviet Math. **29**(4), 1426–1481 (1985). https://doi.org/10.1007/BF02104746

On Complementation of Nondeterministic Finite Automata Without Full Determinization

Lukáš Holík[1,2], Ondřej Lengál[1(✉)], Juraj Major[3], Adéla Štěpková[3], and Jan Strejček[3]

[1] Brno University of Technology, Brno, Czech Republic
ondra.lengal@gmail.com
[2] Aalborg University, Aalborg, Denmark
[3] Masaryk University, Brno, Czech Republic

Abstract. Complementation of finite automata is a basic operation used in numerous applications. The standard way to complement a nondeterministic finite automaton (NFA) is to transform it into an equivalent deterministic finite automaton (DFA) and complement the DFA. The DFA can, however, be exponentially larger than the corresponding NFA. In this paper, we study several alternative approaches to complementation, which are based either on reverse powerset construction or on two novel constructions that exploit a commonly occurring structure of NFAs. Our experiment on a large data set shows that using a different than the classical approach can, in many cases, yield significantly smaller complements.

1 Introduction

Complementation of *finite automata* is an operation with many applications in formal methods. It is used, e.g., in regular model checking [5,6,25], representing extended regular expressions [10,31], to implement negation in automata-based decision procedures for logics such as Presburger arithmetic [17,33] or monadic second order theories like WS1S or MSO(Str) [2,7,11,13,14,16,24], or as the basic underlying operation for testing language inclusion and equivalence over automata. Complementing *deterministic finite automata* (DFAs) is an easy task: it is sufficient to add a single state, direct all missing transitions to this state, and swap accepting and non-accepting states.

In practice, *nondeterministic finite automata* (NFAs) are often favored over DFAs due to their potentially much (up to exponentially) smaller size. The classical approach to complementing NFAs goes through determinization of the input NFA using the *powerset construction* into a DFA and then using DFA complementation. While easy to implement, this approach is prone to cause a blow-up in the number of states, as determinizing an NFA with n states may, in the worst case, result in a DFA with 2^n states [28] and the size of the complement would then also be exponential. Some automata-based algorithms are highly

A. Jeż and J. Otop (Eds.): FCT 2025, LNCS 16106, pp. 221–237, 2026.
https://doi.org/10.1007/978-3-032-04700-7_17

sensitive to the sizes of complement automata, such as decision procedures of certain logics [2,7,11,13,14,16,17,24,33], where the output of the complement may be the basic structure over which another complementation is performed (usually after projection, which can turn a potentially deterministic automaton into a nondeterministic one); for some of the logics, the increase in the size of the complement is the underlying cause of their non-elementary complexity [27]. Due to this, some of the applications tried to avoid complementation altogether, for instance using symbolic techniques [13,14,31]. This is, however, not always possible or feasible, as symbolic techniques often disallow to use, e.g., standard automata reduction techniques (cf., [8,22]), which often have a great impact on the performance of the applications.

Since the lower bounds on the worst-case sizes of deterministic and complement automata are both 2^n [3,20,23,29], one might think that the determinization-based approach is optimal. In practice, this is, however, far from the truth. Consider, e.g., the NFA $\mathcal{A}_2$ accepting the language $\{a,b\}^*.\{a\}.\{a,b\}^2$ shown in Fig. 1 (left). While the (minimal) deterministic complement has 8 states, there exists an NFA with 4 states given in Fig. 1 (right) that is a complement of $\mathcal{A}_2$. This complement was obtained as follows. We reversed $\mathcal{A}_2$, then determinized it (which is easy as $\mathcal{A}_2$ is reverse-deterministic), complemented the output by swapping accepting and non-accepting states, reversed again, and, finally, removed one unreachable state. We can generalize the example above to the family of NFAs $\mathcal{A}_n$ whose languages are $\{a,b\}^*.\{a\}.\{a,b\}^n$ for $n \in \mathbb{N}$. Here, the size of the minimal complement DFA for the NFA $\mathcal{A}_n$ is 2^{n+1}, while the complement NFA constructed by the *reverse powerset* procedure mentioned above has $n+2$ states, the same as $\mathcal{A}_n$. This example is a motivation for a deeper study of NFA complementation.

In this paper, we present several alternative approaches to NFA complementation. Besides the reverse powerset complementation (Sect. 3), we introduce two novel complementation constructions that target NFAs with a particular structure (containing several strongly connected components), common in practice. We first introduce a basic version of the novel constructions on NFAs with a very restricted structure (Sect. 4) and then we briefly present the generalized version of these constructions and the generalized complementation problem that allows to combine these constructions (Sect. 5). Our experimental evaluation (Sect. 6) shows that in a significant number of cases, using an alternative complementation method can give a much smaller complement than the classical construction. Due to the page limit, many parts of the paper are relegated to the technical report [21], including a precise description of the generalized complementation constructions with the corresponding correctness proofs, an example of NFA subclass with a subexponential complement size, implementation details, and additional experiments.

Related Work. While complementation of automata over infinite words is a lively topic (e.g., [1,4,19]), complementation of NFAs seems to be under-researched with not many relevant prior work. The powerset approach to determinization, which is the basic block of the classical complementation, can be traced to Rabin and Scott [28]. Optimizations of the powerset determinization were proposed in [15].

From the theoretical side, the exponential lower bound of NFA complementation was studied with respect to various alphabet sizes (the larger the alphabet size, the easier it is to construct an NFA whose complement is forced to be exponential) in [3,20,23,29].

2 Preliminaries

An *alphabet* Σ is a finite nonempty set of *symbols*. A *word* over Σ is a sequence $u = u_1 \dots u_n$ where $u_i \in \Sigma$ for all $1 \leq i \leq n$, with its *length* n denoted by $|u|$. The *empty word* is denoted by ε. The set of all words over Σ is denoted by Σ^*, and its subsets are *languages* over Σ. The *concatenation* of u with a word $v = u_{n+1} \dots u_m$ is the word $uv = u_1 \dots u_m$. The *reverse* of u is the word $rev(u) = u_n u_{n-1} \dots u_1$. The *concatenation* of languages $L, L' \subseteq \Sigma^*$ is the language $L.L' = \{uv \mid u \in L, v \in L'\}$. The *reverse* of L is the language $rev(L) = \{rev(w) \mid w \in L\}$ and its *complement* is the language $co(L) = \Sigma^* \setminus L$.

Finite Automata. A *nondeterministic finite automaton (NFA)* is defined as a tuple $\mathcal{A} = (Q, \Sigma, \delta, I, F)$, where Q is a finite set of *states*, Σ is an alphabet, $\delta \subseteq Q \times \Sigma \times Q$ is a *transition relation*, $I \subseteq Q$ is a set of *initial states*, and $F \subseteq Q$ is a set of *accepting states* (also called *final states*). We write $q \xrightarrow{a} r \in \delta$ instead of $(q, a, r) \in \delta$. If δ is clear from the context, we write just $q \xrightarrow{a} r$. The *size* of $\mathcal{A}$ is defined as $|\mathcal{A}| = |Q|$. We abuse the notation and use δ also as the function $\delta \colon Q \times \Sigma \to 2^Q$ defined as $\delta(q, a) = \{r \in Q \mid q \xrightarrow{a} r\}$. We also extend the function δ to sets of states $P \subseteq Q$ as $\delta(P, a) = \bigcup_{q \in P} \delta(q, a)$. A *deterministic finite automaton (DFA)* is an NFA with $|I| = 1$ and $|\delta(q, a)| \leq 1$ for all $q \in Q$ and $a \in \Sigma$. A DFA is *complete* if $|\delta(q, a)| = 1$ for all $q \in Q$ and $a \in \Sigma$. A *run* of an NFA $\mathcal{A}$ over a word $v = v_1 v_2 \dots v_n$ is a sequence of transitions $q_0 \xrightarrow{v_1} q_1, q_1 \xrightarrow{v_2} q_2, \dots, q_{n-1} \xrightarrow{v_n} q_n$ with $q_0 \in I$. It is *accepting* if $q_n \in F$. A state that appears in a run over some word $v \in \Sigma^*$ is *reachable*, else it is *unreachable*. A state is *reachable from a state* q if it is reachable in the NFA $(Q, \Sigma, \delta, \{q\}, F)$. $\mathcal{A}$ *accepts* the language $\mathcal{L}(\mathcal{A})$ of all words over Σ for which it has an accepting run. Automata $\mathcal{A}$ and $\mathcal{B}$ are *equivalent* iff $\mathcal{L}(\mathcal{A}) = \mathcal{L}(\mathcal{B})$.

The *reverse* of $\mathcal{A}$ is the NFA $rev(\mathcal{A}) = (Q, \Sigma, rev(\delta), F, I)$, where $rev(\delta) = \{r \xrightarrow{a} q \mid q \xrightarrow{a} r \in \delta\}$. $\mathcal{A}$ is called *reverse-deterministic* iff $rev(\mathcal{A})$ is a DFA. An NFA $\mathcal{C}$ is called a *complement* of $\mathcal{A}$ with respect to an alphabet Λ if $\mathcal{L}(\mathcal{C}) = \Lambda^* \setminus \mathcal{L}(\mathcal{A})$. If Λ is not specified, we assume that $\Lambda = \Sigma$. Given two NFAs $\mathcal{A}_1 = (Q_1, \Sigma, \delta_1, I_1, F_1)$ and $\mathcal{A}_2 = (Q_2, \Sigma, \delta_2, I_2, F_2)$ where $Q_1 \cap Q_2 = \emptyset$, their *union* is the NFA $\mathcal{A}_1 \uplus \mathcal{A}_2 = (Q_1 \cup Q_2, \Sigma, \delta_1 \cup \delta_2, I_1 \cup I_2, F_1 \cup F_2)$. A *strongly connected component (SCC)* of $\mathcal{A}$ is a maximal subset $C \subseteq Q$ in which every state is reachable from every state. Note that the term *component* will be used in a more general sense later.

Forward Powerset Complementation. The standard complementation first uses the *powerset construction* to transform a given NFA $\mathcal{A} = (Q, \Sigma, \delta, I, F)$

into an equivalent complete DFA $det(\mathcal{A}) = (Q', \Sigma, \delta', I', F')$, where $Q' = 2^Q$, $\delta' = \{P \xrightarrow{a} \delta(P, a) \mid P \in Q', a \in \Sigma\}$, $I' = \{I\}$, and $F' = \{P \in Q' \mid P \cap F \neq \emptyset\}$. In the following, we assume that $det(\mathcal{A})$ does not contain unreachable states (only the reachable part of Q' is constructed). This does not improve the upper bound on the size of the DFA, which is still $2^{|Q|}$. Given a complete DFA $\mathcal{D} = (Q, \Sigma, \delta, I, F)$, its complement can be constructed as $co(\mathcal{D}) = (Q, \Sigma, \delta, I, Q \setminus F)$. A complement of an NFA $\mathcal{A}$ can thus be constructed as $co(det(\mathcal{A}))$. We call this construction *forward powerset complementation.*

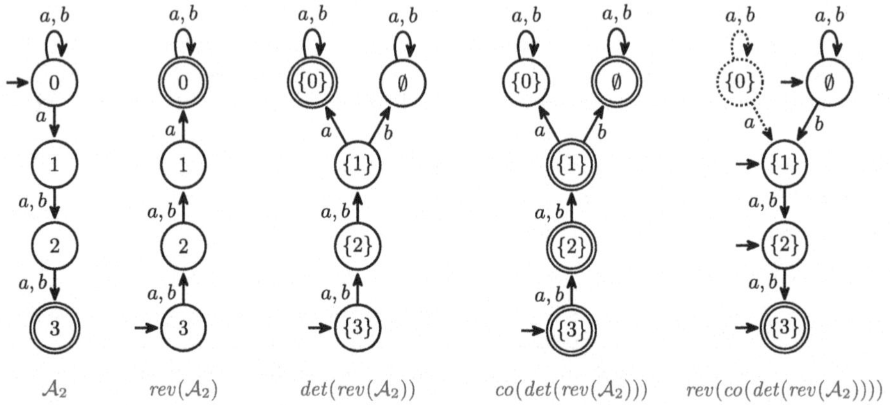

Fig. 1. NFA $\mathcal{A}_2$ accepting language $\{a, b\}^*.\{a\}.\{a, b\}^2$ and all phases of its reverse powerset complementation; the dotted part of the complement automaton is unreachable.

3 Reverse Powerset Complementation

The idea of this approach to complementation is simple: if we reverse an automaton, then complement it (for example using the powerset complementation given above) and reverse again, we obtain the complement of the original automaton. Formally, given an NFA $\mathcal{A}$, its complement can be constructed as $rev(co(det(rev(\mathcal{A}))))$. We call this approach *reverse powerset complementation.* Figure 1 shows all phases of this complementation process on an example automaton.

Contrary to the forward powerset construction, this method can yield a (forward-) nondeterministic automaton, which may be significantly smaller than the minimal deterministic complement. This can be documented by a generalization of the automata in Fig. 1. For any $n \in \mathbb{N}$, there exists an NFA $\mathcal{A}_n$ with $n+2$ states that accepts the language $\{a, b\}^*.\{a\}.\{a, b\}^n$. While the minimal forward powerset complement of $\mathcal{A}_n$ has 2^{n+1} states (intuitively, the DFA must store a bit vector of $n+1$ elements tracking which of the last $n+1$ symbols read were a), the reverse powerset complementation produces an NFA with $n+2$ reachable states. This is due to $\mathcal{A}_n$ being reverse-deterministic. Our experiments in [21] show that the reverse powerset construction outperforms the forward one in many cases.

Heuristic for Forward vs. Reverse Powerset. Naturally, the powerset construction can be efficient in one direction (forward or reverse) while causing a blow-up in the other. To address this, we designed a cheap heuristic to choose the more favorable direction for a given NFA, for cases when running a portfolio is too expensive.

The heuristic sums the sizes of all powerset successors of each state in a given NFA. Formally, for an NFA $\mathcal{A} = (Q, \Sigma, \delta, I, F)$, we define $sc(q) = \{\delta(q, a) \mid a \in \Sigma\}$ for each state $q \in Q$ and compute $powSC(\mathcal{A}) = |I| + \sum_{q \in Q} \sum_{S \in sc(q)} |S|$. We emphasize that if q has the same successors under two different symbols $a, b \in \Sigma$, then the set $\delta(q, a) = \delta(q, b)$ contributes to the sum only once.

Intuitively, a higher value of $powSC(\mathcal{A})$ should indicate a higher number of states produced by the powerset construction applied to $\mathcal{A}$. Hence, comparing $powSC(\mathcal{A})$ and $powSC(rev(\mathcal{A}))$ gives a hint which of the powerset constructions has a greater risk of blow-up. The heuristic's performance is experimentally evaluated in [21].

4 Sequential and Gate Complementation

This section introduces the basic ideas of two techniques called *sequential* and *gate complementation* with use on automata with a specific shape. The general form of these techniques is presented in Sect. 5.

Consider an NFA $\mathcal{A}$ that can be seen as two disjoint NFAs $\mathcal{A}_1 = (Q_1, \Sigma, \delta_1, I_1, \{q_F\})$ and $\mathcal{A}_2 = (Q_2, \Sigma, \delta_2, \{q_I\}, F_2)$ connected with a single transition $q_F \xrightarrow{c} q_I$ for some $c \in \Sigma$, i.e., $\mathcal{A} = (Q_1 \cup Q_2, \Sigma, \delta_1 \cup \delta_2 \cup \{q_F \xrightarrow{c} q_I\}, I_1, F_2)$. The automaton $\mathcal{A}$ accepts the language $L_1.\{c\}.L_2$ where $L_1 = \mathcal{L}(\mathcal{A}_1)$ and $L_2 = \mathcal{L}(\mathcal{A}_2)$. An example of such an NFA $\mathcal{A}$ and the corresponding automata $\mathcal{A}_1$ and $\mathcal{A}_2$ can be found in Fig. 2. The transition $q_F \xrightarrow{c} q_I$ is called the *transfer transition* and automata $\mathcal{A}_1$ and $\mathcal{A}_2$ are referred to as the *front* and the *rear component* of $\mathcal{A}$, respectively.

4.1 Sequential Complementation

Sequential complementation builds a potentially nondeterministic complement $\mathcal{C}$ of $\mathcal{A}$ from a complete DFA $\mathcal{A}'_1 = det(\mathcal{A}_1)$ equivalent to $\mathcal{A}_1$ and an NFA complement $\mathcal{C}_2$ of $\mathcal{A}_2$. The technique is called sequential complementation because we first complement $\mathcal{A}_2$ using an arbitrary complementation approach, and only then we build the complement of $\mathcal{A}$.

The construction is based on the following observation. The language $co(L_1.\{c\}.L_2)$ consists of words w such that, for all pairs of words u, v satisfying $ucv = w$, if $u \in L_1$ then $v \in co(L_2)$. Note that if w does not contain any

c, the condition is trivially satisfied. Therefore, $\mathcal{C}$ will simulate $\mathcal{A}'_1$ and whenever a final state of $\mathcal{A}'_1$ is visited, that is, $\mathcal{C}$ finished reading a prefix $u \in L_1$, a transition under c initiates a new instance of $\mathcal{C}_2$ that will check that the corresponding suffix v is indeed in $co(L_2)$. Automaton $\mathcal{C}$ will accept if each initiated instance of $\mathcal{C}_2$ accepts, meaning that for all possible splittings of the input word w into ucv where $u \in L_1$, it holds that $v \in co(L_2)$.

Formally, let $\mathcal{A}'_1 = (Q'_1, \Sigma, \delta'_1, \{q'_0\}, F'_1)$ be a complete DFA representing the language L_1 and $\mathcal{C}_2 = (\widetilde{Q}_2, \Sigma, \widetilde{\delta}_2, \widetilde{I}_2, \widetilde{F}_2)$ be an NFA representing $co(L_2)$. We construct an NFA $\mathcal{C} = (\widetilde{Q}, \Sigma, \widetilde{\delta}, \widetilde{I}, \widetilde{F})$ representing $co(L_1.\{c\}.L_2)$ as follows.

- $\mathcal{C}$'s states are pairs composed of the current state of $\mathcal{A}'_1$ and states of instances of $\mathcal{C}_2$, $\widetilde{Q} = Q'_1 \times 2^{\widetilde{Q}_2}$.
- For each $a \in \Sigma$, the transition relation $\widetilde{\delta}$ simulates the corresponding transition of $\mathcal{A}'_1$ and arbitrary transitions under a of the running instances of $\mathcal{C}_2$. Moreover, it initiates a new instance of $\mathcal{C}_2$ whenever $\mathcal{A}'_1$ moves from an accepting state by reading c. Formally, for each $(p, \{r_1, \ldots, r_n\}) \in \widetilde{Q}$, $a \in \Sigma$, and transitions $p \xrightarrow{a} q \in \delta'_1$, $r_i \xrightarrow{a} s_i \in \widetilde{\delta}_2$ for all $1 \leq i \leq n$, the transition relation $\widetilde{\delta}$ contains transitions
  - $(p, \{r_1, \ldots, r_n\}) \xrightarrow{a} (q, \{s_1, \ldots, s_n\})$ if $p \notin F'_1$ or $a \neq c$, and
  - $(p, \{r_1, \ldots, r_n\}) \xrightarrow{a} (q, \{s_1, \ldots, s_n\} \cup \{s_0\})$ for all $s_0 \in \widetilde{I}_2$ if $p \in F'_1$ and $a = c$.
- $\mathcal{C}$ starts in the initial state q'_0 of $\mathcal{A}'_1$ with no running instance of $\mathcal{C}_2$, i.e., $\widetilde{I} = \{(q'_0, \emptyset)\}$.
- $\mathcal{C}$ accepts whenever all running instances of $\mathcal{C}_2$ accept, i.e., $\widetilde{F} = Q'_1 \times 2^{\widetilde{F}_2}$.

Theorem 1. *The NFA $\mathcal{C}$ accepts $co(\mathcal{L}(\mathcal{A}))$.*

Proof (Proof (sketch)). Recall that $\mathcal{L}(\mathcal{A}) = L_1.\{c\}.L_2$. First, consider $w \in \mathcal{L}(\mathcal{A})$. Then there are $u \in L_1 = \mathcal{L}(\mathcal{A}'_1)$ and $v \in L_2$ such that $w = ucv$. As $\mathcal{A}'_1$ is deterministic, it has to reach a state $q_f \in F'_1$ after reading u. Hence, $\mathcal{C}$ can reach only states of the form (q_f, R) after reading u. When $\mathcal{C}$ reads c from this state, it reaches a state (q', R'), where R' has to contain some initial state s_0 of $\mathcal{C}_2$. However, $v \in L_2$ implies that $\mathcal{C}_2$ does not accept v. Hence, each state (q'', R'') of $\mathcal{C}$ reached from (q', R') by reading v is not accepting as it cannot satisfy $R'' \subseteq \widetilde{F}_2$. To sum up, $\mathcal{C}$ has no accepting run over $ucv = w$.

Now assume that $w \notin \mathcal{L}(\mathcal{A})$. As $\mathcal{A}'_1$ is deterministic and complete, it has a single run over w. Whenever the run reaches an accepting state over some prefix u of w, we know that $u \in L_1$ and thus the corresponding suffix cannot be of the form $cv \in \{c\}.L_2$ as that would contradict the assumption $w = ucv \notin \mathcal{L}(\mathcal{A})$. In other words, if the prefix u is followed by c, then $\mathcal{C}_2$ has an accepting run over the corresponding suffix v as $v \notin L_2$. We can construct an accepting run of $\mathcal{C}$ over w such that whenever the automaton $\mathcal{A}'_1$ tracked in the first element of the states of $\mathcal{C}$ reaches an accepting state and $\mathcal{C}$ reads c, we add to the second element of the state of $\mathcal{C}$ the initial state of the corresponding accepting run of $\mathcal{C}_2$ and then follow this run in the future transitions of $\mathcal{C}$. After reading the whole w, the second element of the reached state of $\mathcal{C}$ will contain only accepting states of $\widetilde{F}_2$. Thus, the constructed run of $\mathcal{C}$ over w is accepting and $w \in \mathcal{L}(\mathcal{C})$.

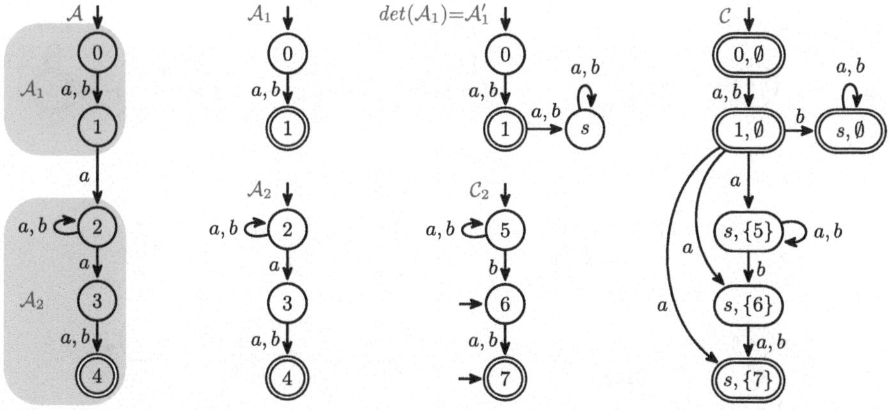

Fig. 2. An NFA $\mathcal{A}$, its front and rear components $\mathcal{A}_1, \mathcal{A}_2$, a complete DFA $\mathcal{A}'_1$ equivalent to $\mathcal{A}_1$, a complement $\mathcal{C}_2$ of $\mathcal{A}_2$, and the complement $\mathcal{C}$ of $\mathcal{A}$ constructed from $\mathcal{A}'_1$ and $\mathcal{C}_2$.

Figure 2 shows sequential complementation of an automaton $\mathcal{A}$ built from $\mathcal{A}_1$ and $\mathcal{A}_2$ connected by the transition $1 \xrightarrow{a} 2$ and illustrates that this complementation can produce nondeterministic results. Moreover, there exist automata for which the sequential complementation produces complements of linear size while both forward and reverse powerset complementations produce exponential complements. Consider the language $L_n = \{a,b\}^n.\{a\}.\{a,b\}^*.\{a\}.\{a,b\}^n$ for any $n \in \mathbb{N}$. There exists an automaton $\mathcal{B}_n$ with $2n+3$ states that accepts L_n (the automaton $\mathcal{B}_1$ is actually the automaton $\mathcal{A}$ in Fig. 2). Analogously to the figure, $\mathcal{B}_n$ can be decomposed into $\mathcal{B}_{n1}$ and $\mathcal{B}_{n2}$ accepting $L_{n1} = \{a,b\}^n$ and $L_{n2} = \{a,b\}^*.\{a\}.\{a,b\}^n$, respectively. Complementing $\mathcal{B}_{n2}$ into $\mathcal{C}_{n2}$ via the reverse powerset and applying sequential complementation yields a complement $\mathcal{C}_n$ with $2n+4$ states. In contrast, complementing $\mathcal{B}_n$ directly with either powerset method leads to an exponential blow-up – both results have $2^{n+1}+n+1$ states due to the loop under a, b and the nondeterminism in the middle of $\mathcal{B}_n$. Moreover, $\mathcal{B}_n$ belongs to an NFA class for which we prove a subexponential upper bound on the sequential complement size in [21]. This upper bound is strictly better compared to forward powerset for this NFA class.

4.2 Gate Complementation

Recall that we consider an automaton $\mathcal{A}$ that can be seen as two disjoint automata $\mathcal{A}_1 = (Q_1, \Sigma, \delta_1, I_1, \{q_F\})$ and $\mathcal{A}_2 = (Q_2, \Sigma, \delta_2, \{q_I\}, F_2)$ connected by a transfer transition $q_F \xrightarrow{c} q_I$ for some $c \in \Sigma$, i.e., $\mathcal{A} = (Q_1 \cup Q_2, \Sigma, \delta_1 \cup \delta_2 \cup \{q_F \xrightarrow{c} q_I\}, I_1, F_2)$. Now we additionally assume that the symbol c of the transfer transition does not appear in any transition of the front component $\mathcal{A}_1$. The transfer transition is then called a *gate* and c is the *gate symbol*. A scheme and an example of an NFA $\mathcal{A}$ with a gate can be found in Fig. 3.

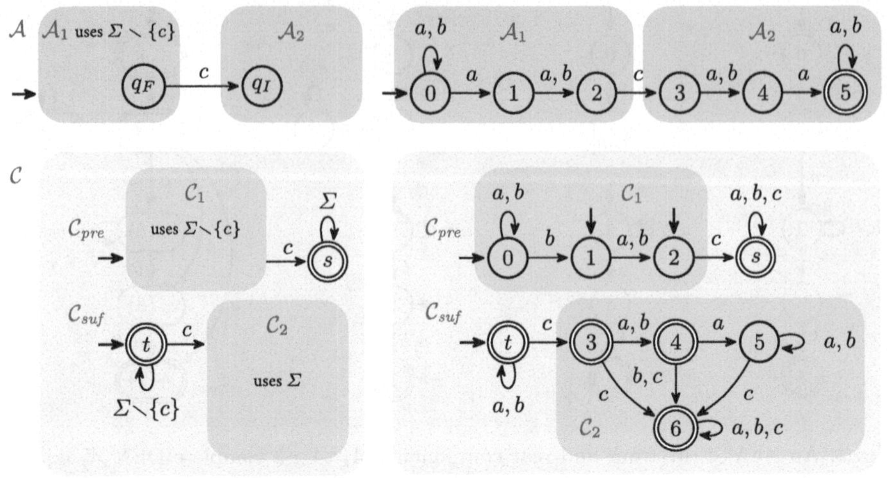

Fig. 3. A generic scheme of an NFA $\mathcal{A}$ with a gate (top left), a scheme of its complement $\mathcal{C}$ (bottom left), an instance of $\mathcal{A}$ for language $\{a, b\}^*.\{a\}.\{a, b\}.\{c\}.\{a, b\}.\{a\}.\{a, b\}^*$ (top right), and its gate complement $\mathcal{C}$ (bottom right).

Gate complementation utilizes the specific properties of $\mathcal{A}$. The automaton $\mathcal{A}$ accepts the language $L_1.\{c\}.L_2$ where $L_1 = \mathcal{L}(\mathcal{A}_1)$ and $L_2 = \mathcal{L}(\mathcal{A}_2)$. A word $w \in \Sigma^*$ belongs to $co(L_1.\{c\}.L_2)$ in the following cases:

1. w does not contain any c,
2. $w = ucv$ where $u \in (\Sigma \setminus \{c\})^*$ and $u \notin L_1$, or
3. $w = ucv$ where $u \in (\Sigma \setminus \{c\})^*$ and $v \notin L_2$.

Let $\mathcal{C}_1 = (\widetilde{Q}_1, \Sigma \setminus \{c\}, \widetilde{\delta}_1, \widetilde{I}_1, \widetilde{F}_1)$ be an arbitrary complement of $\mathcal{A}_1$ with respect to the alphabet $\Sigma \setminus \{c\}$ and $\mathcal{C}_2 = (\widetilde{Q}_2, \Sigma, \widetilde{\delta}_2, \widetilde{I}_2, \widetilde{F}_2)$ be an arbitrary complement of $\mathcal{A}_2$ with respect to the alphabet Σ. Then, the complement $\mathcal{C}$ of $\mathcal{A}$ consists of two parts: $\mathcal{C}_{pre}$ accepting the words according to Case 2 and $\mathcal{C}_{suf}$ accepting the words according to Cases 1 and 3. Schemes of $\mathcal{C}$, $\mathcal{C}_{pre}$, and $\mathcal{C}_{suf}$ can be seen in Fig. 3.

Formally, we set $\mathcal{C} = \mathcal{C}_{pre} \uplus \mathcal{C}_{suf}$ where $\mathcal{C}_{pre} = (\widetilde{Q}_1 \cup \{s\}, \Sigma, \delta_{pre}, \widetilde{I}_1, \{s\})$ and $\mathcal{C}_{suf} = (\widetilde{Q}_2 \cup \{t\}, \Sigma, \delta_{suf}, \{t\}, \{t\} \cup \widetilde{F}_2)$ such that s, t are fresh states,

$$\delta_{pre} = \widetilde{\delta}_1 \cup \{p \xrightarrow{c} s \mid p \in \widetilde{F}_1\} \cup \{s \xrightarrow{a} s \mid a \in \Sigma\}, \text{ and}$$
$$\delta_{suf} = \widetilde{\delta}_2 \cup \{t \xrightarrow{c} p \mid p \in \widetilde{I}_2\} \cup \{t \xrightarrow{a} t \mid a \in \Sigma \setminus \{c\}\}.$$

Theorem 2. *The NFA $\mathcal{C}$ accepts $co(\mathcal{L}(\mathcal{A}))$.*

Proof (Proof (sketch)). Recall that $\mathcal{L}(A) = L_1.\{c\}.L_2$. First, consider $w \in \mathcal{L}(\mathcal{A})$. Then $w = ucv$, where $u \in L_1$ and $v \in L_2$. Because $u \in L_1$, no run of $\mathcal{C}_1$ over u reaches $\widetilde{F}_1$. Hence, no run of $\mathcal{C}_{pre}$ over a word starting with uc reaches s and thus $w \notin \mathcal{L}(\mathcal{C}_{pre})$. The run of $\mathcal{C}_{suf}$ over uc reaches $\widetilde{I}_2$ and it cannot be prolonged into an accepting run over w as $\mathcal{C}_2$ does not accept v. Altogether, we get $w \notin \mathcal{L}(\mathcal{C}_{pre}) \cup \mathcal{L}(\mathcal{C}_{suf}) = \mathcal{L}(\mathcal{C})$.

Now assume that $w \notin \mathcal{L}(\mathcal{A})$. If w does not contain any c, it is accepted by $\mathcal{C}_{suf}$. Let $w = ucv$, where $u \in (\Sigma \setminus \{c\})^*$. As $w \notin \mathcal{L}(\mathcal{A})$, we know that $u \notin L_1$ or $v \notin L_2$. In the former case, $u \in \mathcal{L}(\mathcal{C}_1)$ and thus w is accepted by $\mathcal{C}_{pre}$. In the latter case, $v \in \mathcal{L}(\mathcal{C}_2)$ and thus w is accepted by $\mathcal{C}_{suf}$. To sum up, $w \in \mathcal{L}(\mathcal{C}_{pre}) \cup \mathcal{L}(\mathcal{C}_{suf}) = \mathcal{L}(\mathcal{C})$.

Note that $\mathcal{C}$ has only $|\mathcal{C}_1| + |\mathcal{C}_2| + 2$ states. To see the advantage of this complementation approach, consider $L_n = L_{n1}.\{c\}.L_{n2}$, where $L_{n1} = \{a,b\}^*.\{a\}.\{a,b\}^n$, $L_{n2} = \{a,b\}^n.\{a\}.\{a,b\}^*$, and $n \in \mathbb{N}$. There exists an NFA $\mathcal{B}_n$ with only $2n+4$ states accepting L_n, which can be deconstructed into two components accepting L_{n1} and L_{n2}, respectively (the automaton $\mathcal{B}_1$ is actually the automaton $\mathcal{A}$ in Fig. 3). If we use reverse powerset to complement the front component, forward powerset for the rear one, and combine the outputs by gate complementation, the result $\mathcal{C}_n$ has $2n+7$ states. Complementing the whole NFA $\mathcal{B}_n$ with either forward or reverse powerset causes an exponential blow-up, with both results having $2^{n+1} + n + 2$ states. Sequential complementation (Sect. 4.1) also results in a blow-up (for all possible divisions of $\mathcal{B}_n$ into front and rear components) due to the determinization of the front component and/or tracking possibly many instances of the complement of the rear component.

5 Generalized Complementation Problem

This section briefly generalizes the ideas from Sect. 4. The generalized complementation constructions are still applicable to automata consisting of two components, but there can be an arbitrary number of transfer transitions under various symbols leading from the front to the rear component. The generalized constructions again use complements of the components. These complements can be constructed either by forward or reverse powerset complementation, or by a recursive application of (generalized) sequential or gate complementation.

Due to the potentially recursive application of our complementation constructions and due to the fact that components can be connected by multiple transfer transitions, we need to complement components with many incoming and outgoing transfer transitions. Therefore, we work with automata that generalize initial and final states to multiple sets of entry and exit states. These sets are called *entry* and *exit port sets.* We talk about *port automata* and generalize the complementation problem to these automata as follows.

A *port NFA* or simply a *port automaton* is a tuple $\mathcal{A} = (Q, \Sigma, \delta, \mathcal{I}, \mathcal{F})$, where Q, Σ, and δ are as in an NFA, and $\mathcal{I} = (I_0, \ldots, I_k)$ and $\mathcal{F} = (F_0, \ldots, F_\ell)$ are sequences of subsets of Q called *entry port sets*, resp. *exit port sets.* A *slice* of $\mathcal{A}$ is an NFA with one entry and one exit port set chosen as the initial, resp. final states, i.e., the NFA $\mathcal{A}^{i,j} = (Q, \Sigma, \delta, I_i, F_j)$ for $0 \le i \le k$ and $0 \le j \le \ell$. $\mathcal{A}$ is *deterministic (port DFA)* if all its slices are deterministic (in particular, $|I_i| = 1$ must hold for every $0 \le i \le k$). A port DFA is *complete* if all its slices are complete. A *complement* of $\mathcal{A}$ is a port NFA representing complements of all slices. More precisely, a complement of $\mathcal{A}$ is a port NFA $\mathcal{C} = (\widetilde{Q}, \Sigma, \widetilde{\delta}, \widetilde{\mathcal{I}}, \widetilde{\mathcal{F}})$ with $\widetilde{\mathcal{I}} = (\widetilde{I}_0, \dots, \widetilde{I}_k)$ and $\widetilde{\mathcal{F}} = (\widetilde{F}_1, \dots, \widetilde{F}_\ell)$ such that $\mathcal{L}(\mathcal{C}^{i,j}) = co(\mathcal{L}(\mathcal{A}^{i,j}))$ for each $0 \le i \le k$ and $0 \le j \le \ell$. We call $\widetilde{I}_i$ an *entry port complement* of I_i and $\widetilde{F}_j$ an *exit port complement* of F_j, together shortened to *port complements.*

In the rest of this section, we first generalize the powerset construction to port automata to get a determinization procedure needed in the sequential complementation. With that, we generalize the forward and reverse powerset complementation to port automata. Powerset complementations are applied to components that cannot be recursively complemented by sequential or gate complementation (for example, because they cannot be further decomposed into a front and a rear component). Finally, we outline general versions of sequential and gate complementations.

5.1 Powerset Construction and Complementation for Port Automata

The powerset construction and complementation generalize to port automata as follows. Given a port NFA $\mathcal{A} = (Q, \Sigma, \delta, \mathcal{I}, \mathcal{F})$ with $\mathcal{I} = (I_i)_{0 \le i \le k}$ and $\mathcal{F} = (F_j)_{0 \le j \le \ell}$, the *port powerset construction* produces a port DFA $det(\mathcal{A}) = (Q', \Sigma, \delta', \mathcal{I}', \mathcal{F}')$, where Q' and δ' are defined as in the standard powerset construction (see Sect. 2), $\mathcal{I}' = (\{I_i\})_{0 \le i \le k}$, and $\mathcal{F}' = (\{P \in Q' \mid P \cap F_j \neq \emptyset\})_{0 \le j \le \ell}$. The original $\mathcal{A}$ and $det(\mathcal{A})$ are equivalent, i.e., $\mathcal{L}(\mathcal{A}^{i,j}) = \mathcal{L}(det(\mathcal{A})^{i,j})$ for all $0 \le i \le k$ and $0 \le j \le \ell$. Moreover, $det(\mathcal{A})$ is *complete.* The *complement* of any complete port DFA $\mathcal{D} = (Q', \Sigma, \delta', \mathcal{I}', \mathcal{F}')$ with $\mathcal{F}' = (F'_j)_{0 \le j \le \ell}$ is the port DFA $co(\mathcal{D}) = (Q', \Sigma, \delta', \mathcal{I}', \mathcal{F}'')$ where $\mathcal{F}'' = (Q' \setminus F'_j)_{0 \le j \le \ell}$. The forward powerset complement of a port NFA is, as for non-port NFAs, defined as $co(det(\mathcal{A}))$. The *reverse* of a port NFA $\mathcal{A}$ is the port NFA $rev(\mathcal{A}) = (Q, \Sigma, rev(\delta), \mathcal{F}, \mathcal{I})$. The reverse powerset complement is then constructed as $rev(co(det(rev(\mathcal{A}))))$.

5.2 Generalized Sequential Complementation

We now outline the generalization of sequential complementation to port NFAs, highlighting only the differences from Sect. 4.1 (see [21] for details). Let $\mathcal{A}$ be a component constructed by merging $\mathcal{A}_1$ and $\mathcal{A}_2$. Hence, $\mathcal{A}$ is a port NFA consisting of the front component $\mathcal{A}_1$ and the rear component $\mathcal{A}_2$, both port NFAs, connected by a set of transfer transitions δ_{trans} leading from $\mathcal{A}_1$ to $\mathcal{A}_2$. Unlike Sect. 4.1, we have a set of transfer transitions δ_{trans} instead of a single one, and both components have multiple entry/exit port sets. We first construct $det(\mathcal{A}_1)$ and a complement port NFA $\mathcal{C}_2$ of $\mathcal{A}_2$. Analogously to Sect. 4.1, the constructed complement $\mathcal{C}$ of $\mathcal{A}$ contains states (q, R), where q tracks the only run of $det(\mathcal{A}_1)$ and R tracks runs of $\mathcal{C}_2$ over suffixes of the input word.

The generalized construction closely follows the basic one. It differs in the following:

1. Given a state q of $det(\mathcal{A}_1)$ and a symbol $a \in \Sigma$, let P be the set of states p of $\mathcal{A}_2$ such that $\mathcal{A}$ contains a transfer transition $q'' \xrightarrow{a} p$ from some $q'' \in q$. Whenever $\mathcal{C}$ reaches a state (q, R) with a next on input, it has to check that the rest of the word is not accepted by $\mathcal{A}_2$ starting from any $p \in P$ and thus it spawns a new instance of $\mathcal{C}_2$ for each $p \in P$. Formally, $\mathcal{C}$ has all transitions $(q, R) \xrightarrow{a} (q', R')$ where (1) $q \xrightarrow{a} q'$ is the transition from q under a in $det(\mathcal{A}_1)$, (2) for each $r \in R$, R' contains some s such that $r \xrightarrow{a} s$ is a transition in $\mathcal{C}_2$, and, additionally, (3) for every $p \in P$, R' contains some state s from the port complement of the newly added entry port set $\{p\}$.
2. Since $\mathcal{A}_1$ may have exit ports with transitions leaving $\mathcal{A}$ and $\mathcal{A}_2$ may have entry ports with incoming edges from outside, $\mathcal{C}$ must reject words accepted entirely within either $\mathcal{A}_1$ or $\mathcal{A}_2$. Therefore, if a slice $\mathcal{A}^{i,j}$ has entry ports in $\mathcal{A}_2$, we activate one instance of $\mathcal{C}_2^{i,j}$ at the start: $\mathcal{C}^{i,j}$ has entry ports of the form $(q, \{r\})$, where q is the only entry port in $det(\mathcal{A}_1)^{i,j}$ and r is an entry port in $\mathcal{C}_2^{i,j}$. At the same time, a state (q, R) is an exit port of $\mathcal{C}^{i,j}$ only if q is not an exit port of $det(\mathcal{A}_1)^{i,j}$, ensuring the input word was not accepted in $\mathcal{A}_1^{i,j}$.

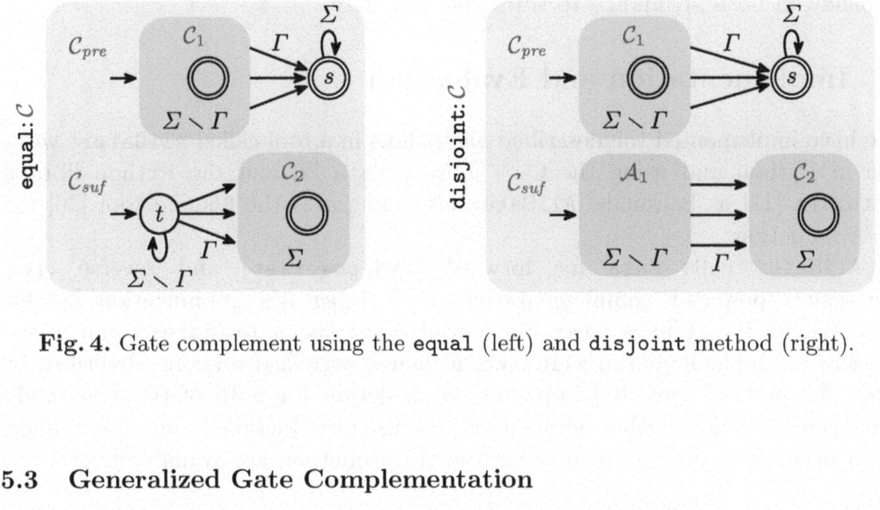

Fig. 4. Gate complement using the `equal` (left) and `disjoint` method (right).

5.3 Generalized Gate Complementation

Here we outline the core ideas of generalized gate complementation as an extension of Sect. 4.2, with full details in [21]. Recall that in Sect. 4.2, the NFA $\mathcal{A}$ was assumed to consist of two components connected by a single gate transition labeled with a gate symbol not occurring in the front component. We now generalize this by allowing arbitrary transitions leading from the front to the rear component and labeled with gate symbols $\Gamma \subseteq \Sigma$ such that $\mathcal{A}_1$ has no transitions under Γ. We call a gate transition labeled with c a c-gate. We also suppose that the input NFA may be a port NFA. We present two variants of the algorithm, `equal` and `disjoint` (illustrated in Fig. 4), each adding further constraints on the input NFA.

The simpler variant `equal` more closely resembles Sect. 4.2. If we write the words read by $\mathcal{A}$ as ucv with $u \in (\Sigma \setminus \Gamma)^*$, $c \in \Gamma$, and $v \in \Sigma^*$, this variant assumes that for every gate symbol $c \in \Gamma$, every two entry ports of $\mathcal{A}_2$ with an incoming c-gate can be reached by reading exactly the same prefixes of the form uc. To recognize all prefixes that cannot be read in $\mathcal{A}_1$ and followed by c, we collect the ports of $\mathcal{A}_1$ with outgoing c-gates into a new exit port set. We do the same for the suffixes that are not accepted in $\mathcal{A}_2$ from states with an incoming c-gate, and give $\mathcal{A}_2$ a new entry port set consisting of those states. These new port sets (one for each gate symbol) are preserved through complementation, and in $\mathcal{C}_{pre}$ and $\mathcal{C}_{suf}$, we can connect their complement port sets in $\mathcal{C}_1$ and $\mathcal{C}_2$ to the states s and t using an appropriate gate symbol, respectively. This way, $\mathcal{C}_{pre}$ accepts words where the first-appearing gate symbol c follows an invalid prefix and $\mathcal{C}_{suf}$ accepts words where c is followed by an invalid suffix.

The `disjoint` variant relaxes the requirement that all $\mathcal{A}_2$'s entry ports with an incoming c-gate must be reached by the same words of the form uc. Instead, $\mathcal{C}_{suf}$ tracks used gates, assuming the complemented port NFA is partitioned as follows: for any $c \in \Gamma$ and any two gates $p \xrightarrow{c} r$, $p' \xrightarrow{c} r'$, if the languages accepted by p and p' in any slice $\mathcal{A}_1^{i,j}$ are not disjoint, then the languages accepted from r and r' in the slice $\mathcal{A}_2^{i,j}$ must be equal. Unlike the `equal` method, $\mathcal{C}_{suf}$ now depends on both $\mathcal{A}_1$ and $\mathcal{C}_2$ to track the used gate.

6 Implementation and Evaluation

We have implemented the described algorithms in a tool called `AliGater`[1] written in Python and using the C++ library `Mata` [9] and the Python library `Automata` [12] as backends. `AliGater` also integrates the `Reduce` tool [26] for NFA reduction.

`AliGater` calls `Mata` for forward (`fwd powerset`) and reverse (`rev powerset`) powerset complementation, and Hopcroft's minimization [22,30] (denoted by the suffix `+ min`). Sequential (`seq`) and gate (`gate`) complementations are implemented in `AliGater` in their generalized versions, described in Sects. 5.2 and 5.3. Selected implementation details for both of these methods are discussed below. Other details (e.g., the use of reductions) and an extended description of `AliGater` settings used in the evaluation are available in [21].

6.1 Implementation of Sequential Complementation

The implementation of sequential complementation first divides the NFA $\mathcal{A}$ into components $\mathcal{A}_1, \ldots, \mathcal{A}_n$. Then $\mathcal{A}_n$ is complemented using forward or reverse powerset complementation, and $\mathcal{A}_1, \ldots, \mathcal{A}_{n-1}$ are determinized. The automaton $det(\mathcal{A}_{n-1})$ is then composed with the complement of $\mathcal{A}_n$ to form a new complemented bottom component; the same process is repeated with every preceding component up to $\mathcal{A}_1$.

We implemented three approaches to divide $\mathcal{A}$ into components:

[1] https://gitlab.fi.muni.cz/xstepkov/aligater.

1. *Deterministic components.* Because the components $\mathcal{A}_1, \ldots, \mathcal{A}_{n-1}$ are determinized during sequential complementation, this partitioning approach tries to avoid determinization and use transfer transitions to cover some nondeterminism in $\mathcal{A}$. We first decompose $\mathcal{A}$ into SCCs and order them by topological ordering. If the first SCC is nondeterministic, we set $\mathcal{A}_1$ to be this SCC. If it is deterministic, we set $\mathcal{A}_1$ to be the first SCC and we repeatedly add previously unused SCCs in the order of the topological ordering (with the corresponding transfer transitions) until a maximal deterministic $\mathcal{A}_1$ is obtained. Note that $\mathcal{A}_1$ may not be connected. The rest of the automaton $\mathcal{A}$ is then recursively divided in the same manner.
2. *Deterministic components + reverse-deterministic bottom component.* Sequential complementation requires $\mathcal{A}_n$ to be complemented, and if it is reverse-deterministic, we can complement it easily by reverse powerset complementation. We compute $\mathcal{A}_n$ as a maximal bottom reverse-deterministic part of $\mathcal{A}$, analogously to the computation of $\mathcal{A}_1$ in the previous approach. The rest of $\mathcal{A}$ is divided in the same way as above.
3. *Min-cut.* $\mathcal{A}$ is divided into two components, where the partition with the fewest transfer transitions from $\mathcal{A}_1$ to $\mathcal{A}_2$ is chosen. To obtain this partition, we construct a directed graph whose vertices are the SCCs of $\mathcal{A}$. The capacity of each edge is the amount of transitions between the SCCs. The function `minimum_cut()` from the `NetworkX` library [18] is then used to compute the minimum cut of this graph. The source is a fresh vertex with edges to all SCCs with no predecessors, the sink is the last SCC in a topological ordering.

6.2 Implementation of Gate Complementation

`AliGater` implements the generalized gate complementation methods. The implementation tries to find all possible partitions of the input NFA $\mathcal{A}$ satisfying the input conditions of any of the methods presented in Sect. 5.3 (cf. [21] for more details). The input conditions are formulated as the language equivalence or disjointness of certain states and evaluated by `Mata`; language equivalence is tested using the *antichains algorithm* [34], language disjointness is checked via an emptiness check on the product automaton. The concrete details can be found in [21], but the main idea is to prefer partitions tagged with `equal` to those tagged with `disjoint`, since they usually deliver smaller results, and that we try to pick a partition where the two components have a similar size.

6.3 Evaluation

All experiments were run on computers with the Intel®Core™i7-8700 CPU. Each complementation algorithm was executed on each input NFA with the timeout of 5 min and the memory limit of 8 GiB. We focus on the size (number of states) of the results and also how often the individual methods were successful (finished within the time and memory limits). TO and MO means that the time or memory limit was reached, respectively.

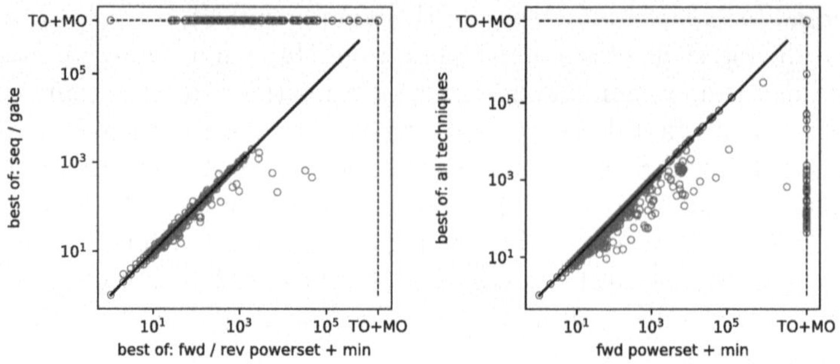

Fig. 5. Comparison of sizes of complements (number of states) given by the best of `fwd powerset + min` and `rev powerset + min` against the best of `seq` and `gate` (left), and `fwd powerset + min` against the best result of all techniques (right).

For the evaluation, we used a total of 9,450 benchmarks from `nfa-bench` [32], which gathers automata benchmarks from diverse applications. Additional details regarding the families of benchmarks are in [21].

Results. We evaluated the performance of the proposed algorithms `seq` and `gate` compared to the best result of `fwd powerset + min` and `rev powerset + min` (Fig. 5, left). The methods `seq` and `gate` are, in general, computationally more intensive than the powerset constructions. Running `seq` often produces large automata unless reduced during the process. The automata structure, however, allows `Reduce` to effectively reduce them, unlike automata generated by powerset constructions. Applicability of `gate` is limited by input conditions (it produced results for 2,577 benchmarks) and evaluating these conditions can be costly. Despite frequent timeouts, these methods sometimes achieve significantly better results than powerset-based methods, suggesting further room for improvement in NFA complementation beyond powerset-based techniques.

We have also evaluated the benefit of using all available techniques (e.g., in a portfolio) against `fwd powerset + min` as the baseline method (Fig. 5, right). This use case is targeted at applications where it pays off to obtain as small automaton for the complement as possible, such as when translating an extended regex into an NFA that will be used millions of times during matching. The results show that the proposed techniques were in many cases able to bring significant benefits, in particular solving a number of cases when `fwd powerset + min` ran out of resources.

7 Conclusion

We have presented, to the best of our knowledge, the first systematic empirical study of NFA complementation approaches. We suggested several novel algorithms for complementation of (subclasses of) NFAs. We carried out an extensive experimental evaluation of the approaches and showed that alternative complementation algorithms can often give a significantly better result than the classic approach (sometimes even in orders of magnitude). We have also suggested a heuristic that helps to select between the classic approach and reverse powerset complementation.

There are still multiple opportunities for improvement, e.g., in the partitioning process for the sequential complementation. Moreover, other structural classes of automata amenable for efficient complementation may exist.

Acknowledgements. We thank the anonymous reviewers for careful reading of the paper and their suggestions that improved its quality. L. Holík and O. Lengál were supported by the Czech Ministry of Education, Youth and Sports ERC.CZ project LL1908, the Czech Science Foundation projects 23-07565S and 25-17934S, and the FIT BUT internal project FIT-S-23-8151. A. Štěpková and J. Strejček were supported by the European Union's Horizon Europe program under the grant agreement No. 101087529 (CHESS).

References

1. Allred, J.D., Ultes-Nitsche, U.: A simple and optimal complementation algorithm for Büchi automata. In: LICS 2018, pp. 46–55. ACM (2018)
2. Ayari, A., Basin, D.: Bounded model construction for monadic second-order logics. In: Emerson, E.A., Sistla, A.P. (eds.) CAV 2000. LNCS, vol. 1855, pp. 99–112. Springer, Heidelberg (2000). https://doi.org/10.1007/10722167_11
3. Birget, J.: Partial orders on words, minimal elements of regular languages and state complexity. Theor. Comput. Sci. **119**(2), 267–291 (1993)
4. Blahoudek, F., Duret-Lutz, A., Strejček, J.: Seminator 2 can complement generalized Büchi automata via improved semi-determinization. In: Lahiri, S.K., Wang, C. (eds.) CAV 2020. LNCS, vol. 12225, pp. 15–27. Springer, Cham (2020). https://doi.org/10.1007/978-3-030-53291-8_2
5. Bouajjani, A., Habermehl, P., Rogalewicz, A., Vojnar, T.: Abstract regular (tree) model checking. Int. J. Softw. Tools Technol. Transf. **14**(2), 167–191 (2012)
6. Bouajjani, A., Jonsson, B., Nilsson, M., Touili, T.: Regular model checking. In: Emerson, E.A., Sistla, A.P. (eds.) CAV 2000. LNCS, vol. 1855, pp. 403–418. Springer, Heidelberg (2000). https://doi.org/10.1007/10722167_31
7. Büchi, J.R.: On a decision method in restricted second-order arithmetic. In: International Congress on Logic, Methodology, and Philosophy of Science, pp. 1–11. Stanford University Press (1962)
8. Bustan, D., Grumberg, O.: Simulation-based minimization. ACM Trans. Comput. Log. **4**(2), 181–206 (2003)
9. Chocholatý, D., et al.: Mata: a fast and simple finite automata library. In: Finkbeiner, B., Kovács, L. (eds.) TACAS 2024. LNCS, vol. 14571, pp. 130–151. Springer, Cham (2024). https://doi.org/10.1007/978-3-031-57249-4_7
10. Chocholatý, D., Havlena, V., Holík, L., Hranička, J., Lengál, O., Síč, J.: Z3-noodler 1.3: shepherding decision procedures for strings with model generation. In: Gurfinkel, A., Heule, M. (eds.) TACAS 2025. LNCS, vol. 15697, pp. 23–44. Springer, Cham (2025). https://doi.org/10.1007/978-3-031-90653-4_2

11. Elgaard, J., Klarlund, N., Møller, A.: MONA 1.x: new techniques for WS1S and WS2S. In: Hu, A.J., Vardi, M.Y. (eds.) CAV 1998. LNCS, vol. 1427, pp. 516–520. Springer, Heidelberg (1998). https://doi.org/10.1007/BFb0028773
12. Evans, C.: CALEB531/automata: a Python library for simulating finite automata, pushdown automata, and Turing machines (2025)
13. Fiedor, T., Holík, L., Janků, P., Lengál, O., Vojnar, T.: Lazy automata techniques for WS1S. In: Legay, A., Margaria, T. (eds.) TACAS 2017. LNCS, vol. 10205, pp. 407–425. Springer, Heidelberg (2017). https://doi.org/10.1007/978-3-662-54577-5_24
14. Fiedor, T., Holík, L., Lengál, O., Vojnar, T.: Nested antichains for WS1S. Acta Informatica **56**(3), 205–228 (2019). https://doi.org/10.1007/s00236-018-0331-z
15. van Glabbeek, R., Ploeger, B.: Five determinisation algorithms. In: Ibarra, O.H., Ravikumar, B. (eds.) CIAA 2008. LNCS, vol. 5148, pp. 161–170. Springer, Heidelberg (2008). https://doi.org/10.1007/978-3-540-70844-5_17
16. Glenn, J., Gasarch, W.: Implementing WS1S via finite automata. In: Raymond, D., Wood, D., Yu, S. (eds.) WIA 1996. LNCS, vol. 1260, pp. 50–63. Springer, Heidelberg (1997). https://doi.org/10.1007/3-540-63174-7_5
17. Habermehl, P., Havlena, V., Hečko, M., Holík, L., Lengál, O.: Algebraic reasoning meets automata in solving linear integer arithmetic. In: Gurfinkel, A., Ganesh, V. (eds.) CAV 2024. LNCS, vol. 14681, pp. 42–67. Springer, Cham (2024). https://doi.org/10.1007/978-3-031-65627-9_3
18. Hagberg, A.A., Schult, D.A., Swart, P.J.: Exploring network structure, dynamics, and function using networkx. In: SciPy 2008, pp. 11 – 15. Pasadena, CA, USA (2008)
19. Havlena, V., Lengál, O., Šmahlíková, B.: Complementation of Emerson-Lei automata. In: Abdulla, P.A., Kesner, D. (eds.) FoSSaCS 2025. LNCS, vol. 15691, pp. 88–110. Springer, Cham (2025). https://doi.org/10.1007/978-3-031-90897-2_5
20. Holzer, M., Kutrib, M.: State complexity of basic operations on nondeterministic finite automata. In: Champarnaud, J.-M., Maurel, D. (eds.) CIAA 2002. LNCS, vol. 2608, pp. 148–157. Springer, Heidelberg (2003). https://doi.org/10.1007/3-540-44977-9_14
21. Holík, L., Lengál, O., Major, J., Štěpková, A., Strejček, J.: On complementation of nondeterministic finite automata without full determinization (technical report). CoRR abs/2507.03439 (2025). https://arxiv.org/abs/2507.03439
22. Hopcroft, J.: An $n \log n$ algorithm for minimizing states in a finite automaton. In: Kohavi, Z., Paz, A. (eds.) Theory of Machines and Computations, pp. 189–196. Academic Press (1971)
23. Jirásková, G.: State complexity of some operations on binary regular languages. Theor. Comput. Sci. **330**(2), 287–298 (2005)
24. Kelb, P., Margaria, T., Mendier, M., Gsottberger, C.: MOSEL: a sound and efficient tool for M2L(Str). In: Grumberg, O. (ed.) CAV 1997. LNCS, vol. 1254, pp. 448–451. Springer, Heidelberg (1997). https://doi.org/10.1007/3-540-63166-6_45
25. Kesten, Y., Maler, O., Marcus, M., Pnueli, A., Shahar, E.: Symbolic model checking with rich assertional languages. Theor. Comput. Sci. **256**(1–2), 93–112 (2001)
26. Mayr, R., Clemente, L.: Advanced automata minimization. In: POPL 2013. ACM (2013)
27. Meyer, A.R.: Weak monadic second order theory of succesor is not elementary-recursive. In: Parikh, R. (ed.) Logic Colloquium. LNM, vol. 453, pp. 132–154. Springer, Heidelberg (1975). https://doi.org/10.1007/BFb0064872
28. Rabin, M.O., Scott, D.S.: Finite automata and their decision problems. IBM J. Res. Dev. **3**(2) (1959)

29. Sakoda, W.J., Sipser, M.: Nondeterminism and the size of two way finite automata. In: STOC 1978, pp. 275–286. ACM (1978)
30. Valmari, A., Lehtinen, P.: Efficient minimization of DFAs with partial transition functions. In: STACS 2008. LIPIcs, vol. 1, pp. 645–656. Dagstuhl (2008)
31. Varatalu, I.E., Veanes, M., Ernits, J.P.: RE#: high performance derivative-based regex matching with intersection, complement, and restricted lookarounds. PACMPL **9**(POPL), 1–32 (2025)
32. VeriFIT: `nfa-bench` Extensive benchmark for reasoning about regular properties (2025). https://github.com/VeriFIT/nfa-bench
33. Wolper, P., Boigelot, B.: An automata-theoretic approach to Presburger arithmetic constraints. In: Mycroft, A. (ed.) SAS 1995. LNCS, vol. 983, pp. 21–32. Springer, Heidelberg (1995). https://doi.org/10.1007/3-540-60360-3_30
34. De Wulf, M., Doyen, L., Henzinger, T.A., Raskin, J.-F.: Antichains: a new algorithm for checking universality of finite automata. In: Ball, T., Jones, R.B. (eds.) CAV 2006. LNCS, vol. 4144, pp. 17–30. Springer, Heidelberg (2006). https://doi.org/10.1007/11817963_5

Hardness of Pre-assignment Problem for Unique Minimum Vertex Cover on Planar Graphs with Maximum Degree 3

Takashi Horiyama, Fumiya Sakamoto, Kazuhisa Seto(✉), and Ryu Suzuki

Hokkaido University, Sapporo, Japan
{horiyama,seto}@ist.hokudai.ac.jp, sakamoto.fumiya.o7@elms.hokudai.ac.jp

Abstract. Horiyama, Kobayashi, Ono, Seto, and Suzuki introduced the pre-assignment model for uniquifying the optimal solutions of a given instance of optimization problems [AAAI 2024]. For example, a pre-assignment of minimum vertex cover C is defined as a vertex set $C_{\mathrm{pre}} = (C_{\mathrm{in}}, C_{\mathrm{ex}})$ such that C includes vertices in C_{in} and excludes ones from C_{ex}. The pre-assignment problem for unique minimum vertex covers (PAU-VC) asks whether there exists a pre-assignment of at most k vertices that induces a unique minimum vertex cover of a given graph. PAU-VC has three pre-assignment models: Include (i.e., $C_{\mathrm{ex}} = \emptyset$), Exclude (i.e., $C_{\mathrm{in}} = \emptyset$), and Mixed. Horiyama et al. showed that PAU-VC is Σ_2^p-complete on general and NP-complete on bipartite graphs under all three models. This paper shows PAU-VC remains Σ_2^p-complete in all three models even if a given graph is restricted to be planar with maximum degree 3.

Keywords: fewest clues problem · pre-assignment problem · uniqueness · minimum vertex cover · planar · maximum degree 3 · Σ_2^p-complete

1 Introduction

The uniqueness of solutions is crucial in combinatorial puzzles such as Sudoku. If an instance of Sudoku has multiple solutions, we cannot logically determine a number in some cell, and then the puzzle is no fun. For this reason, we must construct an instance with a unique solution. It is a natural question to ask how many clues are needed to transform an instance of the puzzle with multiple solutions into one with a unique solution, and what the complexity of computing such clues is. Demaine, Ma, Schvartzman, Waingarten, and Aaronson [8] formulated this problem and named it the Fewest Clues Problem (FCP). For a decision problem $\mathcal{A}$ in NP, FCP $\mathcal{A}$ asks whether there exists a clue of size at most k that ensures a unique solution for an instance $I \in \mathcal{A}$. The paper [8] established the computational intractability of the FCP for puzzles and satisfiability problems

A. Jeż and J. Otop (Eds.): FCT 2025, LNCS 16106, pp. 238–251, 2026.
https://doi.org/10.1007/978-3-032-04700-7_18

(SAT). For example, it was shown that FCP Sudoku, FCP Planar 3SAT, and FCP 3SAT are Σ_2^p-complete, and FCP 2SAT is NP-complete. The class Σ_2^p is a set of languages that can be decided in nondeterministic polynomial-time with an NP-oracle. Other hardness results exist on the complexity of the FCP version of puzzles. The Σ_2^p-completeness of the FCP versions of Picross 3D [17], Rome [10], and Fill-a-Pix [14] has been shown.

Horiyama, Kobayashi, Ono, Seto, and Suzuki [15] extended the notion of clues to pre-assignment for optimization problems, and introduced three pre-assignment models: Include, Exclude, and Mixed models. For the minimum vertex cover problem, the pre-assignment is defined as the pair of sets $C_{\mathrm{pre}} = (C_{\mathrm{in}}, C_{\mathrm{ex}})$, where the set C_{in} must be the vertices included in and the set C_{ex} must be the vertices excluded from the minimum vertex cover. In the pre-assignment models, the Include model is defined as $C_{\mathrm{ex}} = \emptyset$, the Exclude model is defined as $C_{\mathrm{in}} = \emptyset$, and the Mixed model imposes no restriction on C_{in} and C_{ex}. The paper [15] investigated the complexity of computing the pre-assignment for uniquifying the minimum vertex covers of a given graph (PAU-VC). Formally, PAU-VC asks, given a graph G and a non-negative integer k, whether there exists a pre-assignment C_{pre} of size at most k $(= |C_{\mathrm{in}}| + |C_{\mathrm{ex}}|)$ that induces a unique minimum vertex cover of G. The paper [15] showed that PAU-VC is Σ_2^p-complete on general graphs and is NP-complete on bipartite graphs under all three models. It also provided FPT algorithms parameterized by the size of the minimum vertex cover for general graphs and an exponential-time algorithm for trees under all three models. However, it remained open whether PAU-VC on trees is NP-complete. Under the Include model, An, Chang, Cho, Kwon, Lee, Oh, and Shin [2] clarified that it is solvable in polynomial time by providing an FPT algorithm for PAU-VC on bounded clique-width, and also gave linear-time algorithms for PAU-VC on split graphs and unit interval graphs that have no bounded clique-width.

In this paper, we investigate the computational complexity of PAU-VC on planar graphs. We first show the Σ_2^p-completeness of the FCP version of Planar 1-in-3SAT with the restriction that each clause has three positive literals for proving Theorem 1. The proofs use the Σ_2^p-completeness of FCP 1-in-3SAT [8], and the parsimonious reduction in [16]. By refining the parsimonious reduction in [4] and combining the above result, our main contribution is obtained.

Theorem 1. *PAU-VC is Σ_2^p-complete even when a given graph is restricted to be planar with maximum degree 3 under all three models.*

Due to the space limitation, some proofs (marked $\star$) are omitted. They appear in the full version of this paper.

Related Work

Similar concepts to the fewest clues and pre-assignments have been widely studied. In theoretical chemistry, Klein and Randić [18] introduce the notion of *forcing set* to investigate molecular resonance structures, in which edges of a perfect matching correspond to double bonds of a molecule. Harary, Klein, and

Živkovič [12] defined it in terms of graph theory as follows: For a graph G with a perfect matching M, a forcing set is a set of $S \subseteq M$ such that M is the only perfect matching of G inducing S. Afshani, Hatami, and Mahmoodian [1] investigated the complexity of computing a forcing set. They showed that it is NP-complete to determine whether the given bipartite graph of maximum degree 4 has a perfect matching with a forcing set of size at most the given integer k. Vukičević and Trinajstić [21] introduced anti-forcing sets as the opposite concept of forcing sets. A forcing set and an anti-forcing set are the same as pre-assignment with the Include model and the Exclude model, respectively. The forcing set is generalized to graph colorings [13], independent set [19], dominating sets [5], and independent dominating sets [3]. The notion of forcing sets has arisen in various contexts, including the definition of sets [11,20] and critical sets [6,7].

2 Preliminaries

2.1 Graphs and Satisfiability Problems

Let $G = (V(G), E(G))$ be an undirected graph with the vertex set $V(G)$ and the edge set $E(G)$. For $v \in V(G)$, we denote by $N_G(v)$ the set of neighbors of v in G, i.e., $N_G(v) = \{w \in V(G) \mid \{v, w\} \in E(G)\}$. We extend this notation to the vertex set: $N_G(V') = \bigcup_{v \in V'} N_G(v) \setminus V'$. We may omit the subscript G when no confusion arises. A *vertex cover* of G is a vertex set $C \subseteq V(G)$ such that for every edge in G, at least one end vertex is contained in C. The vertex cover problem (VC) is, given a graph G and a non-negative integer k, to determine whether G has a vertex cover C of size at most k. Planar VC is VC on planar graphs. Let x denote a variable. A *literal* is a Boolean variable x (positive literal) or its negation $\bar{x}$ (negative literal). A *clause* is a disjunction of literals, and the length of a clause (for short, clause length) is defined as the number of literals contained in the clause. A CNF formula with $\{x_1, \dots, x_n\}$ is represented as follows:

$$\varphi(x_1, x_2, ..., x_n) = \bigwedge_i \varphi_i = \bigwedge_i \bigvee_j \ell_{i,j},$$

where $\ell_{i,j}$ denotes the j-th literal appearing in the i-th clause φ_i. A k-CNF formula is a CNF formula in which the length of each clause is at most k. The satisfiability problem (SAT) is the problem of determining whether there exists an assignment to the variables that makes a given formula φ true. A *satisfying assignment* is an assignment that makes φ true. Let α_X be an assignment of φ over the variable set X. A partial assignment to $X' \subseteq X$ under α_X is denoted by $\alpha_{X'}$. For example, let $X = \{x_1, x_2, \dots, x_5\}$ and a satisfying assignment $\alpha_X = (1, 0, 1, 0, 1)$. For $X' = \{x_1, x_3, x_4\} \subseteq X$, then $\alpha'_X = (1, *, 1, 0, *)$ where $*$ is unassigned. The formula obtained by applying a partial assignment $\alpha_{X'}$ to φ is denoted by $\varphi|_{\alpha_{X'}}$. We denote $[n] = \{1, 2, \dots, n\}$. Let $X = \{x_1, \dots, x_n\}$ and $\varphi = \varphi_1 \wedge \cdots \wedge \varphi_m$ be a CNF formula over X. A *SAT graph* is the graph constructed from φ as follows:

1. Construct $n + m$ vertices corresponding to the elements in X and Φ.
2. Connect any pair of the vertices x_i ($i \in [n]$) and φ_j ($j \in [m]$) when x_i or $\bar{x}_i$ appears in the clause φ_j.

Figure 1 shows an example of the SAT graph constructed from $\varphi = (x_1 \vee \overline{x_2} \vee x_3)(\overline{x_1} \vee x_4 \vee \overline{x_5})$.

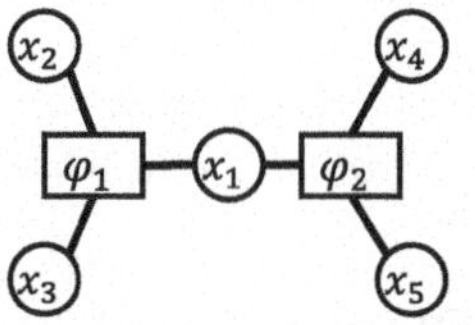

Fig. 1. SAT graph of $\varphi = \varphi_1 \wedge \varphi_2 = (x_1 \vee \overline{x_2} \vee x_3)(\overline{x_1} \vee x_4 \vee \overline{x_5})$

We introduce the various variants of SAT addressed in this paper. One-in-three SAT (1-in-3SAT) is, given a 3-CNF formula with n variables, determining whether there exists a satisfying assignment such that for each clause, exactly one literal is true. Such a satisfying assignment is called a *1-in-3 satisfying assignment.* Planar 1-in-3SAT is 1-in-3SAT whose SAT graph is planar. Planar 1-in-E3SAT is Planar 1-in-3SAT with each clause length exactly 3. Planar Mono 1-in-E3SAT is Planar 1-in-E3SAT, where each clause has only positive literals.

2.2 Fewest Clues Problem and Pre-Assignment Problem

Fewest Clues Problem $\mathcal{A}$ (FCP $\mathcal{A}$) is, given an instance I of a decision problem $\mathcal{A}$ and a non-negative integer k, to determine whether there exists a partial assignment to at most k input variables of I such that I has an exactly one solution consistent with the assignment. The paper [8] shows the following.

Theorem 2 ([8]). *If* $\mathcal{A} \in \text{NP}$, *then FCP* $\mathcal{A} \in \Sigma_2^p$.

Theorem 3 ([8]). *FCP Planar 1-in-3SAT is* Σ_2^p*-complete even when a given formula is satisfiable.*

A pre-assignment problem for a unique optimal solution (PAU) is a natural extension of FCP. Given an instance of an optimization problem $\mathcal{A}$ and a non-negative integer k, PAU-$\mathcal{A}$ asks whether there exists a pre-assignment of size at most k that induces a unique optimal solution. We call such a pre-assignment *feasible.* For the minimum vertex cover problem, the pre-assignment C_{pre} consists of two vertex sets: C_{in} is the set of vertices that must be included in the minimum vertex cover, and C_{ex} is the set of vertices that must be excluded from the minimum vertex cover to induce a unique minimum vertex cover. A pre-assignment for the minimum vertex cover C^* is feasible if the minimum vertex cover C^* satisfying $C_{\text{in}} \subseteq C^* \subseteq V(G) \setminus C_{\text{ex}}$ is unique. The pre-assignment problem for unique minimum vertex cover (PAU-VC) asks, given an undirected graph

G and a non-negative integer k, whether there exists a feasible pre-assignment $C_{\text{pre}} = (C_{\text{in}}, C_{\text{ex}})$ such that $|C_{\text{in}}| + |C_{\text{ex}}| \leq k$. PAU-VC has three pre-assignment models: Include (i.e., $C_{\text{ex}} = \emptyset$), Exclude (i.e., $C_{\text{in}} = \emptyset$), and Mixed. The paper [15] shows the following.

Lemma 1 **([15]).** *In PAU-VC, a graph G has a feasible pre-assignment of size at most k under the Exclude model if and only if G has a feasible pre-assignment of size at most k under the Mixed model.*

Theorem 4 **([15]).** *PAU-VC is Σ_2^p-complete on general and NP-complete on bipartite graphs under Include, Exclude, and Mixed models.*

3 Hardness of FCP Planar 1-in-E3SAT and FCP Planar Mono 1-in-E3SAT

3.1 FCP Planar 1-in-E3SAT

We prove Σ_2^p-hardness of FCP Planar 1-in-E3SAT by providing a polynomial-time reduction from FCP Planar 1-in-3SAT.

We first construct an instance of FCP Planar 1-in-E3SAT from an instance of FCP Planar 1-in-3SAT. Let $\varphi = \varphi_1 \wedge \cdots \wedge \varphi_m$ be a formula of FCP Planar 1-in-3SAT with n variables and m clauses. We assume that φ has no clause of length 1. For each clause φ_i $(1 \leq i \leq m)$ of length 2, let $\varphi_i = (\ell_{i,1} \vee \ell_{i,2})$ where $\ell_{i,1}$ and $\ell_{i,2}$ are literals. By introducing a new variable d_i, we replace φ_i with $\varphi_{i,1} \wedge \varphi_{i,2} = (\ell_{i,1} \vee \ell_{i,2} \vee d_i)(\bar{\ell}_{i,1} \vee \bar{\ell}_{i,2} \vee d_i)$. Note that we omit the conjunction between the clauses for compactness. We obtain the new formula ψ from φ by this operation. Let $D = \{d_{i_1}, d_{i_2}, \ldots, d_{i_{m'}}\}$ $(m' \leq m)$. The number of variables in the formula ψ is $|X| + |D| = n + m' \leq n + m$, and the number of clauses is $m + |D| = m + m' \leq 2m$.

Proposition 1 **($\star$).** *For the formula ψ constructed from the formula φ satisfies:*

(1) The SAT graph of ψ is planar,
(2) For any clause $\varphi_{i,1} \wedge \varphi_{i,2} = (\ell_{i,1} \vee \ell_{i,2} \vee d_i)(\bar{\ell}_{i,1} \vee \bar{\ell}_{i,2} \vee d_i)$, we have $\ell_{i,1} \neq \ell_{i,2}$ and $d_i = 0$ under any 1-in-3 satisfying assignment α of ψ.

Theorem 5. *FCP Planar 1-in-E3SAT is Σ_2^p-complete.*

Proof. It is a well-known fact that Planar 1-in-3SAT is NP-complete [9]. Since Planar 1-in-E3SAT is the restricted version of Planar 1-in-3SAT, Planar 1-in-E3SAT is in NP. And from Theorem 2, it follows that FCP Planar 1-in-E3SAT is restricted in Σ_2^p.

Let φ be a formula with n variables and m clauses, and let ψ be the formula constructed from φ. Since the number of variables and the number of clauses in ψ are at most $n + m$ and at most $2m$, this construction can be done in time polynomial in m.

We prove that (i) and (ii) are equivalent.

(i) There exists a feasible pre-assignment of size at most k for φ.
(ii) There exists a feasible pre-assignment of size at most k for ψ.

(i) $\Rightarrow$ (ii): Let $\alpha_{X'}$ be the feasible pre-assignment for the variable set $X' \subseteq X$ with $|X'| \leq k$ such that $\varphi|_{\alpha_{X'}}$ has a unique 1-in-3 satisfying assignment. From Proposition 1-(2), any 1-in-3 satisfying assignment α_X satisfies the formula ψ' obtained from ψ by assigning 0 to all variables in D. Thus, the unique 1-in-3 satisfying assignment of $\varphi|_{\alpha_{X'}}$ is also a unique 1-in-3 satisfying assignment of $\psi'|_{\alpha_{X'}}$. Therefore, $\alpha_{X'}$ is also a feasible pre-assignment for the 1-in-3 satisfying assignment of ψ.

(ii) $\Rightarrow$ (i): Let $\alpha_{(X' \cup D')}$ be the feasible pre-assignment for the variable set $(X' \cup D') \subseteq (X \cup D)$ with $|X'| + |D'| \leq k$ such that $\psi|_{\alpha_{(X' \cup D')}}$ has a unique 1-in-3 satisfying assignment. For any satisfying assignment of ψ, Proposition 1-(2) shows that all variables in D must be assigned 0. It implies that any pre-assignment to the variables in D is redundant. Therefore, it is sufficient to consider the case $D' = \emptyset$, i.e., $\alpha_{(X' \cup D')} = \alpha_{X'}$. Let ψ' be the formula obtained from ψ by assigning 0 to all variables in D. The unique 1-in-3 satisfying assignment of $\psi'|_{\alpha_{X'}}$ is also a unique 1-in-3 satisfying assignment of $\varphi|_{\alpha_{X'}}$. Therefore, $\alpha_{X'}$ is also a feasible pre-assignment for the 1-in-3 satisfying assignment of φ.

Hence, FCP Planar 1-in-E3SAT is Σ_2^p-hard since FCP Planar 1-in-3SAT is Σ_2^p-hard. □

3.2 FCP Planar Mono 1-in-E3SAT

We provide a polynomial-time reduction from FCP Planar 1-in-E3SAT to FCP Planar Mono 1-in-E3SAT for proving Σ_2^p-hardness. This reduction is based on the polynomial-time parsimonious reduction from Planar 1-in-E3SAT to Planar Mono 1-in-E3SAT in [16].

We first construct an instance of FCP Planar Mono 1-in-E3SAT from an instance of FCP Planar 1-in-E3SAT. Let $X = \{x_1, \cdots, x_n\}$ and $\varphi = \varphi_1 \wedge \cdots \wedge \varphi_m$ be a formula of Planar 1-in-E3SAT over X. Let p ($p \leq 3m$) denote the total number of negative literals that appear in the formula φ.

Let $\varphi_i = (\ell_{i,1} \vee \ell_{i,2} \vee \ell_{i,3})$ and we assume that $\ell_{i,1}$ is a negative literal. We replace the clause φ_i with $\varphi_i' = (y_{i,1} \vee \ell_{i,2} \vee \ell_{i,3})(\bar{\ell}_{i,1} \vee y_{i,1} \vee a_{i,1})(a_{i,1} \vee b_{i,1} \vee c_{i,1})(a_{i,1} \vee d_{i,1} \vee c_{i,1})(b_{i,1} \vee d_{i,1} \vee c_{i,1})$. In the clause φ_i', all literals except for $\ell_{i,2}$ and $\ell_{i,3}$ are positive. By repeating this replacement until p negative literals are removed in φ, we obtain the formula ψ whose all clauses have exactly three positive literals. For each negative literal, we add five new variables and four new clauses. Thus, the total number of variables in ψ is $n + 5p \leq n + 15m$, and the total number of clauses in ψ is $m + 4p \leq 13m$.

Proposition 2 ($\star$). *For the formula ψ constructed from the formula φ satisfies:*

(1) The SAT graph of ψ is planar.
(2) For any clause $\varphi_i' = (y_{i,1} \vee \ell_{i,2} \vee \ell_{i,3})(\bar{\ell}_{i,1} \vee y_{i,1} \vee a_{i,1})(a_{i,1} \vee b_{i,1} \vee c_{i,1})(a_{i,1} \vee d_{i,1} \vee c_{i,1})(b_{i,1} \vee d_{i,1} \vee c_{i,1})$, we have $a_{i,1} = b_{i,1} = d_{i,1} = 0, c_{i,1} = 1$ and $\ell_{i,1} = y_{i,1}$ under any 1-in-3 satisfying assignment α of ψ.

Theorem 6. *FCP Planar Mono 1-in-E3SAT is Σ_2^p-complete.*

Proof. FCP Planar Mono 1-in-E3SAT is in Σ_2^p since Planar Mono 1-in-E3SAT is NP-complete [16] and Theorem 2.

Let φ be a formula with n variables and m clauses, and let ψ be the formula constructed from φ. Since the number of variables and the number of clauses in ψ are at most $n + 15m$ and at most $13m$, this construction can be done in time polynomial in m.

We prove that (i) and (ii) are equivalent.

(i) There exists a feasible pre-assignment of size at most k for φ.
(ii) There exists a feasible pre-assignment of size at most k for ψ

(i) $\Rightarrow$ (ii): Let $\alpha_{X'}$ be the feasible pre-assignment for the variable set $X' \subseteq X$ with $|X'| \leq k$ such that $\varphi|_{\alpha_{X'}}$ has a unique 1-in-3 satisfying assignment. From Proposition 2-(2), any 1-in-3 satisfying assignment α_X satisfies the formula ψ' obtained from ψ by assigning 0 to all $a_{i,j}, b_{i,j}, d_{i,j}$ and 1 to all $c_{i,j}$. In addition, the value of $y_{i,j}$ is forced by that of $\ell_{i,j}$. Thus, the unique 1-in-3 satisfying assignment of $\varphi|_{\alpha_{X'}}$ is also a unique 1-in-3 satisfying assignment of $\psi'|_{\alpha_{X'}}$. Therefore, $\alpha_{X'}$ is also a feasible pre-assignment for the 1-in-3 satisfying assignment of ψ.

(ii) $\Rightarrow$ (i): Let A, B, C, D and Y be a set of variables $a_{i,j}, b_{i,j}, c_{i,j}, d_{i,j}$ and $y_{i,j}$, respectively. We set $Z = A \cup B \cup C \cup D$. Let $\alpha_{(X' \cup Y' \cup Z')}$ be the feasible pre-assignment for the variable set $(X' \cup Y' \cup Z') \subseteq (X \cup Y \cup Z)$ with $|X'| + |Y'| + |Z'| \leq k$ such that $\psi|_{\alpha_{(X' \cup Y' \cup Z')}}$ has a unique 1-in-3 satisfying assignment. For any satisfying assignment of ψ, Proposition 2-(2) shows that all variables in A, B, D must be assigned 0, and all variables in C must be assigned 1. It implies that any pre-assignment to the variables in A, B, C, and D is redundant. Therefore, it is sufficient to consider the case $A' = B' = C' = D' = \emptyset$, i.e., $\alpha_{(X' \cup Y' \cup Z')} = \alpha_{X' \cup Y'}$. Moreover, from $\ell_{i,j} = y_{i,j}$ in Proposition 2-(2), the pre-assignment to $y_{i,j}$ is replaced with the pre-assignment to $\ell_{i,j}$. This replacement preserves the size of the pre-assignment. Thus, we suffice to consider the case $Y' = \emptyset$, i.e., $\alpha_{X' \cup Y'} = \alpha_{X'}$. Let ψ' be the formula obtained from ψ by assigning 0 to all variables in A, B, D and 1 to all variables in C. The unique 1-in-3 satisfying assignment of $\psi'|_{\alpha_{X'}}$ is also a unique 1-in-3 satisfying assignment of $\varphi|_{\alpha_{X'}}$. Therefore, $\alpha_{X'}$ is also a feasible pre-assignment for the 1-in-3 satisfying assignment of φ.

Hence, FCP Planar Mono 1-in-E3SAT is Σ_2^p-hard since FCP Planar 1-in-E3SAT is Σ_2^p-hard. $\square$

4 Hardness of PAU-VC on Planar Graphs with Maximum Degree 3

In this section, we show that PAU-VC on planar graphs with maximum degree 3 is Σ_2^p-complete. Barbanchon and Grandjean [4] provided a polynomial-time parsimonious reduction from Planar Mono 1-in-E3SAT to the minimum vertex cover problem on planar graphs with maximum degree 4. We lower the maximum degree constraint.

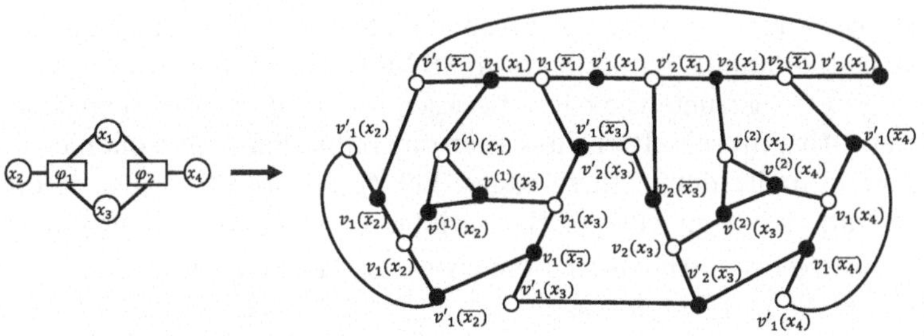

Fig. 2. An example of the graph G_φ. The formula $\varphi = (x_1 \vee x_2 \vee x_3)(x_1 \vee x_3 \vee x_4)$ and four variables are an instance of Planar Mono 1-in-E3SAT. The sixteen black vertices represent the optimal solution C_α and correspond to the satisfying assignment $(x_1, x_2, x_3, x_4) = (1, 0, 0, 0)$ in Planar Mono 1-in-3SAT.

We first construct an instance of PAU-VC on planar graphs with maximum degree 3 from an instance of Planar Mono 1-in-E3SAT. Let $X = \{x_1, \dots, x_n\}$ and $\varphi = \varphi_1 \wedge \cdots \wedge \varphi_m$ be a formula of Planar Mono 1-in-E3SAT over X with m clauses. We assume that φ has no clause of length 1, and each clause has three distinct literals. Note that all clauses have only positive literals. We construct the graph $G_\varphi = (V(G_\varphi), E(G_\varphi))$ from φ. The vertex set $V(G_\varphi)$ consists of two vertex sets V_φ^X and V_φ^Φ, where V_φ^X and V_φ^Φ correspond to the set of variables X and the set of clauses Φ in φ, respectively. Let p_{x_i} is the number of occurrences of x_i in φ. In the order from φ_1 to φ_m, for the j-th occurrence of x_i, the vertex set V_φ^X has four vertices of $v_j(x_i), v_j(\bar{x}_i), v'_j(x_i)$, and $v'_j(\bar{x}_i)$. The vertex set V_φ^Φ has three vertices $v^{(i)}(\ell_{i,1}), v^{(i)}(\ell_{i,2})$, and $v^{(i)}(\ell_{i,3})$ for each clause $\varphi_i = (\ell_{i,1} \vee \ell_{i,2} \vee \ell_{i,3})$. The vertex sets V_φ^X and V_φ^Φ are formally defined as follows:

$$V_\varphi^X := \bigcup_{x_i \in X} \{v_j(x_i), v_j(\bar{x}_i), v'_j(x_i), v'_j(\bar{x}_i) \mid j \in [p_{x_i}]\},$$
$$V_\varphi^\Phi := \bigcup_{\varphi_i \in \varphi} \{v^{(i)}(\ell_{i,1}), v^{(i)}(\ell_{i,2}), v^{(i)}(\ell_{i,3})\}.$$

The edge set $E(G_\varphi)$ consists of four edge sets E_φ^X, E_φ^Φ, $E_\varphi^{X,X'}$, and $E_\varphi^{X,\Phi}$. The edge set E_φ^X has the edges between the vertices corresponding to the variable x_i. The edge set E_φ^Φ has the edges between the vertices corresponding to the clause φ_i. The edge set $E_\varphi^{X,X'}$ has the edges between the vertices corresponding to the different variables x_i and x_j. The edge set $E_\varphi^{X,\Phi}$ has the edges between the vertices corresponding to the variable x_i and the clause φ_j.

The edge set E_φ^X, for each vertex set $\{v_j(x_i), v_j(\bar{x}_i), v'_j(x_i), v'_j(\bar{x}_i) \mid j \in [p_{x_i}]\}$, constructs the cycle as follows:

$$(v'_1(\bar{x}_i), v_1(x_i), v_1(\bar{x}_i), v'_1(x_i), v'_2(\bar{x}_i), \dots v'_2(x_i), \dots, v'_{p_{x_i}}(\bar{x}_i) \dots, v'_{p_{x_i}}(x_i), v'_1(\bar{x}_i))$$

The edge set E_φ^Φ, for each edge set $\{v^{(i)}(\ell_{i,1}), v^{(i)}(\ell_{i,2})$ and $v^{(i)}(\ell_{i,3})\}$, constructs the triangle $(v^{(i)}(\ell_{i,1}), v^{(i)}(\ell_{i,2}), v^{(i)}(\ell_{i,3}), v^{(i)}(\ell_{i,1}))$. The edge set $E_\varphi^{X,\Phi}$

constructs the edges between the variable gadget and the clause gadget in the same way as the famous reduction from 3SAT to the vertex cover problem. The edge set $E_\varphi^{X,X'}$ constructs the edges between a pair of vertices corresponding to different negative literals. It ensures that our reduction is parsimonious.

We now formally define these edges. Let $\ell_{i,k}$ denote the k-th literal in the clause φ_i, and the occurrence of the variable corresponding to $\ell_{i,k}$ in φ_i is its $g(\ell_{i,k})$-th occurrence in the order from φ_1 to φ_m. The edge sets E_φ^X, E_φ^Φ, $E_\varphi^{X,X'}$, and $E_\varphi^{X,\Phi}$ are defined as follows:

$$\begin{aligned}
E_\varphi^X :=& \bigcup_{x_i \in X} \{\{v'_j(\bar{x}_i), v_j(x_i)\}, \{v_j(x_i), v_j(\bar{x}_i)\},\\
&\{v_j(\bar{x}_i), v'_j(x_i)\}, \{v'_j(x_i), v'_{j+1}(\bar{x}_i)\} \mid j \in [p_{x_i} - 1]\}\\
&\cup \bigcup_{x_i \in X} \{\{v'_{p_{x_i}}(\bar{x}_i), v_{p_{x_i}}(x_i)\}, \{v_{p_{x_i}}(x_i), v_{p_{x_i}}(\bar{x}_i)\},\\
&\{v_{p_{x_i}}(\bar{x}_i), v'_{p_{x_i}}(x_i)\}, \{v'_{p_{x_i}}(x_i), v'_1(\bar{x}_i)\}\},\\
E_\varphi^\Phi :=& \bigcup_{\varphi_i \in \varphi} \{\{v^{(i)}(\ell_{i,1}), v^{(i)}(\ell_{i,2})\}, \{v^{(i)}(\ell_{i,2}), v^{(i)}(\ell_{i,3})\}, \{v^{(i)}(\ell_{i,3}), v^{(i)}(\ell_{i,1})\}\},\\
E_\varphi^{X,\Phi} :=& \bigcup_{\varphi_i \in \varphi} \{\{v_{g(\ell_{i,k})}(\ell_{i,k}), v^{(i)}(\ell_{i,k})\} \mid k = 1,2,3\},\\
E_\varphi^{X,X'} :=& \bigcup_{\varphi_i \in \varphi} \{\{v'_{g(\ell_{i,1})}(\bar{\ell}_{i,1}), v_{g(\ell_{i,2})}(\bar{\ell}_{i,2})\}, \{v'_{g(\ell_{i,2})}(\bar{\ell}_{i,2}), v_{g(\ell_{i,3})}(\bar{\ell}_{i,3})\},\\
&\{v'_{g(\ell_{i,3})}(\bar{\ell}_{i,3}), v_{g(\ell_{i,1})}(\bar{\ell}_{i,1})\}\}.
\end{aligned}$$

Figure 2 shows the reduction from the formula $\varphi = (x_1 \vee x_2 \vee x_3)(x_1 \vee x_3 \vee x_4)$ to the graph G_φ. The size of $V(G_\varphi)$ is $|V_\varphi^X| + |V_\varphi^\Phi| \le \sum_{x_i \in X} 4 \cdot |p_{x_i}| + 3m = 4 \cdot 3m + 3m = 15m$. The size of $E(G_\varphi)$ is $|E_\varphi^X| + |E_\varphi^\Phi| + |E_\varphi^{X,X'}| + |E_\varphi^{X,\Phi}| \le \sum_{x_i \in X} 4 \cdot |p_{x_i}| + 3m + 3m + 3m = 21m$. The total number of variable occurrences is equal to the total number of literals in φ, and each clause has three literals ($\sum_{x_i \in X} |p_{x_i}| = 3m$); then the last equality holds.

Proposition 3. *The graph G_φ is planar with maximum degree 3.*

Proof. Since the formula φ is an instance of Planar Mono 1-in-E3SAT, the SAT graph of φ is planar. Each variable gadget forms a cycle, and the clause gadget forms a triangle. These components are also planar. It is easy to see that the edges in $E_\varphi^{X,X'}$ and $E_\varphi^{X,\Phi}$ preserve the planarity of G_φ. Thus, the graph G_ϕ is planar. The vertices in V_φ^X form a cycle, then each vertex v in V_φ^X has at least two edges. Furthermore, v connects to at most one vertex in V_φ^Φ or V_φ^X from our construction $E_\varphi^{X,\Phi}$ and $E_\varphi^{X,X'}$. Thus, the degree of v is at most 3. From a similar argument, the degree of each vertex in V_φ^Φ is also at most 3. □

Proposition 4. *The size of any minimum vertex cover of G_φ is at least 8m.*

Proof. Let C be a minimum vertex cover of G_φ. For each clause $\varphi_i \in \varphi$, C must contain at least two of $v^{(i)}(\ell_{i,1}), v^{(i)}(\ell_{i,2})$, and $v^{(i)}(\ell_{i,3})$ because these three vertices form a triangle. Therefore, C needs at least $2m$ vertices. For each variable $x_i \in X$, $\{v_j(x_i), v_j(\bar{x}_i), v'_j(x_i), v'_j(\bar{x}_i) \mid j \in [p_{x_i}]\}$ forms an cycle of even length. Since the length of the cycle is $4p_{x_i}$, C must contain at least $2p_{x_i}$ vertices from the cycle. Therefore, C must have at least $2m + \sum_{x_i \in X} 2p_{x_i} = 2m + 6m = 8m$. □

Proposition 5. *A vertex cover C of size $8m$ for the graph G_φ satisfies:*

(1) $|C \cap \{v^{(i)}(\ell_{i,1}), v^{(i)}(\ell_{i,2}), v^{(i)}(\ell_{i,3})\}| = 2$ *for every* $\varphi_i \in \varphi$,
(2) For any variable x_i, $C \cap \{v_j(x_i), v_j(\bar{x}_i), v'_j(x_i), v'_j(\bar{x}_i) \mid j \in [p_{x_i}]\}$ *is either* $\{v_j(x_i), v'_j(x_i) \mid j \in [p_{x_i}]\}$ *or* $\{v_j(\bar{x}_i), v'_j(\bar{x}_i) \mid j \in [p_{x_i}]\}$,
(3) $N(V_\varphi^X \cap C) \cap (V_\varphi^\Phi \cap C) = \emptyset$.

Proof. Let C be a vertex cover of size 8 m for the graph G_φ.

(1) We can easily see that it holds from the proof of Proposition 4.
(2) Suppose (2) is false, i.e., $C \cap \{v_j(x_i), v_j(\bar{x}_i), v'_j(x_i), v'_j(\bar{x}_i) \mid j \in [p_{x_i}]\}$ is neither $\{v_j(x_i), v'_j(x_i) \mid j \in [p_{x_i}]\}$ nor $\{v_j(\bar{x}_i), v'_j(\bar{x}_i) \mid j \in [p_{x_i}]\}$ for some variable $x_i \in X$. Note that C contains at least one of these two sets since $\{v_j(x_i), v_j(\bar{x}_i), v'_j(x_i), v'_j(\bar{x}_i) \mid j \in [p_{x_i}]\}$ forms an even cycle. Without loss of generality, assume $\{v_j(x_i), v'_j(x_i) \mid j \in [p_{x_i}]\} \subset C$. In this case, there exists a vertex $v \in \{v_j(\bar{x}_i), v'_j(\bar{x}_i) \mid j \in [p_{x_i}]\}$ such that $v \in C$. This implies $|C \cap V_\varphi^X| \geq 2\sum_{x \in X} p_x + 1 = 6m+1$, and hence we have $|C| = |C \cap V_\varphi^\Phi| + |C \cap V_\varphi^X| \geq 8m+1$, which contradicts $|C| = 8m$. Thus, condition (2) holds.
(3) Assume $N(V_\varphi^X \cap C) \cap (V_\varphi^\Phi \cap C) \neq \emptyset$ for a contradiction. Then, for some i and k, there exists an edge e both endpoints of $\{v^{(i)}(\ell_{i,k}), v_{g(\ell_{i,k})}(\ell_{i,k})\} \in E^{X,\Phi}$ are covered by C. For the simplicity of the proof, we prove the case that $\varphi_1 = (x_1 \vee x_2 \vee x_3)$ and the edge $e = \{v^{(1)}(x_1), v_1(x_1)\}$. In this case, $\ell_{1,1} = x_1, \ell_{1,2} = x_2, \ell_{1,3} = x_3$, and $g(\ell_{1,1}) = g(\ell_{1,2}) = g(\ell_{1,3}) = 1$. Since the vertices $v^{(1)}(x_1)$ and $v_1(x_1)$ are covered by C, from condition (2), the vertices $v_1(\bar{x}_1)$ and $v'_1(\bar{x}_1)$ are not in C. To cover the edge $\{v'_1(\bar{x}_1), v_1(\bar{x}_2)\} \in E^{X,X'}$, C must include the vertex $v_1(\bar{x}_2)$. Then, from condition (2), $v_1(x_2) \notin C$. Thus, C must include the vertex $v^{(1)}(x_2)$ to cover the edge $\{v_1(x_2), v^{(1)}(x_2)\} \in E^{X,\Phi}$. By a similar argument, we can show that $v^{(1)}(x_3) \in C$. Therefore, the vertices $v^{(1)}(x_1), v^{(1)}(x_2)$, and $v^{(1)}(x_3)$ are in C. It contradicts condition (1). The above arguments hold for an arbitrary case; hence, condition (3) holds. □

Lemma 2. *There exists a polynomial-time parsimonious reduction from Planar Mono 1-in-E3SAT to the minimum vertex cover problem on planar graphs with maximum degree 3.*

Proof. Let $\varphi = \varphi_1 \wedge \cdots \wedge \varphi_m$ be a formula of Planar Mono 1-in-E3SAT over $X = \{x_1, \ldots, x_n\}$, and G_φ be the graph constructed from φ. The number of vertices and the number of edges in G_φ are at most $15m$ and at most $21m$, respectively. Thus, this construction can be done in time polynomial in m.

Let S_φ be the set of 1-in-3 satisfying assignments of the instance φ and $C(G_\varphi)$ be the set of the vertex covers of size $8m$ in G_φ. We construct a bijection ϕ such that $\alpha \in S_\varphi$ if and only if $\phi(\alpha) \in C(G_\varphi)$.

We first show that the mapping ϕ is injective. For a 1-in-3 satisfying assignment α, let $C_\alpha = \phi(\alpha)$. For each variable x_i and $j \in [p_{x_i}]$, if x_i is assigned 1 by α, we include the vertices $v_j(x_i)$ and $v'_j(x_i)$ in C_α, otherwise include the vertices $v_j(\bar{x}_i)$ and $v'_j(\bar{x}_i)$ in C_α. These vertices can cover all edges in E^X_φ. The total number of vertices included in C_α is $2\sum_{x_i \in X} p_{x_i} = 6m$. Since α is 1-in-3 satisfying assignment, for any i, exactly one literal of $\ell_{i,1}, \ell_{i,2}$ and $\ell_{i,3}$ is true in each clause φ_i. Hence, for any i, the edges $\{v'_{g(\ell_{i,1})}(\bar{\ell}_{i,1}), v_{g(\ell_{i,2})}(\bar{\ell}_{i,2})\}, \{v'_{g(\ell_{i,2})}(\bar{\ell}_{i,2}), v_{g(\ell_{i,3})}(\bar{\ell}_{i,3})\}$, and $\{v'_{g(\ell_{i,3})}(\bar{\ell}_{i,3}), v_{g(\ell_{i,1})}(\bar{\ell}_{i,1})\} \in E^{X,X'}$ are covered. Furthermore, C_α also covers exactly one edge among $\{v^{(i)}(\ell_{i,1}), v_{g(\ell_{i,1})}(\ell_{i,1})\}, \{v^{(i)}(\ell_{i,2}), v_{g(\ell_{i,2})}(\ell_{i,2})\}$, and $\{v^{(i)}(\ell_{i,3}), v_{g(\ell_{i,3})}(\ell_{i,3})\}$. Without loss of generality, we assume that the edge $\{v^{i}(\ell_{i,1}), v_{g(\ell_{i,1})}(\ell_{i,1})\}$ is covered. To cover the other two edges, C_α must include $v^{(i)}(\ell_{i,2})$ and $v^{(i)}(\ell_{i,3})$. Thus, C_α covers all the edges in E^Φ_φ and $E^{X,\Phi}_\varphi$, and $2m$ vertices in V^Φ_φ are added in C_α. Note that for any pair of different satisfying assignments α_1 and α_2 of the formula φ, $C_{\alpha_1} \neq C_{\alpha_2}$ holds. Therefore, C is a vertex cover of $8m$ and ϕ is injective.

We next show that the mapping ϕ is surjective. Let C be a vertex cover of size $8m$ for G_φ. We define an assignment α_C such that $\alpha_C(x_i) = 1$ (resp. $\alpha_C(x_i) = 0$) if and only if $v_1(x_i) \in C$ (resp. $v_1(\bar{x}_i) \in C$) for every variable x_i. Since this assignment is derived from C, which is consistent with Proposition 5-(2), there can be no multiple assignments for a single variable. From Proposition 5-(1) and Proposition 5-(3), for each clause $\varphi_i \in \varphi$, exactly one literal in φ_i is true under α_C. Therefore, the mapping ϕ is surjective. □

We now show the Σ^p_2-completeness of PAU-VC on planar graphs with maximum degree 3 under the Include and Exclude models. Theorem 1 immediately follows from Lemma 1, 3 and 4.

Lemma 3. *PAU-VC on planar graphs with maximum degree 3 is Σ^p_2-complete under the Include model.*

Proof. By Lemma 2, we construct a planar graph G_φ with maximum degree 3 from the instance φ of FCP Planar Mono 1-in-E3SAT. We show that (i) and (ii) are equivalent.

(i) There exists a feasible pre-assignment of size at most k for φ.
(ii) There exists a feasible pre-assignment of size at most k for the minimum vertex covers of G_φ.

(i) ⇒ (ii): Let $\alpha_{X'}$ be a feasible pre-assignment for φ on a set $X' \subseteq X$ with $|X'| \leq k$. A pre-assignment $C_{\text{pre}} = (C_{\text{in}}, \emptyset)$ is constructed as follows. If $x_i \in X'$ is assigned true, then $v_1(x_i) \in C_{\text{in}}$, otherwise $v_1(\bar{x}_i) \in C_{\text{in}}$. Since ϕ is a bijection, C_{pre} induces a unique vertex cover C^* of size $8m$ in G_φ, and C^* is a minimum vertex cover due to Proposition 4. Thus, (i) ⇒ (ii) holds.

(ii) $\Rightarrow$ (i): Let $C_{\text{pre}} = (C_{\text{in}}, \emptyset)$ be a feasible pre-assignment for the minimum vertex covers of G_φ with $|C_{\text{in}}| \leq k$. Let C^* be the unique minimum vertex cover such that $C_{\text{in}} \subseteq C^*$. Note that C_{pre} satisfies Proposition 5. We consider the following three cases. Note that the cases (a) and (b) occur simultaneously.

(a) There exists a vertex $v^{(i)}(\ell_{i,k}) \in C_{\text{in}}$ for some $i \in [m]$ and $k \in \{1, 2, 3\}$.
(b) There exists a vertex $v'_j(x_i) \in C_{\text{in}}$ or $v'_j(\bar{x}_i) \in C_{\text{in}}$ for some $j \in [p_{x_i}]$.
(c) $C_{\text{in}} \subseteq \bigcup_{x_i \in X} \{v_j(x_i), v_j(\bar{x}_i) \mid j \in [p_{x_i}]\}$

We first show that the vertex in cases (a) and (b) can be replaced with a vertex in $\bigcup_{x_i \in X} \{v_j(x_i), v_j(\bar{x}_i) \mid j \in [p_{x_i}]\}$.

(a) From Proposition 5-(3), $v^{(i)}(\ell_{i,k}) \in C^* \Rightarrow v_{g(\ell_{i,k})}(\ell_{i,k}) \notin C^*$, and from Proposition 5-(2), $v_{g(\ell_{i,k})}(\ell_{i,k}) \notin C^* \Rightarrow v_{g(\ell_{i,k})}(\bar{\ell}_{i,k}) \in C^*$. Hence, $v^{(i)}(\ell_{i,k}) \in C^* \Rightarrow v_{g(\ell_{i,k})}(\bar{\ell}_{i,k}) \in C^*$. From Proposition 5-(2), $v_{g(\ell_{i,k})}(\bar{\ell}_{i,k}) \in C^* \Rightarrow v_{g(\ell_{i,k})}(\ell_{i,k}) \notin C^*$. To cover the edge $\{v^{(i)}(\ell_{i,k}), v_{g(\ell_{i,k})}(\ell_{i,k})\}$, $v_{g(\ell_{i,k})}(\ell_{i,k}) \notin C^* \Rightarrow v^{(i)}(\ell_{i,k}) \in C^*$. Hence, $v_{g(\ell_{i,k})}(\bar{\ell}_{i,k}) \in C^* \Rightarrow v^{(i)}(\ell_{i,k}) \in C^*$. Therefore, $v^{(i)}(\ell_{i,k}) \in C^* \Leftrightarrow v_{g(\ell_{i,k})}(\bar{\ell}_{i,k}) \in C^*$. Thus, we can replace C_{in} with $(C_{\text{in}} \setminus \{v^{(i)}(\ell_{i,k})\}) \cup \{v_{g(\ell_{i,k})}(\bar{\ell}_{i,k})\}$ without changing the size of C_{in}.
(b) If $v'_j(x_i) \in C_{\text{in}}$, then $v_j(x_i) \in C^*$ follows from Proposition 5-(2). We can replace C_{in} with $(C_{\text{in}} \setminus \{v'_j(x_i)\}) \cup \{v_j(x_i)\}$ without changing the size of C_{in}. If $v'_j(\bar{x}_i) \in C_{\text{in}}$, then $v_j(\bar{x}_i) \in C^*$ follows from Proposition 5-(2). We can replace C_{in} with $(C_{\text{in}} \setminus \{v'_j(\bar{x}_i)\}) \cup \{v_j(\bar{x}_i)\}$ without changing the size of C_{in}.

From the above argument, we only consider a pre-assignment C_{in} of the case (c). If $v_j(x_i) \in C_{\text{in}}$ for some $j \in [p_{x_i}]$, assign true to x_i, and if $v_j(\bar{x}_i) \in C_{\text{in}}$, assign false to x_i. Proposition 5-(2) ensures that, for every variable x_i, C^* contains either $v_j(x_i)$ or $v_j(\bar{x}_i)$ not both. By Lemma 2, there is a one-to-one correspondence between the satisfying assignment of φ and the vertex covers of size $8m$ in G_φ. From these facts, the pre-assignment of φ converted from C_{pre} induces a unique satisfying assignment of φ. □

The following lemma can be proven by similar arguments to Lemma 3.

Lemma 4. *PAU-VC on planar graphs with maximum degree 3 is Σ_2^p-complete under the Exclude model.*

Proof. By Lemma 2, we construct a planar graph G_φ with maximum degree 3 from the instance φ of FCP Planar Mono 1-in-E3SAT. We show that (i) and (ii) are equivalent.

(i) There exists a feasible pre-assignment of size at most k for φ.
(ii) There exists a feasible pre-assignment of size at most k for the minimum vertex covers of G_φ.

(i) $\Rightarrow$ (ii): Let $\alpha_{X'}$ be a feasible pre-assignment for φ on a set $X' \subseteq X$ with $|X'| \leq k$. We construct a pre-assignment $C_{\text{pre}} = (\emptyset, C_{\text{ex}})$ as follows. If $x_i \in X'$ is assigned true, then $v_1(\bar{x}_i) \in C_{\text{ex}}$. If $x_i \in X'$ is assigned false, then $v_1(x_i) \in C_{\text{ex}}$.

Since ϕ is a bijection, the pre-assignment C_{pre} induces a unique vertex cover of size $8m$ in the graph G_φ.

(ii) $\Rightarrow$ (i): Let $C_{\text{pre}} = (\emptyset, C_{\text{ex}})$ be a feasible pre-assignment of size at most k for G_φ. Let C^* be the unique minimum vertex cover such that $C^* \subseteq V(G_\varphi) \setminus C_{\text{ex}}$.

We consider the following three cases. Note that the cases (a) and (b) occur simultaneously.

(a) There exists a vertex $v^{(i)}(\ell_{i,k}) \in C_{\text{ex}}$ for some $i \in [m]$ and $k \in \{1, 2, 3\}$.
(b) There exists a vertex $v'_j(x_i) \in C_{\text{ex}}$ or $v'_j(\bar{x}_i) \in C_{\text{ex}}$ for some $j \in [p_{x_i}]$.
(c) $C_{\text{ex}} \subseteq \bigcup_{x_i \in X} \{v_j(x_i), v_j(\bar{x}_i) \mid j \in [p_{x_i}]\}$

We first show that the vertex in cases (a) and (b) can be replaced with a vertex in $\bigcup_{x_i \in X} \{v_j(x_i), v_j(\bar{x}_i) \mid j \in [p_{x_i}]\}$.

(a) From the case in Lemma 3, $v^{(i)}(\ell_{i,k}) \in C^* \Leftrightarrow v_{g(\ell_{i,k})}(\bar{\ell}_{i,k}) \in C^*$. Then, $v^{(i)}(\ell_{i,k}) \notin C^* \Leftrightarrow v_{g(\ell_{i,k})}(\bar{\ell}_{i,k}) \notin C^*$. Thus, C_{ex} can be replaced with $(C_{\text{ex}} \setminus \{v^{(i)}(\ell_{i,k})\}) \cup \{v_{g(\ell_{i,k})}(\bar{\ell}_{i,k})\}$ without changing the size of C_{ex}.
(b) Assume $v'_j(x_i) \in C_{\text{ex}}$ for some j. Since $v'_j(x_i) \notin C^* \Leftrightarrow v_j(x_i) \notin C^*$ due to Proposition 5-(2), we can replace C_{ex} with $(C_{\text{ex}} \setminus \{v'_j(x_i)\}) \cup \{v_j(x_i)\}$ without changing the size of C_{ex}. A similar argument holds if $v'_j(\bar{x}_i) \in C_{\text{ex}}$ for some j. Thus, we can replace C_{ex} with $(C_{\text{ex}} \setminus \{v'_j(\bar{x}_i)\}) \cup \{v_j(\bar{x}_i)\}$ without changing the size of C_{ex}.

From the above argument, we only consider a pre-assignment C_{ex} of the case (c). If $C_{\text{ex}} \subseteq \bigcup_{x_i \in X} \{v_j(x_i), v_j(\bar{x}_i) \mid j \in [p_{x_i}]\}$, then by the same arguments as the Include model, there exists a feasible pre-assignment for φ of size k. $\square$

Acknowledgments. This work was supported by JSPS KAKENHI Grant Numbers 20H05964 (TH), 22H00513 (KS), 23K24806 (TH), and 24K02898 (KS).

References

1. Afshani, P., Hatami, H., Mahmoodian, E.S.: On the spectrum of the forced matching number of graphs. Australas. J Comb. **30**, 147–160 (2004)
2. An, S., et al.: Pre-assignment problem for unique minimum vertex cover on bounded clique-width graphs. In: Proceedings of the Thirty-Ninth AAAI Conference on Artificial Intelligence, AAAI 2025, pp. 26886–26894 (2025)
3. Armada, C.L., Canoy, S.R., Jr.: Forcing independent domination number of a graph. Eur. J. Pure Appl. Math. **12**(4), 1371–1381 (2019)
4. Barbanchon, R., Grandjean, E.: Local problems, planar local problems and linear time. In: Bradfield, J. (ed.) CSL 2002. LNCS, vol. 2471, pp. 397–411. Springer, Heidelberg (2002). https://doi.org/10.1007/3-540-45793-3_27
5. Chartrand, G., Gavlas, H., Vandell, R.C., Harary, F.: The forcing domination number of a graph. J. Comb. Math. Comb. Comput. **25**, 161–174 (1997)
6. Cooper, J., Donovan, D., Seberry, J.: Secret sharing schemes arising from Latin squares. Bull. Inst. Combin. Appl. **12**, 33–43 (1994)

7. Cooper, J., Kirkpatrick, A.: Critical sets for sudoku and general graph colorings. Discret. Math. **315–316**, 112–119 (2014)
8. Demaine, E.D., Ma, F., Schvartzman, A., Waingarten, E., Aaronson, S.: The fewest clues problem. Theor. Comput. Sci. **748**, 28–39 (2018)
9. Dyer, M.E., Frieze, A.M.: Planar 3dm is NP-complete. J. Algorithms **7**(2), 174–184 (1986)
10. Goergen, K., Fernau, H., Oest, E., Wolf, P.: All paths lead to Rome. CoRR, abs/2207.09439 (2022)
11. Gray, K.: On the minimum number of blocks defining a design. Bull. Aust. Math. Soc. **41**(1), 97–112 (1990)
12. Harary, F., Klein, D.J., Živkovič, T.P.: Graphical properties of polyhexes: perfect matching vector and forcing. J. Math. Chem. **6**(1), 295–306 (1991)
13. Harary, F., Slany, W., Verbitsky, O.: On the computational complexity of the forcing chromatic number. SIAM J. Comput. **37**(1), 1–19 (2007)
14. Higuchi, Y., Kimura, K.: NP-completeness of fill-a-pix and σ_2^p-completeness of its fewest clues problem. IEICE Trans. Fundam. Electron. Commun. Comput. Sci. **102-A**(11), 1490–1496 (2019)
15. Horiyama, T., Kobayashi, Y., Ono, H., Seto, K., Suzuki, R.: Theoretical aspects of generating instances with unique solutions: pre-assignment models for unique vertex cover. In: Proceedings of the Thirty-Eighth AAAI Conference on Artificial Intelligence, AAAI 2024, pp. 20726–20734 (2024)
16. Hunt III, H.B., Marathe, M.V., Radhakrishnan, V., Stearns, R.E.: The complexity of planar counting problems. SIAM J. Comput. **27**(4), 1142–1167 (1998)
17. Kimura, K., Kamehashi, T., Fujito, T.: The fewest clues problem of picross 3D. In: Proceedings of the 9th International Conference on Fun with Algorithms, FUN 2018. LIPIcs, vol. 100, pp. 25:1–25:13 (2018)
18. Klein, D.J., Randić, M.: Innate degree of freedom of a graph. J. Comput. Chem. **8**(4), 516–521 (1987)
19. Larson, C., Van Cleemput, N.: Forcing independence. Croatia Chem. Acta **86**(4), 469–475 (2013)
20. Mahmoodian, E.S.: Defining sets and uniqueness in graph colorings: a survey. J. Stat. Plann. Inference **73**(1), 85–89 (1998)
21. Vukičević, D., Trinajstić, N.: On the anti-forcing number of benzenoids. J. Math. Chem. **42**(3), 575–583 (2007)

Combinatorics of Palindromes

Michael Itzhaki(✉)

Bar-Ilan University, Ramat Gan, Israel
michaelitzhaki@gmail.com

Abstract. We investigate the structure and reconstruction complexity of Manacher arrays. First, we establish a combinatorial lower bound, proving that the number of rooted tandem repeat trees with $n + 1$ genes exceeds the number of distinct Manacher arrays of length n. Second, we introduce a graph-theoretic framework that associates a graph to each Manacher array, where every proper vertex coloring yields a string consistent with the array. Finally, we analyze a reconstruction algorithm by I et al. (SPIRE 2010), showing that it simultaneously achieves a globally minimal alphabet size, uses at most $\log_2(n-1) + 2$ distinct symbols, and can be adapted to produce reconstructions over arbitrary alphabets when possible. Our results also resolve an open problem posed by the original authors. Together, these findings advance the combinatorial understanding of Manacher arrays and open new directions for string reconstruction under structural constraints.

Keywords: Palindrome · Combinatorics · Manacher Array

1 Introduction

Palindromes—strings that read identically forward and backward—arise naturally in fields such as mathematics, computer science, and bioinformatics. Their study sheds light on structural symmetry in strings and has implications for sequence alignment, data compression, and formal language theory [8,9,16–18,22,24]. In computational settings, palindromes are central to problems ranging from error detection in coding theory to modeling self-replicating biological patterns. A classical linear-time algorithm for computing all maximal palindromes in a string is due to Manacher [19], with further variants addressing the longest palindromic substring [5,14], palindromic length [3,11], efficient indexing [21], and sublinear-time detection [4].

Tandem Duplication Trees provide a combinatorial model of genome evolution through repeated duplications of adjacent segments [2]; see also [6,13]. We uncover a structural connection between Manacher arrays and rooted tandem duplication trees.

Relations between strings and graphs have been explored before [1,23]; we extend this programme by attaching a graph to every Manacher array and analyzing its vertex colorings. This perspective yields a reconstruction framework for Manacher arrays based solely on the vertex coloring of the graph.

A. Jeż and J. Otop (Eds.): FCT 2025, LNCS 16106, pp. 252–266, 2026.
https://doi.org/10.1007/978-3-032-04700-7_19

In this work, we establish a combinatorial link between Manacher arrays and duplication trees, introduce a graph-based reconstruction framework via vertex colorings, and prove that a classical algorithm achieves minimal alphabet size. Our results resolve an open problem and deepen the combinatorial understanding of palindromic structures and Manacher arrays.

Our Contributions.

- **Combinatorial Lower Bound.** We prove a surprising combinatorial bound relating Manacher arrays and rooted tandem repeat trees (Theorem 1): the number of rooted tandem repeat trees with $n+1$ genes is greater than or equal to the number of distinct Manacher arrays of length $2n-1$. The combinatorial problem of counting the number of unique Manacher arrays is motivated by the literature [10].
- **Graph-Theoretic Reconstruction.** We introduce a novel graph construction for Manacher arrays (Theorem 2), where:
  - Every Manacher array can be represented by a graph.
  - Every proper coloring of the graph yields a unique string corresponding to the given Manacher array.
- **Minimal Reconstruction Algorithm.** We analyze the algorithm of I et al. [10] that reconstructs a string from its Manacher array and prove the following (Theorem 3):
  - It outputs a string with the globally minimal alphabet size, achieving the given Manacher array.
  - The number of distinct symbols used is at most $\log(n-1)+2$, and there exists a tight example for every n.
  - The algorithm can be modified to generate a string with any desired alphabet size (when such a string exists).

2 Preliminaries

An ordered sequence of characters is a *string*. A string S of length $|S| = n$ is the sequence $S[1]S[2]\dots S[n]$ where $\forall 1 \le i \le n,\ S[i] \in \Sigma$. The set Σ is called the *Alphabet* of S. The *empty string* with $|S| = 0$ is denoted as ε. A *substring* of a string is the string $S[i..j]$ where $1 \le i \le j \le n$ and it is formed from the characters of S starting at index i and ending at index j, i.e., $S[i..j] = S[i]S[i+1]\dots S[j]$. If $j < i$, $S[i..j]$ is the empty string ε. A prefix of S is a substring $S[1..i]$, and a suffix of S is a substring $S[i..n]$. We say that a character c *occurs* in S if and only if $\exists i$ s.t. $S[i] = c$, and denote $c \in S$.

For an integer $i \in \mathbb{N}$, we denote as S^i the concatenation of a string to itself i times, i.e., $S^1 = S$, and $S^i = S^{i-1} \cdot S$, for any $i \ge 2$. A string $S = p^i p'$ is called *periodic*, where $i \ge 2$ and p' is a prefix of p. The substring p is called the *period* or *factor* of S. A string with two periods p, q has a period of length $\gcd(|p|, |q|)$ (Fine & Wilf [7]).

We call S a *palindrome* or *palindromic* if $\forall 1 \leq i \leq n$, $S[i] = S[n-i+1]$. For example, $S =$ `level`, or $S =$ **deed** are palindromes. However, $S = abcbaa$ is not a palindrome. A substring $P = S[i..j]$ is called a *maximal palindrome* of S if P is a palindrome, and $S[i-1..j+1]$ is either undefined or not a palindrome. The center of P is $c_P = \frac{i+j}{2}$ and its radius is $r_P = \lceil \frac{j-i}{2} \rceil$.

There are $2n-1$ possible centers in S and each has exactly one corresponding maximal palindrome. An array of all maximal palindromes can be found in linear time, e.g., using the Manacher algorithm [19]. The that retains the length of the maximal palindrome for every given center is known as the *Manacher array*, also referred to in the literature as the *palindromic structure* of S.

Definition 1 (Manacher array). *Let $S[1..n]$ be a string of length n. The* Manacher array *of S is the array $\mathsf{A}[1..2n-1]$ defined as follows:*

Each position i corresponds to a center between characters of S:

- *If i is odd, $i = 2k-1$, then $\mathsf{A}[i]$ is the maximum integer $r \geq 0$ such that $S[k-r..k+r]$ is a palindrome, and $1 \leq k-r \leq k+r \leq n$.*
- *If i is even, $i = 2k$, then $\mathsf{A}[i]$ is the maximum integer $r \geq 0$ such that $S[k-r+1..k+r]$ is a palindrome, and $1 \leq k-r+1 \leq k+r \leq n$.*

In other words, $\mathsf{A}[2k]$ is the radius of the longest palindrome centered between $S[k]$ and $S[k+1]$, and $\mathsf{A}[2k-1]$ is the radius of the longest palindrome centered at $S[k]$.

Lemma 1 (Folklore, [11]). *Let P be a palindrome. p is a palindromic suffix of P iff $|P| - |p|$ is a period of P.*

Definition 2. *Let S be a string. An index m is said to be* palindromically dependent *iff there exists a palindromic substring $p = S[i..j]$ such that $c_p < m \leq j$.*

The *Zimin word* z_n is a recursively defined sequence of words, where:

$$z_1 = w_1 \quad \text{and} \quad z_{n+1} = z_n w_{n+1} z_n,$$

with each w_i being a distinct non-empty variable. The *Zimin degree* of a word w is the largest number k such that w can be written as z_k (i.e., w is a substitution instance of z_k). A Zimin word of degree k has a length of at least $2^k - 1$.

Definition 3 (Palindromic Zimin Word). *Let Z_1^P be any arbitrary palindrome, and let Z_k^P be the pattern of all palindromes that match $Z_{k-1}^P P_k Z_{k-1}^P$ where P_k is an arbitrary palindrome, and subsequent choices of P_i satisfy $P_i \neq P_k$. A word that matches the structure Z_k^P is a* Palindromic Zimin Word.

Since every word that follows the structure Z_k^P also follows Z_{k-1}^P, we say that S follows (or matches) Z_k^P if it follows Z_k^P and does not follow Z_{k+1}^P[1].

[1] This pattern is also known as the "ABACABA pattern".

Tandem Duplication Trees. Let T be a rooted binary tree and let $v \in T$ be a node. We write $\mathrm{par}(v)$ for the parent of v, $\mathrm{lc}(v)$ for its left child, and $\mathrm{rc}(v)$ for its right child. If v is the root, then $\mathrm{par}(v)$ is undefined; if v is a leaf, then $\mathrm{lc}(v)$ and $\mathrm{rc}(v)$ are undefined.

We now define the *Tandem Duplication Tree* and *Rooted Tandem Duplication Tree*, data structures motivated by biological evolutionary constructs.

Definition 4 (Tandem-Duplication Event). *Let Γ be a universe of gene identifiers, and let $A = \{g_1, \ldots, g_m\}$ be an ordered sequence with $g_j \in \Gamma$. Pick any contiguous block of indices:*

$$B = \{i, i+1, \ldots, i+\ell-1\}, \quad (1 \leq i \leq i+\ell-1 \leq m)$$

Replace $A[i..i+\ell-1]$ by:

$$\{\,\mathrm{lc}(g_i), \ldots, \mathrm{lc}(g_{i+\ell-1}),\ \mathrm{rc}(g_i), \ldots, \mathrm{rc}(g_{i+\ell-1})\,\},$$

where for every $g_j \in B$ the symbols $\mathrm{lc}(g_j)$ and $\mathrm{rc}(g_j)$ are distinct new *genes. Such an operation is referred to as* tandem-duplication event *or simply a* duplication event*. A block B of size $k = \ell$ is called an ℓ*-duplication.

Using the latter definition, we define a *rooted duplication tree* as a tree that conserves all duplication events on a single gene.

Definition 5 (Rooted Duplication Tree). *Let $A_0 = \{1\}$ be the original gene, and let A be the final genes array that results from consecutive tandem duplication events on the original array A_0.*

The history of all events is stored in a rooted binary tree T:

- *the root represents the ancestral gene 1;*
- *each internal node corresponds to one duplication event, its left (and right) subtree containing all $\mathrm{lc}(\cdot)$ (resp. $\mathrm{rc}(\cdot)$) descendants created by that event;*
- *leaves are the genes present in the final array.*

An example can be found at Fig. 1.

Lemma 2 (Rooted Duplication Tree's count [13]). *Let r_n be the number of distinct (non-isomorphic) rooted duplication trees with n leaves. The value r_n follows the recurrence formula:*

$$r_n = \begin{cases} 1 & n = 1, 2 \\ \sum_{k=1}^{\lfloor (n+1)/3 \rfloor} (-1)^{k+1} \binom{n+1-2k}{k} r_{n-k} & n \geq 3 \end{cases}$$

Using Stirling's approximation [15], it can be shown that $r_n = o(6.75^n)$.

Graph Theory. We use standard graph-theoretic terminology. A graph $G = (V, E)$ consists of a finite set of vertices V and a set of edges $E \subseteq V \times V$, where all edges are undirected and loop-free. A *proper vertex coloring* is a function $\psi : V \to \Sigma$ such that $\psi(u) \neq \psi(v)$ for every edge $(u, v) \in E$; the set Σ is referred to as the set of colors. The minimal number of colors required for a proper coloring of G is its chromatic number, denoted $\chi(G)$.

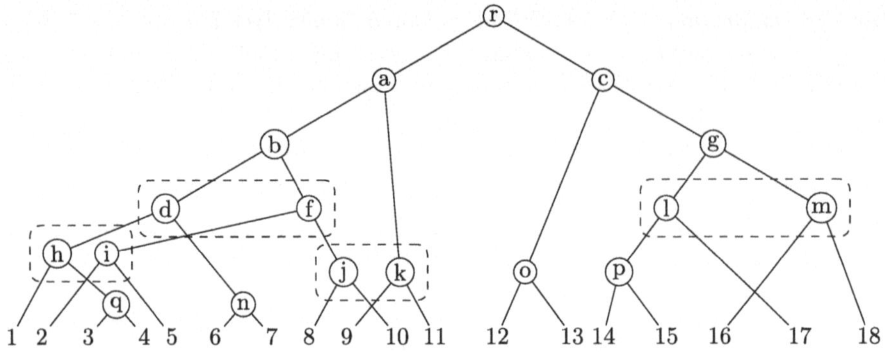

Fig. 1. Rooted tandem duplication tree. Duplication events with more than one gene are marked with a dashed rectangle. The leaves are labeled and positioned in order.

3 Combinatorial Complexity of the Manacher Array

In this section, we discuss the combinatorial complexity of the Manacher array, i.e., we attempt to find the number ρ_n of distinct Manacher arrays generated from strings of length n.

The number of distinct strings of length n with alphabet Σ is $|\Sigma|^n$. However, the number of distinct Manacher arrays corresponding to these strings is smaller. We show an upper and lower bound for ρ_n. The lower bound is $\Omega\left(3^n\right)$, while the upper bound is $\mathcal{O}\left(r_{n+1}\right)$, the number of rooted duplication trees with $n+1$ leaves.

Lemma 3. *Let ρ_n denote the number of distinct Manacher arrays corresponding to strings of length n. Then $\rho_n = \Omega(3^n)$.*

The Lemma follows from Lemma 2 in [10], where they prove that a string over a ternary alphabet can be reconstructed from its Manacher array, up to a permutation.

We now proceed to prove the upper bound with rooted tandem repeat trees.

Theorem 1. *Let ρ_n denote the number of distinct Manacher arrays corresponding to strings of length n, and let r_n denote the number of rooted tandem duplication trees with n leaves. Then $\rho_n \leq r_{n+1}$.*

Throughout the proof, we use an auxiliary combinatorial structure, denoted as the *counter array.* The array is defined as follows:

Definition 6 (Counter array). *Let $A = (a_1, \ldots, a_n)$ be an array of length n of integers satisfying:*

$$1 \leq a_i \leq i \quad \text{and} \quad a_{i+1} \geq a_i - 1 \qquad (1 \leq i < n).$$

We call A a counter array *and denote by σ_n the number of such arrays.*

The proof process of Theorem 1 consists of three parts; In the first part of the proof (Lemma 4), we prove that $\sigma_n = r_{n+1}$. In the second part (Lemma 6), we show a compact representation of the Manacher array and prove that the Manacher array can be retrieved from it. In the third and last part (Lemma 7), we show that the number of compact representations corresponding to strings of length n is exactly σ_n, establishing $\rho_n \leq \sigma_n = r_{n+1}$, as claimed.

3.1 Rooted Duplication Trees

In this subsection, we prove the following lemma:

Lemma 4 (Duplication trees and counter arrays). *The number of rooted duplication trees with n leaves equals the number of counter arrays of length n. That is,*

$$r_{n+1} = \sigma_n.$$

The proof is primarily combinatorial and requires several additional definitions related to rooted duplication trees. We refer to these trees simply as *duplication trees.*

Let T be a duplication tree. All non-leaf nodes in T have left and right children. We refer to the ordered list of genes at the leaves of the duplication as the *leaves array* of T. In Fig. 1, the leaves array of T is the numbers from 1 to 18.

Given a duplication tree T with leaves array $A = \{g_1, g_2, \ldots, g_m\}$, a duplication event is a continuous indices array $B = \{i, i+1, \ldots, i+\ell-1\}$. Applying the event B to the array results in:

$$B(A) = \{g_1, g_2, \ldots, g_{i-1}, lc(g_i), lc(g_{i+1}), lc(g_{i+\ell-1}), rc(g_i), \ldots, rc(g_{i+\ell-1}), \ldots, g_m\}$$

Let $A = \{g_1, \ldots, g_m\}$ be the leaves array. We define a strict total order $\prec$ on the leaves by $g_i \prec g_j \iff i < j$ for $1 \leq i, j \leq m$; thus $g_1 \prec g_2 \prec \cdots \prec g_m$. When needed, we write $g_i \preceq g_j$ to mean $i \leq j$.

Partial Events Ordering. We define a partial order between two duplication events, B_1 and B_2. We say that B_1 is smaller than B_2 if $\max B_1 < \min B_2$. Note that the order is not necessarily defined; let $B_1 = \{1, 2\}, B_2 = \{2, 3\}$, then $B_1 \not\prec B_2$ and $B_2 \not\prec B_1$. However, if $B_1 = \{6\}, B_2 = \{1, 2, 3\}$, then $B_2 \prec B_1$. We say that $B_1 \preceq B_2$ if and only if $B_2 \not\prec B_1$. The relation $\preceq$ is defined between every pair of events.

We can now prove a key lemma that also results in an algorithm to decompose a duplication tree into a unique list of duplication events.

Lemma 5 (Unique event decomposition). *Any rooted duplication tree with n leaves admits a* unique *ordered list* $(B_1 \preceq B_2 \preceq \cdots \preceq B_k)$ *of duplication events such that applying the events in that order recreates the tree's leaves.*

Proof (Proof sketch of Lemma 5). Induct on n.

Low Nodes. A node is *low* if its two children are leaves. Two low nodes lie in the *same* event iff their four child leaves interleave $\mathrm{lc}(u) \prec \mathrm{lc}(v) \prec \mathrm{rc}(u) \prec \mathrm{rc}(v)$ or its mirror.

Last Event. Partition the low nodes by this interleaving rule; discard any block whose leaves are *not* contiguous (such a block cannot be produced by the *last* event). Of the remaining event blocks, choose the rightmost block in the ordering B^*.

Recurse. Remove the right-copy leaves of B^* to obtain a tree $\widetilde{T}$ with fewer leaves. By the induction hypothesis $\widetilde{T}$ decomposes uniquely as $(B_1, \ldots, B_{k-1})$; append B^* to finish.

Uniqueness. The block selection step is deterministic, so B^* is unique; the induction hypothesis fixes the rest. □

The full proof can be found in the full version of the paper [12].

Next, we prove that there is a bijection between rooted duplication trees and counter arrays, as stated at the beginning of the subsection at Lemma 4.

Proof. Encoding events as a counter array. Let T be a duplication tree with events $B_1 \preceq \cdots \preceq B_k$ and $B_i = (b_i, \ldots, b_i + \ell_i - 1)$. Encode B_i by the strictly decreasing sequence $(b_i + \ell_i - 1, \ldots, b_i)$ and concatenate the k sequences to obtain $A = (a_1, \ldots, a_n)$. Inside an event, the drop is exactly 1; between events, we have $a_{j+1} \geq a_j$[2], so $a_{j+1} \geq a_j - 1$ for all j. While encoding B_i there are exactly $1 + \sum_{t<i} \ell_t$ genes, hence every symbol written satisfies $1 \leq a_j \leq j$. Thus, A is a counter array, and its length is $n - 1 = \sum_{i=1}^{k} \ell_i$.

Decoding a Counter Array. Conversely, scan a counter array A from left to right. Start a new event whenever the next entry fails to drop by 1. If entry a_j starts a new event, the number of available genes is exactly j. Therefore, an entry that starts an event and the subsequent entries within the event are valid gene indices when read, hence every recovered event is legal. The two procedures are inverses, establishing a bijection between rooted duplication trees with n genes and counter arrays of length $n - 1$. Hence $r_{n+1} = \sigma_n$. □

3.2 Compact Manacher Array

At the beginning of the section, we proved that a Manacher array requires $\Omega(n)$ bits to represent. We show a folklore representation of the Manacher array that requires $\Theta(n)$ bits[3]. Later, we show that this representation is equivalent to the counter array. We conclude by showing that the representation is not tight—meaning that some counter arrays do not represent a valid Manacher array.

[2] Because the events are ordered.

[3] Arseny M. Shur, private communication.

Lemma 6. *The Manacher array A of a string S of length n can be represented using $\mathcal{O}(n)$ bits.*

Proof (Proof sketch for Lemma 6). For each position i let c_i be the center of the maximal palindromic suffix of $S[1..i]$. Define $b_1 := 1$ and $b_i := 2(c_i - c_{i-1})$ for $i \geq 2$. Because each suffix extends the previous one or ends earlier, the sequence $C = (c_i)$ is non-decreasing; hence $\mathsf{B} = (b_i)$ is a monotone sequence of non-negative integers.

Space. Since $c_i \leq n$ and $2c_i = \sum_{j \leq i} b_j$,

$$\sum_{i=1}^{n} b_i \leq 2n.$$

Encode B in unary, separating consecutive values by a single 0. This uses at most $(n-1)$ zeros and $\leq 2n$ ones, i.e. $\leq 3n - 1 = \mathcal{O}(n)$ bits.

Reconstruction. From the unary code we recover B, prefix-sum to obtain each c_i, and hence every center. A right-to-left sweep (mirroring the standard Manacher update) then assigns the correct radius to each center in $\mathcal{O}(n)$ time.

Thus, the Manacher array admits an $\mathcal{O}(n)$-bit representation. □

The full proof can be found in the full version of the paper [12].

Lemma 7. *The compact representation as defined in Lemma 6 is equivalent to the counter array.*

The proof constructs the centers array from the compact representation and shows that the centers array can be transformed into the counter array. The full proof is in the full version of the paper [12].

Corollary 1. *The value ρ_n is less than or equal to the number of distinct compact representations from Lemma 6.*

And with that, we conclude the proof of Theorem 1. □

4 Restriction Graphs

In this section, we prove the following theorem:

Theorem 2. *Let A be a valid Manacher array. There exists a graph $G = (V, E)$, such that every proper coloring of G with k colors yields a string with Manacher array A and exactly k different alphabet symbols, and every string with Manacher array A yields a proper graph coloring.*

A similar framework considering only equality dependency relations in strings was presented by Gawrychowski et al. [23], where they seek the biggest possible alphabet to satisfy the given set of dependencies. To the contrary, our approach incorporates both equality and inequality dependencies, but restricts the dependencies to a specific, non-arbitrary set.

Throughout the proof, we refer to a variant of the Manacher array of a string S as the *palindromic fingerprint*, or simply *fingerprint* of S.

Definition 7. *The* palindromic fingerprint F *of a string* S *is a set of all maximal palindromes in* S.

A pair (i, j) *is included in the fingerprint* F *if and only if* $S[i..j]$ *is a maximal palindrome. Zero-length maximal palindromes are* $(i, i-1)$.

The length of a fingerprint F*, denoted* $|\mathsf{F}|$*, is equivalent to the length of the underlying string* $|S|$.

A string T *is a* reconstruction *of* F *if the set of maximal palindromes of* T *results in the set* F*, and* $|T| = |\mathsf{F}|$.

For example, denote F as the fingerprint of string S with length 12 and a maximal palindrome at $S[2..7]$. Then $(2, 7) \in \mathsf{F}$ and $|\mathsf{F}| = 12$. Additionally, S is a reconstruction of F.

We present the *restriction graph* G of a palindromic fingerprint F, which is the graph that is described at Theorem 2.

The equivalence between restriction graph coloring and palindromic fingerprint reconstruction gives us a straightforward bound on the number of distinct alphabet characters required to reconstruct a given fingerprint - the smallest possible number is the chromatic number of the graph $\chi(G)$. The highest possible number is the number of nodes in G, $|V|$.

In this section, we provide a detailed description of the restriction graph. First, we define an auxiliary graph, called the *equality graph*.

Definition 8 (Equality Graph). *The equality graph of a fingerprint* F *of length* n *denoted as* $G_{=}(\mathsf{F}) := (V, E)$ *is an undirected graph defined as follows:* $V = \{i\}_{i=1}^{n}$*, and* $(i, j) \in E$ *if and only if* (i, j) *are palindromically dependent (recall from Definition 2: If* i, j *are palindromically dependent, then in every reconstruction* T *of* F*,* $T[i] = T[j]$*).*

The following follows from the definition of $G_{=}(\mathsf{F})$.

Observation 8. *Let* $A_1, A_2, \ldots, A_\ell$ *be the connected components of* $G_{=}(\mathsf{F})$*. Every reconstruction* T *of* F *satisfies* $S[i] = S[j]$ *if and only if:*

$$\exists k \quad s.t. \quad i, j \in A_k$$

We now define the restriction graph of F.

Definition 9 (Restriction Graph). *Let* $G' = G_{=}(\mathsf{F}) = (V', E')$ *be the equality graph of* F.

The ***restriction graph*** $G = G(\mathsf{F}) := (V, E)$ *is an undirected graph defined as follows:*

- *The vertex set V consists of the connected components of G', i.e.,*

$$V = \{A_k\}_{k=1}^{\ell}$$

- *The edge set E consists of edges between connected components that are constrained by palindromic properties in F.*
Informally, if in every possible reconstruction T of F, $T[i_1] \neq T[i_2]$, then we draw an edge between the vertex A_k^1 that contains i_1 and the vertex A_k^2 that contains i_2.
Formally,

$$E = \{(A_{k_1}, A_{k_2}) \mid \exists i \in A_{k_1}, \exists j \in A_{k_2} \text{ such that } (i+1, j-1) \in \mathsf{F}\}.$$

An example of the restriction graph can be found at Sect. 4.

The following two observations can hint to the relation between graph coloring and fingerprint reconstruction:

Observation 9. *Let $G = (V, E)$ be a restriction graph for a palindromic fingerprint F. The vertices $V = (v_1, v_2, ...)$ form a partition of $[n] = \{1, 2, \ldots, n\}$.*

Observation 10. *Let T be a reconstruction of F, and let $G = (V, E)$ be the restriction graph of F. If $(A_{k_1}, A_{k_2}) \in E$, then:*

$$\forall i \in A_{k_1}, \forall j \in A_{k_2} \qquad T[i] \neq T[j].$$

We proceed to prove the equivalence "coloring $\iff$ reconstruction" (Lemma 11).

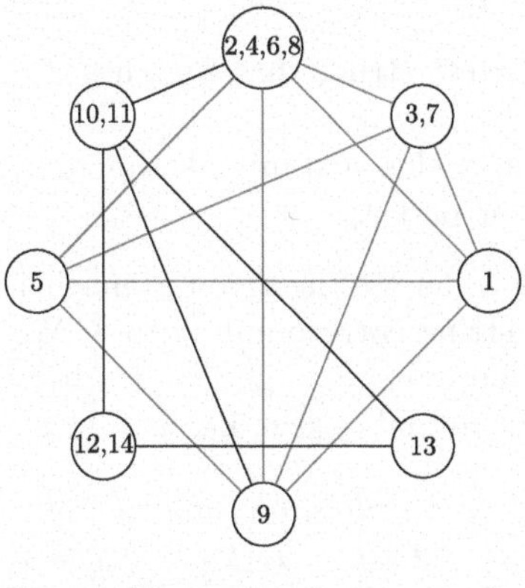

Fig. 2. The restriction graph G of $S = \texttt{41213121566757}$.

Lemma 11. *Let $\psi : V \to \Sigma$ be a coloring of the restriction graph $G(\mathsf{F}) = (V, E)$ with $n = |\mathsf{F}|$. Define another function $\psi' : [n] \to \Sigma$:*

$$\forall A_k \in V \quad \forall i \in A_k \qquad \psi'(i) := \psi(A_k)$$

A string T can be reconstructed as

$$T = \psi'(1)\psi'(2)\ldots\psi'(n),$$

And the fingerprint of T is F. Conversely, given a string T with fingerprint F, define a coloring function ψ:

$$\psi(A_k) := T[i] \quad \textit{For some } i \in A_k$$

The function ψ is a proper coloring of $G(\mathsf{F})$.

The proof is in the full version of the paper [12].

And the latter proof completes the proof for Theorem 2. □

5 Alphabet Size Bounds

Logarithmic bounds on the alphabet size in palindromic-equivalent structures are well established. However, a finer combinatorial analysis reveals even more specific structural constraints. In this section, we present results that clarify the relationship between a fingerprint's alphabet size and its underlying combinatorial structure.

Theorem 3. *Let* A *be a Manacher array, and let* S *be a lexicographically minimal string with the Manacher array* A, *i.e.,* a reconstruction *of* A. *Assume the alphabet of* S *is* $\Sigma = \{\sigma_1, \sigma_2, \dots\}$, *and that the lexicographic ordering is* $\sigma_i < \sigma_{i+1}$ *for every* i.

1. *The first occurrence of a character* σ_i, $i \geq 3$ *is preceded by a substring that follows the pattern* Z^P_{i-2}.
2. *Any subsequent occurrences of a character* σ_i *(where* $i \geq 3$*) are either:*
   - *At a palindromically dependent index (recall Definition 2), or*
   - *Preceded by a substring that follows the pattern* Z^P_{i-2}.
3. *The string* S *is constructed using a globally minimal alphabet.*

The above theorem implies the following:

Corollary 2 (Alphabet Size). *Let* $\alpha : \mathbb{N} \to \mathbb{N}$ *be the function such that* $2\alpha(k) - 1$ *is the minimal length of a Manacher array that cannot be realized by any string over an alphabet of fewer than* k *distinct symbols.*

Then,

$$\alpha(k) = \begin{cases} 1, & \text{if } k = 1, \\ 2^{k-2} + 1, & \text{if } k \geq 2. \end{cases}$$

The corollary is achieved by plugging the shortest string that matches Z^P_k into the theorem.
We acknowledge that [10] claims their algorithm yields a minimal alphabet; however, this claim is not rigorously proven in their paper.

Proof of Theorem 3. Throughout the proof, let A denote the given Manacher array, and let S be the lexicographically minimal string corresponding to A. We assume that S contains more than three distinct characters, i.e., $|\Sigma_S| \geq 4$, where Σ_S represents the alphabet of S. We fix k to be the alphabet size, i.e., $k = |\Sigma_S|$.

Observation 12. *Let* σ_i *be a character at a palindromically-independent index* m, *and let* $s_m = S[1..m]$. *For every character* σ_j *with* $j < i$, *there exists a palindrome* p_j *such that* $\sigma_j p_j \sigma_i$ *is a suffix of* s_m.

This observation is an alternative formulation of the requirement that the string S is lexicographically minimal.

Observation 13. *Let* P *be a periodic palindrome.* P *can be written as* $(q_0 q_1)^i q_0$, *where* q_0 *and* q_1 *are palindromes and* $i \geq 2$.

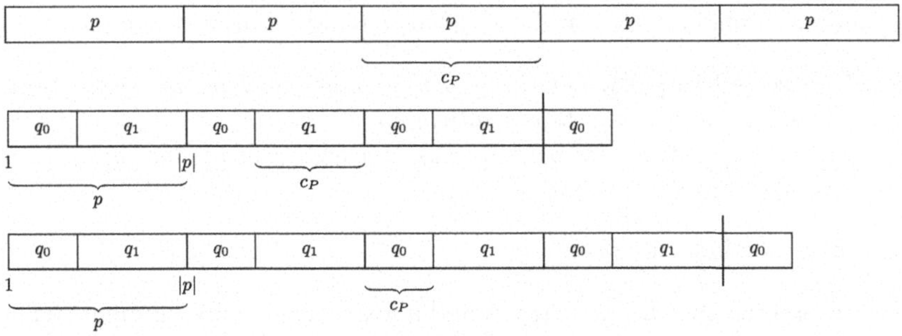

Fig. 3. Demonstration of Observation 13. c_P is the center of the periodic palindrome.

A demonstration of this observation can be found at Fig. 3.

Lemma 14. *Let σ_i be a character at a palindromically independent index m, and let $s_m = S[1..m]$. Let $\mathcal{P} = \{p_1, p_2, \ldots, p_{i-1}\}$ be the set of palindromes as defined in Observation 12, where each $p \in \mathcal{P}$ is the shortest possible. For any distinct pair of palindromes p_j and $p_{j'}$ from $\mathcal{P}$, where $|p_j| < |p_{j'}|$, it holds that $2|p_j| < |p_{j'}|$. The longest palindrome p_j matches the pattern Z^P_{i-2}.*

Proof. Let us assume without loss of generality that the set $\mathcal{P} = \{p_1, p_2, \ldots, p_{i-1}\}$ is sorted, i.e., for every pair p_j, p_{j+1} it holds that $|p_j| > |p_{j+1}|$.

Assume to the contrary there exists a triplet (σ_i, p_j, p_{j+1}) that contradicts the lemma, meaning:

1. $2|p_j| \geq |p_{j+1}|$
2. $\sigma_j p_j \sigma_i$ is a suffix of s_m. The same condition would hold for σ_{j+1} and p_{j+1}, respectively.

First, let's rule out the possibility of $2|p_j| = |p_{j+1}|$. If $2|p_j| = |p_{j+1}|$, then $p_{j+1}[|p_j|] = p_{j+1}[|p_j| + 1] = \sigma_j$, and by symmetry of p_j, the last character of p_j also equals σ_j, and therefore the minimal palindrome p_j is the empty string ε. However, we assumed $2|p_j| = |p_{j+1}| = 0$, contradiction.

Let us now assume $2|p_j| > |p_{j+1}|$. When two palindromic suffixes are of similar length, the longer palindrome has a period of length at most $|p_{j+1}| - |p_j|$. Therefore, let q be the minimal periodic factor, and rewrite p_j as $(q_0 q_1)^i q_0$ Observation 13.

We know that:

1. The character preceding p_j in S' is σ_j.
2. $|p_{j+1}| \geq |p_j| + 1$, and therefore $a_j p_j$ is a suffix of p_{j+1}, which is a periodic string with factor q.

Therefore, $q_1[1] = \sigma_j$, and p_j has a prefix $q_0 \sigma_j$, and since a periodic factor must occur at least twice, $|q_0| < |p_j|$, contradicting the minimality of p_j. If q_1 is empty, it follows that $q_0[1] = a_j$ and p_j can be replaced with the empty string ε.

The string p_j has p_{j-1} as a non-overlapping prefix and suffix. Therefore, since $p_1 = \varepsilon$, the string p_{i-1} matches the pattern Z^P_{i-2}. □

Next, we need to prove that the lexicographically minimal string has the globally minimal alphabet size. To do so, let us cite an additional lemma:

Lemma 15 (Proved in [20]). *Let w and u be strings that have the same Manacher array, and let i and j be integers satisfying $1 \leq i < j \leq |w| = |u| = n+1$. If $w[i+1..n]$ and $w[j+1..n]$ are palindromes and $w[i] = w[j]$, then $u[i+1..n]$ and $u[j+1..n]$ are palindromes and $u[i] = u[j]$.*

We can now conclude the proof.

Proof. Let us assume that for an array A and a lexicographically minimal string S, there exists a string S' with a strictly smaller alphabet.

Looking at the algorithm by [10] and the proof of Lemma 14, a new character is introduced to the reconstruction only if no existing character can satisfy the left-to-right reconstruction of the array A. Therefore, if $|\Sigma_S| > |\Sigma_{S'}|$ then there exists a minimal index j such that $\sigma_i = S[j]$, and $\sigma_j \notin \Sigma_{S'}$. In Lemma 14, we proved that the new character σ_i is preceded by $(i-1)$ maximal palindromic suffixes, each of them preceded by a different alphabet symbol. However, since a new character was not introduced at $S'[j]$, at least two palindromic suffixes were preceded by the same character, contradicting Lemma 15. □

With that, Theorem 3 is proved. □

The last lemma for this section combines Theorem 3 and Theorem 2 to solve an open problem proposed by [10].

Theorem 4 (Open problem (3) in [10]). *Given a set of maximal palindromes F and a predefined parameter k, we can find a reconstruction of F that contains exactly k characters, if possible.*

The proof creates an optimal coloring using the algorithm from [10], updates the coloring to include exactly k colors, and concludes by making a string from the updated coloring. The proof can be found in the full version of the paper [12].

6 Future Work

Several natural questions remain open. Most notably, the exact number of valid Manacher arrays of length n is unknown, and a direct combinatorial characterization of these arrays would significantly deepen our understanding. It is also unclear whether efficient uniform sampling of such arrays is possible. Finally, extending the reconstruction and combinatorial framework to approximate palindromes or palindromes containing wildcards remains a largely unexplored avenue.

Acknowledgments. I thank Professor Ron Adin for his guidance and valuable assistance with this problem.

This study was partially funded by ISF grant No. 168/23.

References

1. Bannai, H., Inenaga, S., Shinohara, A., Takeda, M.: Inferring strings from graphs and arrays. In: Rovan, B., Vojtáš, P. (eds.) MFCS 2003. LNCS, vol. 2747, pp. 208–217. Springer, Heidelberg (2003). https://doi.org/10.1007/978-3-540-45138-9_15
2. Benson, G.: Tandem repeats finder: a program to analyze DNA sequence. Nucleic Acids Res. **27**(2), 573–580 (1999)
3. Borozdin, K., Kosolobov, D., Rubinchik, M., Shur, A.: Palindromic length in linear time. In: CPM 2017, vol. 78, pp. 23:1–23:12 (2017)
4. Charalampopoulos, P., Pissis, S., Radoszewski, J.: Longest palindromic substring in sublinear time. In: CPM 2022, vol. 223, pp. 20:1–20:9 (2022)
5. Dan, G.: Algorithms on stings, trees, and sequences: computer science and computational biology. SIGACT News **28**(4), 41–60 (1997)
6. Eichler, E.: Recent duplication, domain accretion and the dynamic mutation of the human genome. Trends Genet. **17**(11), 661–669 (2001)
7. Fine, N., Wilf, H.: Uniqueness theorems for periodic functions. Proc. Amer. Math. Soc. **16**, 109–114 (1965)
8. Fuglsang, A.: Distribution of potential type II restriction sites (palindromes) in prokaryotes. Biochem. Biophys. Res. Commun. **310**(2), 280–285 (2003)
9. Gelfand, M., Koonin, E.: Avoidance of palindromic words in bacterial and archaeal genomes: a close connection with restriction enzymes. Nucleic Acids Res. **25**, 2430–2439 (1997)
10. I, T., Inenaga, S., Bannai, H., Takeda, M.: Counting and verifying maximal palindromes. In: Chavez, E., Lonardi, S. (eds.) SPIRE 2010. LNCS, vol. 6393, pp. 135–146. Springer, Heidelberg (2010). https://doi.org/10.1007/978-3-642-16321-0_13
11. I, T., Sugimoto, S., Inenaga, S., Bannai, H., Takeda, M.: Computing palindromic factorizations and palindromic covers on-line. In: Kulikov, A.S., Kuznetsov, S.O., Pevzner, P. (eds.) CPM 2014. LNCS, vol. 8486, pp. 150–161. Springer, Cham (2014). https://doi.org/10.1007/978-3-319-07566-2_16
12. Itzhaki, M.: Combinatorics of palindromes (2025). https://arxiv.org/abs/2507.13671
13. Jialiang, Y., Louxin, Z.: On counting tandem duplication trees. Mol. Biol. Evol. **21**(6), 1160–1163 (2004)
14. Johan, J.: The derivation of on-line algorithms, with an application to finding palindromes. Algorithmica **11**(2), 146–184 (1994)
15. Knuth, D.E., Graham, R.L., Patashnik, O.: Concrete Mathematics: A Foundation for Computer Science, 2nd edn. Addison-Wesley (1994)
16. Knuth, D., Morris, J., Pratt, V.: Fast pattern matching in strings. SIAM J. Comput. **6**, 323–350 (1977)
17. Lenz, A., Wachter-Zeh, A., Yaakobi, E.: Duplication-correcting codes. Des. Codes Crypt. **87**(2), 277–298 (2019)
18. Lisnic, B., Svetec, I., Saric, H., Nikolic, I., Zgaga, Z.: Palindrome content of the yeast Saccharomyces cerevisiae genome. Curr. Genet. **47**, 289–297 (2005)
19. Manacher, G.: A new linear-time "on-line" algorithm for finding the smallest initial palindrome of a string. J. ACM **22**(3), 346–351 (1975)
20. Nagashita, S., I, T.: PalFM-index: FM-index for palindrome pattern matching. In: CPM 2023, vol. 259, pp. 23:1–23:15 (2023)
21. Rubinchik, M., Shur, A.: EERTREE: an efficient data structure for processing palindromes in strings. In: Combinatorial Algorithms, pp. 321–333 (2016)

22. Srivastava, S., Robins, H.: Palindromic nucleotide analysis in human T cell receptor rearrangements. PLoS ONE **7**(12), e52250 (2012)
23. Walen, P.: Universal reconstruction of a string. Theor. Comput. Sci. **812**, 174–186 (2020)
24. Zvi, G., Joel, S.: Recognizing certain repetitions and reversals within strings. In: 17th Annual Symposium on Foundations of Computer Science (SFCS 1976), pp. 236–252 (1976)

Complexity of Positive Influence Domination on Partial Grids

Enrico Iurlano(✉) and Günther R. Raidl

Algorithms and Complexity Group, TU Wien, Favoritenstraße 9-11/192-01, 1040 Vienna, Austria
{eiurlano,raidl}@ac.tuwien.ac.at

Abstract. The Positive Influence Domination (PID) problem asks to find a minimum cardinality subset of influencers among the vertices of an undirected graph such that at least half the number of neighbors of each vertex are influencers. The problem's underlying model can be used to determine an economical way of promoting (and keeping) good habits in society by interpreting vertices as individuals, neighbors as social contacts, and influencers as, for example, healthy eaters. We show that the problem is NP-hard even when restricted to planar subcubic graphs. The same result turns out to apply for the so-called double total domination problem, which exhibits a similar behavior on this graph class. We use this insight to derive NP-hardness of PID on the class of induced partial grids via a technique relying on orthogonal graph drawing. Finally, we derive bounds on the size of optimal solutions for arbitrarily dimensioned grids.

Keywords: Positive influence domination · NP-hardness · Partial grids

1 Introduction

The *Positive Influence Domination* (PID) problem [22] asks one to find a minimum cardinality subset of *influencers* of the vertex set of a given undirected graph such that for each vertex, the influencers in its neighborhood constitute a (not necessarily strict) majority. The problem, introduced in 2009 by Wang et al. [22], can be illustrated as follows: Suppose a public authority aims to establish a positive habit in society, e.g., the routine of eating healthily. Further assume that healthy eaters risk becoming unhealthy eaters when there are more unhealthy than healthy eaters in their circle of contacts. At this point the authority might wonder what is the minimum number of unhealthy eaters in the society who would need to be "turned" into healthy ones such that healthy eaters remain unaffected by the aforementioned social rebound effect.

While the latter application scenario is meaningful in graphs modeling a social network, the whole question can also be seen as a resource distribution problem in more general graphs. Here sufficient locally available neighboring

A. Jeż and J. Otop (Eds.): FCT 2025, LNCS 16106, pp. 267–280, 2026.
https://doi.org/10.1007/978-3-032-04700-7_20

resources must be provided as economically as possible. For practical applications like broadcasting problems, related problems (some considering also propagation aspects) have been studied on, e.g., unit disk graphs [5] but also more regular graph structures, e.g., hypercubes [1]. Optimal solutions on grid graphs are the subject of study for several variants [20] of the classical dominating set problem [10]. Remarkably, for the latter problem itself, although studied extensively, the process of finding and certifying optimal solutions was completed only in 2011 by Gonçalves et al. [10]. Motivated by these works, we point out properties on (partial) grids for the variant of positive influence domination.

Several closely related combinatorial optimization problems on graphs have been proposed in the form of *monopoly problems* [17], *α-domination* [8], and *signed domination* [7], to mention a few examples. Considering the aspect of temporal propagation of influence, the *Target Set Selection* problem [13] has also been studied. One can find comprehensive comparisons [4,9] of many such problems. This class of domination-type problems has been studied mainly from a theoretical viewpoint so far. For any scalar $\alpha \in (0,1]$—steering the "strength" of a majority—the α-domination problem was shown to be NP-hard [8]. Its slight generalization *vector domination* [11] was shown to be fixed-parameter tractable on graphs excluding cycles of length 4 and on graphs of bounded degeneracy [19].

The more constrained problem *total vector domination* [4], which includes PID as special case, was shown to be solvable in polynomial time on graphs of bounded branchwidth and bounded treewidth [12]—the same was shown to hold for vector domination. Appearing under various names in the literature, *k-tuple total domination* (see, e.g., [14]) is a particularly uniform special case of total vector domination, for which NP-hardness on a special case of chordal graphs was shown [14]; NP-hardness was also shown on bipartite and split graphs [18].

Concerning complexity theoretic results on PID itself, it is shown in [23] that PID is APX-hard; this is also true more generally when, instead of a majority, at least a fixed proportion of neighbors is required to consist of influencers [4]. Recently, NP-hardness for the PID problem on bipartite and split graphs has been established [24]. NP-hardness was shown for the first time in [23] by a reduction from vertex cover (via a proof affirming this for graphs with a maximum degree of 6). The latter work is also our main motivation to identify more restrictive graph classes for which the hardness still applies.

The paper is organized as follows. After providing relevant notation in Sect. 2, in Sect. 3 we show that PID and 2-tuple total domination are NP-hard problems on planar subcubic graphs. The result is lifted to NP-hardness of PID on induced partial grids in Sect. 4. Bounds on complete grids are given in Sect. 5, and Sect. 6 concludes the paper.

2 Notation and Preliminaries

For two integers r and s, we use the notation $[r:s] := \{r, r+1, \ldots, s-1, s\}$ and $[s] := [1:s]$. For $a, b \in \mathbb{Z}$, $b \neq 0$, denote by $\mathrm{mod}(a,b)$ the binary modulo operator (returning the unique number $r \in [0:b-1]$ satisfying $a = kb + r$ for some $k \in \mathbb{Z}$).

In our discussion, we assume graphs $G = (V, E)$ to be simple and undirected. For a vertex $v \in V$ we denote its (open) neighborhood of adjacent vertices by $N(v)$ and its degree by $\deg(v) := |N(v)|$. A graph $G = (V, E)$ with $\max_{v\in V} \deg(v) \leq k$ is called *k-subregular*—or *subcubic* for $k = 3$. If further $\min_{v\in V} \deg(v) = k$, it is called *k-regular*—or *cubic* for $k = 3$. We say that G is k-connected, if G remains connected when fewer than k vertices are removed. Denote by $\mathcal{G}_3^{\text{plan}}$ the class of *planar cubic graphs* and by $\mathcal{G}_{\leq 3}^{\text{plan}}$ the class of *planar subcubic graphs.*

For the following we define a length-r path as a graph $P_r = (V, E)$ with $V = [r]$ and $E = \{\{i, i+1\} : i \in [r-1]\}$. Denote by $P_m \square P_n$ the Cartesian product of P_m and P_n, henceforth also called *complete* $m \times n$ *grid (graph).* Its vertices are given by all pairs in $[m] \times [n]$; two such vertices are defined as adjacent if their first coordinates or second coordinates form an adjacent pair in their respective path of origin P_m or P_n. Let us think of $P_m \square P_n$ as embedded in the plane, with vertices on integer-valued coordinates and unit-length straight, axes-parallel edges; tacitly we assume the vertices to be oriented as the indices of an $m \times n$ matrix. By a *partial grid* we refer to an arbitrary (not necessarily induced) subgraph of a grid, in other words, a graph resulting from the removal of a nonnegative number of vertices and edges from $P_m \square P_n$ for some specific $m, n \in \mathbb{N}$; here, a vertex-removal automatically removes all incident edges. A *unit disk graph* has vertices identifiable with points in the plane such that two vertices are adjacent if and only if the two compact unit disks centered at the respective points have nonempty intersection.

Although many of the subsequent variants of domination/independence are typically defined via subsets of vertices, let us state them for easier usability—see later Theorem 5—in a binary-encoded form: For a graph $G = (V, E)$, a function $f : V \to \{0, 1\}$ is *independent* if $f(v) + f(w) \leq 1$ for all $\{v, w\} \in E$. A function $f : V \to \{0, 1\}$ is said to be

- *positive influence dominating* [22] if $\sum_{w\in N(v)} f(w) \geq \lceil \deg(v)/2 \rceil$ for $v \in V$,
- *k-tuple total dominating* (for some $k \in \mathbb{N}$) if $\sum_{w\in N(v)} f(w) \geq k$ for $v \in V$,
- *double total dominating* if it is 2-tuple total dominating.

When f is positive influence dominating we will often abbreviate this by saying that f is *PID-feasible* or a *PID-function.* For such binary functions f let us call the value $f(v)$ the *weight* or *label* of $v \in V$ and $\sum_{v\in V} f(v)$ the *(cumulative) weight* of f. Given a graph $G = (V, E)$ and $k \in \mathbb{N}$, the problem of determining the existence of an independent $g : V \to \{0, 1\}$ with $\sum_{v\in V} g(v) \geq k$ is known as the independent set problem. The problem of determining the existence of $f : V \to \{0, 1\}$ with $\sum_{v\in V} f(v) \leq k$ being (i) positive influence dominating is denoted as PosInflDom(G, k); (ii) double total dominating is denoted as DblTotalDom(G, k). For fixed G, the minimum $k \in \mathbb{N}$ for which such a feasible function exists, is denoted as $\gamma_{\text{PID}}(G)$ for (i), and $\gamma_{\times 2,\text{t}}(G)$ for (ii).

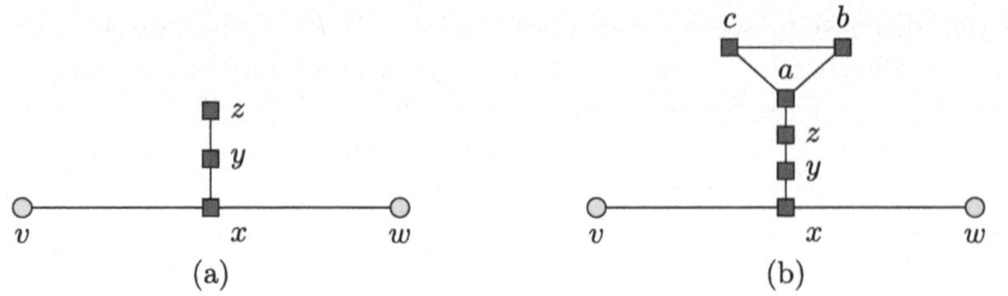

Fig. 1. Modifications of an edge $e = \{v, w\}$ used in the proof of (a) Theorem 2; and (b) Theorem 3. Subscripts e are omitted for the vertices x, y, z, a, b, and c.

3 Hardness on Planar Subcubic Graphs

The results of the current and the next section are inspired by the works [6] and [25] which address the hardness of the so-called *signed dominating set* problem. In the latter problem, a minimum-weight function $f : V \to \{-1, 1\}$ must be found such that $f(v) + \sum_{w \in N(v)} f(w) \geq 1$ for $v \in V$. Although signed domination does not allow a zero-label (i.e., "neutral") and each vertex' own label contributes to meeting its threshold, we still obtain equivalent hardness results for positive influence domination. Unlike Damaschke [6] in 2001, we can now rely on a stronger NP-hardness result that permits assuming 2-connectedness. The latter is stated in the subsequent and underpins our later derived reductions.

Theorem 1 ([16, **Theorem 4.1(a)**]). *The maximum independent set problem is NP-hard when restricted to planar* 2*-connected cubic graphs.*

One can verify that Theorem 2 and Theorem 3 remain valid even when restricted to connected graphs. We omitted this observation which also follows from the fact that both concerned problems can always be solved independently on all connected components.

Consider the planarity-preserving, polynomial-time operation

$$\mathsf{SplApd} : \mathcal{G}^{\text{plan}}_{\leq 3} \to \mathcal{G}^{\text{plan}}_{\leq 3}$$

replacing each edge $e = \{v, w\} \in E$ of $G = (V, E)$ with new edges $\{v, x_e\}$, $\{w, x_e\}$, $\{x_e, y_e\}$, $\{y_e, z_e\}$ on top of fresh vertices x_e, y_e, z_e; see Fig. 1a. Define $\mathcal{SP}^{\text{plan}}_{\leq 3}$ as the class of all graphs that are—up to isomorphy—contained in $\{\mathsf{SplApd}(G) : G \in \mathcal{G}^{\text{plan}}_3\}$. Note that by detecting the vertices at (shortest-path) distance at most 2 from the leaves of $G' \in \mathcal{SP}^{\text{plan}}_{\leq 3}$, we can always determine the underlying cubic graph, even when no explicit specification of the fresh vertices is provided.

Theorem 2. *When restricted to the class of planar subcubic graphs, the problem* PosInflDom(G, k) *is NP-complete.*

Proof. More strongly we prove that the problem is NP-complete on $\mathcal{SP}^{\text{plan}}_{\leq 3}$, directly implying NP-completeness on $\mathcal{G}^{\text{plan}}_{\leq 3}$. The reduction is from maximum independent set on planar 2-connected cubic graphs; see Theorem 1. Let $G = (V, E)$ be a planar 2-connected cubic graph and let $G' := \mathsf{SplApd}(G)$, denoted as $G' = (V', E')$.

The problem is in NP, as a guessed function can be checked for feasibility in linear time (in the size of G).

We claim that G has an independent $g : V \to \{0,1\}$ with $\sum_{v \in V} g(v) \geq k$ iff G' has a PID-feasible $f : V' \to \{0,1\}$ with $\sum_{v' \in V'} f(v') \leq k'$ for some k' depending on G and k. In fact, if an independent g with

$$\sum_{v \in V} g(v) \geq k \tag{1}$$

exists, then the function

$$f(v') := \begin{cases} 1 - g(v') & \text{if } v' \in V, \\ 1 & \text{if } v' \in \{x_e, y_e : e \in E\}, \\ 0 & \text{if } v' \in \{z_e : e \in E\}, \end{cases}$$

is positive influence dominating on G': each vertex in V has three 1-labeled x-neighbors; all vertices in $\{x_e, y_e, z_e : e \in E\}$ have one 1-labeled neighbor; and, finally, the absence of an assigned 1-label by g to one of v and w (according to the independence of g) guarantees at least one additional 1-label assigned by f to one of the neighbors v or w of $x_{\{v,w\}}$.

We conclude that (below obtaining estimate (2) via (1))

$$\begin{aligned} \sum_{v' \in V'} f(v') &= \sum_{v \in V} f(v) + \sum_{v' \in V' \setminus V} f(v') = |V| - \sum_{v \in V} g(v) + \sum_{v' \in V' \setminus V} f(v') \\ &\leq |V| + \sum_{v' \in V' \setminus V} f(v') - k = |V| + 2|E| - k =: k'. \end{aligned} \tag{2}$$

The converse proof direction requires us to show that for a PID-function f on G' the circumstance $\sum_{v' \in V'} f(v') \leq k'$ allows constructing an independent g on G of cumulative weight at least $k = |V| + 2|E| - k'$. We need the intermediate observation that each positive influence dominating $f : V' \to \{0,1\}$ can be updated on $V' \setminus V$ in a weight-preserving—or even weight-reducing—manner such that for each $e \in E$ we eventually have more canonically $f(x_e) = f(y_e) = 1$ and $f(z_e) = 0$.

This simple claim can be justified as follows: We recognize that always $f(y_e) = 1$ due to $\deg(z_e) = 1$. In case now $f(z_e) = 1$, we can overwrite $f(z_e) \leftarrow 0$ and if not already $f(x_e) = 1$, we update $f(x_e) \leftarrow 1$ to maintain the influence threshold of y_e—the cumulative weight does not increase by doing so. On the other hand, if initially $f(z_e) = 0$, necessarily we have already $f(x_e) = 1$.

We can therefore assume that a PID-feasible f with $\sum_{v' \in V'} f(v') \leq k'$, further satisfying the above canonicity, exists. Setting $g := 1 - f(\cdot)$ yields independence for g, as $f(v)+f(w)\geq 1$ implies $g(v)+g(w)\leq 1$. Furthermore, we have

$$\sum_{v\in V} g(v) = |V| - \sum_{v\in V} f(v) = |V| - \sum_{v\in V} f(v) - \sum_{v'\in V'\setminus V} f(v') + \sum_{v'\in V'\setminus V} f(v')$$
$$= |V| - \sum_{v'\in V'} f(v') + 2|E| \geq |V| - k' + 2|E|.$$

□

Corollary 1. *PosInflDom(G,k) is NP-complete on planar subcubic* bipartite *graphs.*

Proof. The vertices of $G' \in \mathcal{SP}^{\text{plan}}_{\leq 3}$ with an underlying cubic graph $G=(V,E)$ are bi-partitioned by $V \cup \{y_e : e \in E\}$ and $\{x_e, z_e : e \in E\}$. □

Similarly, we obtain the following result.

Theorem 3. *DblTotalDom(G,k) is NP-complete even when restricted to planar subcubic graphs.*

Proof. We fall back on a leaves-free class of subcubic graphs resulting from the replacement of the edges of a cubic graph by the gadgets described in Fig. (1b) yielding six additional fresh vertices per edge. Analogously to the proof of Theorem 2, we can here carry out the argumentation with $k' := |V| + 6|E| - k$ and $f(v') := 1 - g(v')$ if $v' \in V$ and otherwise, if $v' \in \{x_e, y_e, z_e, a_e, b_e, c_e : e \in E\}$, $f(v') := 1$. For the converse proof direction the degree of freedom for the choice of f-values is now considerably lower than in the last proof, as, to be feasible for DblTotalDom, all fresh vertices necessarily must be 1-labeled. □

4 Hardness on Induced Partial Grids

In this section we lift the result in Theorem 2 to the class of induced partial grids. We start with the following preparatory observation.

Lemma 1. *Consider an undirected path $P = (v, x_1, x_2, \dots, x_{4\ell-1}, x_{4\ell}, w)$ for some $\ell \in \mathbb{N}\cup\{0\}$ and $f : \{v,w\} \cup \{x_j : j = 1,\dots,4\ell\} \to \{0,1\}$ such that each vertex x_j, $j = 1,\dots,4\ell$, has at least one 1-labeled neighbor. Suppose further that $f(v) = f(w) = 1$. Then, $\sum_{j=1}^{4\ell} f(x_j) \geq 2\ell$, where the lower bound 2ℓ is attainable.*

Proof. Examine f on the induced subpaths $(x_1,x_2,x_3,x_4), \dots, (x_{4\ell-3}, x_{4\ell-2}, x_{4\ell-1}, x_{4\ell})$. By exhaustion it turns out that each of them, independently of the f-assignment on their neighboring paths, must have a cumulative f-weight of at least 2. Thus, $\sum_{j=1}^{4\ell} f(x_j) \geq 2\ell$. This weight is attainable by setting $f(x_j) := 1$, when $\mathrm{mod}(j-1,4) \in \{1,2\}$; and $f(x_j) := 0$, otherwise. □

We anticipate an auxiliary result, suppressing its proof following from a simple canonization strategy comparable to the proof of Theorem 2.

Lemma 2. *Let $k \in \mathbb{N} \cup \{0\}$. Further, let $G = (V, E)$ be a subcubic graph, and $L_e \in \mathbb{N} \cup \{0\}$ be divisible by 4, for $e \in E$. Consider $G' = (V', E')$ resulting from replacing each $e = \{v, w\} \in E$ by an undirected path $(v, p_e^1, \ldots, p_e^{L_e}, w)$ of length $L_e + 1$, i.e., the former edge e is split by L_e fresh vertices. Let $U = \{(0,1,1,0), (0,0,1,0), (0,0,1,1), (0,1,1,1)\}$. If there exists a PID-feasible f' with $\sum_{v \in V'} f'(v) = K$, then there exists a PID-feasible $f'' : V' \to \{0,1\}$ satisfying $(f''(v), f''(p_e^1), f''(p_e^{L_e}), f''(w)) \notin U$, for $e = \{v, w\} \in E$, and $\sum_{v' \in V'} f''(v') \leq K$.*

We will consider inter-vertex paths as in Lemma 1 for any two adjacent vertices v and w. The paths' lengths will again be zero or multiples of 4 but can vary depending on the edge $e = \{v, w\}$. The following lemma affirms that, after inserting a number of splitting vertices divisible by 4 on each edge, a PID-function of the same quality can be achieved (up to an additional constant originating from the number of splitting vertices). The core argument of the proof is that the labels on the splitting vertices can be chosen such that the constellations of neighboring labels remains invariant around all original vertices. The fresh labels can further be feasibly chosen such that the weight does not exceed the half of the count of splitting vertices.

Lemma 3. *Let $k \in \mathbb{N} \cup \{0\}$. Further, let $G = (V, E)$ be a subcubic graph, and $L_e \in \mathbb{N} \cup \{0\}$ be divisible by 4, for $e \in E$. Consider $G' = (V', E')$ resulting from replacing each $e = \{v, w\} \in E$ by an undirected path $(v, p_e^1, \ldots, p_e^{L_e}, w)$ of length $L_e + 1$, i.e., the former edge e is split by L_e fresh vertices. Then, the following assertions are equivalent.*

(i) A PID-feasible f on G with $\sum_{v \in V} f(v) \leq k$ exists.
(ii) A PID-feasible f' on G' with $\sum_{v \in V'} f'(v) \leq k + \sum_{e \in E} L_e/2$ exists.

Proof. On all undirected paths we will consider the values of f'. Formally, as each undirected path can be traversed in two directions, and as labels have to be assigned to the vertices on the path, let us encode the labels via $F'_{(v,w)} \in \{0,1\}^{L_{\{v,w\}}}$ determining hence the chronological ordering of the vertex labels (excluding those of v and w) encountered when traversing the path from v to w in G'. Implicitly we assume that $F'_{(w,v)}$ is automatically coherently specified by the reversal of $F'_{(v,w)}$. With a slight abuse of notation, by writing $F'_{\{v,w\}}$ we refer to $F'_{(v,w)}$.

(i) $\Longrightarrow$ (ii): Starting from a feasible f on G, defining $f'(v) := f(v)$, for $v \in V$, the idea is then, for all $e = \{v, w\} \in E$, to choose the remaining f'-values on $(v, p_e^1, \ldots, p_e^{L_e}, w)$ ensuring feasibility, having cumulative weight $L_e/2$, and satisfying

$$f'(v) = f'(p_e^{L_e}) \text{ and } f'(p_e^1) = f'(w). \tag{3}$$

The latter property requires for vertices $v \in V$, seen as vertices from G', that the neighboring label constellation around v corresponds to the neighboring label constellation around v according to f on G.

The choice of the following length-L_e labeled paths ensures such a behavior (note that together with the labeled endpoints $f(v)$, $f(w)$ PID-feasibility is guaranteed on such a path, i.e., at least one 1-labeled neighbor is present for the path's vertices):

$$F'_{(v,w)} := \begin{cases} (0,1,1,0,\dots,0,1,1,0,\dots,0,1,1,0) & \text{if } f(v)=0 \wedge f(w)=0, \\ (1,1,0,0,\dots,1,1,0,0,\dots,1,1,0,0) & \text{if } f(v)=0 \wedge f(w)=1, \\ (0,0,1,1,\dots,0,0,1,1,\dots,0,0,1,1) & \text{if } f(v)=1 \wedge f(w)=0, \\ (1,0,0,1,\dots,1,0,0,1,\dots,1,0,0,1) & \text{if } f(v)=1 \wedge f(w)=1. \end{cases}$$

The definition of f' on V results hence from the older values of f on V while on the paths' fresh vertices the labels are inferrable from the definition of F'_e according to the above case distinction, which also shows that a weight of only $\sum_{e \in E} L_e/2$ is added to the cumulative weight of f on V.

(ii) $\Longrightarrow$ (i): Due to Lemma 2 we can not only assume f' to be PID-feasible but also without loss of generality $(f'(v), f'(p_e^1), f'(p_e^{L_e}), f'(w)) \notin U$. For all edges $e = \{v, w\}$, up to symmetry breaking (reversed quadruples) we can therefore only have $(f'(v), f'(p_e^1), f'(p_e^{L_e}), f'(w)) \in T \dot\cup W$ with $T := \{(1,1,1,1), (0,1,0,1), (0,0,0,0)\}$ and $W := \{(1,0,1,1), (1,0,0,1), (0,0,0,1)\}$. The ultimate goal will be to modify f' towards a more useful version f'' being as well PID-feasibile, having a weight not greater than f' on G' and satisfying for all edges $\{v, w\} \in E$ the special property

$$f''(v) = f''(p_e^{L_e}) \text{ and } f''(p_e^1) = f''(w). \tag{4}$$

Initially set $f'' := f'$. Immediately we notice that all edges whose behavior is captured by one of the scenarios in T satisfies (4); no updates will hence be needed.

On the other hand, an update leaving the values of f'' unchanged on v and w suffices for the scenarios covered by W:

$$F''_e \leftarrow \begin{cases} (1,0,0,1,\dots,1,0,0,1) & \text{if } (f'(v), f'(p_e^1), f'(p_e^{L_e}), f'(w)) = (1,0,1,1), \\ (1,0,0,1,\dots,1,0,0,1) & \text{if } (f'(v), f'(p_e^1), f'(p_e^{L_e}), f'(w)) = (1,0,0,1), \\ (1,1,0,0,\dots,1,1,0,0) & \text{if } (f'(v), f'(p_e^1), f'(p_e^{L_e}), f'(w)) = (0,0,0,1). \end{cases} \tag{5}$$

Note that the updated values on $p_e^1, \dots, p_e^{L_e}$ meet the bound $L_e/2$ (see Lemma 1) and yield an on-par or better weight than the initial values of f'. After carrying out the updates on all edges, we end up with a labeling f'' fulfilling (4). Therefore, if we contract the inter-vertex paths to the original edges of G, the constellation of neighboring f''-labels will remain unchanged for all vertices $v \in V$. Consequently, the restriction $f := f''|_V$ is the claimed existing labeling. As the weight of f does not include an additional weight $L_e/2$ per edge e, we finally conclude

$$\sum_{v \in V} f(v) \leq \sum_{v' \in V'} f''(v) - \sum_{e \in E} L_e/2 \leq \sum_{v' \in V'} f'(v) - \sum_{e \in E} L_e/2 \leq k.$$

□

Next, let us show how the previous result is related to partial grids. Recall that a *simple rectilinear polyline* (SRP) is an injectively parameterizable polygonal chain made up of axes-parallel line segments. However, for all what follows, we assume a strengthened definition of SRP requiring each of its atomic line segments to be of *integer-valued length.* The argumentation relies on the following auxiliary Lemma due to Valiant [21].

Lemma 4 (adapted from [21, Theorem 2]). *There is a polynomial-time algorithm taking as input a planar 4-subregular graph G and returning a planar embedding of G into the plane using $O(|V|^2)$ area such that embedded vertices possess integer-valued coordinates, and embedded edges are SRPs.*

Theorem 4. *The problem* PosInflDom(G, k) *is NP-complete on the class of induced partial grids.*

Proof. From the proof of Theorem 1 we know that PosInflDom(G, k) is NP-complete on $\mathcal{SP}_{\leq 3}^{\text{plan}}$. We will rely on a reduction from the latter.

Starting from $G' = (V', E') \in \mathcal{SP}_{\leq 3}^{\text{plan}}$ consider also its underlying cubic graph $G = (V, E)$, i.e., $\mathsf{SplApd}(G) = G'$. The first goal will be to find a particular embedding for G' by adapting an embedding for G. This is accompanied by an illustration in Fig. 2 and works as follows:

Using Lemma 4, planarly draw in $\mathbb{R}^2$ the embedding of G such that embedded vertices have integer-valued coordinates and all embedded edges are SRPs; to keep the result polynomially-sized, make sure the embedded vertices and edges are bounded as in Lemma 4. Afterwards, subject the drawing in $\mathbb{R}^2$ to the geometric scaling $(x, y) \mapsto (8x, 8y)$. For each edge $\{v, w\}$ of G, prolong by two units the associated (now scaled) SRP $\rho_{\{v,w\}}$ via the following procedure consisting of two steps:

Step 1. Fix an arbitrary endpoint $p = p_{\{v,w\}} \in \{v, w\}$ of the edge and consider the length-4 line subsegment $s(p, \{v, w\}; 2, 6)$ of $\rho_{\{v,w\}}$ starting at geodesic distance 2 from p and ending at geodesic distance 6 from p. Let us translate $s(p, \{v, w\}; 2, 6)$ by one unit in direction orthogonal to the extension direction of $s(p, \{v, w\}; 2, 6)$ itself. The now modified $\rho_{\{v,w\}}$ consists of three disconnected components, which we join by two unit-length line segments, yielding our definitive form of $\rho_{\{v,w\}}$. The length of the original SRP $\rho_{\{v,w\}}$ was hence artificially increased by two units by performing a local detour geometrically reminding of a rectangular U-turn. This guarantees that all present SRPs are now of length congruent 2 modulo 8. The absence of crossing edges is clearly maintained.

Step 2. Determine on $\rho_{\{v,w\}}$ the point at geodesic distance 5 from p (according to the geometry of the curve $\rho_{\{v,w\}}$ itself). Let us regard it as the embedded "splitting" vertex $x_{\{v,w\}}$ of G'. Then, append to the $x_{\{v,w\}}$-embedding a fresh length-2 straight path, meant to host the embeddings of $y_{\{v,w\}}$ and $z_{\{v,w\}}$, in a way ensuring that one coordinate-entry of $y_{\{v,w\}}$ is divisible by 8. Eventually, we have found an embedding with SRPs for all edges of G'.

Note that for each edge $e = \{v, w\}$ of G, the number of lattice points of $\mathbb{Z} \times \mathbb{Z}$ which are covered by the SRP of G' connecting $x_{\{v,w\}}$ and v is divisible

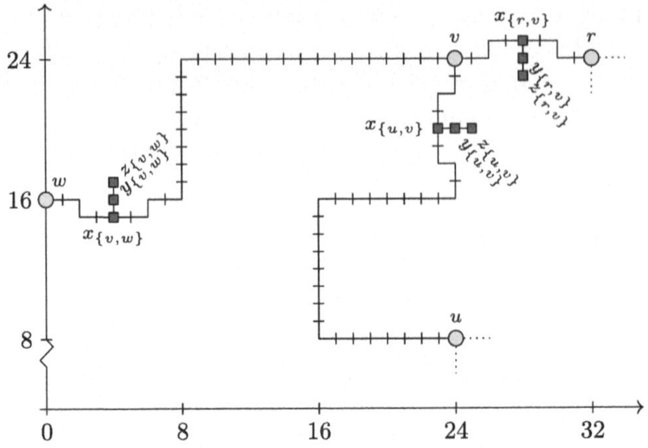

Fig. 2. Up to the detours (visible as lengthy U-turns) and the paths between the quadratic vertices, the drawing shows an initial SRP embedding already scaled by a factor of 8. The Euclidean distance between u and v, e.g., is assumed to have increased eightfold from initially 2 to eventually 16. The splitting vertices x_e, $e \in E$, lie—with respect to the associated SRP—at geodesic distance 5 from an endpoint.

by 4 when this number renounces counting the embedded endpoints $x_{\{v,w\}}$ and v. The same is true for the SRP connecting $x_{\{v,w\}}$ and w. On these $x_{\{v,w\}}$-incident SRPs let us consider these intermediate $\mathbb{Z} \times \mathbb{Z}$ points as embedded fresh subdivisors of the combinatorial edges of G; call this new graph $\tilde{G}$. In particular, as they lie at Manhattan distance 1, no combinatorial subdivision of the edges $\{x_e, y_e\}, \{y_e, z_e\}$ occurs. $\tilde{G}$ is a (subcubic planar) induced partial grid and has the form of the transformed graphs from Lemma 3 with

$$4 \leq \min\{L_{\{v,x_e\}}, L_{\{w,x_e\}}\} \equiv \max\{L_{\{v,x_e\}}, L_{\{w,x_e\}}\} \equiv 0 \pmod 4,$$
$$0 = L_{\{x_e,y_e\}} = L_{\{y_e,z_e\}} \equiv 0 \pmod 4,$$

for all $e = \{v, w\} \in E$. Thus, by Lemma 3, the existence of a feasible f for PosInflDom(G, k) is equivalent to the existence of a feasible f' on the induced partial grid $\tilde{G}$ of weight at most $k + \sum_{e=\{v,w\} \in E} L_{\{v,x_e\}}/2 + L_{\{w,x_e\}}/2$. □

The latter result has implications for the class of unit disk graphs on which domination-type problems are often considered [5].

Corollary 2. *The problem PosInflDom(G, k) remains NP-complete on unit disk graphs.*

5 Bounds on Complete Grids

This section derives lower and upper bounds on complete grids for γ_{PID} and $\gamma_{\times 2,\mathrm{t}}$. For this graph class the requirements of $\gamma_{\times 2,\mathrm{t}}$ are strictly more constrained than those of γ_{PID}. In fact, tighter lower bounds for $\gamma_{\times 2,\mathrm{t}}$ will be found.

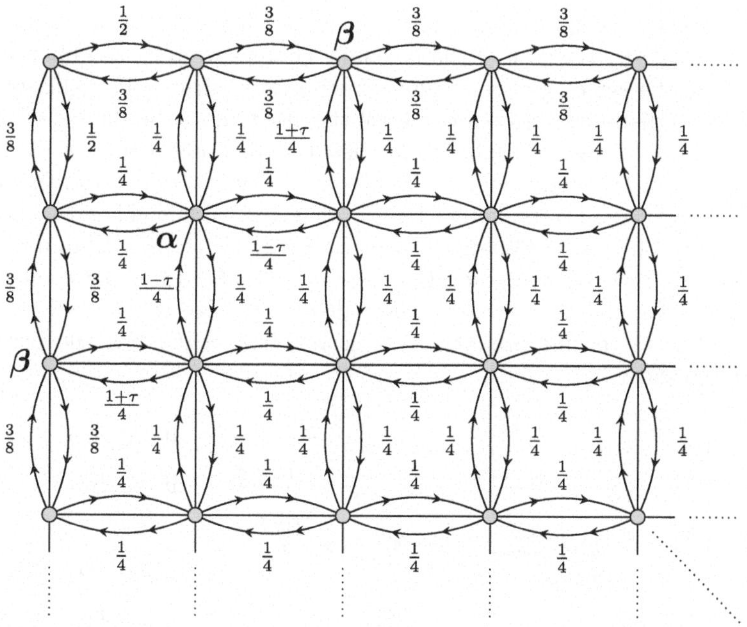

Fig. 3. The top-left area of an $m \times n$ grid subject to the discharging process. Coefficients attached to the arrows indicate the proportion of transferred charge. Vertices in the set A or B from the proof of Theorem 6 are marked by α or β, respectively. Exemplarily assuming that v is the upper β-marked vertex, the situation $f(v) = 1$ implies that the two neighbors of v on the grid's upper boundary receive additional charge of 3/8 whereas the interior neighbor just 1/4; if $f(v) = 0$, a zero-charge transfer to neighbors applies.

Theorem 5. *For $m, n \geq 6$ we have*

$$\frac{mn}{2} + \frac{m+n}{4} - \frac{3}{2} \leq \gamma_{\mathrm{PID}}(P_m \Box P_n) \leq \frac{mn}{2} + \frac{m+n}{2} + \frac{3 - |\operatorname{mod}(n,4) - 1|}{2}. \quad (6)$$

Proof. As any double total dominating function on grids is automatically positive influence dominating, the upper bound of [15] for $\gamma_{\times 2,\mathrm{t}}$ is valid for γ_{PID}, too; the latter is stated as the right-hand side of (6). For $\gamma_{\times 2,\mathrm{t}}$ it constitutes the currently tightest upper bound with exception of the constellations $(\operatorname{mod}(m,4), \operatorname{mod}(n,4)) \in \{(0,0),(1,1)\}$, where it worsens the older bound [3] by 1.

Let us prove the lower bound by using the so-called *discharging method*, arguably best known due to the proof of Appel and Haken [2] of the Four Color Theorem. We transform a PID-feasible $f : [m] \times [n] \to \{0,1\}$ to a function $g : [m] \times [n] \to \mathbb{Q}$ in a weight-preserving manner, i.e., $\sum_{v \in [m] \times [n]} f(v) = \sum_{v \in [m] \times [n]} g(v)$ and then prove the bound for g: To obtain g—simultaneously for all vertices—we entirely redistribute the f-value of each vertex, according to a specific convex combination, among its neighbors. Adopting, conversely, a passive perspective, this means that the initial "charge" $f(v)$ of a vertex v is entirely replaced by the sum of the incoming charges from its neighborhood.

Let A be the set of vertices having precisely two degree-4 neighbors (lying at Manhattan distance 2 from a corner of the grid) and B be the set of vertices having only one degree-4 neighbor and being at distance 2 from a corner; for both see Fig. 3. In the following, we directly state the function g resulting from a tailored discharging process sketched in Fig. 3. Apart from fetching it from Fig. 3, more formally, the proportionality-scalar of the charge-transfer from a neighbor w to v can be read off the respective case for v in (7) from the coefficient of the unique summand associated to the index w. Denote by $D_t := \{v \in [m] \times [n] : \deg(v) = t\}$ specific preimages of the function $\deg(\cdot)$. For the parameter $\tau \in \{0, 1\}$ consider

$$g^{\tau}(v) := \begin{cases} \sum_{w \in N(v) \cap D_3} \frac{f(w)}{4} + \sum_{w \in N(v) \cap D_4} \frac{(1-\tau) f(w)}{4} & \text{if } v \in A, \\ \sum_{w \in N(v) \cap D_3} \frac{3f(w)}{8} + \sum_{w \in N(v) \cap D_4} \frac{(1+\tau) f(w)}{4} & \text{if } v \in B, \\ \sum_{w \in N(v) \cap D_2} \frac{f(w)}{2} + \sum_{w \in N(v) \cap D_3} \frac{3f(w)}{8} + & \\ \quad \sum_{w \in N(v) \cap D_4} \frac{f(w)}{4} & \text{if } v \in (D_2 \cup D_3) \setminus B, \\ \sum_{w \in N(v) \cap D_4} \frac{f(w)}{4} & \text{if } v \in D_4 \setminus A. \end{cases} \tag{7}$$

In the current proof, we are interested in the discharging process associated to $g := g^0$, i.e., $\tau = 0$ (the choice $\tau = 1$ concerns the proof of the later stated Theorem 6). As f is a PID-function, we have

$$\begin{aligned} \sum_{v \in [m] \times [n]} g(v) &= \sum_{v \in D_4} g(v) + \sum_{v \in D_3} g(v) + \sum_{v \in D_2} g(v) \geq \frac{1}{2} |D_4| + \frac{5}{8} |D_3| + \frac{3}{8} |D_2| \\ &= \frac{mn}{2} + \frac{m+n}{4} - \frac{3}{2}. \end{aligned}$$

Here, the three sums have been estimated from below one by one with the following justification: By feasibility of f, every $v \in D_4$ has at least two neighbors with an f-value of 1; this simply allows to conclude that $g(v) \geq 1/4 + 1/4$, for $v \in D_4$. The estimate for each vertex in D_3 holds due to the fact, that again two neighbors with an f-value of 1 must exist leading in the lowest possible case to an inflow of charge $1/4 + 3/8 = 5/8$. The charge inflow for a corner must be at least $3/8$.

Repeating the proof of Theorem 5—this time using $g := g^1$ defined as in (7)—we almost verbatim obtain the following result.

Theorem 6. *For $m, n \geq 6$ we have*

$$\frac{mn}{2} + \frac{m+n}{4} + 1 \leq \gamma_{\times 2,t}(P_m \Box P_n) \leq \frac{mn}{2} + \frac{m+n}{2} + \frac{3 - |\mathrm{mod}(n,4) - 1|}{2}. \tag{8}$$

Despite the simplicity of the discharging approach, the bound (8) on $\gamma_{\times 2,t}$ slightly strengthens—the difference is precisely 1—the lower bound given in [15]. However, the more striking insight lies here in the possibility to give a considerably shorter proof for the lower bound than in the latter work.

6 Conclusion

We have shown that PosInflDom(G, k) and DblTotalDom(G, k) remain NP-complete problems even when restricted to planar subcubic graphs. The employed reductions in Sect. 3 lose certain structural properties raising two open questions: Is it possible to determine a small k—perhaps even $k = 3$—for which the hardness result of Theorem 2 still applies for k-subregular graphs in the setting of 2-*connected* planarity? Can we adapt the reductions to work for 3-regular instead of 3-subregular graphs?

Using a geometry-accented argument, we observed that NP-completeness for PID is inherited by the class of induced partial grids. Concerning the complete $m \times n$ grids we showed how to derive a lower bound for γ_{PID} by a short argument, which applied to the setting of $\gamma_{\times 2,t}$ slightly tightens the known lower bound.

Acknowledgments. We thank Johannes Varga for fruitful exchange. This work is partially supported by the program VGSCO [10.55776/W1260-N35] of the Austrian Science Fund (FWF).

Disclosure of Interests. The authors have no competing interests.

References

1. Alon, N.: Transmitting in the n-dimensional cube. Discret. Appl. Math. **37**, 9–11 (1992)
2. Appel, K., Haken, W.: Every planar map is four colorable. I: discharging. Illinois J. Math. **21**, 429–490 (1977)
3. Bermudo, S., Jalemskaya, D.L., Sigarreta, J.M.: Total 2-domination in grid graphs. Utilitas Math. **110**, 151–173 (2019)
4. Cicalese, F., Milanič, M., Vaccaro, U.: Hardness, approximability, and exact algorithms for vector domination and total vector domination in graphs. In: Owe, O., Steffen, M., Telle, J.A. (eds.) FCT 2011. LNCS, vol. 6914, pp. 288–297. Springer, Heidelberg (2011). https://doi.org/10.1007/978-3-642-22953-4_25
5. Clark, B.N., Colbourn, C.J., Johnson, D.S.: Unit disk graphs. Discret. Math. **86**(1–3), 165–177 (1990)
6. Damaschke, P.: Minus domination in small-degree graphs. Discret. Appl. Math. **108**(1–2), 53–64 (2001)
7. Dunbar, J., Hedetniemi, S., Henning, M., Slater, P.: Signed domination in graphs. In: Alavi, Y., Schwenk, A. (eds.) Graph Theory, Combinatorics, and Algorithms—Proceedings of the Seventh Quadrennial International Conference on the Theory and Applications of Graphs, pp. 311–322. Wiley, New York (1995)
8. Dunbar, J.E., Hoffman, D.G., Laskar, R.C., Markus, L.R.: α-domination. Discret. Math. **211**(1–3), 11–26 (2000)

9. Fernau, H., Rodríguez-Velázquez, J.A.: A survey on alliances and related parameters in graphs. Electron. J. Graph Theory Appl. **2**(1), 70–86 (2014)
10. Gonçalves, D., Pinlou, A., Rao, M., Thomassé, S.: The domination number of grids. SIAM J. Discret. Math. **25**(3), 1443–1453 (2011)
11. Harant, J., Pruchnewski, A., Voigt, M.: On dominating sets and independent sets of graphs. Comb. Probab. Comput. **8**(6), 547–553 (1999)
12. Ishii, T., Ono, H., Uno, Y.: (Total) vector domination for graphs with bounded branchwidth. In: Pardo, A., Viola, A. (eds.) LATIN 2014. LNCS, vol. 8392, pp. 238–249. Springer, Heidelberg (2014). https://doi.org/10.1007/978-3-642-54423-1_21
13. Kempe, D., Kleinberg, J., Tardos, É.: Maximizing the spread of influence through a social network. In: Proceedings of the 9th ACM SIGKDD International Conference on Knowledge Discovery and Data Mining, pp. 137–146. Association for Computing Machinery (2003)
14. Lan, J.K., Chang, G.J.: On the algorithmic complexity of k-tuple total domination. Discret. Appl. Math. **174**, 81–91 (2014)
15. Li, L., Yue, J., Zhang, X.: Double total domination number of Cartesian product of paths. AIMS Math. **8**(4), 9506–9519 (2023)
16. Mohar, B.: Face covers and the genus problem for apex graphs. J. Combin. Theory Ser. B **82**(1), 102–117 (2001)
17. Peleg, D.: Local majorities, coalitions and monopolies in graphs: a review. Theoret. Comput. Sci. **282**(2), 231–257 (2002)
18. Pradhan, D.: Algorithmic aspects of k-tuple total domination in graphs. Inf. Process. Lett. **112**(21), 816–822 (2012)
19. Raman, V., Saurabh, S., Srihari, S.: Parameterized algorithms for generalized domination. In: Yang, B., Du, D.-Z., Wang, C.A. (eds.) COCOA 2008. LNCS, vol. 5165, pp. 116–126. Springer, Heidelberg (2008). https://doi.org/10.1007/978-3-540-85097-7_11
20. Rao, M., Talon, A.: The 2-domination and Roman domination numbers of grid graphs. Discrete Math. Theor. Comput. Sci. **21**(1) (2019)
21. Valiant, L.G.: Universality considerations in VLSI circuits. IEEE Trans. Comput. **100**(2), 135–140 (1981)
22. Wang, F., Camacho, E., Xu, K.: Positive influence dominating set in online social networks. In: Du, D.-Z., Hu, X., Pardalos, P.M. (eds.) COCOA 2009. LNCS, vol. 5573, pp. 313–321. Springer, Heidelberg (2009). https://doi.org/10.1007/978-3-642-02026-1_29
23. Wang, F., et al.: On positive influence dominating sets in social networks. Theoret. Comput. Sci. **412**(3), 265–269 (2011)
24. Wang, H., Wang, N., Zhang, Y., Xiao, J., Sun, Y.: Positive influence domination in graphs. Bull. Malay. Math. Sci. Soc. **45**(5), 1975–1985 (2022)
25. Zheng, Y., Wang, J., Feng, Q.: Kernelization and lower bounds of the signed domination problem. In: Fellows, M., Tan, X., Zhu, B. (eds.) AAIM/FAW -2013. LNCS, vol. 7924, pp. 261–271. Springer, Heidelberg (2013). https://doi.org/10.1007/978-3-642-38756-2_27

On the Parameterized Complexity of Cosecure Domination

D. Karthika[1(✉)], R. Muthucumaraswamy[1], V. P. Abidha[2], Pradeesha Ashok[3], Sriram Bhyravarapu[4], Sayani Das[5], Saket Saurabh[6], Ayush Sawlani[7], and Vikash Tripathi[8]

[1] Sri Venkateswara College of Engineering, Sriperumbudur, Kanchipuram, India
{2022pm0001,msamy}@svce.ac.in

[2] National Institute of Technology, Calicut, Kattangal, India
abidhavp@nitc.ac.in

[3] International Institute of Information Technology - Bangalore, Bangalore, India
pradeesha@iiitb.ac.in

[4] Indian Institute of Technology Guwahati, Guwahati, India
sriramb@iitg.ac.in

[5] Mahindra University, Hyderabad, India
sayani.das@mahindrauniversity.edu.in

[6] The Institute of Mathematical Sciences, Chennai, India
saket@imsc.res.in

[7] ThoughtSpot, Bangalore, India
ayush.sawlani@thoughtspot.com

[8] Indian Institute of Technology Mandi, Mandi, Himachal Pradesh, India
vikash@iitmandi.ac.in

Abstract. For a graph G, a set $S \subseteq V(G)$ is a cosecure dominating set if S is a dominating set of G and for each $v \in S$ there exists a vertex $v' \in N(v) \setminus S$ such that $(S \cup \{v'\}) \setminus \{v\}$ is also a dominating set. Given a graph G, Cosecure Domination (Co-SDS, in short) asks to find a minimum cosecure dominating set for G. The problem is W[2]-hard parameterized by the solution size. Kusum and Pandey (Theor. Comput. Sci., 2024) showed that Co-SDS is linear time solvable on graphs of bounded treewidth and cliquewdith by using monadic second order logic. In this paper, we present explicit algorithms for the problem with respect to structural parameters like modular width, treewidth, distance to cluster, distance to co-cluster, and vertex cover number. Here, we want to emphasize that the running times of all our FPT algorithms are single exponential or only slightly super-exponential. We further show that Co-SDS when parameterized by vertex cover number, does not admit a polynomial kernel unless coNP $\subseteq$ NP/poly.

Keywords: Domination · Cosecure Domination · Fixed Parameter Tractability · Modular Width · Tree Width · Kernelization

The fourth author acknowledges the support from Anusandhan National Research Foundation under the grant MTR/2023/001078, and the ninth author acknowledges the support from IIT Mandi Seed Grant IITM/SG/VT/142.

A. Jeż and J. Otop (Eds.): FCT 2025, LNCS 16106, pp. 281–294, 2026.
https://doi.org/10.1007/978-3-032-04700-7_21

1 Introduction

The DOMINATION problem and its variants hold significant theoretical and practical interest in graph theory. For a graph $G = (V, E)$, a set $S \subseteq V(G)$ is a *dominating set* of G if each vertex $v \in V(G) \setminus S$ has a neighbor in S. Given a graph G and an integer k, DOMINATION asks if there exists a dominating set of size at most k. This problem has been extensively studied in the literature, see for instance, [1–3]. To make it more applicable and adaptive to various real-world problems, several variants of the dominating set problem have been defined and extensively studied, including Connected Dominating Set [1,2,4], Total Dominating Set [5–7], Independent Dominated Set [8,9], Roman Domination [10,11], and many more [12,13]. These variants explore different aspects of dominating sets and their applications in diverse contexts. In this paper, we focus on the algorithmic study of one such variant of the domination problem, defined below.

Definition 1 (Cosecure dominating set [14,15]**).** *A set $S \subseteq V(G)$ is a* cosecure dominating set *of G if S is a dominating set of G and for each $v \in S$, there exists a vertex $w \in N(v) \setminus S$ such that $(S \cup \{w\}) \setminus \{v\}$ is also a dominating set. We call the vertex w a* covertex *of v.*

Given a graph G and an integer k, COSECURE DOMINATION (CO-SDS, in short) asks whether there exists a cosecure dominating set of G of size at most k. The motivation for the problem comes from security in various networks. Consider, for example, a communication network where nodes represent critical systems or servers, and edges denote direct communication links essential for maintaining connectivity and information flow. The entire network can be effectively "secured" by installing specialized security features at a selected subset of nodes. These features are designed to protect not only the node in which they are installed but also the nodes that are directly connected to it. In this context, a dominating set ensures comprehensive network coverage and the establishment of a co-secure dominating set provides backup flexibility to mitigate such threats.

COSECURE DOMINATION was introduced by Arumugam et al. [14], in 2014. The problem was shown to be NP-complete for several graph classes including bipartite graphs, chordal graphs, planar graphs, chordal bipartite graphs, and circle graphs [14,16]. It was shown that the problem is NP-complete for doubly chordal graphs, star-convex bipartite graphs, and comb-convex bipartite graphs [17], grid graphs, supergrid graphs, unit disk, and unit square graphs [18]. In contrast, efficient polynomial time algorithms are known for the problem on co-graphs, chain graphs [16,19], proper interval graphs [15], bounded tree-width graphs, and bounded clique-width graphs [17]. The complexity of the problem is well explored in the realm of approximation as well [18–20].

While significant progress has been made in understanding the problem from the perspective of classical complexity, approximability, and restricted graph classes, the problem has not been explored much from the viewpoint of parameterized complexity. In this work, we study the CO-SDS from the perspective of parameterized algorithms. We now present the results of the paper.

Our Results: It was shown that COSECURE DOMINATION is W[2]-hard parameterized by the solution size [19]. Kusum and Pandey [17] showed that the problem is fixed parameter tractable (FPT) when parameterized by treewidth or cliquewidth [17] by formulating the problem in monadic second order logic. Since the problem can be expressed in MSO_1, the problem is FPT parameterized by the cliquewidth of the input graph. Since cliquewidth generalizes (see Fig. 1) the parameters modular width, treewidth, distance to cluster and distance to co-cluster numbers, etc., the problem is FPT when parameterized by any of the above parameters. In this paper, we present efficient parameterized algorithms with respect to these parameters.

In Sect. 3, we obtain an algorithm running in time $3^{\mathsf{mw}} n^{\mathcal{O}(1)}$ where mw is the modular width of the input graph. The algorithm is non-trivial and we present it in full in this paper. We also consider the problem on graphs of bounded treewidth and give an FPT algorithm parameterized by treewidth. We then consider dense graphs that have unbounded treewidth. In particular, graphs with bounded distance to cluster and bounded distance to co-cluster and provide efficient parameterized algorithms. Due to space constraints, we give the algorithms full version of the paper. The results are stated in Sect. 4. Lastly, we consider the problem from a kernelization perspective and show that CO-SDS does not admit a polynomial kernel parameterized by the vertex cover number unless coNP $\subseteq$ NP/poly, which is presented in Sect. 5.

Since distance to cluster generalizes vertex cover number, the problem does not admit a polynomial kernel parameterized by distance to cluster. It will be interesting to obtain a polynomial kernel for the distance to clique parameter, which is a restricted case of distance to cluster parameter. DOMINATION admits $4^{\mathsf{cw}} n^{\mathcal{O}(1)}$, $2^{\mathsf{mw}} n^{\mathcal{O}(1)}$, $3^{\mathsf{dc}} n^{\mathcal{O}(1)}$, and $3^{\tau} n^{\mathcal{O}(1)}$ time algorithms, where τ, cw and dc denote the treewidth, cliquewidth and distance to cluster of the input graph [21,22]. Lower bounds for the parameters have also been obtained in the above papers. It will be interesting to obtain efficient algorithms for the parameters discussed in the paper. Due to space constraints, the proofs of the results marked (★) are presented in the extended version of the paper.

2 Preliminaries

Let $G = (V, E)$ be a simple connected graph with $|V| = n$ and $|E| = m$. For $v \in V(G)$, $N_G(v)$ (respectively, $N_G[v]$) denotes the open (respectively, closed) neighborhood of v in G. The subscript is dropped when it is clear from the context. For $k \in \mathbb{N}$, $[k] = \{1, 2, \ldots, k\}$. A vertex $v \in V$ is called *cosecurely dominated* if either v contains two solution vertices as neighbors, or v has a solution neighbor w, w has a covertex w' such that $w' \in N(v)$. Consider sets A, B, and C such that $A \subseteq B$. If $f : B \mapsto C$ is a function then the *restriction* of f to A, denoted as $f|_A$, is a function $f|_A : A \mapsto C$ defined as $f|_A(x) = f(x)$ for all $x \in A$. For the basic terminology of parameterized complexity, and graph theory, we refer the reader to [23] and [24].

Modular Decomposition: A *modular decomposition tree* of a graph G represents the construction of G in a hierarchical manner where every node of the tree corresponds to one of the following operations.

(O1) An isolated vertex; The corresponding nodes in the modular decomposition tree are the leaf nodes.
(O2) The *disjoint union* of two graphs G_1 and G_2, denoted by $G_1 \otimes G_2$, is the graph with vertex set $V(G_1) \cup V(G_2)$ and edge set $E(G_1) \cup E(G_2)$.
(O3) The *complete join* of two graphs G_1 and G_2, denoted by $G_1 \oplus G_2$, is the graph with vertex set $V(G_1) \cup V(G_2)$ and edge set $E(G_1) \cup E(G_2) \cup \{\{v, u\} : v \in V(G_1) \text{ and } u \in V(G_2)\}$.
(O4) The *substitution operation* with respect to some graph G' with vertices $v_1, \ldots, v_w$, and using graphs $G_1, \ldots, G_w$ constructs the graph with vertex set $\bigcup_{1 \leq i \leq w} V(G_i)$ and edge set $\bigcup_{1 \leq i \leq w} E(G_i) \cup \{(u, v) : u \in V(G_i), v \in V(G_j), v_i v_j \in E(G') \text{ and } i \neq j\}$. Every graph G_i, $1 \leq i \leq w$ is called a *module* in the resulting graph.

Every node in a modular decomposition tree of G corresponds to a subgraph of G and the root corresponds to G itself. The number of labels of an operation is the number of graphs on which the operation is performed. The *width* of a modular decomposition is the maximum number of labels used over all the operations used in the construction of G. The *modular width* of G, denoted by mw, is the minimum width over all of its modular decompositions [25].

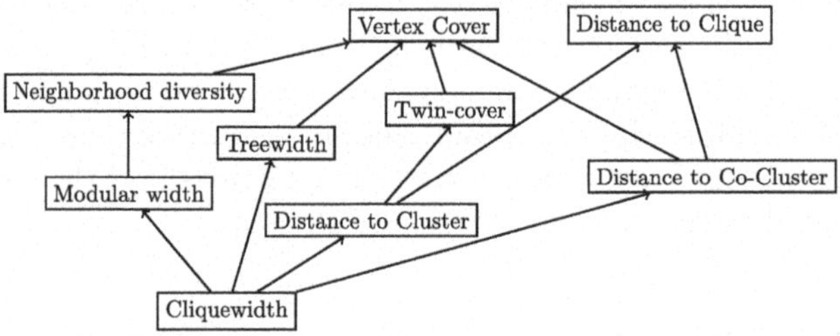

Fig. 1. Hierarchy of parameters. An arrow from parameter a to parameter b indicates that $a \leq f(b)$ for all graphs, where f is a computable function.

3 Modular Width

In this section, we present an explicit algorithm for Cosecure Domination parameterized by modular width. We formally state the result below.

Theorem 1. Cosecure Domination *can be solved in time $3^{\mathsf{mw}} n^{\mathcal{O}(1)}$, where* mw *is the modular width of the input graph on n vertices.*

Let (G, k) be an instance of the problem and $\mathcal{T}$ be a modular decomposition tree of G of width mw. Let r be the root node of $\mathcal{T}$. Since G is connected, the operation performed at r is either complete join (O3) or substitution (O4). Since a complete join can be represented by a substitution operation, without loss of generality, we assume that the operation at r is a substitution operation. Let $\mathcal{V} = \{V_1, V_2, \ldots, V_q\}$ be the modules corresponding to this operation at r, where $2 \leq q \leq \mathsf{mw}$. Notice that each vertex of G is contained in exactly one of the modules of $\mathcal{V}$. We say a module V_i is *adjacent* to V_j, where $i \neq j$, if each vertex in V_i is adjacent to every vertex in V_j. The set $\mathsf{adj}(V_j)$ contains the set of modules adjacent to V_j, for each j. Let us denote the closed neighborhood of V_j by $\mathsf{adj}^+(V_j)$ that contains V_j and the modules adjacent to V_j.

Let S be a hypothetical solution of (G, k). We now present an overview of our algorithm.

- We partition the modules of $\mathcal{V}$ based on the number of solution vertices they contain. In particular, whether they contain 0, 1, or at least two solution vertices. Based on this, we identify a module $V_i \in \mathcal{V}$ is "critical" if it contains a non-solution vertex say $w \in V_i$ such that $|N(w) \cap S| = 1$. We call the module that contains the vertex $N(w) \cap S$ as the "critical module" of V_i.
- To identify the critical modules, we consider the domination numbers of the modules and the number of solution vertices in each of the modules. Further, we also identify the "placeholder" (a module) of the covertex for each solution vertex. Towards this, we run the $2^{\mathsf{mw}} n^{O(1)}$ algorithm for DOMINATION [26] apriori to obtain the domination numbers of each of the modules of $\mathcal{V}$.
- The rest of the algorithm is to check whether there exists a cosecure dominating set that satisfies the solution vertices in each module and their placeholders.

We introduce some definitions that are required for our algorithm. Let $f : \mathcal{V} \to \{0, 1, 2\}$ be a *solution function* that captures the number of solution vertices in each module of $\mathcal{V}$. We say a cosecure dominating set $S \subseteq V(G)$ *respects* f if for each $V_i \in \mathcal{V}$,

$$f(V_i) = \begin{cases} 0 & \text{if } |S \cap V_i| = 0, \\ 1 & \text{if } |S \cap V_i| = 1, \\ 2 & \text{otherwise.} \end{cases}$$

The *domination number* of a graph G is the size of a smallest sized dominating set of G. For each $V_i \in \mathcal{V}$, let $d(V_i)$ denote the domination number of $G[V_i]$. For each i, let $n_i = \sum_{j : V_j \in \mathsf{adj}(V_i)} f(V_j)$. Next, we define a critical module. Intuitively, a module is critical V_j if it contains exactly one solution vertex and there exists a neighboring module V_i such that each vertex in V_i has exactly one solution neighbor outside V_i. Therefore, the covertex of the solution vertex in V_j may be important to V_i. We give the definition of critical module and an observation.

Definition 2 (critical module). *We say a module V_j is* critical *if there exists some module V_i such that $V_i \in \mathsf{adj}(V_j)$, $f(V_j) = 1$, $n_i = 1$ and $f(V_i) < d(V_i)$. We also say V_j is critical for V_i and denote $\mathsf{crit}(V_j) = \{V_i \mid V_j \text{ is critical for } V_i\}$.*

Observation 1. *For each module V_i, we have $|\{V_j \mid V_i \in \mathsf{crit}(V_j)\}| \leq 1$.*

For each critical module V_j, we identify a module in the closed neighborhood of V_j (V_j and the modules adjacent to V_j) which we call the *cosecure module* of V_j that contains the covertex of the solution vertex in V_j. In Sect. 3.1, we give a procedure to compute a cosecure module for each critical module V_i. In Sect. 3.2, we perform some checks to ignore "invalid" functions f for which there is no cosecure dominating set that respects f. In Sect. 3.3, we construct a minimum sized set that respects f. As a final step, we show that the constructed set is a cosecure dominating set in Lemma 3.

3.1 Identifying the Covertex Module for Each Critical Module

Consider a critical module V_j and let $S \cap V_j = \{w\}$. Let $\mathsf{crit}^+(V_j) = \mathsf{crit}(V_j) \cup \{V_j\}$. We obtain a partition of $\mathsf{crit}(V_j)$ as follows: $A_j = \{V_i \in \mathsf{crit}(V_j) \mid d(V_i) - f(V_i) = 1\}$ and $B_j = \{V_i \in \mathsf{crit}(V_j) \mid d(V_i) - f(V_i) \geq 2\}$. For any module $V_p \in \mathsf{adj}(V_j) \setminus \mathsf{crit}(V_j)$, either n_p is 0 or at least 2, or $d(V_p) \leq f(V_p) \leq 2$. For the former case, we have $n_p = 2$ as $f(V_j) = 1$. For the latter case, since $d(V_p) \geq 1$, we have $f(V_p) \in \{1, 2\}$. Thus, V_p has at least $d(V_p)$ and at most two vertices to cosecurely dominate V_p, and therefore the number of solution vertices in V_p is fixed by the function f. Therefore we do not include V_p in this process.

In the following, we compute the "best" choice for the cosecure module from $\mathsf{adj}^+(V_j)$, in terms of the cost associated with the covertex of the solution vertex in V_j being in $V_i \in \mathsf{adj}(V_j)$. Before moving to the discussion, we would like to recall that for each $V_p \in \mathsf{crit}(V_j)$ we have $n_p = 1$ and all other modules $V_{p'} \in \mathsf{adj}(V_p) \setminus \{V_j\}$ has $f(V_{p'}) = 0$. Let $Y_j = A_j \cup B_j$. For a critical module V_j and a module $V_i \in \mathsf{adj}^+(V_j)$, we formally define $\mathsf{cost}(j, i)$ below.

For each $V_i \in \mathsf{adj}^+(V_j)$ we compute $\mathsf{cost}(j, i)$ which stores the size of a smallest sized vertex set D from the vertices in $Y_j \cup \{V_j, V_i\}$ such that

(a) for each $V_p \in Y_j \setminus \{V_j, V_i\}$, if $f(V_p) = 2$, then $|D \cap V_p| \geq 2$, else if $f(V_p) = c$ where $c \in \{0, 1\}$, then $|D \cap V_p| = c$.
(b) if $i = j$, then $|D \cap V_j| = 1$. Further, if $n_j = 0$ then D dominates the vertices in the modules $Y_j \cup \{V_j\}$, otherwise D dominates the vertices in the modules Y_j.
(c) else if $i \neq j$, then $|D \cap V_j| = 0$. If $V_i \notin Y_j$ then $|D \cap V_i| = 1$, else $|D \cap V_i| = d(V_i)$. Further, D dominates the vertices in the modules Y_j.

Computing $\mathsf{cost}(j, i)$: Consider a module $V_i \in \mathsf{adj}(V_j)$. Let $Y_j^i = \{V_p \in Y_j \mid V_p \in \mathsf{adj}(V_i)\}$.

- If $V_i \in \mathsf{adj}^+(V_j) \setminus Y_j$
If $i \neq j$, then from Definition 2, either $n_i \geq 2$ or $d(V_i) \leq f(V_i) \leq 2$. In both the cases, any optimum solution contains at most two vertices from V_i. Thus

depending on $f(V_i)$, we check whether there exists at least $f(V_i)+1$ many vertices in V_i. If not, we set $\mathsf{cost}(j,i)=\infty$. Notice that the extra vertex is a placeholder for the covertex of the solution vertex in V_j.
When $i=j$, the basic requirement is that $|V_j|>1$ and $G[V_j]$ is not an independent set. Otherwise we set $\mathsf{cost}(j,j)=\infty$. We have an additional check based on the value of n_j. If $n_j=0$, there is an additional requirement that there exist two vertices $w_1,w_2\in V_j$ such that $N[w_1]=N[w_2]=V_j$. Otherwise, we set $\mathsf{cost}(j,j)=\infty$. We compute the cost as follows:

$$\mathsf{cost}(j,i)=1+\sum_{V_p\in(A_j\cup B_j)\setminus Y_j^i} d(V_p)+\sum_{V_p\in Y_j^i} f(V_p).$$

– Else if $V_i\in Y_j$.

$$\mathsf{cost}(j,i)=d(V_i)+\sum_{V_p\in(A_j\cup B_j)\setminus(Y_j^i\cup\{V_i\})} d(V_p)+\sum_{V_p\in Y_j^i\setminus\{V_i\}} f(V_p).$$

Once we compute the values of $\mathsf{cost}(j,i)$ for each $V_i\in\mathsf{adj}^+(V_j)$, we are ready to identify the covertex module, denoted by V_{j^*}, for V_j. We assign $V_{j^*}\in\mathsf{adj}^+(V_j)$ to be the module such that $\mathsf{cost}(j,j^*)\le\mathsf{cost}(j,p)$ for all $V_p\in\mathsf{adj}^+(V_j)$. If there are multiple such j^* then arbitrarily choose one of them. If $\mathsf{cost}(j,j^*)=\infty$ then we conclude that the guess f is invalid and we move to the next guess. We would like to highlight that due to Observation 1, each V_i features in the cost computation corresponding to V_j.

Claim 1 (★). *The computed value* $\mathsf{cost}(j,i)$ *is correct.*

We have the following lemma and the proof follows from the above discussion.

Lemma 1. *Let S be a cosecure dominating set that respects f and let V_j be a critical module. Then there exists a cosecure dominating set S' of size $|S|$ and the covertex module of the vertex in V_j is the module V_{j^*}.*

3.2 Invalid Functions

For the rest of the proof, we work with a fixed solution function f. Before constructing a cosecure dominating set, we perform some checks to find out if f cannot be realized to a cosecure dominating set.

Definition 3 (detecting invalid solution functions). *We say a solution function f is* invalid *if one of the following conditions holds:*

- **(d0)** *there is a critical module V_i and $\mathsf{cost}(i,i^*)=\infty$ where V_{i^*} is the covertex module for V_i.*
- **(d1)** *there is a module V_i such that $f(V_i)<d(V_i)$, $f(V_i)\in\{0,1\}$ and $n_i=0$,*
- **(d2)** *there is a module V_i such that $\mathsf{crit}(V_i)=\emptyset$, $f(V_i)=1$, $n_i=1$, $V_i\in\mathsf{crit}(V_p)$ for some p, cosecure module of V_p is V_i, $G[V_i]$ is an independent set and $|N(w)|=1$, for all $w\in V_i$.*

(d3) *there is a module* V_i *such that* $f(V_i) = 2$, $n_i \geq 2$, $|N(w) \setminus V_i| \leq n_i$, *for some* $w \in V_i$, *and there is no set* $Z_i = \{w_1, w_2\} \subseteq V_i$ *such that the following hold:* $N(w_1) \cap (V_i \setminus \{w_2\}) \neq \emptyset$ *and* $N(w_2) \cap (V_i \setminus \{w_1\}) \neq \emptyset$.

(d4) *there is a module* V_i *such that* $f(V_i) = 2$, $n_i = 1$, $V_i \in \mathsf{crit}(V_p)$ *for some* p, *the cosecure module for* V_p *is* V_ℓ *where* $\ell \neq i$, $V_\ell \notin \mathsf{adj}(V_i)$, *there exists a degree-zero vertex in* $G[V_i]$ *and* $|N(w) \setminus V_i| \leq n_i$ *for all* $w \in V_i$.

(d5) *there is a module* V_i *such that* $f(V_i) = 2$, $n_i = 1$, $V_i \in \mathsf{crit}(V_p)$ *for some* p, *the cosecure module for* V_p *is* V_i, *there exists at least two degree-zero vertices in* $G[V_i]$ *and* $|N(w) \setminus V_i| \leq n_i$ *for all* $w \in V_i$.

(d6) *there is a module* V_i *such that* $f(V_i) = 1$, $n_i = 1$, $\mathsf{crit}(V_i) = \emptyset$, $V_i \notin \mathsf{crit}(V_p)$ *for any* p, $G[V_i]$ *is an independent set, and* $|N(w)| = 1$ *for all* $w \in V_i$.

(d7) *there is a module* V_i *such that* $f(V_i) = 2$, $n_i = 1$, $V_i \notin \mathsf{crit}(V_p)$ *for any* p, $|N(w) \setminus V_i| \leq n_i$, *for all* $w \in V_i$, *and there is no set* $Z_i \subseteq V_i$ *such that the following hold:* $|Z_i| = 2$, Z_i *is a dominating set of* $G[V_i]$ *and for each vertex* $w \in Z_i$ *we have* $N(w) \cap (V_i \setminus Z_i) \neq \emptyset$.

(d8) *there is a module* V_i *such that* $f(V_i) = 1$, $n_i \geq 2$, $\mathsf{crit}(V_i) = \emptyset$, $G[V_i]$ *is an independent set and* $|N(w) \setminus V_i| \leq n_i$, *for all* $w \in V_i$.

Lemma 2 (⋆). *There exists no cosecure dominating set* S *that respects* f *and satisfying any of the conditions (d0)–(d8).*

3.3 Constructing a Cosecure Dominating Set

For each valid solution function f we check whether there exists a cosecure dominating set S that respects f. For each V_i, if $f(V_i) = 0$ then we do not include any vertex from V_i to S. Thus we consider the cases when $f(V_i) = 1$ and $f(V_i) = 2$. We argue based on the following cases.

- $n_i = 0$.
  If $f(V_i) = 1$ then there exists a universal vertex, say w, in $G[V_i]$ because of the definition of f and (d1) is not satisfied. Thus we include w in the set S. We now consider the case when $f(V_i) = 2$. Let $Z_i \subseteq V_i$ be a minimum sized dominating set of $G[V_i]$. We include Z_i to the set S.
- $n_i = 1$ and $f(V_i) = 1$.
  Let $V_j \in \mathsf{adj}(V_i)$ and $f(V_j) = 1$. We consider the case when $\mathsf{crit}(V_i) \neq \emptyset$ and $\mathsf{crit}(V_i) = \emptyset$. For the case when $\mathsf{crit}(V_i) \neq \emptyset$, let the covertex module of V_i be V_{i^*}. We consider the following cases.
  - $\mathsf{crit}(V_i) \neq \emptyset$ and $i^* \neq i$.
    Arbitrarily pick a vertex from V_i and include it in the set S.
  - $\mathsf{crit}(V_i) \neq \emptyset$ and $i^* = i$.
    * There exists a module V_ℓ such that $V_i \in \mathsf{crit}(V_\ell)$.
      This implies that $\ell = j$. Otherwise $n_i \geq 2$, a contradiction to our assumption. If the covertex module of V_j is not V_i, then arbitrarily pick a vertex $w \in V_i$ such that $N(w) \cap V_i \neq \emptyset$ and include it in S. Else the covertex module of V_j is V_i. Let $Z_i \subseteq V_i$ be a minimum sized dominating of $G[V_i]$. Since the covertex module of V_j is V_i, we get that

$|Z_i| = 2$. Let $Z_i = \{w_1, w_2\}$. Since $G[V_i]$ is not an independent set, either $N(w_1) \cap V_i \neq \emptyset$ or $N(w_2) \cap V_i \neq \emptyset$. Without loss of generality, let $N(w_1) \cap V_i \neq \emptyset$. Include w_1 to the set S.

* There exists no module V_ℓ such that $V_i \in \mathsf{crit}(V_\ell)$.
  Then from Definition 2, we have that $f(V_i) \geq d(V_i)$ and thus $d(V_i) = 1$. There exists a universal vertex, say w, in V_i and include w in S.

- $\mathsf{crit}(V_i) = \emptyset$.
  * There exists a module V_ℓ such that $V_i \in \mathsf{crit}(V_\ell)$.
    This implies that $\ell = j$. Otherwise $n_i \geq 2$, a contradiction to our assumption. If the covertex module of V_j is not V_i, then arbitrarily pick a vertex $w \in V_i$ and include it in the set S.
    Else the covertex module of V_j is V_i. Since (d2) is not met, we either have that $G[V_i]$ is not an independent set or $|N(w)| > 2$. Because of the case assumption, we get that $d(V_i) = 2$. Let $Z_i \subseteq V_i$ be a minimum sized dominating set of $G[V_i]$. Let $Z_i = \{w_1, w_2\}$. Notice that if $G[V_i]$ is an independent set then $|V_i| = 2$ as $d(V_i) = 2$. We arbitrarily pick a vertex from Z_i to the set S. Otherwise, either $N(w_1) \cap V_i \neq \emptyset$ or $N(w_2) \cap V_i \neq \emptyset$. Without loss of generality, let $N(w_1) \cap V_i \neq \emptyset$. Then we include w_1 to the set S.
  * There exists no module V_ℓ such that $V_i \in \mathsf{crit}(V_\ell)$.
    Then from Definition 2, we have that $f(V_i) \geq d(V_i)$ and thus $d(V_i) = 1$. There exists a universal vertex, say w, in V_i and include w in S.

– $n_i = 1$ and $f(V_i) = 2$.
Let $V_j \in \mathsf{adj}(V_i)$ and $f(V_j) = 1$. From the case assumption, we have that $\mathsf{crit}(V_i) = \emptyset$. We have the following cases.

- There exists a module V_ℓ such that $V_i \in \mathsf{crit}(V_\ell)$.
  This implies that $\ell = j$. Otherwise $n_i \geq 2$, a contradiction to our assumption. Let us assume that the covertex module of V_j, say V_p, is not V_i. We consider the case when $V_p \notin \mathsf{adj}(V_i)$. Let $Z_i \subseteq V_i$ be a minimum sized dominating set of $G[V_i]$. We include Z_i to the set S. If $V_p \in \mathsf{adj}(V_i)$, arbitrarily pick two vertices from V_i to the set S.
  Now we consider the case when covertex module of V_j is V_i. Let $Z_i \subseteq V_i$ be a minimum sized dominating set of $G[V_i]$. We include $Z_i \setminus \{w\}$ to the set S, where w is an arbitrary vertex from Z_i.
- There exists no module V_ℓ such that $V_i \in \mathsf{crit}(V_\ell)$.
  Then from Definition 2, we have that $f(V_i) \geq d(V_i)$ and thus $d(V_i) \leq 2$. From (d7), we have the following cases:
  * $|N(w) \setminus V_i| > n_i$, for all $w \in V_i$.
    If $d(V_i) = 1$, there exists a universal vertex w in V_i and include w to the set S. Arbitrarily pick another vertex from V_i to the set S.
    Else if $d(V_i) = 2$. Let $Z_i \subseteq V_i$ be a minimum sized dominating of $G[V_i]$. Include the set Z_i to the set S.
  * $|N(w) \setminus V_i| \leq n_i$, for all $w \in V_i$.
    Then there is a set $Z_i \subseteq V_i$ such that $|Z_i| = 2$, Z_i is a dominating set of $G[V_i]$ and for each vertex $w \in Z_i$ we have $N(w) \cap (V_i \setminus Z_i) \neq \emptyset$. Pick the vertices in Z_i to the set S.

- $n_i \geq 2$ and $f(V_i) = 1$.
  Because of the case assumption, there is no module V_ℓ such that $V_i \in \mathsf{crit}(V_\ell)$. We have the following cases.
  - $\mathsf{crit}(V_i) \neq \emptyset$.
    Let the covertex module of V_i be V_{i^*}. We consider the following cases. If $i^* \neq i$, arbitrarily pick a vertex w from V_i and include in the set S. Else if $i^* = i$, then arbitrarily pick a vertex w from V_i and include in the set S. Otherwise, there exists a vertex $w \in V_i$ such that $N(w) \cap V_i \neq \emptyset$. If there are many such vertices, we arbitrarily pick a vertex to S.
  - $\mathsf{crit}(V_i) = \emptyset$.
    From (d8) we have one of the following cases. If $G[V_i]$ is not an independent set, then there exists a vertex $w \in V_i$ such that $N(w) \cap V_i \neq \emptyset$. Include an arbitrary vertex that satisfies this condition to the set S. Otherwise we have $|N(w) \setminus V_i| > n_i$ for all $w \in V_i$. Then arbitrarily pick a vertex from V_i and include to the set S.
- $n_i \geq 2$ and $f(V_i) = 2$.
  Because of the case assumption, there is no module V_ℓ such that $V_i \in \mathsf{crit}(V_\ell)$. Further $\mathsf{crit}(V_i) = \emptyset$. From (d3) we have the following. If $|N(w) \setminus V_i| > n_i$, for all $w \in V_i$ then arbitrarily pick two vertices from V_i and include to the set S. Else if $|N(w) \setminus V_i| \leq n_i$, for all $w \in V_i$, then the number of vertices of degree at least one in $G[V_i]$ is at least three. Then there exist two vertices $w, w' \in V_i$ such that $N(w) \cap (V_i \setminus \{w'\}) \neq \emptyset$ and $N(w') \cap (V_i \setminus \{w\}) \neq \emptyset$. Include w and w' to S.

The construction of S is complete. We show that S is a cosecure dominating set.

Lemma 3 (★). *S is a cosecure dominating set of G.*

We obtain a cosecure dominating set for each valid function f, and return a set with the minimum sized cosecure domination set over all such sets. The correctness follows from Lemma 3 and the above discussion. This completes the proof of Theorem 1.

4 Other Parameterized Algorithms

We give dynamic programming based explicit algorithms for the problem parameterized by treewidth, distance to cluster and distance to co-cluster.

Theorem 2 (★). Cosecure Domination *can be solved in time $\tau^{\mathcal{O}(\tau)} \cdot n^{\mathcal{O}(1)}$ where τ represents the treewidth of the input graph on n vertices.*

Let $G = (V, E)$ be a graph and $\mathcal{T} = (T, \mathcal{X})$ be a nice tree decomposition of G of width τ. Note that, for a graph with treewidth τ, we can compute a nice tree decomposition of width at most 5τ in $2^{\mathcal{O}(\tau)} n^{\mathcal{O}(1)}$ time [23,27]. Below, we present a sketch of the algorithm.

Sketch of the Algorithm: We design a dynamic programming algorithm that solves a set of sub-problems at every node of the tree decomposition. These sub-problems essentially asks for the existence of partial solutions that are of a particular "type" and the type is defined by functions that are part of the sub-problem. There are two functions here: The first is a coloring function on the vertices in the bag, where every color represents how a vertex is dominated by the partial solution and also how a vertex will be dominated by any solution that extends the partial solution. The second function stores information about the position of the covertex, for every vertex in the bag that is also part of the partial solution. More precisely, the coloring function tells whether a vertex is in the partial solution and if not, whether it is dominated $0, 1$ or ≥ 2 times by the partial solution. Moreover, for vertices that are dominated exactly one time, the function further tells whether it is dominated once or more in the final solution. Thus there are at most 5^{τ} possible coloring functions at any node. For every vertex that is part of the partial solution (according to the coloring function), the second function tells whether its covertex is a vertex that has already been forgotten, or if it is yet to be seen or if it is in the bag. In the third case, the function further tells the actual vertex that is its covertex. Thus there are at most τ^{τ} functions of the second type.

The algorithm solves sub-problems using recurrences specific to the type of the node in the tree decomposition. Note that, to answer a subproblem, we need to verify that a vertex's domination status matches that given by the coloring function. Secondly, we need to verify the "covertex property" : Let a vertex u is uniquely dominated by a vertex v and w is assigned as the covertex of v. We need to verify that u and w are adjacent. The covertex property is checked as a pre-processing step at every node. We now give an overview of the recurrences for each type of node. For an **Introduce Node** t, we consider cases based on the color of the introduced vertex, v. If the coloring function specifies that v is in the partial solution, the answer is given by the sub-problems at the child node of t where the domination status of the neighbors of v is appropriately modified. Otherwise, we verify the domination status of v as per the color assigned to it. For a **Forget Node** t, let v be the vertex being forgotten. We need to confirm that the vertex is cosecurely dominated since v cannot be dominated later. Hence, we consider the solutions to sub-problems at the child node, where each sub-problem assigns different colors to v and select the minimum sized partial solution. Finally, for the **Join node**, the solution to a sub-problem can be obtained by combining the solutions to sub-problems at its two chid nodes. Here, at the child nodes, we consider only those sub-problems that are *compatible* with the sub-problem that we are solving. There are at most c^{τ} such pairs of sub-problems, where c is a constant.

Theorem 3 (★). *Given a graph G,* Cosecure Domination *can be solved in* $t^{\mathcal{O}(t)} \cdot n^{\mathcal{O}(1)}$ *time where t is the distance to cluster number of G.*

Theorem 4 (★). *Given a graph G,* Cosecure Domination *can be solved in time* $2^{\mathcal{O}(t^2)} n^{\mathcal{O}(1)}$ *where t is the co-cluster number of G.*

5 No Poly Kernel Parameterized by Vertex Cover

In this section, we show that there is no polynomial kernel for the problem parameterized by vertex cover number. Formally, we state the theorem below.

Theorem 5. Cosecure Domination *parameterized by the vertex cover number does not admit polynomial kernel, unless* $\mathsf{NP} \subseteq \mathsf{coNP/poly}$.

We will give a reduction from Set Cover to Cosecure Domination. Let $(\mathcal{U}, \mathcal{F})$ be an instance of Set Cover, where $\mathcal{U}$ is a universe of n elements and $\mathcal{F}$ is a family of m subsets of $\mathcal{U}$. Let $\mathcal{U} = \{x_1, x_2, \ldots, x_n\}$ and $\mathcal{F} = \{F_1, F_2, \ldots, F_m\}$. We will construct a graph G as follows:

- For each $x_i \in \mathcal{U}$, add a vertex s_i and for each $F_j \in \mathcal{F}$, add a vertex t_j.
- Add $k+8$ vertices w_1, w_2, w_3, w_4, $u_1, u_2, \ldots, u_{k+4}$.
- For all $i \in [n]$ and $j \in [m]$, add an edge $s_i t_j$, if $x_i \in F_j$.
- For all $i \in [n]$, add edges $w_2 s_i$ and for all $j \in [m]$, add edges $w_3 t_j$ and $w_4 t_j$.
- For all $i \in [k+4]$, add the edges $w_3 u_i$ and $w_4 u_i$. Finally add the edge $w_1 w_2$.

Let $S = \{s_i \mid i \in [n]\}$ and $T = \{t_j \mid j \in [m]\}$. This completes the construction of the graph G. An illustration of the graph is given in Fig. 2. We now prove a few lemmas.

Lemma 4. *If D is a cosecure dominating set of G of size $k+3$, then $\{w_3, w_4\} \subseteq D$.*

Proof. Since there are $k+4$ vertices in $\{u_1, u_2, \ldots, u_{k+4}\}$ and these vertices only have w_3 and w_4 as neighbors, at least one of w_3 and w_4 is in D otherwise there exists a vertex in $\{u_1, u_2, \ldots, u_{k+4}\}$ that is not dominated. If both w_3 and w_4 belong to D, the claim holds. Without loss of generality, let $w_3 \in D$ and $w_4 \notin D$. Since $|D| = k+3$ and $w_3 \in D$, there exists at least two vertices, say x and y, from $\{u_1, u_2, \ldots, u_{k+4}\}$ that do not belong to D. Further, if the covertex vertex of w_3 is x (or y) then y (resp. x) is not cosecurely dominated as $w_4 \notin D$ and $xy \notin E(G)$. This is a contradiction to the fact that D is a cosecure dominating set. Thus $\{w_3, w_4\} \subseteq D$. □

Lemma 5 (★). *Let D be a cosecure dominating set of G of size $k+3$. Then G has a cosecure dominating set D' of size at most $|D|$ such that $w_2 \in D'$, $S \cap D' = \emptyset$, and $|N(s_i) \cap D'| \geq 2$ for each $s_i \in S$.*

Lemma 6. *$(\mathcal{U}, \mathcal{F})$ has a set cover of size k if and only if G has a cosecure dominating set of size $k+3$.*

Proof. Let $X \subseteq \mathcal{F}$ be a set cover of size k for $(\mathcal{U}, \mathcal{F})$. Let $Y = \{w_2, w_3, w_4\} \cup \{t_j : F_j \in X\}$. Notice that $Y \cap S = \emptyset$. We claim that Y is a cosecure dominating set of G. The vertices w_2, w_3 and w_4 dominate all other vertices in G. Thus Y is a dominating set. Since X is a set cover, each vertex in S is adjacent to at least one vertex in $T \cap Y$. It is easy to see that, for each $y \in V(G) \setminus (Y \cup \{w_1\})$, we have $|N[y] \cap Y| \geq 2$. Thus, for any vertex in $y \in Y \setminus \{w_2\}$, an arbitrary neighbor

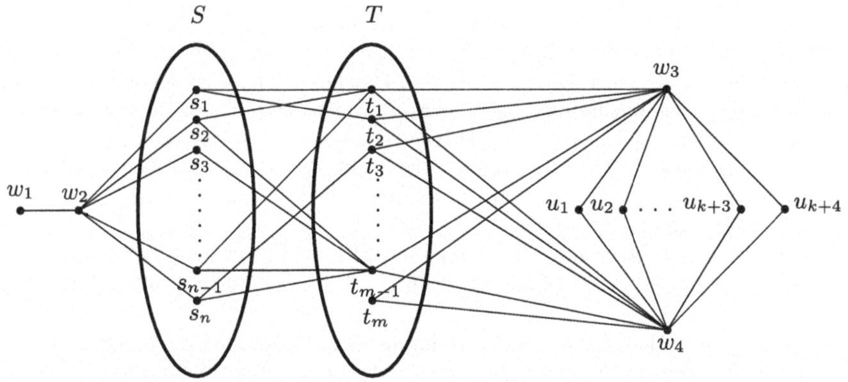

Fig. 2. Suppose $\mathcal{F}$ contains the family $F_1 = \{x_1, x_2, x_{n-1}\}$ where $F_1 \in \mathcal{F}$ and $x_1, x_2, x_{n-1} \in \mathcal{U}$ then add edges t_1s_1,t_1s_2,t_1s_{n-1} where $t_1 \in T$ and $\{s_1, s_2, s_{n-1}\} \subseteq S$ which illustrates the construction of the graph.

of y can be its covertex and the covertex of w_2 is w_1. Thus Y is a cosecure dominating set.

Conversely, let D be a cosecure dominating set of G of size $k+3$ that satisfies the conditions of Lemma 5. By Lemma 4, we have $\{w_3, w_4\} \subseteq D$. Hence, $|D \cap T| = k$. Let $H = \{F_i \mid t_i \in D \cap T\}$. For all $s_i \in S$, we have $|N(s_i) \cap D| \geq 2$, by Lemma 5. Since $w_2 \in D$, we have that $|N(s_i) \cap (D \setminus \{w_2\})| \geq 1$, for each $s_i \in S$. Thus H is a set cover of $(\mathcal{U}, \mathcal{F})$ of size k. □

It is easy to see that $T \cup \{w_2, w_3, w_4\}$ is a vertex cover of G. Thus $\mathsf{vc}(G) \leq |\mathcal{F}| + 3$. Since Set Cover does not admit a polynomial kernel parameterized by $|\mathcal{F}|$, unless $\mathsf{NP} \subseteq \mathsf{coNP/poly}$ [23], we get our desired result.

References

1. Butenko, S., Cheng, X., Oliveira, C.A., Pardalos, P.M.: A new heuristic for the minimum connected dominating set problem on ad hoc wireless networks. In: Recent Developments in Cooperative Control and Optimization, pp. 61–73 (2004)
2. Du, D.-Z., Wan, P.-J.: Connected Dominating Set: Theory and Applications, vol. 77. Springer, Heidelberg (2012)
3. Haynes, T.W., Hedetniemi, S.T., Henning, M.A. (eds.): Structures of Domination in Graphs. Developments in Mathematics, vol. 66. Springer, Cham (2021)
4. Guha, S., Khuller, S.: Approximation algorithms for connected dominating sets. Algorithmica **20**, 374–387 (1998)
5. Cockayne, E.J., Dawes, R.M., Hedetniemi, S.T.: Total domination in graphs. Networks **10**(3), 211–219 (1980)
6. Yuan, F., Li, C., Gao, X., Yin, M., Wang, Y.: A novel hybrid algorithm for minimum total dominating set problem. Mathematics **7**(3), 222 (2019)
7. Zhu, J.: Approximation for minimum total dominating set. In: Proceedings of the 2nd International Conference on Interaction Sciences: Information Technology, Culture and Human, pp. 119–124 (2009)

8. Irving, R.W.: On approximating the minimum independent dominating set. Inf. Process. Lett. **37**(4), 197–200 (1991)
9. Bourgeois, N., Croce, F.D., Escoffier, B., Paschos, V.T.: Fast algorithms for min independent dominating set. Discrete Appl. Math. **161**(4–5), 558–572 (2013)
10. Chambers, E.W., Kinnersley, B., Prince, N., West, D.B.: Extremal problems for roman domination. SIAM J. Discrete Math. **23**(3), 1575–1586 (2009)
11. Fernau, H.: Roman domination: a parameterized perspective. Int. J. Comput. Math. **85**(1), 25–38 (2008)
12. Haynes, T.W., Hedetniemi, S.T., Henning, M.A. (eds.): Topics in Domination in Graphs, vol. 64. Springer, Cham (2020). https://doi.org/10.1007/978-3-030-51117-3
13. Haynes, T.W., Hedetniemi, S.T., Henning, M.A.: Domination in Graphs: Core Concepts. Springer, Cham (2023). https://doi.org/10.1007/978-3-031-09496-5
14. Arumugam, S., Ebadi, K., Manrique, M.: Co-secure and secure domination in graphs. Util. Math. **94**, 167–182 (2014)
15. Zou, Y.-H., Liu, J.-J., Chang, S.-C., Hsu, C.-C.: The co-secure domination in proper interval graphs. Discret. Appl. Math. **311**, 68–71 (2022)
16. Kusum, Pandey, A.: Cosecure domination: hardness results and algorithms. In: Hsieh, S.Y., Hung, L.J., Lee, C.W. (eds.) IWOCA 2023. LNCS, vol. 13889. Springer, Cham (2023). https://doi.org/10.1007/978-3-031-34347-6_21
17. Kusum, Pandey, A.: Some new algorithmic results on co-secure domination in graphs. Theoret. Comput. Sci. **992**, 13 (2024). Paper No. 114451
18. Wang, C.-X., Yang, Y., Xu, S.-J.: The algorithm and complexity of co-secure domination in geometric intersection graphs. Comput. Appl. Math. **44**(1), 22 (2025)
19. Kusum, Pandey, A.: Complexity results on cosecure domination in graphs. In: Bagchi, A., Muthu, R. (eds.) CALDAM 2023. LNCS, vol. 13947, pp. 335–347. Springer, Cham (2023). https://doi.org/10.1007/978-3-031-25211-2_26
20. Panda, B.S., Rana, S., Mishra, S.: On the complexity of co-secure dominating set problem. Inf. Process. Lett. **185**, 106463 (2024)
21. Bodlaender, H.L., van Leeuwen, E.J., van Rooij, J.M.M., Vatshelle, M.: Faster algorithms on branch and clique decompositions. In: Hliněný, P., Kučera, A. (eds.) MFCS 2010. LNCS, vol. 6281, pp. 174–185. Springer, Heidelberg (2010). https://doi.org/10.1007/978-3-642-15155-2_17
22. Goyal, D., Jacob, A., Kumar, K., Majumdar, D., Raman, V.: Parameterized complexity of dominating set variants in almost cluster and split graphs. J. Comput. Syst. Sci. 103631 (2025)
23. Cygan, M., et al.: Parameterized Algorithms, vol. 5. Springer, Cham (2015). https://doi.org/10.1007/978-3-319-21275-3
24. Diestel, R.: Graph Theory. Graduate Texts in Mathematics, vol. 173, 4th edn. Springer, Cham (2012)
25. Gajarský, J., Lampis, M., Ordyniak, S.: Parameterized algorithms for modular-width. In: Gutin, G., Szeider, S. (eds.) IPEC 2013. LNCS, vol. 8246, pp. 163–176. Springer, Cham (2013). https://doi.org/10.1007/978-3-319-03898-8_15
26. Románek, M.: Parameterized algorithms for modular-width (2016)
27. Bodlaender, H.L., Drange, P.G., Dregi, M.S., Fomin, F.V., Lokshtanov, D., Pilipczuk, M.: SIAM J. Comput. **45**(2), 317–378 (2016)

Simultaneous Contact Representations of Planar Graphs

Jan Kratochvíl[1] and Melanie Reihl[2](✉)

[1] Department of Applied Mathematics, Faculty of Mathematics and Physics, Charles University, Prague, Czech Republic
honza@kam.mff.cuni.cz

[2] Technische Universität Berlin, Berlin, Germany
Melanie.Reihl@gmx.de

Abstract. Contact representations of planar graphs by triangles or – in the bipartite case – by vertical and horizontal segments in the plane can be constructed in linear time. It is also known that in both of these cases, deciding whether a representation of an induced subgraph of an input graph can be extended into a representation of the entire graph is NP-hard. We complement these results by showing that simultaneous representation of two graphs is also NP-hard (both for triangle contact graphs, and for grid contact ones). We show this by a unified reduction via simultaneous contact representations of so-called x-y-stretchable sets of convex sets, which implies NP-hardness of the simultaneous representation problem for many more classes of geometric contact graphs.

1 Introduction

We consider simple undirected graphs and their intersection and contact representations by geometric objects in the plane. A *graph* is a pair $G = (V, E)$ where V is its set of vertices and $E \subseteq \binom{V}{2}$ its set of edges, viewed as two-element subsets of the vertex set. We use standard terminology and notation. In particular, the vertex set of a graph G is denoted by $V(G)$, its edge set by $E(G)$. A graph is called *planar* if it can be drawn in the plane without edge crossings.

1.1 Geometric Intersection and Contact Graphs

Intersection graphs of geometric objects in the plane are widely studied both for their interesting combinatorial properties and for their practical motivation. In an *intersection representation* of a graph, vertices are represented by sets (typically from some well-specified universe) and two vertices are adjacent (in the graph) if and only if the corresponding sets have a nonempty intersection. Perhaps the oldest among intersection-defined classes of graphs are *interval*

J. Kratochvíl—This author acknowledges support of Czech Science Foundation through a research grant GAČR 23-04949X.

A. Jeż and J. Otop (Eds.): FCT 2025, LNCS 16106, pp. 295–308, 2026.
https://doi.org/10.1007/978-3-032-04700-7_22

graphs [4,19], intersection graphs of intervals on a real line, which find natural applications in scheduling, but also in areas as diverse as social sciences or biology [38], in particular in DNA sequencing [44]. Other intersection-defined classes that attracted the interest of researchers in algorithmic graph theory are circle graphs [5,41], circular arc graphs [33,43], permutation graphs [13], trapezoid graphs [11] and many others. The reader is further referred to monographs [6,34]. Many of these classes allow elegant structural characterizations and are recognizable in polynomial time. And many of them allow polynomial-time algorithms for basic optimization problems which are known to be NP-complete for general graphs. For instance, both MAXIMUM CLIQUE and MAXIMUM INDEPENDENT SET problems are polynomial-time solvable in *interval filament graphs*, a class of graphs that includes many of the above mentioned ones [18]. Another intersection-defined class of graphs that allows MAXIMUM INDEPENDENT SET be solved in polynomial time is the class of *grid intersection graphs* (intersection graphs of vertical and horizontal segments in the plane, with any two segments of the same direction being disjoint. Grid intersection graphs have been studied in connection to graphs of boxicity 2 (i.e., intersection graphs of axes-parallel rectangles) [2]. Grid intersection graphs, as well as graphs of boxicity 2, are hard (NP-complete) to recognize [30].

A *contact representation* of a graph is an intersection representation in which the objects representing the vertices of the graph may share boundary points, but their interiors are disjoint (in the applications we refer about, the objects are always *arc-connected*, which means that any two points are connected by a simple Jordan arc that belongs to the object). A graph that has a contact representation by objects from a set $\mathcal{C}$ is called a $\mathcal{C}$*-contact graph.* Classes of contact graphs such as *disk-contact graphs*, *triangle-contact graphs*, *segment-contact graphs* or *grid-contact graphs* (i.e., contact graphs of vertical and horizontal segments in the plane, see an illustrative example in Fig. 1) have been studied [15,16,21,22,29]. A further condition is usually imposed on the objects of a contact representation, namely, that no three objects touch in the same point. Under this assumption, the contact graph of such a representation is always planar. It is somewhat surprising that the converse is also true. It has been shown that every planar graph is a contact graph of triangles (even of isosceles ones) [16], and every bipartite planar graph is a grid-contact one [15,20]. Intersection and contact representations are an important means of visualizing graphs. The simpler a representation, the better it is intercepted by the user. From this point of view are the contact representations of planar graphs by simple and symmetric objects highly desirable.

1.2 REPEXT and SIMREP Paradigms

It is a natural question whether a partial intersection (or contact) representation of a graph can be extended to a corresponding representation of the entire graph. For a class of graphs $\mathcal{A}$, this question is referred to as REPEXT($\mathcal{A}$). More formally, the input of REPEXT($\mathcal{A}$) is a graph and an $\mathcal{A}$-representation of some of its vertices. The question is whether objects representing the remaining vertices

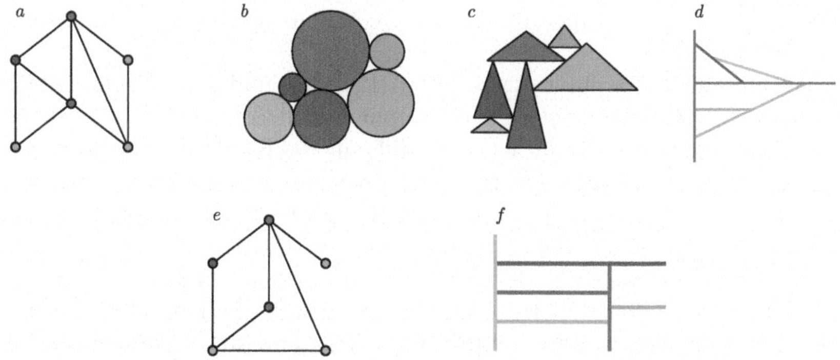

Fig. 1. (a) A planar graph, with (b) a disk, (c) a triangle and (d) a segment contact representation. (e) A bipartite planar graph, with (f) a grid contact representation.

can be added to create an $\mathcal{A}$-representation of the entire input graph, without changing the objects representing the pre-represented vertices. The study of RepExt for geometric intersection graphs was initiated in [28], and it is somewhat surprising that for most intersection-defined classes of graphs that are recognizable in polynomial time, RepExt is polynomial-time decidable as well (today, RepExt is known to be polynomial-time solvable for interval graphs, for unit and proper interval graphs [27], for circle graphs [9], for permutation and function graphs [26] and for trapezoid graphs [32]). Thus, the only intersection-defined class that obeys the "architects' rule" is the class of circular arc graphs, for which the RepExt problem is NP-complete [14]. The authors of [8] show that RepExt is NP-hard for contact graphs of disks, isosceles triangles, and vertical/horizontal segments, while these classes of contact graphs are recognizable in linear time; cf. more details in the next subsection.

A related problem to RepExt is SimRep, the question of whether several graphs have a *simultaneous representation*, that is, a representation of the common part of the graphs that extends to representations of each of them. As outlined in the next paragraph, this question is studied mainly in the case of *sunflower configurations*, which is the case when the intersections of all pairs of input graphs coincide, and this intersection is the common part of the input graphs (note that if the input consists of two graphs, then these are in a sunflower configuration per se). Equivalently, for a class of objects $\mathcal{C}$ and a collection of graphs $G_1, \ldots, G_k$, such that $V_0 = V(G_i) \cap V(G_j)$ for all $0 \leq i < j \leq k$ and $G_0 = G_i[V_0]$ for every $i = 1, 2, \ldots, k$, a simultaneous $\mathcal{C}$-representation of $G_1, \ldots, G_k$ is a mapping $R : \bigcup_{i=1}^{k} V(G_i) \to \mathcal{C}$ such that the restriction R_i of R to $V(G_i)$ is an intersection representation of G_i for every $i = 1, 2, \ldots, k$. A simultaneous contact representation is defined in a similar way. For an intersection- or contact-defined class of graphs $\mathcal{A}$, we denote by SimRep($\mathcal{A}$) the problem of whether a collection of input graphs in sunflower configuration allows a simultaneous $\mathcal{A}$-representation (with the number k of the graphs being a part of the input), and for every fixed k, we denote by k-SimRep($\mathcal{A}$) the problem of whether a collection of k input graphs in sunflower configuration allows a simultaneous $\mathcal{A}$-representation.

From the complexity point of view, SIMREP is less understood than REPEXT. However, in all the cases where the computational complexity of both problems is known, SIMREP is at least as difficult as REPEXT. For example, it is proven in [24] that SIMREP(interval graphs) is polynomial-time solvable for $k = 2$, and it is not difficult to argue that REPEXT(interval graphs) polynomially reduces to 2-SIMREP(interval graphs), thus providing an alternative proof of polynomial-time solvability of REPEXT(interval graphs). Recently, Rutter and Stumpf [40] showed that SIMREP(interval graphs) is polynomial-time solvable (i.e., even when k is part of the input). Other polynomial-time solvable problems include SIMREP(proper interval graphs) and SIMREP(unit interval graphs), as demonstrated in [39], SIMREP(permutation graphs) [35], and SIMREP(function graphs), following from an easy reduction to simultaneous transitive orientation of graphs [25,36] as shown in [42]. In contrast, SIMREP(circular arc graphs) is NP-complete [42]. In [10], the understanding of the interplay of the representation extension and simultaneous representation is identified as one of the current challenges of graph drawing.

Another well-known problem related to SIMREP is the *Simultaneous Embedding with Fixed Edges (SEFE)* problem, which asks whether it is possible to simultaneously draw a set of planar graphs $G_i = (V, E_i)$, such that each vertex is placed at the same point in all drawings and every shared edge $e \in E_i \cap E_j$ is represented by the same curve in all graphs. The SEFE problem for three graphs has been shown NP-complete [1,17]. For $k = 2$, it was proven that the SEFE problem can be solved in quadratic time, provided both graphs are biconnected and the common subgraph is connected [3]. However, for two arbitrary graphs, the complexity of the SEFE problem remains an open question. A related problem *Simultaneous Geometric Embedding (SGE)* requires that the input graphs are drawn on the same set of points as vertices, edges are drawn as straight-line segments and the drawing of each of the graphs is crossing-free. This problem is known to be NP-hard already for two graphs [12].

1.3 Contact Representations of Planar Graphs

It can be easily seen that triangle-free contact graphs of arc-connected sets in the plane are planar [31]. A classical example is *coin graphs*, i.e., contact graphs of disks in the plane. The opposite is also true. The circle packing theorem of Koebe [29] stating that every planar graph is a contact graph of disks in the plane has been rediscovered several times. One of its strengthenings gives a simultaneous contact representation of a planar graph and its dual [7].

Using the concept of canonical orderings of vertices in planar triangulations, the authors of [16] showed that every planar graph is a contact graph of isosceles triangles (with the further restriction on the representation that the bases of the triangles are horizontal and all triangles are pointed upward).

It is shown in [15] that every planar bipartite graph is a grid-contact graph. The same result is implicitly proven in [20].

Since planar graphs are recognizable in linear time [23], it follows that disk contact graphs, isosceles triangle contact graphs, and grid contact graphs can be recognized in linear time. Moreover, triangle-contact and grid-contact representations can be constructed in linear time.

The authors of [8] have explored representations of planar graphs described above from the point of view of the representation extension approach. They showed that the problems RepExt(disk contact graphs) and RepExt(isosceles triangle contact graphs) are NP-hard and that RepExt(grid contact graphs) is NP-complete.

In his talk at the Shonan conference in 2020 [10], J. Kratochvíl reported on NP-hardness of 3-SimRep(grid-contact graphs) and posed as an open problem the complexity of 2-SimRep(grid-contact graphs). The computational complexity of k-SimRep(isosceles triangle contact graphs) stayed unknown for arbitrary k. The aim of this paper is to answer these two questions.

1.4 Our Results

We prove the following.

Theorem 1. *The problem 2-SimRep(grid-contact graphs) is NP-complete.*

Theorem 2. *The problem 2-SimRep(isosceles triangle contact graphs) is NP-hard.*

In both cases, we prove the results by a reduction from the so-called Betweenness Problem (for a definition of this problem, see the next section). We prove the results in a broader setting of so-called *x-y-stretchable* sets.

We denote the horizontal and vertical stretching of the plane by factor a by T_x^a and T_y^a, respectively. Formally, these are the linear transformations of the plane given by $T_x^a(x, y) = (ax, y)$ and $T_y^a(x, y) = (x, ay)$.

Definition 1. *A set $\mathcal{C}$ of bounded subsets of $\mathbb{R}^2$ is **x-y-stretchable** if it is closed under translations, as well as under horizontal and vertical stretchings T_x^a and T_y^b for every $a > 0$ and $b > 0$. Furthermore, $\mathcal{C}$ must contain at least one set A with $x_{max}(A) \neq x_{min}(A)$ and one set B with $y_{max}(B) \neq y_{min}(B)$.*

Both the set of vertical and horizontal segments, and the set of isosceles triangles (even in the restricted case of upward-pointed triangles with vertical axes of symmetry) are x-y-stretchable. Thus the following theorem is a unifying generalization of Theorems 1 and 2.

Theorem 3. *Let $\mathcal{C}$ be an x-y-stretchable set of convex subsets of $\mathbb{R}^2$. Then 2-SimRep($\mathcal{C}$-contact graphs) is NP-hard.*

Many other natural set systems are x-y-stretchable, some examples are listed in the following corollary.

Corollary 1. *SimRep is NP-hard for contact representations by convex sets, by axes-aligned rectangles, by ellipses, by triangles, by right-angled triangles with axes-aligned perpendicular sides.*

The proofs are presented in the next section. We first introduce the Betweenness Problem and a reduction from it to our problem. Then we show how infeasible instances of the Betweenness problem produce instances without any simultaneous representation by contacts of convex sets. Next, we demonstrate that feasible instances lead to grid-contact representations and, finally, also by sets from any x-y-stretchable set.

2 Simultaneous Representation of x-y-Stretchable Convex Contact Graphs Is NP-Hard

We prove Theorems 1, 2, and 3 through a unified reduction from the Betweenness problem.

2.1 Betweenness

The problem Betweenness [37] is defined as follows. Given a finite set S and a finite set T of triples of elements in S, decide whether there exists a total ordering $<$ of S such that for all triples $(\rho, \psi, \varphi) \in T$ either $\rho < \psi < \varphi$ or $\rho > \psi > \varphi$. These triples are called *betweenness constraints*. The problem is also known as the Total Ordering Problem, which is shown NP-complete by Opatrny in [37].

2.2 Reduction from Betweenness – Construction of Graphs G_1 and G_2

Let (S, T), where $S = \{1, ..., l\}$ and $T = \{T_1, ..., T_k\}$, be an instance of Betweenness. In this notation, the triple T_j is represented as $(\rho_j, \psi_j, \varphi_j)$. We use i for elements of S, and χ when referring to any of ρ, ψ or φ in cases where the distinction between them is not important and the statement applies to all three. With that, every χ_j equals an i in S.

We construct two bipartite graphs G_1 and G_2 which share a common subgraph G_0. The nodes of the shared graph G_0 will be denoted by Greek letters. With a slight abuse of notation, convex sets and the corresponding nodes will be referred to by the same names. The graphs are described by means of grid contact representations.

We start by placing two nodes α and β as vertical segments and two nodes γ and δ as horizontal segments, in a way that they create a large rectangle. Then we place the horizontal *row* segments $r_{1,0}, r_{1,1}, r_{2,0}, r_{2,1}, \ldots, r_{l,0}, r_{l,1}$ across the rectangle in G_1, and we place the vertical *column* segments c_j, $j = 1, \ldots, k+1$ across the rectangle in G_2. For each betweenness constraint $T_j = (\rho_j, \psi_j, \varphi_j)$, we place horizontal segments ω^j_ρ, ω^j_ψ and ω^j_φ so that ω^j_χ (for $\chi \in \{\rho, \psi, \varphi\}$) with

$\chi_j = i$ is placed in the intersection of the region between row segments $r_{i,0}$ and $r_{i,1}$ and the region between columns c_j and c_{j+1}. The segment ω_χ^j is connected to $r_{i,0}$ and $r_{i,1}$ with two vertical segments $v_{\chi_j,j,h}$ in G_1. In G_2, we add two vertical segments u, one touching ω_ρ^j and ω_ψ^j, the other one touching ω_ψ^j and ω_φ^j. These segments enforce the correct order of the ω segments corresponding to the betweennes constraint T_j. Simplified illustrations of the graphs G_1 and G_2 are provided in Figs. 2 and 3, respectively.

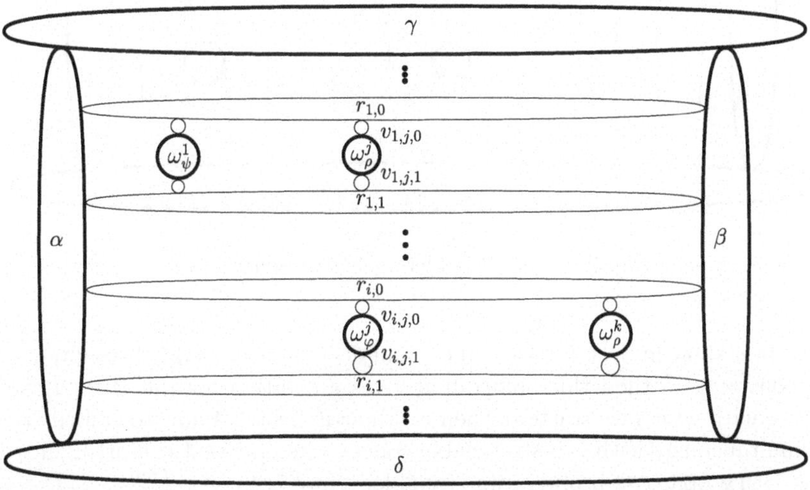

Fig. 2. G_1 for $\psi_1, \rho_j = 1$ and $\varphi_j, \rho_k = i$ and bolder ovals representing G_0

Formally, we set

$G_0 := (V^1 \bigcup V^2, E^1)$ with

$V^1 := \{\omega_\chi^j \mid \chi \in \{\rho, \psi, \varphi\}, j \in \{1, ..., k\}\}$,

$V^2 := \{\alpha, \beta, \gamma, \delta\}$

$E^1 := \{\{\alpha, \gamma\}, \{\alpha, \delta\}, \{\beta, \gamma\}, \{\beta, \delta\}\}$,

$G_1 := (V^1 \bigcup V^2 \bigcup V_1^1 \bigcup V_1^2, E^1 \bigcup E_1^1 \bigcup E_1^2)$ with

$V_1^1 := \{r_{i,h} \mid i \in \{1, ..., l\}, h \in \{0, 1\}\}$

$V_1^2 := \{v_{\chi_j,j,h} \mid \chi \in \{\rho, \psi, \varphi\}, j \in \{1, ..., k\}, h \in \{0, 1\}\}$

$E_1^1 := \{\{\alpha, r_{i,h}\}, \{\beta, r_{i,h}\} \mid i \in \{1, ..., l\}, h \in \{0, 1\}\}$

$E_1^2 := \{\{\omega_\chi^j, v_{\chi_j,j,h}\}, \{r_{\chi_j,h}, v_{\chi_j,j,h}\} \mid \chi \in \{\rho, \psi, \varphi\}, j \in \{1, ..., k\}, h \in \{0, 1\}\}$, and

$G_2 := (V^1 \bigcup V^2 \bigcup V_2^1 \bigcup V_2^2, E^1 \bigcup E_2^1 \bigcup E_2^2)$ with

$V_2^1 := \{c_j \mid j \in \{1, ..., k+1\}\}$

$V_2^2 := \{u_{j,\rho\psi}, u_{j,\psi\varphi} \mid j \in \{1, ..., k\}\}$

$E_2^1 := \{\{c_j, \gamma\}, \{c_j, \delta\}, \{c_j, \omega_\chi^{j-1}\}, \{c_j, \omega_\chi^j\} | j \in \{1, ..., k+1\}, \chi \in \{\rho, \psi, \varphi\}\}$

$E_2^2 := \{\{\omega_\rho^j, u_{j,\rho\psi}\}, \{\omega_\psi^j, u_{j,\rho\psi}\}, \{\omega_\psi^j, u_{j,\psi\varphi}\}, \{\omega_\varphi^j, u_{j,\psi\varphi}\} | j \in \{1, ..., k\}\}$.

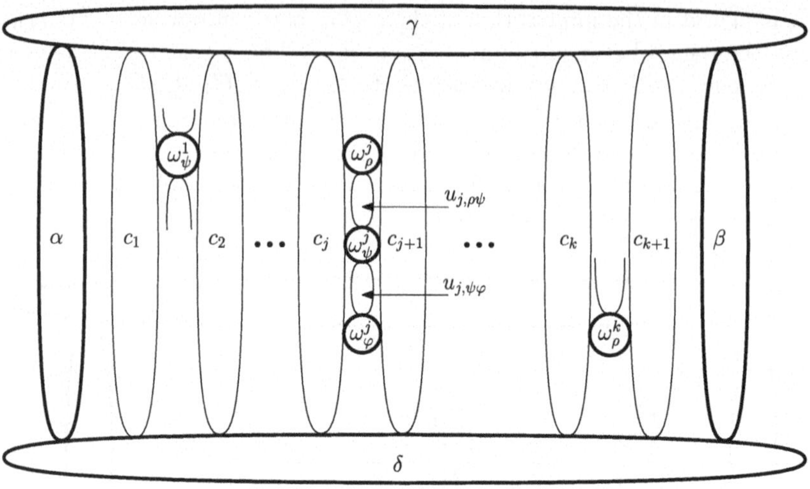

Fig. 3. G_2, with bolder ovals representing G_0

We first show in Sect. 2.3 that, if G_1 and G_2 can be embedded simultaneously with convex sets, then this embedding gives a solution for the BETWEENNESS instance with which we started. Then we show in Sect. 2.4 how to embed G_1 and G_2 simultaneously with x-y-stretchable convex sets, in case the starting instance of the BETWEENNESS problem was feasible.

2.3 Necessity for Convex Sets

Suppose G_1 and G_2 were constructed for (S, T) and suppose that we have a simultaneous contact representation by convex sets. Let Q_i be the quadrilateral spanned by the convex hull of contact points of α, β, $r_{i,0}$ and $r_{i,1}$, for each $i \in \{1, ..., l\}$. If two convex sets contact along a segment, an arbitrary but fixed point on the segment can be selected.

Lemma 1. *All convex sets ω_χ^j with $\chi_j = i$ are contained in Q_i.*

Proof. Assume, for the sake of contradiction, that one of the convex sets ω_χ^j is not contained within Q_i.

In G_1, the nodes $\{\alpha, \beta, (v_{\chi_j,j,0}, \omega_\chi^j, v_{\chi_j,j,1})\}, \{r_{i,0}, r_{i,1}\}$ form a subdivision of $K_{2,3}$. In any planar embedding of $K_{2,3}$, it generally holds that not all three vertices of the tripartition can lie on the outer face. Since $\omega_\chi^j \notin Q_i$, it cannot be the internal vertex. Therefore, the only face incident to α and β is bounded by $\alpha, r_{i,0}, \beta, r_{i,1}$. Since γ and δ touch both α and β, they lie in this face and, consequently, also in Q_i, as shown in Fig. 4a.

If we now consider the graph G_2, we observe from its definition that c_j is adjacent to γ, δ, and ω_χ^j. Therefore, it must touch all three sets. Since c_j must touch ω_χ^j, which is not contained in Q_i, it follows that c_j cannot lie in the face enclosed by $\alpha, \gamma, \beta, \delta$. However, c_j does not touch α, β nor the interior of the face enclosed by $\alpha, \gamma, \beta, \delta$. But together they divide the quadrilateral Q_i into two regions, with γ and δ lying on opposite sides. Therefore, c_j touches γ, crosses the interior of one of the sides of the quadrilateral Q_i, then touches ω^j_χ, cuts through the opposite side of Q_i, and finally touches δ. However, for a convex set c_j, this is not possible, even when Q_i is not convex, as shown in Fig. 4b. Therefore, the assumption that $\omega^j_\chi \notin Q_i$ leads to a contradiction, and it follows that each convex set ω^j_χ must indeed lie within Q_i. □

Lemma 2. *Every ω^j_χ is contained in the region surrounded by $\gamma, c_j, \delta, c_{j+1}$.*

Proof. Analogously to the proof of Lemma 1. □

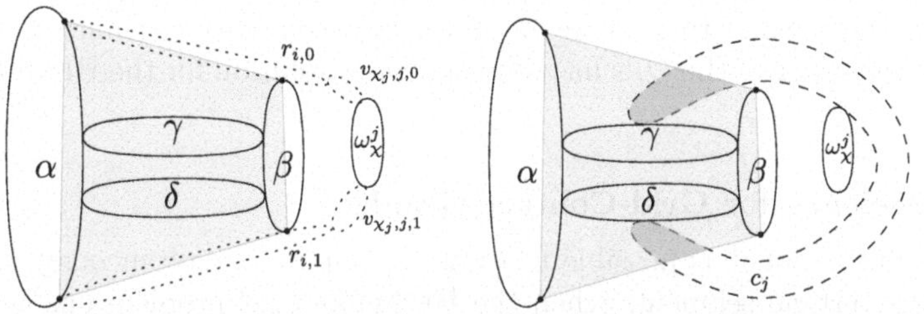

(a) G_1: Q_i contains γ and δ (b) G_2: c_j cannot be convex

Fig. 4. Illustration to the proof of Lemma 1. G_1 and G_2 are represented in dotted lines. The shared graph G_0 is represented in continues lines. The darkest area (in red), $Q_{\chi_j} \cap c_j$, cannot be convex. (Color figure online)

Using Lemma 1 and the property of the contact graph G_1 that the interiors of two sets must not intersect, the following lemma can be easily proven. A detailed proof will appear in the full version of the paper.

Lemma 3. *The quadrilaterals Q_i, for $i = 1, ..., l$, are pairwise disjoint.*

The *order of the Q_i's* now follows their sequence along α, starting from outside the convex hull of α and β, then moving along the boundary of α counterclockwise. This defines a total order $<$ on $S = \{1, ..., l\}$, with $x, y \in S$ where $x < y$ if Q_x appears before Q_y in the counterclockwise traversal.

Lemma 4. *If G_1, G_2 have a simultaneous contact representation, the order of Q_i is a proper solution for the* Betweenness *problem from which we are reducing.*

Proof. From Lemma 3, we know that the sets Q_i are pairwise disjoint. Furthermore from Lemma 1 we know all ω^j_χ are contained in the corresponding Q_i. Therefore if we follow the boundary of the convex set c_j from γ to δ, we know that if it touches multiple ω^j_χ then it can only touch them in the order of the Q_i's, or the reversed order.

From Lemma 2, we know that the convex sets $\omega^j_\psi, \omega^j_\rho$ and ω^j_φ are located within the region $\gamma, c_j, \delta, c_{j+1}$. Furthermore, by definition $\omega^j_\psi, \omega^j_\rho$ and ω^j_φ touch c_j. From this, we can conclude that the boundary of c_j intersects $\omega^j_\psi, \omega^j_\rho$ and ω^j_φ along a path following the boundary from γ to δ in the order of the Q_i's. We now look at the planar subgraph of G_2 induced by the convex sets $\omega^j_\rho, \omega^j_\psi, \omega^j_\varphi, u_{j,\rho\psi}, u_{j,\psi\varphi}, c_j, c_{j+1}, \gamma$ and δ (See Fig. 5). It is easy to see that ω^j_ψ has to lie in the middle, between ω^j_ρ and ω^j_φ.

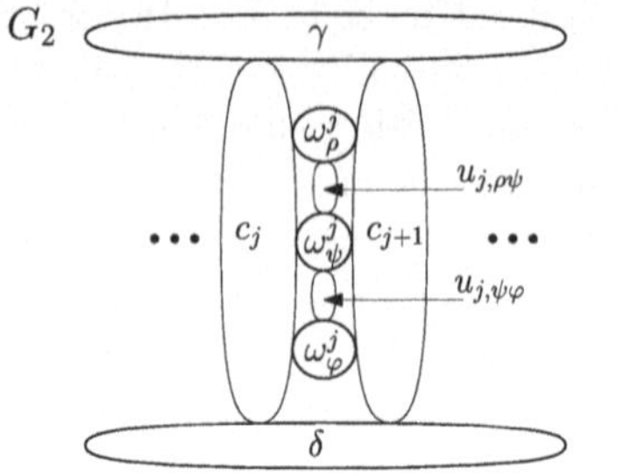

Fig. 5. Ensuring that ω^j_ψ is between the other two

So it has also to lie in the middle in the order of the Q_i's. Since this is true for all the betweenness constraints T_j with $j \in \{1, ..., k\}$, the order of the Q_i's has to be a proper solution for the Betweenness Problem. □

2.4 Sufficiency for Grid-Contact Graphs

First, we show that it is possible to draw G_1 and G_2 simultaneously with horizontal and vertical segments when the Betweenness problem has a solution. In Sect. 2.5, we then extend this result to x-y-stretchable convex sets.

Lemma 5. *Let* (S, T) *with* $T = \{T_1, ..., T_k\}$, $S = \{1, ..., l\}$ *be an instance of* Betweenness *with a solution* $(\pi_1, ...\pi_l)$. *Then* G_1, G_2 *have a simultaneous grid contact representation.*

Proof. Let γ and δ be two horizontal segments, starting and ending with the same x values. The vertical segments $\alpha, c_1, c_2, \ldots, c_k, c_{k+1}, \beta$, touching both γ and δ, appear in this order, following the boundary of γ counterclockwise, with α and β on the outer face. Similarly, the horizontal segments $\gamma, r_{\pi_1,0}, r_{\pi_1,1}, ..., r_{\pi_l,0}, r_{\pi_l,1}, \delta$, each of them touching both α and β, appear in this order, following the boundary of α clockwise, with γ and δ on the outer face. Note that c_j and $r_{\pi_i,h}$ cross, which is allowed since they are not in the same graph. Now there is a region surrounded by $c_j, r_{\pi_i,0}, c_{j+1}, r_{\pi_i,1}$ for all $j \in \{1, ..., k\}$ and $i \in \{1, ..., l\}$. We place ω^j_χ with $\chi_j = \pi_i$ in this region as a horizontal segment connecting c_j and c_{j+1} and two vertical segments $v_{\pi_i,j,0}$ connecting $r_{\pi_i,0}$ and ω^j_χ and the segments $v_{\pi_i,j,1}$ connecting $r_{\pi_i,1}$ and ω^j_χ. The vertical segments $u_{j,\rho\psi}$ connect ω^j_ρ and ω^j_ψ, and the vertical segments $u_{j,\psi\varphi}$ connect ω^j_ψ and ω^j_φ. Thereby $u_{j,\rho\psi}$ does not cross ω^j_φ since the order of the rows is a solution for the Betweenness problem, therefore ω^j_φ is not between ω^j_ρ and ω^j_ψ. The same is true for $u_{j,\psi\varphi}$. □

2.5 Sufficiency for x-y-Stretchable Convex Sets

Lemma 6. *Let* (S, T) *with* $T = \{T_1, ..., T_k\}$, $S = \{1, ..., l\}$ *be a feasible instance of* Betweenness. *Then* G_1, G_2 *allow a simultaneous contact representation with any* x-y-*stretchable set.*

Proof. Sketch of. This representation can be obtained by replacing each horizontal segment with a flat, wide object and each vertical segment with a tall, narrow object at each step of Lemma 5. Details of the proof will be presented in the full version of the paper. □

Our proof cannot be adjusted for arbitrary arc-connected sets. But it is easy to show that SEFE and SIMREP of contact graphs of arc-connected sets are polynomial-time equivalent. Hence, results for SEFE naturally extend to this SIMREP problem; in particular, the NP-hardness of SEFE for three graphs implies the same for SIMREP of contact graphs of arc-connected sets [1,17]. Given the extensive research over several years regarding the complexity status of SEFE for two graphs, the problem appears to lack an immediate or straightforward solution.

2.6 Wrap-Up of the Proofs

We have described a reduction from the BETWEENNESS Problem, constructing a pair of graphs G_1 and G_2 from an instance (S, T). Both graphs G_1 and G_2 are bipartite.

Consider first grid contact representations. By Lemma 4, if (S, T) is an infeasible instance of BETWEENNESS, G_1 and G_2 do not allow a simultaneous contact representation by convex sets, and therefore also not by vertical and horizontal segments (as both types of segments are convex sets). On the other hand, by Lemma 5, if (S, T) is a feasible instance of BETWEENNESS, G_1 and G_2 do have a simultaneous contact representation by vertical and horizontal segments. This proves Theorem 1.

Next consider an x-y-stretchable set $\mathcal{C}$ of convex sets. Since all sets of $\mathcal{C}$ are convex, an infeasible (for BETWEENNESS) instance (S, T) gives rise to a pair G_1, G_2 of graphs that do not allow a simultaneous $\mathcal{C}$-contact representation, while it follows from Proposition 5 that a simultaneous $\mathcal{C}$-contact representation exists if (S, T) is feasible. This concludes the proof of Theorem 3.

Theorem 2 is a direct consequence of Theorem 3, since the set of all isosceles triangles (with horizontal bases and pointed upward) is x-y-stretchable.

3 Conclusion and Open Problems

We have proved NP-hardness of 2-SIMREP($\mathcal{C}$-contact graphs) for several classes $\mathcal{C}$ of convex sets in the plane. A relevant question is for which of these classes the SIMREP problem belongs to NP. This is the case of grid-contact graphs (and more generally of contact and intersection graphs of axes-aligned rectangles), since the correctness of a simultaneous representation only depends on the linear orders of the coordinates; the exact values of the coordinates are not needed.

Problem 1: For which other x-y-stretchable sets $\mathcal{C}$ is the SIMREP Problem of simultaneous representations of $\mathcal{C}$-contact graphs also in the class NP?

Problem 2: In particular, is 2-SIMREP(isosceles triangle contact graphs) in NP?

In this paper, we only considered simultaneous representations by contacts of x-y-stretchable sets. Even less is known for non-stretchable classes.

Problem 3: What is the complexity of SIMREP for contact graphs non-stretchable classes of convex sets? For instance, for contact graphs of equilateral triangles (axes-aligned, or arbitraty)?

Problem 4: What is the complexity of SEFE with 2 graphs and 2-SIMREP for arbitrary arc-connected sets?

References

1. Angelini, P., Da Lozzo, G., Neuwirth, D.: Advancements on SEFE and partitioned book embedding problems. Theoret. Comput. Sci. **575**, 71–89 (2015). https://doi.org/10.1016/j.tcs.2014.11.016
2. Bellantoni, S., Ben-Arroyo Hartman, I., Przytycka, T., Whitesides, S.: Grid intersection graphs and boxicity. Discret. Math. **114**(1–3), 41–49 (1993). https://doi.org/10.1016/0012-365X(93)90354-V
3. Bläsius, T., Rutter, I.: Simultaneous PQ-ordering with applications to constrained embedding problems. ACM Trans. Algorithms **12**(2) (2015). https://doi.org/10.1145/2738054
4. Booth, K.S., Lueker, G.S.: Testing for the consecutive ones property, interval graphs, and graph planarity using PQ-tree algorithms. J. Comput. Syst. Sci. **13**(3), 335–379 (1976). https://doi.org/10.1016/S0022-0000(76)80045-1
5. Bouchet, A.: Reducing prime graphs and recognizing circle graphs. Comb. **7**(3), 243–254 (1987). https://doi.org/10.1007/BF02579301
6. Brandstaedt, A., Le, V.B., Spinrad, J.P.: Graph classes: a survey. Soc. Ind. Appl. Math. (1999). https://doi.org/10.1137/1.9780898719796
7. Brightwell, G.R., Scheinerman, E.R.: Representations of planar graphs. SIAM J. Discret. Math. **6**(2), 214–229 (1993). https://doi.org/10.1137/0406017
8. Chaplick, S., Dorbec, P., Kratochvíl, J., Montassier, M., Stacho, J.: Contact representations of planar graphs: extending a partial representation is hard. In: Kratsch, D., Todinca, I. (eds.) WG 2014. LNCS, vol. 8747, pp. 139–151. Springer, Cham (2014). https://doi.org/10.1007/978-3-319-12340-0_12
9. Chaplick, S., Fulek, R., Klavík, P.: Extending partial representations of circle graphs. J. Graph Theory **91**(4), 365–394 (2019). https://doi.org/10.1002/JGT.22436
10. Chaplick, S., Itoh, T., Liotta, G., Ma, K.L., Rutter, I.: Trends and perspectives in graph drawing and network visualization. In: NII Shonan Meeting Report No. 171 (2020)
11. Dagan, I., Golumbic, M.C., Pinter, R.Y.: Trapezoid graphs and their coloring. Discret. Appl. Math. **21**(1), 35–46 (1988). https://doi.org/10.1016/0166-218X(88)90032-7
12. Estrella-Balderrama, A., Gassner, E., Jünger, M., Percan, M., Schaefer, M., Schulz, M.: Simultaneous geometric graph embeddings. In: Graph Drawing, 15th International Symposium, GD 2007. LNCS, vol. 4875, pp. 280–290. Springer (2007). https://doi.org/10.1007/978-3-540-77537-9_28

13. Even, S., Pnueli, A., Lempel, A.: Permutation graphs and transitive graphs. J. ACM **19**(3), 400–410 (1972). https://doi.org/10.1145/321707.321710
14. Fiala, J., Rutter, I., Stumpf, P., Zeman, P.: Extending partial representations of circular-arc graphs. In: Graph-Theoretic Concepts in Computer Science - 21st International Symposium, WG 2022. LNCS, vol. 13453, pp. 230–243. Springer (2022). https://doi.org/10.1007/978-3-031-15914-5_17
15. de Fraysseix, H., Ossona de Mendez, P., Pach, J.: Representation of planar graphs by segments. Intuitive Geom. **63** (2001)
16. de Fraysseix, H., Ossona de Mendez, P., Rosenstiehl, P.: On Triangle contact graphs. Comb., Probab. Comput. **3**(2), 233–246 (1994). https://doi.org/10.1017/S0963548300001139
17. Gassner, E., Jünger, M., Percan, M., Schaefer, M., Schulz, M.: Simultaneous graph embeddings with fixed edges. In: Fomin, F.V. (ed.) WG 2006. LNCS, vol. 4271, pp. 325–335. Springer, Heidelberg (2006). https://doi.org/10.1007/11917496_29
18. Gavril, F.: Maximum weight independent sets and cliques in intersection graphs of filaments. Inf. Process. Lett. **73**(5–6), 181–188 (2000). https://doi.org/10.1016/S0020-0190(00)00025-9
19. Golumbic, M.C.: Algorithmic aspects of intersection graphs and representation hypergraphs. Graphs Comb. **4**(1), 307–321 (1988). https://doi.org/10.1007/BF01864170
20. Hartman, I., Newman, I., Ziv, R.: On grid intersection graphs. Discret. Math. **87**(1), 41–52 (1991). https://doi.org/10.1016/0012-365X(91)90069-E
21. Hliněný, P.: Classes and recognition of curve contact graphs'. J. Comb. Theory B **74**(1), 87–103 (1998). https://doi.org/10.1006/jctb.1998.1846
22. Hliněný, P.: Contact graphs of line segments are NP-complete. Discret. Math. **235**(1-3), 95–106 (2001). https://doi.org/10.1016/S0012-365X(00)00263-6
23. Hopcroft, J.E., Paul, W.J., Valiant, L.G.: On Time versus Space and Related Problems. In: 16th Annual Symposium on Foundations of Computer Science FOCS 1975, pp. 57–64. IEEE Computer Society (1975). https://doi.org/10.1109/SFCS.1975.23
24. Jampani, K.R., Lubiw, A.: Simultaneous interval graphs. In: Algorithms and Computation - 21st International Symposium, ISAAC 2010. LNCS, vol. 6506, pp. 206–217. Springer (2010). https://doi.org/10.1007/978-3-642-17517-6_20
25. Jampani, K.R., Lubiw, A.: The simultaneous representation problem for chordal, comparability and permutation graphs. J. Graph Algorithms Appl. **16**(2), 283–315 (2012). https://doi.org/10.7155/jgaa.00259
26. Klavík, P., Kratochvíl, J., Krawczyk, T., Walczak, B.: Extending partial representations of function graphs and permutation graphs. In: Algorithms - 20th Annual European Symposium, ESA 2012. LNCS, vol. 7501, pp. 671–682. Springer (2012). https://doi.org/10.1007/978-3-642-33090-2_58
27. Klavík, P.: Extending partial representations of proper and unit interval graphs. Algorithmica **77**(4), 1071–1104 (2017). https://doi.org/10.1007/s00453-016-0133-z
28. Klavík, P., Kratochvíl, J., Vyskocil, T.: Extending partial representations of interval graphs. In: Theory and Applications of Models of Computation - 8th Annual Conference, TAMC 2011. LNCS, vol. 6648, pp. 276–285. Springer (2011). https://doi.org/10.1007/978-3-642-20877-5_28
29. Koebe, P.: Kontaktprobleme der konformen Abbildung. Ber. Verh. Sächs. Akad. Leipzig **88**, 141–164 (1936)

30. Kratochvíl, J.: A special planar satisfiability problem and a consequence of its NP-completeness. Discret. Appl. Math. **52**(3), 233–252 (1994). https://doi.org/10.1016/0166-218X(94)90143-0
31. Kratochvíl, J.: Intersection graphs of noncrossing arc-connected sets in the plane. In: Graph Drawing, Symposium on Graph Drawing, GD '96. LNCS, vol. 1190, pp. 257–270. Springer (1996). https://doi.org/10.1007/3-540-62495-3_53
32. Krawczyk, T., Walczak, B.: Extending partial representations of trapezoid graphs. In: Graph-Theoretic Concepts in Computer Science - 43rd International Workshop, WG 2017. LNCS, vol. 10520, pp. 358–371. Springer (2017). https://doi.org/10.1007/978-3-319-68705-6_27
33. McConnell, R.M.: Linear-time recognition of circular-arc graphs. Algorithmica **37**(2), 93–147 (2003). https://doi.org/10.1007/s00453-003-1032-7
34. McKee, T.A., McMorris, F.R.: Topics in intersection graph theory. Soc. Ind. Appl. Math. (1999)
35. Münch, M., I., R., P., S.: Partial and simultaneous transitive orientations via modular decompositions. In: 33rd International Symposium on Algorithms and Computation, ISAAC 2022, LIPIcs, pp. 1–16. Schloss Dagstuhl – Leibniz-Zentrum für Informatik (2022)
36. Münch, M., Rutter, I., Stumpf, P.: Partial and simultaneous transitive orientations via modular decompositions. Algorithmica **86**(4), 1263–1292 (2024). https://doi.org/10.1007/s00453-023-01188-y
37. Opatrny, J.: Total ordering problem. SIAM J. Comput. **8**(1), 111–114 (1979). https://doi.org/10.1137/0208008
38. Roberts, F.: Applications of Combinatorics and Graph Theory to the Biological and Social Sciences, The IMA Volumes in Mathematics and its Applications, vol. 17. Springer (1989)
39. Rutter, I., Strash, D., Stumpf, P., Vollmer, M.: Simultaneous representation of proper and unit interval graphs. In: Algorithms - 20th Annual European Symposium, ESA 2019. LIPIcs, vol. 144, pp. 1–15. Schloss Dagstuhl – Leibniz-Zentrum für Informatik (2019). https://doi.org/10.4230/LIPIcs.ESA.2019.80
40. Rutter, I., Stumpf, P.: Simultaneous representation of interval graphs in the sunflower case. In: 31st Annual European Symposium on Algorithms - ESA 2023. LIPIcs, vol. 274, pp. 1–15. Schloss Dagstuhl - Leibniz-Zentrum für Informatik (2023). https://doi.org/10.4230/LIPICS.ESA.2023.90
41. Spinrad, J.P.: Recognition of circle graphs. J. Algorithms **16**(2), 264–282 (1994). https://doi.org/10.1006/jagm.1994.1012
42. Stumpf, P.F.: Partial representation extension and simultaneous representation of intersection graphs. Phd thesis, University of Passau, Passau, Germany (2023)
43. Tucker, A.C.: An efficient test for circular-arc graphs. SIAM J. Comput. **9**(1), 1–24 (1980). https://doi.org/10.1137/0209001
44. Zhang, P., et al.: An algorithm based on graph theory for the assembly of contigs in physical mapping of DNA. Comput. Appl. Biosci. **10**(3), 309–317 (1994). https://doi.org/10.1093/bioinformatics/10.3.309

Partially Ordered Sets Corresponding to the Partition Problem

Susumu Kubo(✉)

General Education Center, Tottori University of Environmental Studies, Tottori, Japan
s-kubo@kankyo-u.ac.jp, boku446@gmail.com

Abstract. The partition problem is a well-known NP-complete problem. We focus on its optimization version. This problem has been extensively studied from various perspectives over time.

We propose two partially ordered sets (posets) corresponding to the partition problem and present a novel methodology for solving the problem. The first poset is order-isomorphic to a well-known poset whose structure is related to solutions of the subset sum problem, while the second is a subposet of the first and plays a crucial role in this paper. We first show several properties of the two posets, such as size, height, and width (the size of the largest antichain, i.e., the largest set of pairwise incomparable elements). Both widths are the same and $\Theta(2^n/n^{3/2})$ for n congruent to 0 or 3 modulo 4, which indicates the hardness of the partition problem. We then prove that the initial candidate solutions are the elements of the second poset, whose size is $2^n - 2\binom{n}{\lfloor n/2 \rfloor}$. Since a partition corresponds to two elements of the poset, the number of initial candidate partitions is half of that, i.e., $2^{n-1} - \binom{n}{\lfloor n/2 \rfloor}$. We finally prove that candidate solutions can be reduced based on the partial order. In particular, we give several polynomially solvable cases by considering the minimal and maximal elements of the second poset.

Our approach offers a useful tool for structural analysis of the partition problem.

Keywords: Partial order · Partition problem · Combinatorial optimization

1 Introduction

The partition problem is formulated as follows: given positive integers $c_1, \ldots, c_n$, the objective is to decide whether there exists a subset S of $\{1, \ldots, n\} =: [n]$ such that $\Delta(S) = 0$, where

$$\Delta(S) := \sum_{i \in S} c_i - \sum_{i \in [n] \setminus S} c_i. \tag{1}$$

In this paper, we focus on its optimization version: to find a subset S of $[n]$ that minimizes the absolute value $|\Delta(S)|$. Without loss of generality, we index the

A. Jeż and J. Otop (Eds.): FCT 2025, LNCS 16106, pp. 309–321, 2026.
https://doi.org/10.1007/978-3-032-04700-7_23

integers in non-increasing order ($c_1 \geq c_2 \geq \cdots \geq c_n$). We further assume that $c_n \geq 0$ for simplicity in the subsequent proofs, although originally $c_n > 0$.

The partition problem is one of Karp's 21 NP-complete problems [11] and is also one of Garey and Johnson's six basic NP-complete problems [5]. Therefore, the optimization version is known to be NP-hard. It has important applications in task scheduling [3,21]. In fact, it is equivalent to the two-identical parallel-machine scheduling problem to minimize the makespan.

The problem has been investigated from various perspectives. Several optimal algorithms give exact solutions in time exponential in n [5,8,12,17]. Moreover, a variety of heuristic and metaheuristic algorithms have been developed [1,4,9, 10,16]. An "easy-hard" phase transition was observed and the solution structure has been studied from the point of view of statistical mechanics [2,6,7,15]. An algebraic expression was presented based on max-plus algebra [13].

Furthermore, the number of solutions to the subset sum problem, a special case of which is the partition problem, has been studied. The subset sum problem with positive inputs is to decide whether there exists a subset S of $[n]$ such that

$$\sum_{i \in S} c_i = t,$$

where t is a given constant (the target). Let $N(\{c_1, ..., c_n\}, t)$ denote the number of solutions. It is known that

$$N\left(\{c_1, ..., c_n\}, t\right) \leq N\left([n], \left\lfloor \frac{n(n+1)}{4} \right\rfloor\right)$$

[14,19]. In fact, $N\left([n], \lfloor n(n+1)/4 \rfloor\right)$ is Sequence A025591 in the On-Line Encyclopedia of Integer Sequences [22] and is $\Theta(2^n/n^{3/2})$ for n congruent to 0 or 3 modulo 4 [20]. The inequality is established using the properties of a partially ordered set (poset).

The poset is a well-known poset, denoted by $M(n)$ in Stanley's paper [19, Subsection 4.1.2]. The elements of $M(n)$ are all the subsets of $[n]$. Let A and B be two subsets of $[n]$ with elements $a_1, a_2, \ldots, a_j$ and $b_1, b_2, \ldots, b_k$, respectively, and assume that the elements of A and B are ordered in decreasing order. The partial order is defined as follows: $A \leqslant B$ if and only if $j \leq k$ and $a_i \leq b_i$ for $i \in [j]$.

To the best of our knowledge, no previous work has applied posets to the partition problem from an optimization perspective.

1.1 Main Results

To define two posets that correspond to the partition problem, we first consider the set $\mathbb{R}^n$ of n- dimensional vectors of real numbers. Let $\mathbf{x} = (x_1, \ldots, x_n) \in \mathbb{R}^n$ and $\mathbf{y} = (y_1, \cdots y_n) \in \mathbb{R}^n$, and define a partial order on the set: $\mathbf{x} \preceq \mathbf{y}$ if and only if $x_1 + \cdots + x_k \leq y_1 + \cdots + y_k$ for all $k \in [n]$. In other words, $\mathbf{x} \preceq \mathbf{y}$ means that any entry of the cumulative sum of $\mathbf{x}$ is less than or equal to the corresponding one of $\mathbf{y}$.

Let us define a poset $P(n)$ that corresponds to the partition problem. Let $P(n)$ be the set of all n-dimensional vectors with entries in $\{1, -1\}$. The size of $P(n)$ is of course 2^n. The partial order is defined to be the restriction of $\preceq$ on $P(n)$, denoted by $\preceq_P$. The poset is order-isomorphic to $M(n)$ under the mapping f from $P(n)$ to $M(n)$ defined by the following rule: Given $\mathbf{v} = (v_1, \ldots, v_n) \in P(n)$, we define $f(\mathbf{v})$ by $i \in f(\mathbf{v})$ if and only if $v_{n+1-i} = 1$.

Example 1. The Hasse diagrams of $M(5)$ and $P(5)$ are shown in Figs. 1 and 2, respectively. We can see that the two posets are order-isomorphic under f. For example, $f((\ 1, -1,\ 1, -1, -1)) = \{5, 3\}$. Note that a partial order is shown horizontally due to space limitations, though it is usually shown vertically. The horizontal position indicates rank (See Subsect. 2.1).

Remark 1. An element of $P(n)$ can be also regarded as a path of the random walk on the integer number line which starts at 0 and at each step moves $+1$ or -1. The relation $\mathbf{v} \preceq_P \mathbf{w}$ means that the path corresponding to $\mathbf{v}$ does not "exceed" the one corresponding to $\mathbf{w}$.

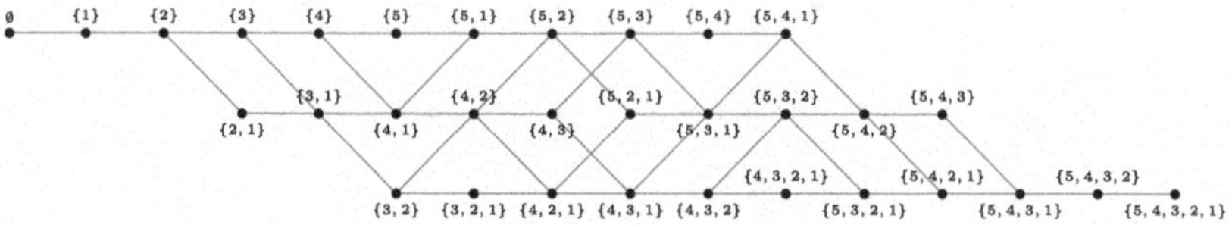

Fig. 1. The Hasse diagram of $M(5)$.

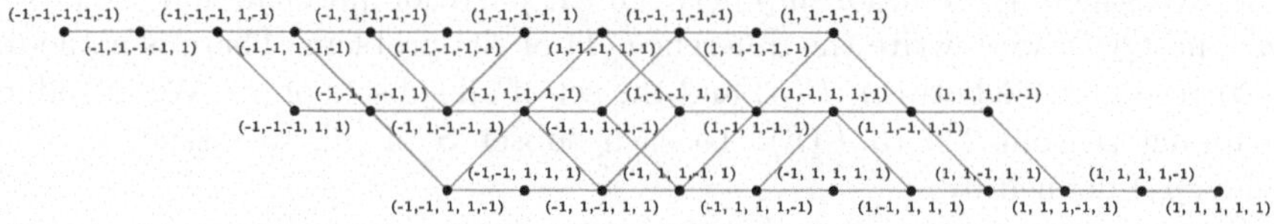

Fig. 2. The Hasse diagram of $P(5)$.

Let us define another poset $Q(n)$ to be the subset of $P(n)$ obtained by deleting from $P(n)$ the elements that are less than or greater than the zero vector, that is, $Q(n) = P(n) \setminus \{\mathbf{v} \in P(n) \colon \mathbf{v} \preceq \mathbf{0} \text{ or } \mathbf{0} \preceq \mathbf{v}\}$, where $\mathbf{0} = (0, \ldots, 0)$. The partial order is defined to be the restriction of $\preceq_P$ on $Q(n)$, denoted by $\preceq_Q$. The poset $Q(n)$ is obtained by truncating both the upper and lower parts of $P(n)$. The size of $Q(n)$ is proved to be $2^n - 2\binom{n}{\lfloor n/2 \rfloor}$.

Example 2. Let us consider $Q(n)$ for $n \leq 6$. We can easily see that $Q(1) = Q(2) = \emptyset$ and $Q(3) = \{(1, -1, -1), (-1, 1, 1)\}$, the elements of which are incomparable. The Hasse diagrams of $Q(4), Q(5)$ and $Q(6)$ are shown in Fig. 3.

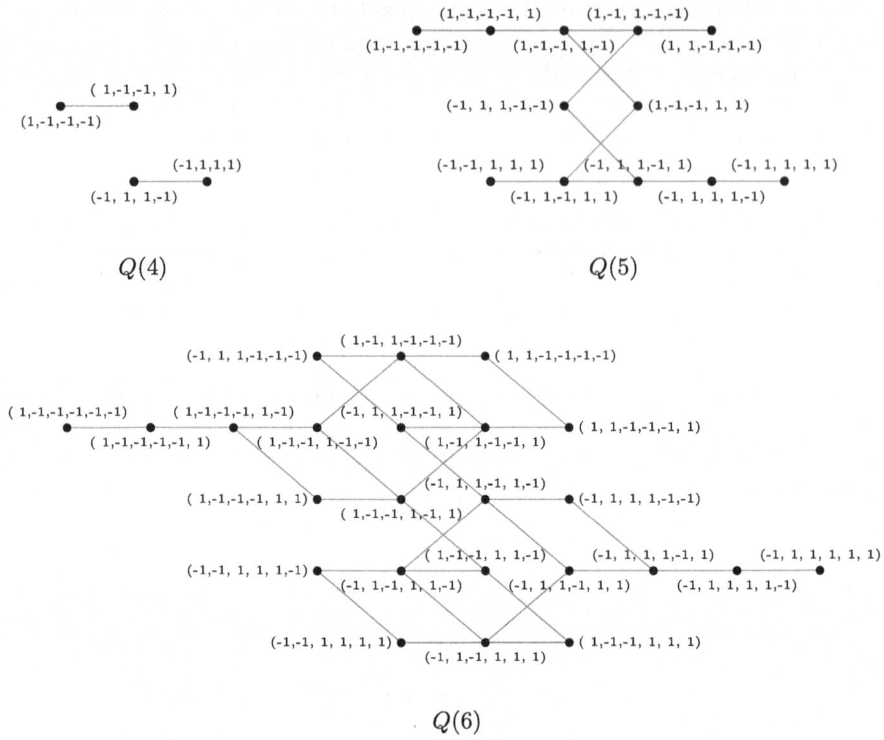

Fig. 3. The Hasse diagrams of $Q(4), Q(5)$ and $Q(6)$.

To investigate the relationship between the partition problem and the posets $P(n)$ and $Q(n)$, we rewrite the difference (1) of the problem. There is a one-to-one correspondence between $P(n)$ and the set of all subsets of $[n]$. We introduce a mapping $\mathbf{p}$ from $2^{[n]}$ to $P(n)$. To each subset S of $[n]$, we assign $\mathbf{p}(S) = (p_1, \ldots, p_n)$ defined by

$$p_i = \begin{cases} 1 & \text{if} \quad i \in S, \\ -1 & \text{if} \quad i \notin S. \end{cases}$$

Let $\mathbf{c} = (c_1, c_2, \ldots, c_n)$. Then

$$\Delta(S) = \mathbf{p}(S) \cdot \mathbf{c},$$

where the dot indicates the usual inner product. Note that $\mathbf{p}([n] \setminus S) = -\mathbf{p}(S)$ and hence $\Delta([n] \setminus S) = -\mathbf{p}(S) \cdot \mathbf{c}$, the absolute value of which is equal to $|\Delta(S)|$. A pair of two elements $\mathbf{p}(S)$ and $-\mathbf{p}(S)$ corresponds to the partition $\{S,\ [n] \setminus S\}$. We can show that we have only to consider $Q(n)$ for solving the partition problem. The *initial candidate solutions* are defined as subsets that cannot be ruled out a priori based on general properties of the problem. This set, called the solution space, is determined independently of any algorithm. A *candidate solution* is a subset that remains under consideration at a given stage of the optimization process. It is an element of the search space.

Theorem 1 (Characterization of Initial Candidate Solutions). *The initial candidate solutions to the partition problem are precisely the subsets corresponding to the elements of* $Q(n)$.

The number of initial candidate partitions is half of the size of $Q(n)$, i.e., $2^{n-1} - \binom{n}{\lfloor n/2 \rfloor}$, since a partition comprises a subset and its complement. Furthermore, candidate solutions can be reduced based on the partial order of $Q(n)$.

Theorem 2 (Order-Based Pruning of Candidates). *Given an instance* $\mathbf{c}$ *of the partition problem, for some subset* S *of* $[n]$ *satisfying* $\mathbf{p}(S) \in Q(n)$ *the following holds:*

(1) If $\mathbf{p}(S) \cdot \mathbf{c} \geq 0$, *then* $|\mathbf{p}(S) \cdot \mathbf{c}| \leq |\mathbf{p}(S') \cdot \mathbf{c}|$ *for any* S' *that satisfies* $\mathbf{p}(S') \preceq_Q -\mathbf{p}(S)$ *or* $\mathbf{p}(S) \preceq_Q \mathbf{p}(S')$.
(2) If $\mathbf{p}(S) \cdot \mathbf{c} \leq 0$, *then* $|\mathbf{p}(S) \cdot \mathbf{c}| \leq |\mathbf{p}(S') \cdot \mathbf{c}|$ *for any* S' *that satisfies* $\mathbf{p}(S') \preceq_Q \mathbf{p}(S)$ *or* $-\mathbf{p}(S) \preceq_Q \mathbf{p}(S')$.

Based on this result, we provide several polynomially solvable cases by considering the minimal and maximal elements of $Q(n)$.

Let $\ell = \lfloor (n-1)/2 \rfloor$. For $k \in \{0, 1, \ldots \ell\}$ we denote by $\mathbf{m}_k$ the element of $Q(n)$

$$(\underbrace{1, \ldots, 1}_{k}, \underbrace{-1, \ldots, -1}_{k+1}, 1, \ldots, 1).$$

Thus, $-\mathbf{m}_k$ stands for the element of $Q(n)$

$$(\underbrace{-1, \ldots, -1}_{k}, \underbrace{1, \ldots, 1}_{k+1}, -1, \ldots, -1).$$

In fact, $-\mathbf{m}_k$ and $\mathbf{m}_k$ are the minimal and maximal elements, respectively.

Theorem 3 (Polynomially Solvable Case for Minimal Elements). *Given an instance* $\mathbf{c}$ *of the partition problem, if* $-\mathbf{m}_k \cdot \mathbf{c} \geq 0$ *for some* $k \in \{0, 1, \ldots \ell\}$, *then the subset* S *such that* $\mathbf{p}(S) = -\mathbf{m}_k$ *is an optimal solution and the optimal value is* $-\mathbf{m}_k \cdot \mathbf{c}$.

Using the adding and swapping operators $\mathcal{A}^{(k)}$ and $\mathcal{S}^{(j,k)}$ defined in Sect. 2, which map an element to another, we immediately obtain the following corollary for the maximal elements.

Corollary 1 (Polynomially Solvable Case for Maximal Elements). *Given an instance* $\mathbf{c}$ *of the partition problem, if* $\mathbf{m}_k \cdot \mathbf{c} \geq 0$, $\left(\mathbf{m}_k - \mathcal{S}^{(k,k+1)}(-\mathbf{m}_k)\right) \cdot \mathbf{c} \leq 0$ *(applicable for* $k \neq 0$*), and* $\left(\mathbf{m}_k - \mathcal{A}^{(n)}(-\mathbf{m}_k)\right) \cdot \mathbf{c} \leq 0$ *(applicable for* $k \neq (n-1)/2$*) for some* $k \in \{0, 1, \ldots \ell\}$, *then the subset* S *such that* $\mathbf{p}(S) = \mathbf{m}_k$ *is an optimal solution and the optimal value is* $\mathbf{m}_k \cdot \mathbf{c}$.

Theorem 3 is a generalization of the obvious fact that the subset $\{1\}$ is an optimal solution (with optimal value $c_1-(c_2+\cdots+c_n)$ when $c_1-(c_2+\cdots+c_n) \geq 0$), corresponding to the minimal element $(1,-1,\ldots,-1)$. But it is difficult to find out this corollary without the poset $Q(n)$. It may be possible to derive similar conditions for other elements. We leave this as a problem for future work.

Our results are very simple once we realize them. However, the realization was the hard part. A key point is to use values $\{1,-1\}$ instead of $\{1,0\}$ when defining the posets. Another is to avoid arbitrarily halving candidate solutions by assumptions such as requiring every subset to contain 1. We believe that our approach provides a useful tool for structural analysis of the partition problem.

1.2 Organization of the Paper

The paper is organized as follows. In Sect. 2 we review basic poset concepts and the properties of $P(n)$ and $Q(n)$. In Sect. 3 we explore the relationship between the partition problem and these posets, and we prove our three main theorems.

Due to space limitations, some proofs are omitted.

2 Two Posets Corresponding to the Partition Problem

2.1 Preliminaries

In this subsection, we review some standard facts on posets and give some properties of the poset $M(n)$.

Let $(P, \preceq)$ be a finite poset. We simply denote the poset by P and define the relation $\prec$ by $x \prec y$ if and only if $x \preceq y$ and $x \neq y$. An element y *covers* x if $x \prec y$ and there is no $z \in P$ such that $x \prec z \prec y$. A subset of P is a *chain* if it is totally ordered, and a subset of P is an *antichain* if no pair of elements in it is comparable. The *height* (*width*) of P is the largest size of a chain (antichain, resp.) in P. An element of P is a *maximal* (*minimal*) *element* if it is not less (greater, resp.) than any other element. An element of P is the *greatest* (*least*) *element* if it is greater (less, resp.) than any other element.

A poset P is *graded*[1] if it is equipped with a *rank function* ρ from P to $\mathbb{Z}$ that satisfies:

(1) if $x \prec y$, then $\rho(x) < \rho(y)$;
(2) if y covers x, then $\rho(y) = \rho(x) + 1$.

An element $x \in P$ has *rank* i if $\rho(x) = i$. By setting the minimum value of the rank function to zero, any graded poset P can be partitioned into $P = \bigcup_{i=0}^{r} P_i$, where $P_i = \{x \in P\colon \rho(x) = i\}$. We denote the size of P_i by s_i. The poset P is *rank-symmetric* if $s_i = s_{r-i}$ for each i. The poset P is *rank-unimodal* if $s_0 \leq s_1 \leq \cdots \leq s_i \geq s_{i+1} \geq \cdots \geq s_r$ for some i, and P is *Sperner* if the width of P is equal to $\max_i\{s_i\}$. The width is lower bounded by $\max_i\{s_i\}$ since every P_i

[1] There are several competing definitions for graded posets.

is an antichain. In a Sperner poset, the largest one among $P_0, P_1, \ldots, P_r$ provides an antichain of maximum size.

A poset P is a *lattice* if any two elements $x, y \in P$ have a least upper bound and a greatest lower bound, denoted by $x \vee y$ and $x \wedge y$ respectively. A lattice L is *distributive* if $x \vee (y \wedge z) = (x \vee y) \wedge (x \vee z)$ for each $x, y, z \in L$.

The poset $M(n)$ has the following properties [18]:

- The rank function with minimum rank 0 is the sum of elements of a subset. The maximum rank is $n(n+1)/2$.
- The poset $M(n)$ is rank-symmetric, rank-unimodal and Sperner. Therefore, its width is equal to the size of the middle rank, which is $N([n], \lfloor n(n+1)/4 \rfloor)$.
- Its height is $n(n+1)/2 + 1$.
- The poset $M(n)$ is a distributive lattice with the greatest element $[n]$ and the least element $\emptyset$.

2.2 Properties of the Two Posets

We will provide the properties of $P(n)$ and $Q(n)$. We first need to become familiar with some elementary properties of the partial order $\preceq$.

Proposition 1. *For* $\mathbf{x}, \mathbf{y}, \mathbf{z} \in \mathbb{R}^n$*, the following holds:*

(1) $\mathbf{x} \preceq \mathbf{y} \Longleftrightarrow \mathbf{x} + \mathbf{z} \preceq \mathbf{y} + \mathbf{z}$.
(2) $\mathbf{x} \preceq \mathbf{y}$ *and* $\alpha \geq 0 \Longrightarrow \alpha\mathbf{x} \preceq \alpha\mathbf{y}$.
(3) $\mathbf{x} \preceq \mathbf{y}$ *and* $\alpha \leq 0 \Longrightarrow \alpha\mathbf{y} \preceq \alpha\mathbf{x}$.

Proof. Straightforward. □

In order to examine the poset $P(n)$, we define two operators. Let us denote the ith entry of a vector $\mathbf{v}$ by v_i hereafter.

Definition 1 (Addition operator). *For* $k \in [n]$*, the addition operator* $\mathcal{A}^{(k)}$ *takes the input* $\mathbf{v} \in P(n)$ *with* $v_k = -1$ *and sets the* k*th entry to* 1*, that is,*

$$\left(\mathcal{A}^{(k)}\mathbf{v}\right)_i = \begin{cases} v_k + 2 = 1 & \text{if } i = k \\ v_i & \text{if } i \neq k. \end{cases}$$

For other elements $\mathbf{v}$*, the output* $\mathcal{A}^{(k)}\mathbf{v}$ *is not defined.*

Definition 2 (Swap operator). *For* $j, k \in [n]$ *with* $j < k$*, the swap operator* $\mathcal{S}^{(j,k)}$ *takes the input* $\mathbf{v} \in P(n)$ *with* $v_j = -1$ *and* $v_k = 1$*, and swaps* v_j *and* v_k*, that is,*

$$\left(\mathcal{S}^{(j,k)}\mathbf{v}\right)_i = \begin{cases} v_k = 1 & \text{if } i = j \\ v_j = -1 & \text{if } i = k \\ v_i & \text{if } i \notin \{j, k\}. \end{cases}$$

For other elements $\mathbf{v}$*, the output* $\mathcal{S}^{(j,k)}\mathbf{v}$ *is not defined.*

We can show some relationships between the operators and the partial order $\preceq_P$.

Proposition 2. *For $\mathbf{v}, \mathbf{w} \in P(n)$, the relation $\mathbf{v} \prec_P \mathbf{w}$ holds if and only if $\mathbf{w}$ is obtained from $\mathbf{v}$ by applying addition and/or swap operators.*

Proposition 3. *For $\mathbf{v}, \mathbf{w} \in P(n)$, the element $\mathbf{w}$ covers $\mathbf{v}$ if and only if $\mathbf{w}$ is obtained from $\mathbf{v}$ by applying exactly one of the following operators: $\mathcal{A}^{(n)}$ or one of $\mathcal{S}^{(1,2)}, \mathcal{S}^{(2,3)}, \dots, \mathcal{S}^{(n-1,n)}$.*

Proposition 4. *The poset $P(n)$ is order-isomorphic to $M(n)$.*

Proof. We need to show that $\mathbf{v} \preceq_P \mathbf{w}$ if and only if $f(\mathbf{v}) \leqslant f(\mathbf{w})$ for each $\mathbf{v}$ and $\mathbf{w}$ in $P(n)$.

Assume that $\mathbf{v} \preceq_P \mathbf{w}$. From Proposition 2 it follows that $\mathbf{w}$ is obtained from $\mathbf{v}$ by applying addition and/or swap operators. An addition operator increases 1's of the input by one and a swap operator shifts one of 1's of the input to the left. Now let $\mathbf{v}$ have m 1's. Then $\mathbf{w}$ has at least m 1's, which implies that $f(\mathbf{w})$ has at least as many elements as $f(\mathbf{v})$. The m left-most 1's of $\mathbf{w}$ do not lie more to the right than the 1's of $\mathbf{v}$, which implies that in decreasing order the ith element of $f(\mathbf{w})$ is not less than that of $f(\mathbf{v})$ for $i \in [m]$. Hence, $f(\mathbf{v}) \leqslant f(\mathbf{w})$.

Conversely, assume that $f(\mathbf{v}) \leqslant f(\mathbf{w})$. Let $f(\mathbf{v}) = \{a_1, \dots, a_j\}$ and $f(\mathbf{w}) = \{b_1, \dots, b_k\}$ in decreasing order. Then $j \leq k$, which means that $\mathbf{w}$ has at least as many 1's as $\mathbf{v}$, and $a_i \leq b_i$ for $i \in [j]$, which means that the j left-most 1's of $\mathbf{w}$ do not lie more to the right than the 1's of $\mathbf{v}$. By the properties of addition and swap operators, $\mathbf{w}$ is obtained from $\mathbf{v}$ by applying addition and/or swap operators, that is, $\mathbf{v} \preceq_P \mathbf{w}$. □

The poset $P(n)$ has the following symmetry.

Proposition 5. *For $\mathbf{v}, \mathbf{w} \in P(n)$, the relation $\mathbf{v} \prec_P \mathbf{w}$ holds if and only if $-\mathbf{w} \prec_P -\mathbf{v}$.*

Proof. For each $\mathbf{v} \in P(n)$, it holds that $-\mathbf{v} \in P(n)$. The statement follows immediately from Proposition 1 (3). □

It follows from Proposition 4 and its proof and Proposition 5 that the poset $P(n)$ has the following properties:

- The rank function with minimum rank 0 is
$$\frac{1}{2}\left(\mathbf{v} \cdot \mathbf{c}_0 + \frac{n(n+1)}{2}\right) \tag{2}$$
for $\mathbf{v} \in P(n)$, where $\mathbf{c}_0 = (n, n-1, \dots, 2, 1)$.
- The poset $P(n)$ is rank-symmetric, rank-unimodal, and Sperner.
- Its width is $N([n], \lfloor n(n+1)/4 \rfloor)$.
- Its height is $n(n+1)/2 + 1$.
- The poset $P(n)$ has a certain symmetry: $\mathbf{v} \prec_P \mathbf{w} \Longleftrightarrow -\mathbf{w} \prec_P -\mathbf{v}$.

- The poset $P(n)$ is a distributive lattice. The greatest element and the least one are respectively $(1, \dots, 1)$ and $(-1, \dots, -1)$.

We investigate the other poset $Q(n)$. Hereafter, let $n \geq 3$. We will show some properties of $Q(n)$.

Let $R_+(n) = \{\mathbf{v} \in P(n) \colon \mathbf{0} \prec \mathbf{v}\}$ and $R_-(n) = \{\mathbf{v} \in P(n) \colon \mathbf{v} \prec \mathbf{0}\}$. We can see that the subposets $R_+(n)$ and $R_-(n)$ are sublattices of $P(n)$. The least element of $R_+(n)$ is $(1, -1, 1, -1, \dots)$, the rank of which in $P(n)$ is $n(n+2)/4$ if n is even and $(n+1)^2/4$ if n is odd. The elements of $R_+(n)$ are not in the middle rank of $P(n)$. This is true for $R_-(n)$ by the symmetry of $P(n)$. Therefore, we obtain the next proposition.

Proposition 6. *The width of $Q(n)$ is the same as that of $P(n)$.*

For the size of $Q(n)$, the next proposition holds.

Proposition 7. *The size of $Q(n)$ is $2^n - 2\binom{n}{\lfloor n/2 \rfloor}$.*

Proof. By definition, $\#Q(n) = \#P(n) - \#R_+(n) - \#R_-(n)$, where $\#$ indicates the size of a set. Since $\#P(n) = 2^n$ and $\#R_+(n) = \#R_-(n)$, it is sufficient to show that $\#R_+(n) = \binom{n}{\lfloor n/2 \rfloor}$. Let us regard the elements of $R_+(n)$ as the paths of the random walk. The size is the number of n-step paths from 0 never going below 0. Let X_n be this count, with $X_0 = 1$. Then we have

$$X_n = \begin{cases} 2X_{n-1} - C_{(n-1)/2} & \text{if } n \text{ is odd} \\ 2X_{n-1} & \text{if } n \text{ is even,} \end{cases}$$

where C_m is the m-th Catalan number. The sequence X_n is known to be $\binom{n}{\lfloor n/2 \rfloor}$. □

The poset $Q(n)$ has a similar symmetry.

Proposition 8. *For $\mathbf{v}, \mathbf{w} \in Q(n)$, the relation $\mathbf{v} \prec_Q \mathbf{w}$ holds if and only if $-\mathbf{w} \prec_Q -\mathbf{v}$.*

Proof. For each $\mathbf{v} \in Q(n)$, we show that $-\mathbf{v} \in Q(n)$. Suppose, contrary to our claim, that $-\mathbf{v} \notin Q(n)$. Then $-\mathbf{v} \prec \mathbf{0}$ or $\mathbf{0} \prec -\mathbf{v}$, which contradicts $\mathbf{v} \in Q(n)$. The statement follows immediately from Proposition 1 (3). □

Proposition 9. *The poset $Q(n)$ is graded and one of the rank functions is the same as the rank function (2) of $P(n)$.*

Proof. Denote the rank function of $P(n)$ by ρ. We show that a rank function of $Q(n)$ is ρ. Since $Q(n)$ is a subposet of $P(n)$, if $\mathbf{v} \prec_Q \mathbf{w}$ then $\rho(\mathbf{v}) < \rho(\mathbf{w})$. When $\mathbf{w}$ covers $\mathbf{v}$ in $Q(n)$, suppose, contrary to our claim, that $\rho(\mathbf{w}) \geq \rho(\mathbf{v}) + 2$. Then there exists some $\mathbf{u} \in P(n) \setminus Q(n)$ such that $\mathbf{v} \prec_P \mathbf{u} \prec_P \mathbf{w}$ and $\rho(\mathbf{u}) = \rho(\mathbf{v}) + 1$. Since $\mathbf{u}$ is not in $Q(n)$, $\mathbf{u} \prec \mathbf{0}$ or $\mathbf{0} \prec \mathbf{u}$. If $\mathbf{u} \prec \mathbf{0}$ then $\mathbf{v} \prec \mathbf{0}$. If $\mathbf{0} \prec \mathbf{u}$ then $\mathbf{0} \prec \mathbf{w}$. Either case is a contradiction. □

Proposition 10. *The maximal (minimal) elements of* $Q(n)$ *are* $\mathbf{m}_0$, $\mathbf{m}_1$, $\ldots, \mathbf{m}_\ell$ $(-\mathbf{m}_0, -\mathbf{m}_1, \ldots, -\mathbf{m}_\ell$, *resp.).*

Proposition 11. *The height of* $Q(n)$ *is*

$$\frac{n(n-1)}{2} - \left(n - \frac{3}{2}\ell\right)(\ell+1) + 1 \qquad \text{for } n \le 7, \text{ and}$$

$$\frac{(n-2)(n-3)}{2} + 1 \qquad \text{for } n \ge 8.$$

In summary, we have the following properties of $Q(n)$:

- The rank function with minimum rank 0 is

$$\frac{1}{2}\left(\mathbf{v} \cdot \mathbf{c}_0 + \frac{n(n+1)}{2}\right) - n$$

for $\mathbf{v} \in Q(n)$.
- Its size is $2^n - 2\binom{n}{\lfloor n/2 \rfloor}$.
- The poset $Q(n)$ is rank-symmetric and Sperner.
- Its width is the same as that of $P(n)$, i.e., $N([n], \lfloor n(n+1)/4 \rfloor)$.
- Its height is $n(n-1)/2 - (n - 3\ell/2)(\ell+1) + 1$ for $n \le 7$ and $(n-2)(n-3)/2+1$ for $n \ge 8$.
- The poset $Q(n)$ also has the same symmetry as $P(n)$: $\mathbf{v} \prec_Q \mathbf{w} \Longleftrightarrow -\mathbf{w} \prec_Q -\mathbf{v}$.
- The minimal elements and the maximal ones are respectively $-\mathbf{m}_0, -\mathbf{m}_1, \ldots, -\mathbf{m}_\ell$ and $\mathbf{m}_0, \mathbf{m}_1, \ldots, \mathbf{m}_\ell$.

It appears difficult to determine whether $Q(n)$ is rank-unimodal. Empirically, $Q(n)$ is rank-unimodal for $n \le 21$.

3 Relationship Between the Partition Problem and the Posets

We will discuss the relationship between the partition problem and the posets $P(n)$ and $Q(n)$. We first relate the difference $\mathbf{p}(S) \cdot \mathbf{c}$ in the partition problem with the partial orders $\preceq_P$ and $\preceq_Q$.

Proposition 12. *Given two subsets* S *and* S' *of* $[n]$, *the inequality* $\mathbf{p}(S) \cdot \mathbf{c} \le \mathbf{p}(S') \cdot \mathbf{c}$ *holds for any instance* $\mathbf{c}$ *if and only if* $\mathbf{p}(S) \preceq_P \mathbf{p}(S')$.

A similar statement holds for the poset $Q(n)$.

Corollary 2. *Given two subsets* S *and* S' *of* $[n]$ *with* $\mathbf{p}(S), \mathbf{p}(S') \in Q(n)$, *the inequality* $\mathbf{p}(S) \cdot \mathbf{c} \le \mathbf{p}(S') \cdot \mathbf{c}$ *holds for any instance* $\mathbf{c}$ *if and only if* $\mathbf{p}(S) \preceq_Q \mathbf{p}(S')$.

Next, we identify subsets that can be ignored in solving the partition problem.

Lemma 1. *If a subset S of $[n]$ has $\mathbf{p}(S) \prec \mathbf{0}$ or $\mathbf{0} \prec \mathbf{p}(S)$, then there exists some subset S' of $[n]$, not equal to S or $[n] \setminus S$, such that $|\mathbf{p}(S') \cdot \mathbf{c}| \leq |\mathbf{p}(S) \cdot \mathbf{c}|$ for any instance $\mathbf{c}$.*

Thus, Lemma 1 implies that we need not consider elements of $P(n) \setminus Q(n)$. We then show that, in general, all elements of $Q(n)$ must be considered.

Lemma 2. *Let a subset S of $[n]$ satisfy $\mathbf{p}(S) \in Q(n)$. Then there exists some instance $\mathbf{c}$ such that $|\mathbf{p}(S) \cdot \mathbf{c}| < |\mathbf{p}(S') \cdot \mathbf{c}|$ for any S' that satisfies $\mathbf{p}(S') \in Q(n)$ and is not equal to S or $[n] \setminus S$.*

We thus can prove Theorem 1.

Proof of Theorem 1. This follows immediately from Lemmas 1 and 2. □

We will consider a specific instance $\mathbf{c}$. We can prune candidate solutions based on the partial order.

Proposition 13. *Given an instance $\mathbf{c}$, for some subset S of $[n]$, the following holds:*

(1) If $\mathbf{p}(S) \cdot \mathbf{c} \geq 0$, then $|\mathbf{p}(S) \cdot \mathbf{c}| \leq |\mathbf{p}(S') \cdot \mathbf{c}|$ for any S' such that $\mathbf{p}(S') \preceq_P -\mathbf{p}(S)$ or $\mathbf{p}(S) \preceq_P \mathbf{p}(S')$.
(2) If $\mathbf{p}(S) \cdot \mathbf{c} \leq 0$, then $|\mathbf{p}(S) \cdot \mathbf{c}| \leq |\mathbf{p}(S') \cdot \mathbf{c}|$ for any S' such that $\mathbf{p}(S') \preceq_P \mathbf{p}(S)$ or $-\mathbf{p}(S) \preceq_P \mathbf{p}(S')$.

Proof. We show the first statement. From Proposition 12 it follows that $\mathbf{p}(S') \cdot \mathbf{c} \leq -\mathbf{p}(S) \cdot \mathbf{c}$ or $\mathbf{p}(S) \cdot \mathbf{c} \leq \mathbf{p}(S') \cdot \mathbf{c}$. Hence, $\mathbf{p}(S') \cdot \mathbf{c} \leq -\mathbf{p}(S) \cdot \mathbf{c} \leq 0$ or $0 \leq \mathbf{p}(S) \cdot \mathbf{c} \leq \mathbf{p}(S') \cdot \mathbf{c}$. Either case implies that $|\mathbf{p}(S) \cdot \mathbf{c}| \leq |\mathbf{p}(S') \cdot \mathbf{c}|$.

In the same manner, we can prove the second. □

We can show that a similar statement, which is Theorem 2, holds for $Q(n)$.

Proof of Theorem 2. It immediately follows from Proposition 13. □

Remark 2. The width of $Q(n)$ is $\Theta(2^n/n^{3/2})$ for n congruent to 0 or 3 modulo 4. The large width indicates the hardness of the partition problem. That is, this exponential width precludes a polynomial-time exhaustive search algorithm based on the partial order.

We will give some polynomially solvable cases by considering the minimal and maximal elements of $Q(n)$. To this end, we show the following property of the minimal elements of $Q(n)$.

Lemma 3. *Given any subset S of $[n]$, each minimal element of $Q(n)$ lies below either $\mathbf{p}(S)$ or $-\mathbf{p}(S)$ with respect to $\preceq_Q$.*

Using this lemma, we can prove Theorem 3 and Corollary 1.

Proof of Theorem 3. By Lemma 3, we have $-\mathbf{m}_k \preceq_Q \mathbf{p}(S)$ or $-\mathbf{m}_k \preceq_Q -\mathbf{p}(S)$ for any $\mathbf{p}(S) \in Q(n)$. Applying Theorem 2 (1), it follows that $|\mathbf{m}_k \cdot \mathbf{c}| \leq |\mathbf{p}(S) \cdot \mathbf{c}|$. Since $-\mathbf{m}_k \cdot \mathbf{c} \geq 0$, the optimal value is $-\mathbf{m}_k \cdot \mathbf{c} = |\mathbf{m}_k \cdot \mathbf{c}|$. □

Proof of Corollary 1. Rewriting the three inequalities, we have

$$0 \leq \mathbf{m}_k \cdot \mathbf{c} \leq \mathcal{S}^{(k,k+1)}(-\mathbf{m}_k) \cdot \mathbf{c}, \quad 0 \leq \mathbf{m}_k \cdot \mathbf{c} \leq \mathcal{A}^{(n)}(-\mathbf{m}_k) \cdot \mathbf{c}. \tag{3}$$

Note that $\mathcal{S}^{(k,k+1)}(-\mathbf{m}_k) \cdot \mathbf{c} \geq 0$ and $\mathcal{A}^{(n)}(-\mathbf{m}_k) \cdot \mathbf{c} \geq 0$. The only elements covering the minimal element $-\mathbf{m}_k$ are $\mathcal{S}^{(k,k+1)}(-\mathbf{m}_k)$ for $k \neq 0$ and $\mathcal{A}^{(n)}(-\mathbf{m}_k)$ for $k \neq (n-1)/2$. By Theorem 2 (1) and Lemma 3, we deduce that the candidate solutions are reduced to six elements: $\pm\mathbf{m}_k$, $\pm\mathcal{S}^{(k,k+1)}(-\mathbf{m}_k)$, and $\pm\mathcal{A}^{(n)}(-\mathbf{m}_k)$. The inequalities (3) imply that the subsets S with $\mathbf{p}(S) = \pm\mathbf{m}_k$ are optimal solutions and the optimal value is $\mathbf{m}_k \cdot \mathbf{c}$. □

References

1. Alidaee, B., Glover, F., Kochenberger, G.A., Rego, C.: A new modeling and solution approach for the number partitioning problem. J. Appl. Math. Decis. Sci. **2**, 113–121 (2005). https://doi.org/10.1155/JAMDS.2005.113
2. Borgs, C., Chayes, J., Pittel, B.: Phase transition and finete-size scaling for the integer partitioning problem. Random Struct. Algorithms **19**, 247–288 (2001). https://doi.org/10.1002/rsa.10004
3. Coffman, E.G., Jr., Lueker, G.S.: Probabilistic Analysis of Packing and Partitioning Algorithms. Wiley, Chichester (1991)
4. Fuksz, L., Pop, P.C.: A hybrid genetic algorithm with variable neighborhood search approach to the number partitioning problem. In: Pan, J.S., Polycarpou, M.M., Woźniak, M., de Carvalho, A.C.P.L.F., Quintián, H., Corchado, E. (eds.) 8th International Conference on Hybrid Artificial Intelligence Systems, pp. 649–658 (2013). https://doi.org/10.1007/978-3-642-40846-5_65
5. Garey, M.R., Johnson, D.S.: Computers and intractability: a guide to the theory of NP-completeness. W. H. Freeman, New York (1979)
6. Gent, I.P., Walsh, T.: Phase transitions and annealed theories: number partitioning as a case study. In: Wahlster, W. (ed.) ECAI 96: 12th European Conference on Artificial Intelligence, pp. 170–174. Wiley, Chichester (1996)
7. Hartmann, A.K., Mann, A., Radenbach, W.: Solution-space structure of (some) optimization problems. J. Phys. Conf. Ser. **95**, 012011 (2008). https://doi.org/10.1088/1742-6596/95/1/012011
8. Horowitz, E., Sahni, S.: Computing partitions with applications to the knapsack problem. J. ACM **21**, 277–292 (1974). https://doi.org/10.1145/321812.321823
9. Johnson, D.S., Aragon, C.R., McGeoch, L.A., Schevon, C.: Optimization by simulated annealing: an experimental evaluation; part II, graph coloring and number partitioning. Oper. Res. **39**, 378–406 (1991). https://doi.org/10.1287/opre.39.3.378
10. Karmarkar, N., Karp, R.M.: The differencing method of set partitioning. Technical Report, Computer Science Division, University of California, Berkeley 82/113 (1982). http://www2.eecs.berkeley.edu/Pubs/TechRpts/1983/6353.html
11. Karp, R.M.: Reducibility among combinatorial problems. In: Miller, R.E., Thatcher, J.W., Bohlinger, J. (eds.) Complexity of Computer Computations, pp. 85–103. Springer, Boston (1972). https://doi.org/10.1007/978-1-4684-2001-2_9
12. Korf, R.E.: A complete anytime algorithm for number partitioning. Artif. Intell. **106**, 181–203 (1998). https://doi.org/10.1016/S0004-3702(98)00086-1, https://doi.org/10.1016/S0004-3702(98)00086-1

13. Kubo, S., Nishinari, K.: An algebraic expression of the number partitioning problem. Discrete Appl. Math. **285**, 283–296 (2020). https://doi.org/10.1016/j.dam.2020.04.020
14. Lindström: Conjecture on a theorem similar to Sperner's. In: Guy, R., Hanani, H., Sauer, N., Schönheim, J. (eds.) Combinatorial Structures and Their Applications, p. 241. Gordon and Breach, New York (1970)
15. Mertens, S.: Phase transition in the number partitioning problem. Phys. Rev. Lett. **81**, 4281–4284 (1998). https://doi.org/10.1103/PhysRevLett.81.4281
16. Ruml, W., Ngo, J.T., Marks, J., Shieber, S.M.: Easily searched encodings for number partitioning. J. Optim. Theory Appl. **89**, 251–291 (1996). https://doi.org/10.1007/BF02192530
17. Schroeppels, R., Shamir, A.: A $T = O(2^{n/2})$, $S = O(2^{n/4})$ algorithm for certain NP-complete problems. SIAM J. Comput. **10**, 456–464 (1981). https://doi.org/10.1137/0210033
18. Stanley, R.P.: Weyl groups, the hard Lefschetz theorem, and the Sperner property. SIAM J. Algebraic Discrete Methods **1**, 168–184 (1980). https://doi.org/10.1137/0601021
19. Stanley, R.P.: Some applications of algebra to combinatorics. Discret. Appl. Math. **34**, 241–277 (1991). https://doi.org/10.1016/0166-218X(91)90089-F
20. Sullivan, B.D.: On a conjecture of Andrica and Tomescu. J. Integer Sequences **16**, Article 13.3.1 (2013)
21. Tsai, L.H.: Asymptotic analysis of an algorithm for balanced processor scheduling. SIAM J. Comput. **21**, 59–64 (1992). https://doi.org/10.1137/0221007
22. Wilson, D.W.: Sequence A025591. The On-Line Encyclopedia of Integer Sequences (2023). https://oeis.org/A025591

Disjointness, Inclusion, and Regularity of ω-Rational Trace Languages – Extended Abstract –

Dietrich Kuske(✉)

Technische Universität Ilmenau, Ilmenau, Germany
dietrich.kuske@tu-ilmenau.de

Abstract. This paper studies the disjointness, the inclusion, and the regularity problem for nonterminating behaviors of concurrent systems modelled using Mazurkiewicz traces. The main finding is a trichotomy that, for every independence alphabet, determines the complexity of the problems (almost) completely. Noteworthy, for all the problems, the classes are the same. While these classes already appeared in the study of terminating behaviors, the trichotomy obtained here is more pronounced as the complexities vary from decidable via low levels of the arithmetical hierarchy to low levels of the analytical hierarchy (that do not feature in the result for terminating behaviors).

1 Introduction

Free partially commutative monoids were first studied by Cartier and Foata in combinatorics [4]. Later, Mazurkiewicz introduced them as trace monoids into computer science as a means for the description of the behavior of certain concurrent systems [22]. More recently, traces have also been used to model multi-buffer systems [15] and multi-pushdown systems [17,18]. Graph monoids are a more general notion [19,20] and are used in valence systems [10,32] to model systems with multiple partially blind counters and pushdowns. Starting with [7,9,11,12], interest has extended to infinite traces as they allow to model nonterminating behaviors of concurrent systems.

The idea behind traces is that concurrent processes synchronize via shared actions. Then sequences of basic actions, i.e., words, form interleavings of the executions of these processes. Words that are interleavings of one and the same concurrent execution are considered equivalent. Formally, the concurrency in a system is given by an independence alphabet that describes which pairs of actions are executed by disjoint sets of sequential processes and can therefore commute in interleaving executions. As a consequence, questions like "Do two concurrent systems allow the same concurrent executions?" translate into "Does every interleaving execution of one system have an equivalent interleaving execution of the other system?" or "Are the closures of the sets of interleaving executions under the equivalence of words equal?" In other words, one is interested in the closure of a language under this equivalence.

A. Jeż and J. Otop (Eds.): FCT 2025, LNCS 16106, pp. 322–335, 2026.
https://doi.org/10.1007/978-3-032-04700-7_24

For regular sets of finite words, questions of this type (in particular, disjointness, inclusion, and regularity of closures) have been studied in detail. Here, e.g., non-inclusion asks for the existence of some finite object (a word) with some decidable property (to belong to the first regular set, but not to the closure of the second). Hence inclusion belongs to Π_1^0, the first universal level of the arithmetical hierarchy. Differently, regularity asks for the existence of a finite object (a finite automaton) such that all finite objects (words) satisfy some decidable property (to belong to the closure if, and only if, they are accepted by the automaton). Hence, this problem belongs to Σ_2^0, the second existential level of the arithmetical hierarchy. These easy observations do not say that inclusion or regularity cannot be expressed by simpler means, e.g., are decidable or belong Π_1^0. Using two properties of independence alphabets (transitivity and diagonality), borders between decidability and Π_1^0-hardness have been identified for disjointness [1] (diagonal vs. non-diagonal) and for inclusion [2] and regularity [29] (transitive vs. non-transitive). While disjointness and inclusion belong to Π_1^0, the precise complexity of regularity remained open. Section 3 settles this question: regularity is Σ_2^0-complete for every non-transitive independence alphabet. Computationally, this says that neither the regularity nor the non-regularity of the closure of some regular set is semi-decidable.

The main part of this paper deals with nonterminating behaviours, i.e., ω-words and Büchi-automata. To express disjointness of closures in this setting, it does not suffice to talk about finite objects, but one needs to make statements about all ω-words. As a consequence, disjointness, inclusion, and regularity now belong to Π_1^1 and Π_2^1, i.e., the first two universal levels of the analytical hierarchy. The question, again, is whether simpler means suffice to characterise, e.g., those Büchi-automata that accept some ω-language with regular closure. Also in this case, transitivity and diagonality characterise the situation completely: (1) for transitive independence alphabets, disjointness and inclusion are decidable while regularity belongs to Π_1^0, (2) for non-transitive diagonal independence alphabets, disjointness belongs to Π_1^0, inclusion is complete for Π_1^0, and regularity is complete for Σ_2^0, and (3) for non-diagonal independence alphabets, disjointness is complete for Π_1^1 and inclusion and regularity both are hard for Π_1^1. Thus, diagonal independence alphabets allow to express disjointness etc. referring to finite objects only while this is not possible for any non-diagonal independence alphabet. Furthermore, transitive and diagonal independence alphabets behave much the same in the finite and the infinite setting (it is still open whether regularity for the transitive case and disjointness for the diagonal case are decidable).

The following section introduces all the necessary concepts for this work, Sect. 3 deals with finite words, ω-words are handled thereafter. Section 4 presents our results for transitive independence alphabets. The decidability proofs for disjointness and inclusion are similar to the corresponding arguments from [2] for finite words. For the decidability of the regularity problem for finite words, Sakarovitch uses minimal deterministic automata which are not available for ω-words. As a consequence, we are only able to prove semi-decidability. The new results for diagonal independence alphabets can be found in Sect. 5. Technically, the most interesting finding is that non-disjointness and non-inclusion are necessarily witnessed by some ultimately periodic word; this proof is inspired by Sakarovitch's bipartite automata [28,29]. Non-diagonal independence alphabets are the topic of Sect. 6. The main results are analytical lower bounds for all problems. Constructions as in [1] (that were also used in [19,20,33]) yield this lower bound for the inclusion and the disjointness problem. For the regularity problem, this proof technique is merged with the basic idea of the undecidability proof from [29].

2 Preliminaries

2.1 Words, Automata

Let Γ be an *alphabet*, i.e., a finite non-empty set. The set of finite words over Γ is denoted Γ^*, Γ^ω denotes the set of ω-words over Γ, and $\Gamma^\infty = \Gamma^* \cup \Gamma^\omega$. For $x \in \Gamma^\infty$, the *length* of x is denoted $|x| \in \mathbb{N} \cup \{\omega\}$. For $A \subseteq \Gamma$, the *projection* $\mathrm{proj}_A(x)$ of x to A is obtained by deleting all letters from x that do not belong to A; $|\alpha|_A$ is the length of $\mathrm{proj}_A(\alpha)$. For two words or ω-words $x, y \in \Gamma^\infty$, the *shuffle* $x \sqcup\!\sqcup\, y \subseteq \Gamma^\infty$ is the set of all words or ω-words $x_0 y_0\, x_1 y_1\, x_2 y_2 \cdots$ with $x_i, y_i \in \Gamma^*$ where $x = x_0 x_1 x_2 \cdots$ and $y = y_0 y_1 y_2 \cdots$; $K \sqcup\!\sqcup\, L$ is the union of all shuffles $x \sqcup\!\sqcup\, y$ with $x \in K$ and $y \in L$.

A *nondeterministic finite automaton* or *nfa* is a tuple $\mathcal{A} = (Q, \Gamma, S, T, F)$ where Q is a finite set of "states", Γ an alphabet, $S, F \subseteq Q$ sets of "initial" and "final states", respectively, and $T \subseteq Q \times \Gamma \times Q$ the transition relation. For $p, q \in Q$ and $u \in \Gamma^*$, we write $p \xrightarrow{u} q$ if there is some u-labeled path from p to q. A word $u \in \Gamma^*$ is *accepted* by $\mathcal{A}$ if there are $p \in S$ and $q \in F$ with $p \xrightarrow{u} q$; $L^*(\mathcal{A})$ denotes the language of words accepted by $\mathcal{A}$.

An ω-word α is *accepted* by $\mathcal{A}$ if there are words $u_0, u_1, \dots \in \Gamma^+$ with $\alpha = u_0\, u_1 \cdots$ and states $p \in S$ and $q \in F$ with $p \xrightarrow{u_0} q$ and $q \xrightarrow{u_i} q$ for all $i \geq 1$. The set of ω-words accepted by $\mathcal{A}$ is $L^\omega(\mathcal{A})$. A language or ω-language L is *regular* if there is some nfa $\mathcal{A}$ with $L = L^*(\mathcal{A})$ or $L = L^\omega(\mathcal{A})$. The classes of regular languages and ω-languages, respectively, are effectively closed under Boolean operations, shuffle, and projection (cf. [27,30]).

Convention. If we are mainly interested in the ω-language $L^\omega(\mathcal{A})$ accepted by the nfa $\mathcal{A}$, we speak of a *Büchi-automaton.*

2.2 Traces

An *independence alphabet* is a pair (Γ, I) of an alphabet Γ and an irreflexive and symmetric *independence relation* $I \subseteq \Gamma^2$. The complement of I is the *dependence relation* $D = \Gamma^2 \setminus I$. IndA denotes the set of all independence alphabets.

The independence alphabet (Γ, I) is *complete* (*transitive*, resp.) if the reflexive closure of I equals Γ^2 (is transitive, resp.). TransA denotes the set of all transitive independence alphabets. An independence alphabet (Γ, I) is *diagonal* if, for any four distinct letters $a, b, c, d \in \Gamma$ with $(a, b), (b, c), (c, d) \in I$, also

$(a,c) \in I$ or $(b,d) \in I$. This is the case iff (Γ, I) is the comparability graph of an order tree iff it is P_4- and C_4-free [31]. This is equivalent to saying that the finite graph can be constructed from singleton graphs by disjoint union and addition of new nodes that are connected to all existing ones. The set of all diagonal independence alphabets is denoted DiagA.

Let (Γ, I) be an independence alphabet and $x, y \in \Gamma^\infty$. We call x and y *equivalent* [11] (denoted $x \sim y$) if, for all (not necessarily distinct) letters $a, b \in \Gamma$ with $(a,b) \in D$, we have $\mathtt{proj}_{\{a,b\}}(x) = \mathtt{proj}_{\{a,b\}}(y)$.[1] As an example, suppose $(a,b),(b,c) \in D$ and $(a,c) \in I$. Then $acbac \sim cabac \not\sim cbaac$ and $(ac)^\omega \sim ac\, acc\, ac^3\, ac^4 \cdots$.

For $L \subseteq \Gamma^\infty$, the *closure* is the set

$$[L] = \{y \in \Gamma^\infty \mid \exists x \in L \colon x \sim y\}$$

of all (ω-)words y that are equivalent to some element of L. The set $L \subseteq \Gamma^\infty$ is *closed* if $L = [L]$. If $L = \{x\}$ is a singleton, the closure $[L]$ of L equals the equivalence class $[x]$ of x wrt. $\sim$.

In this paper, we will regularly be concerned with closures of regular sets and with closed regular sets. While every closed regular set is the closure of some regular set (namely itself), the converse implication does not hold: with $(a,b) \in I$, $[(ab)^*] = \{u \colon |u|_a = |u|_b\}$ is the closure of some regular set, but not regular.

Remark. A *trace* [22] is an equivalence class $[u]$ for some word $u \in \Gamma^*$, i.e., an element of the quotient $\Gamma^*/_\sim$. Similarly, an *infinite trace* is an element of $\Gamma^\omega/_\sim$. Let $\odot \in \{*, \omega\}$. Fundamental results (cf. [8,12]) state that a set T of (infinite) traces is rational iff $T = \{[u] \mid u \in L\}$ for some regular set $L \subseteq \Gamma^\odot$. Dually, a set of (infinite) traces T is recognizable if $L = \{u \in \Gamma^\odot \mid [u] \in T\} = \bigcup_{t \in T} t$ is regular. Identifying the set of equivalence classes $\{[u] \mid u \in L\}$ with their union $[L]$, we see that rational and recognizable sets of (infinite) traces correspond to closures of regular $\odot$-languages and to closed regular $\odot$-languages, respectively.

Thus, our results on closures of regular sets and on closed regular sets could be phrased as results on rational and recognizable sets of (infinite) traces.

2.3 Decision Problems

This paper deals with three decision problems related to the closure of regular languages and ω-languages, respectively. For instance, the inclusion problem $\mathrm{IncP}_*(\Gamma, I)$ asks whether $[L^*(\mathcal{A})] \subseteq [L^*(\mathcal{B})]$ for two given nfas $\mathcal{A}$ and $\mathcal{B}$. Considering $\mathcal{A}$ and $\mathcal{B}$ as Büchi-automata, $\mathrm{IncP}_\omega(\Gamma, I)$ asks for $[L^\omega(\mathcal{A})] \subseteq [L^\omega(\mathcal{B})]$.

Let $\odot \in \{*, \omega\}$, $\mathrm{A} \subseteq \mathrm{IndA}$ a set of independence alphabets, $(\Gamma, I) \in \mathrm{A}$ an independence alphabet from A, and $\mathcal{A}$ and $\mathcal{B}$ nfas over Γ.

- Disjointness problem: $(\Gamma, I, \mathcal{A}, \mathcal{B}) \in \mathrm{DisjP}_\odot(\mathrm{A})$ iff $[L^\odot(\mathcal{A})] \cap [L^\odot(\mathcal{B})] = \emptyset$.
- Inclusion problem: $(\Gamma, I, \mathcal{A}, \mathcal{B}) \in \mathrm{IncP}_\odot(\mathrm{A})$ iff $[L^\odot(\mathcal{A})] \subseteq [L^\odot(\mathcal{B})]$.
- Regularity problem: $(\Gamma, I, \mathcal{A}) \in \mathrm{RegP}_\odot(\mathrm{A})$ iff $[L^\odot(\mathcal{A})]$ is regular.

[1] Two other, equivalent, definitions of $\sim$ have been formulated by Gastin [11].

The $*$-versions of these problems have been considered extensively; the known results can be read from Table 1 (see below for the definitions of Π_1^0 etc.)

For instance, the Π_1^0-completeness result for $\mathrm{DisjP}_*(\mathrm{IndA})$ is to be read as "The uniform problem $\mathrm{DisjP}_*(\mathrm{IndA})$ is in Π_1^0 and, for every non-diagonal independence alphabet (Γ, I), the non-uniform problem $\mathrm{DisjP}_*(\Gamma, I)$ is hard for Π_1^0."; the other two statements of completeness have to be read analogously.

Table 1. Finitary languages

	transitive	diagonal	general
disjointness	decidable ([2])	decidable ([1])	Π_1^0-complete ([1])
inclusion	decidable ([2])	Π_1^0-complete ([16])	
regularity	decidable ([29])	Σ_2^0-complete (Theorem 1) Π_1^0-hard by [29]	

2.4 Arithmetical and Analytical Hierarchy

We follow the definitions of the arithmetical and analytical hierarchy from [25]. The class $\Sigma_n^0 \subseteq 2^{\mathbb{N}}$ is the collection of all sets $A \subseteq \mathbb{N}$ of the form

$$A = \{x \in \mathbb{N} \mid (\mathbb{N}, +, \cdot) \models \exists y_1 \, \forall y_2 \cdots Q y_n^1, \ldots, y_n^m : \varphi(x, y_1, \ldots, y_n^1, \ldots, y_n^m)\},$$

where $Q = \forall$ (resp. $Q = \exists$) if n is even (resp. odd) and φ is a quantifier-free formula using addition and multiplication; Π_n^0 is the class of complements of Σ_n^0-sets.

As usual in computability theory, a Gödel numbering of all finite objects of interest allows to quantify over, say, nfas or words as well. A set is semi-decidable iff it is in Σ_1^0 [21] and co-semi-decidable iff it is in Π_1^0.

The class $\Sigma_n^1 \subseteq 2^{\mathbb{N}}$ is the collection of all sets $A \subseteq \mathbb{N}$ of the form

$$A = \{x \in \mathbb{N} \mid (\mathbb{N}, +, \cdot) \models \exists f_1 \, \forall f_2 \cdots Q f_n : \varphi(x, f_1, \ldots, f_n)\}, \tag{1}$$

where Q is as before and φ is a first-order formula over the signature containing $+$, $\cdot$, and the function symbols $f_1, \ldots, f_n$ that stand for functions $\mathbb{N} \to \mathbb{N}$. The class Π_n^1 is the class of all complements of Σ_n^1 sets.

It is useful to note that a formula as in (1) with number quantifiers added anywhere in the function quantifier prefix still defines a set from Σ_n^1, e.g., the set of natural numbers satisfying $\forall y \, \exists f_1 \, \exists z \, \forall f_2 : \varphi(x, y, z, f_1, f_2)$ belongs to Σ_2^1.

Now we can formulate our results on the ω-versions of our decision problems that are collected in Table 2.

Table 2. ω-languages

	transitive	diagonal	general
disjointness	decidable (Theorem 2)	in Π_1^0 (Cor. 2)	Π_1^1-complete (Theorem 8)
inclusion	decidable (Theorem 2)	Π_1^0-complete ([16] & Cor. 2)	Π_1^1-hard and in Π_2^1 (Theorem 8)
regularity	in Σ_1^0 (Thm. 3)	Σ_2^0-complete (Theorem 1 & 5)	Π_1^1-hard and in Π_2^1 (Theorem 8)

3 Finitary Languages

In this section, we show Σ_2^0-completeness of the regularity problem for languages. Sakarovitch showed that $\mathrm{RegP}_*(\mathrm{TransA})$ of all transitive independence alphabets is decidable and $\mathrm{RegP}_*(\Gamma, I)$ is Π_1^0-hard for any non-transitive independence alphabet. Hence we have to prove both, the lower and the upper bound Σ_2^0, the lower bound being more difficult and interesting (for the upper bound, see the introduction).

To this aim, we extend Post's correspondence problem PCP as follows. An instance of the *extended Post correspondence problem* is a tuple (A, B, α, β, L) where A and B are disjoint alphabets, α and β are homomorphisms from A^* to B^*, and $L \subseteq A^+$ is a regular language of non-empty words. A *solution* of the instance (A, B, α, β, L) is a word $u \in L$ from L with $\alpha(u) = \beta(u)$. Let $\mathrm{ePCP}_{\mathrm{fin}}$ denote the set of all instances with finitely many solutions.[2] Minimal modifications of any text-book proof of the undecidability of PCP (e.g., [14]) yield Σ_2^0-hardness of $\mathrm{ePCP}_{\mathrm{fin}}$. The rest of the proof of the lower bound in the following theorem makes only moderate adjustments to Sakarovitch's proof of Π_1^0-hardness [29].

Theorem 1. *The regularity problem* $\mathrm{RegP}_*(\mathrm{IndA})$ *of all independence alphabets is* Σ_2^0*-complete.*

4 Transitive Independence Alphabets

In this section, we do not prove any lower bounds, but only decidability and semi-decidability (cf. Table 2).

We will first prove that the inclusion problem is decidable for transitive independence alphabets.

Any transitive independence alphabet (Γ, I) is the disjoint union of n complete independence alphabets (Γ_k, I_k). We partition the set of all ω-words into those that are "mono-alphabetic at infinity" and those that are "multi-alphabetic at infinity": $\mathrm{MonoInf} := \bigcup_{k\in[n]} \Gamma^* \Gamma_k^\omega$ and $\mathrm{MultiInf} := \Gamma^\omega \setminus \mathrm{MonoInf}$.

2 Note that any instance with $L = A^+$ and at least one solution $u \in L = A^+$ has infinitely many solutions, namely u^k for all $k \geq 1$.

Consequently, $[L^\omega(\mathcal{A})] \subseteq [L^\omega(\mathcal{B})]$ for Büchi-automata $\mathcal{A}$ and $\mathcal{B}$ iff

$$[L^\omega(\mathcal{A})] \cap \text{MonoInf} \subseteq [L^\omega(\mathcal{B})] \text{ and} \tag{2}$$
$$[L^\omega(\mathcal{A})] \cap \text{MultiInf} \subseteq [L^\omega(\mathcal{B})]\,. \tag{3}$$

To decide inclusion (2), it suffices to decide it with $\Gamma^*\Gamma_k^\omega$ in place of MonoInf (for all $k \in [n]$). Any word from this set splits uniquely into a prefix from $\Gamma^* \setminus \Gamma^*\Gamma_k$ and an ω-word over Γ_k. This observation allows to reduce the problem to the finitary version (that was solved in [2]) and complete independence alphabets.

The inclusion (3) is more involved and uses some notions that will be introduced next. Finite words from Γ_k^+ for any $k \in [n]$ are called *mono-alphabetic.* The characteristic $\chi(u)$ of such a mono-alphabetic word $u \in \Gamma_k^+$ collects the information how equivalent words $v \sim u$ can behave in the automata $\mathcal{A}$ and $\mathcal{B}$ (in that sense, it generalises transition profiles, cf. [3]) as well as the alphabet u belongs to: $\chi(u) = (A, A^F, B, B^F, k)$ with

- $A = \{(p,q) \in Q_\mathcal{A}^2 \mid \exists u' \sim u \colon p \xrightarrow{u'}_\mathcal{A} q\}$
- $A^F = \{(p,r) \in Q_\mathcal{A}^2 \mid \exists u'u'' \sim u, q \in F_\mathcal{A} \colon p \xrightarrow{u'}_\mathcal{A} q \xrightarrow{u''}_\mathcal{A} r\}$
- $B = \{(p,q) \in Q_\mathcal{B}^2 \mid \exists u' \sim u \colon p \xrightarrow{u'}_\mathcal{B} q\}$
- $B^F = \{(p,r) \in Q_\mathcal{B}^2 \mid \exists u'u'' \sim u, q \in F_\mathcal{B} \colon p \xrightarrow{u'}_\mathcal{B} q \xrightarrow{u''}_\mathcal{B} r\}$

Let X denote the set of all characteristics of mono-alphabetic words. For a tuple $C \in X$, let $\chi^{-1}(C) = \{v \text{ mono-alphabetic} \mid \chi(v) = C\} \subseteq \Gamma_k^+$.

The following lemma says that X is decidable uniformly in (Γ, I), $\mathcal{A}$, and $\mathcal{B}$. Its proof is based on [2] and therefore on the transitivity of (Γ, I).

Lemma 1. *Given a disjoint union (Γ, I) of complete independence alphabets (Γ_k, I_k) for $k \in [n]$, Büchi-automata $\mathcal{A}$ and $\mathcal{B}$ over Γ, sets $A, A^F \subseteq Q_\mathcal{A}^2$ and $B, B^F \subseteq Q_\mathcal{B}^2$, and $k \in [n]$, it is decidable whether there exists a word $u \in \Gamma_k^+$ with $\chi(u) = (A, A^F, B, B^F, k)$.*

Proof. For $p, q \in Q_\mathcal{A}$, let K_{pq} denote the set of words $u' \in \Gamma_k^+$ with $p \xrightarrow{u'}_\mathcal{A} q$. The language K_{pq}^F is defined similarly with the additional requirement that the path $p \xrightarrow{u'}_\mathcal{A} q$ visits some accepting state. Then the set of words u with $\chi(u) = (A, A^F, B, B^F, k)$ is a Boolean combination of the closures of the regular languages K_{pq}, K_{pq}^F and languages defined similarly wrt. the nfa $\mathcal{B}$. From [2], we know that the emptiness of such Boolean combinations is decidable since the sub-alphabets (Γ_k, I_k) are transitive (in this case, even complete). □

Remark 1. Later, we will also use the characteristic as defined above for independence alphabets (Γ, I) that are disjoint unions of *not necessarily transitive* independence alphabets (Γ_k, I_k) for $k \in [n]$. In that case, the above proof does not work. Even more, the decidability is unlikely to hold since already for a diagonal alphabet (Γ_k, I_k), the non-emptiness of the difference of the closures of two regular languages is undecidable [16].

Any ω-word $\alpha \in \text{MultiInf}$ splits uniquely into maximal mono-alphabetic words, i.e., $\alpha = u_1 u_2 \cdots$ with $u_i \in \Gamma_{k_i}^+$ and $k_i \neq k_{i+1}$ for all $i \geq 1$. Hence α can be abstracted by the sequence $\chi(u_1)\,\chi(u_2) \cdots \in X^\omega$ of characteristics and the ω-languages of $\mathcal{A}$ and $\mathcal{B}$ by ω-languages over X. This abstraction actually characterises the *closures* of $L^\omega(\mathcal{A})$ and $L^\omega(\mathcal{B})$:

Lemma 2. *From Büchi-automata $\mathcal{A}$ and $\mathcal{B}$ and the set of characteristics X, one can construct Büchi-automata $\mathcal{A}_\chi$ and $\mathcal{B}_\chi$ over X such that $[L^\omega(\mathcal{A})] \cap \text{MultiInf}$ is the disjoint union of all ω-languages $\chi^{-1}(C_1)\,\chi^{-1}(C_2) \cdots \subseteq \Gamma^\omega$ for $C_1\,C_2 \cdots \in L^\omega(\mathcal{A}_\chi)$ and similarly for $\mathcal{B}$ and $\mathcal{B}_\chi$.*

Now the following characterisation of the inclusion (3) is an easy consequence.

Corollary 1. $[L^\omega(\mathcal{A})] \cap \text{MultiInf} \subseteq [L^\omega(\mathcal{B})]$ *if, and only if,* $L^\omega(\mathcal{A}_\chi) \subseteq L^\omega(\mathcal{B}_\chi)$.

In particular the inclusion problem is therefore decidable.

Theorem 2. *The inclusion problem* $\text{IncP}_\omega(\text{TransA})$ *as well as the disjointness problem* $\text{DisjP}_\omega(\text{TransA})$ *for transitive independence alphabets are decidable.*

Proof. Since (Γ, I) is transitive, one can compute the set X of characteristics and the Büchi-automata $\mathcal{A}_\chi$ and $\mathcal{B}_\chi$. Hence the inclusion $L^\omega(\mathcal{A}_\chi) \subseteq L^\omega(\mathcal{B}_\chi)$ can be decided as well (cf. [30, Theorem 2.3]). By Cor. 1, it holds iff the inclusion (3) holds. Since validity of the inclusion (2) is decidable as well, the claim regarding $\text{IncP}_\omega(\text{TransA})$ holds.

For disjointness, one can argue analogously, using appropriate (and not too complicated) modifications of the results presented so far. □

For the regularity problem, membership in Σ_1^0 can be shown by expressing, for a Büchi-automaton $\mathcal{A}$, the existence of some Büchi-automaton $\mathcal{B}$ with $L^\omega(\mathcal{B}) = [L^\omega(\mathcal{A})]$. This equality holds iff $[L^\omega(\mathcal{A})] = [L^\omega(\mathcal{B})]$ (decidable by Theorem 2) and $L^\omega(\mathcal{B}) = [L^\omega(\mathcal{B})]$ (decidability by [24,26]). Hence one obtains the following result.

Theorem 3. *The regularity problem* $\text{RegP}_\omega(\text{TransA})$ *for all transitive independence alphabets is in* Σ_1^0.

5 Diagonal Independence Alphabets

The claimed lower bounds Π_1^0 and Σ_2^0 (cf. Table 2) follow from the corresponding lower bounds for the $*$-version of the respective decision problem. As in the previous section, this extended abstract concentrates on the inclusion problem; the disjointness problem can be handled along very similar lines, and the result on the regularity problem follows as before.

To show that the inclusion problem $\text{IncP}_\omega(\text{DiagA})$ is in Π_1^0, we will show that its complement is semi-decidable. The crucial insight is that, assuming $[L^\omega(\mathcal{A})] \not\subseteq [L^\omega(\mathcal{B})]$, the difference $[L^\omega(\mathcal{A})] \setminus [L^\omega(\mathcal{B})]$ contains some ultimately periodic word. While this is a purely existential statement, we can then use a decision algorithm for the questions "$uv^\omega \in [L^\omega(\mathcal{A})]$?" and "$uv^\omega \in [L^\omega(\mathcal{B})]$?" (see below). Thus, the semi-algorithm for non-inclusion can search for a pair (u, v) of nonempty words with $uv^\omega \in [L^\omega(\mathcal{A})] \setminus [L^\omega(\mathcal{B})]$.

We start with the decidability result that holds for all independence alphabets, its proof uses that $[u\, v^\omega]$ is regular for any nonempty words u and v [13, Theorem 3.22].

Observation 1. *Given an independence alphabet (Γ, I), some Büchi-automaton $\mathcal{A}$ over Γ, and two non-empty words $u, v \in \Gamma^+$, it is decidable whether $uv^\omega \in [L^\omega(\mathcal{A})]$. Furthermore, if this is the case, there are nonempty words $x, y \in \Gamma^+$ with $uv^\omega \sim xy^\omega \in L^\omega(\mathcal{A})$.*

We next show that non-inclusions are witnessed by ultimately periodic words. As this proof is by induction on the size of the independence alphabet (Γ, I), we say that (Γ, I) *guarantees witnesses* if, for any two Büchi-automata $\mathcal{A}$ and $\mathcal{B}$ with $[L^\omega(\mathcal{A})] \not\subseteq [L^\omega(\mathcal{B})]$, there are $u, v \in \Gamma^+$ with $uv^\omega \in [L^\omega(\mathcal{A})] \setminus [L^\omega(\mathcal{B})]$.

The following two propositions show that adding an independent letter as well as taking the disjoint union of independence alphabets preserves this property.

Proposition 1. *Let (Γ, I) be an independence alphabet and $a \in \Gamma$ with $\{a\} \times (\Gamma \setminus \{a\}) \subseteq I$. If $(\Gamma \setminus \{a\}, I)$ guarantees witnesses, then so does (Γ, I).*

Proof. Set $\Delta = \Gamma \setminus \{a\}$. Suppose K and L are regular ω-languages over Γ with $\alpha \in [K] \setminus [L]$ and let $\varkappa = |\alpha|_a$ denote the number of occurrences of the letter a in α. The projection K' of $K \cap (\Delta^\infty \sqcup a^\varkappa)$ to Δ is regular and similarly for L'. Using $\{a\} \times (\Gamma \setminus \{a\}) \subseteq I$, one gets $\text{proj}_a(\alpha) \in [K'] \setminus [L']$. Now the claim follows since K' and L' are regular languages over Δ and (Δ, I) guarantees witnesses. □

Proposition 2. *Let (Γ_1, I_1) and (Γ_2, I_2) be disjoint independence alphabets that guarantee witnesses. Then $(\Gamma_1 \cup \Gamma_2, I_1 \cup I_2)$ guarantees witnesses.*

Proof. Suppose K and L are the closures of $L^\omega(\mathcal{A})$ and $L^\omega(\mathcal{B})$, respectively, and $\alpha \in K \setminus L$. First, assume α to be mono-alphabetic at infinity. Then $\alpha = u\,\beta$ with $\beta \in \Gamma_k^\omega$ for some $k \in [2]$. Then β can be replaced by some vw^ω since (Γ_k, I_k) guarantees witnesses. Hence $uv\,w^\omega \in K \setminus L$. If, alternatively, α is multi-alphabetic at infinity, we can refer to the automata $\mathcal{A}_\chi$ and $\mathcal{B}_\chi$ from the previous section (defined analogously based on the partition of Γ into the sub-alphabets Γ_1 and Γ_2). Here, we only need the existence of these automata. By Corollary 1, $L^\omega(\mathcal{A}_\chi) \not\subseteq L^\omega(\mathcal{B}_\chi)$ since $\alpha \in K \setminus L$. Hence there is an ultimately periodic word in $L^\omega(\mathcal{A}_\chi) \setminus L^\omega(\mathcal{B}_\chi)$ and therefore, again by Corollary 1, in $K \setminus L$. □

Using the characterisation of diagonal independence alphabets from [31], by induction, one obtains the following existential statement.

Theorem 4. *Every diagonal independence alphabet guarantees witnesses.*

As a consequence, the inclusion problem for diagonal independence alphabets is co-semi-decidable, the claim regarding the disjointness problem is shown similarly, with a few more technical difficulties; it is omitted for lack of space.

Corollary 2. *The inclusion problem* $\mathrm{IncP}_\omega(\mathrm{DiagA})$ *and the disjointness problem* $\mathrm{DisjP}_\omega(\mathrm{DiagA})$ *for diagonal independence alphabets are in* Π^0_1.

As in the transitive case, we can infer an upper bound for the regularity problem, the matching lower bound follows from the finite case.

Theorem 5. *The regularity problem* $\mathrm{RegP}_\omega(\mathrm{DiagA})$ *for all diagonal independence alphabets is* Σ^0_2*-complete.*

6 General Independence Alphabets

The upper bounds Π^1_1 and Π^1_2 are relatively easy consequences of the definition of these problems, but the lower bounds Π^1_1 are more demanding and interesting. We present the ideas for the inclusion and the regularity problem that both use nondeterministic Turing machines with Büchi-acceptance [6], two-counter automata with Büchi-acceptance, and an idea from [1] to encode computations of two-counter automata by words over a non-diagonal independence alphabet.

The starting point is that the emptiness problem for nondeterministic Turing machines with Büchi-acceptance is complete for Π^1_1 ([5, Prop. 3.1] together with [6, Theorems 8.2 and 8.6]). Then, we translate these machines into machines with counters. A *2-counter automaton with Büchi-acceptance* is just a Büchi-automaton $\mathcal{M}$ over the alphabet $\mathrm{Op} = \{\mathrm{inc}_k, \mathrm{dec}_k, \mathrm{zero}_k \mid k \in [2]\}$. Hence any such $\mathcal{M}$ accepts an ω-language of sequences of "counter operations" from Op. We next define what we mean by "valid sequence". Let $\alpha = a_0 a_1 \cdots \in \mathrm{Op}^\omega$ with $a_i \in \mathrm{Op}$ for all $i \geq 0$. For $n \geq 0$ and $k \in [2]$, let

$$\mathrm{value}_k(\alpha, n) = |a_0 a_1 \cdots a_{n-1}|_{\mathrm{inc}_k} - |a_0 a_1 \cdots a_{n-1}|_{\mathrm{dec}_k}$$

denote the value of the counter k after executing the first n operations of α. The ω-word α as above is *valid* if, for all $n \geq 0$, $\mathrm{value}_k(\alpha, n) = 0$ whenever $a_n = \mathrm{zero}_k$ and $\mathrm{value}_k(\alpha, n) > 0$ whenever $a_n = \mathrm{dec}_k$. Let $\mathrm{Val} \subseteq \mathrm{Op}^\omega$ denote the set of all valid ω-words.

Recall that Minsky machines, i.e., deterministic 2-counter automata modeled as deterministic finite automata, can simulate deterministic Turing machines [23]. Similarly, nondeterministic 2-counter automata with Büchi-acceptance can simulate nondeterministic Turing machines with Büchi-acceptance. Hence we obtain

Theorem 6. *The set of Büchi-automata* $\mathcal{M}$ *over* Op *with* $L^\omega(\mathcal{M}) \cap \mathrm{Val} = \emptyset$ *is complete for* Π^1_1.

Before we derive the lower bounds for all non-diagonal independence alphabets, we first prove them for two typical examples, so called *standard alphabets* (Σ, I_i) for $i \in \{1, 2\}$: Set $\Sigma = \mathrm{Op} \cup \{b, c, d\}$ with $\mathrm{Op} \times \{b\} \subseteq I_i$, $(b, c), (c, d) \in I_i$, $\mathrm{Op} \times (\mathrm{Op} \cup \{c\}) \subseteq D_i$, $(b, d) \in D_i$, $\mathrm{Op} \times \{d\} \subseteq I_1$, and $\mathrm{Op} \times \{d\} \subseteq D_2$.

In the following, let (Σ, I) be any of these two standard alphabets (Σ, I_1) or (Σ, I_2). We will construct two regular ω-languages K and L such that the intersection of their closures corresponds to the set of valid sequences (cf. first item in Prop. 3 below). To this aim, we will ensure that $\mathrm{proj}_{\mathrm{Op}\cup\{c\}}([K]\cap[L])$ is the set of ω-words of the form $a_0\, c^{m_1}\, a_1\, c^{m_2} \cdots$ with $a_i \in \mathrm{Op}$ and, for all $i \geq 0$,

$$m_{i+1} = 2^{\mathrm{value}_1(a_0 a_1 \cdots a_i)} \cdot 3^{\mathrm{value}_2(a_0 a_1 \cdots a_i)} \cdot 5^i$$

as well as $2 \not| m_{i+1}$ and $3 \not| m_{i+1}$ if $a_i = \mathrm{zero}_1$ and $a_i = \mathrm{zero}_2$, respectively.

So let $K_0 = \mathrm{Op}\, d\, (c\, b)^+$ and $K = b\, K_0^\omega$. Then $\alpha \in [K]$ if, and only if,

- $\mathrm{proj}_{\{b,d\}}(\alpha) = b\, db^{m_1}\, db^{m_2}\, db^{m_3} \cdots,$
- $\mathrm{proj}_{\mathrm{Op}\cup\{c\}}(\alpha) = a_0 c^{m_1}\, a_1 c^{m_2}\, a_2 c^{m_3} \cdots,$
- and $\mathrm{proj}_{\mathrm{Op}\cup\{d\}} \in \begin{cases} \mathrm{Op}^\omega \sqcup d^\omega & \text{if } I = I_1 \\ (\mathrm{Op}\, d)^\omega & \text{if } I = I_2 \end{cases}$

for some positive integers m_i and some operations $a_i \in \mathrm{Op}$. The central point here is that the blocks no. i of occurrences of b and of c, resp., have the same length m_i.

A second language is similar, but its definition might look a bit frightening:

$$\begin{aligned} L_0 = &\,\mathrm{inc}_1\,(b\,c^5\,c^5)^+\,d \cup \mathrm{inc}_2\,(b\,c^5\,c^5\,c^5)^+\,d \\ &\cup \mathrm{dec}_1\,(b\,b\,c^5)^+\,d \cup \mathrm{dec}_2\,(b\,b\,b\,c^5)^+\,d \\ &\cup \mathrm{zero}_1\,(b\,c^5\,b\,c^5)^*\,b\,c^5\,d \cup \mathrm{zero}_2\,(b\,c^5\,b\,c^5\,b\,c^5)^*\,\{b\,c^5\,b\,c^5, b\,c^5\}\,d \end{aligned}$$

and $L = L_0^\omega$. Then, e.g., $\mathrm{inc}_1\, b^k c^\ell\, d \in [L_0]$ iff $2\cdot 5\cdot k = \ell$, and $\mathrm{zero}_2\, b^k c^\ell\, d \in [L_0]$ iff $5 \cdot k = \ell \notin 3 \cdot \mathbb{N}$. The idea is that the numbers k and ℓ of copies of b and c shall encode values of the two counters and the step counter before and after the execution of the respective counter operation. More precisely, if (c_1, c_2) and (d_1, d_2) are the counter values before and after the operation $a \in \mathrm{Op}$ and the step counter is i, then the numbers k and ℓ are $2^{c_1} \cdot 3^{c_2} \cdot 5^i$ and $2^{d_1} \cdot 3^{d_2} \cdot 5^{i+1}$, respectively. Note that, no matter what a is, at most one counter can decrease by at most one, the other remaining the same, hence $2^{c_1} \cdot 3^{c_2} \leq 2^{d_1} \cdot 3^{d_2} \cdot 3$. But this implies $k < \ell$.

Since K_0 and L_0 are regular, also K and L are regular. Similarly to the above definitions, one can construct two more regular ω-languages K' and L'. The central properties of these four regular ω-languages are listed as follows, the increase of the encodings k and ℓ ensures the second property.

Proposition 3.
- $\mathrm{Val} = \mathrm{proj}_{\mathrm{Op}}\big([K]\cap[L]\big)$,
- *no ω-word from $[K]\cap[L]$ is ultimately periodic,*
- $\Sigma^\omega \setminus [K] = [K']$, *and* $\Sigma^\omega \setminus [L] = [L']$.

Now let $\mathcal{M}$ be some Büchi-automaton on the alphabet Op and set $H = L^\omega(\mathcal{M}) \sqcup \{b, c, d\}^\infty$. Then the following are equivalent:

(i) $L^\omega(\mathcal{M}) \cap \mathrm{Val} = \emptyset$. (iii) $[\Sigma^\omega \setminus H \cup K' \cup L'] = \Sigma^\omega$.
(ii) $[H \cap K] \cap [L] = \emptyset$. (iv) $[\Sigma^\omega \setminus H \cup K' \cup L']$ is regular.

The equivalence of (i) and (ii) follows from Prop. 3, the proof of the implication (iv)$\Rightarrow$(iii) is the most interesting one. For this, note that $[\Sigma^\omega \setminus H \cup K' \cup L']$ is the complement of $[H \cap K] \cap [L] \subseteq [K] \cap [L]$. Since $[K] \cap [L]$ contains no ultimately periodic word, they all belong to $[\Sigma^\omega \setminus H \cup K' \cup L']$. But Σ^ω is the only regular ω-language containing all ultimately periodic words.

It is important that in (ii), (iii), and (iv), we talk about closures of effectively regular ω-languages $H \cap K$, L, and $\Sigma^\omega \setminus H \cup K' \cup L'$. This allows to reduce the problem from Theorem 6 to the disjointness problem because of (i) $\iff$ (ii), the inclusion problem because of (i) $\iff$ (iii) (with $L^\omega(\mathcal{A}) = \Sigma^\omega$), and the regularity problem because of (i) $\iff$ (iv).

Theorem 7. *Let (Σ, I) be a standard independence alphabet. Then the problems* $\mathrm{DisjP}_\omega(\Sigma, I)$, $\mathrm{IncP}_\omega(\Sigma, I)$, *and* $\mathrm{RegP}_\omega(\Sigma, I)$ *are hard for* Π^1_1.

Finally, let (Γ, I) be some non-diagonal independence alphabet, and choose $a, b, c, d \in \Gamma$ with $(a, b), (b, c), (c, d) \in I$, but $(a, c), (b, d) \in D$. Then for one of the standard independence alphabets (Σ, I), one can define an injection $f\colon \Sigma \to \{a, b, c, d\}^*$ that, when applied letter-wise to ω-words, has the following properties.

(a) $f\colon \Sigma^\omega \to \Gamma^\omega$ is injective.
(b) $\alpha \sim \beta \iff f(\alpha) \sim f(\beta)$ for all $\alpha, \beta \in \Sigma^\omega$.
(c) $[L] \subseteq \Sigma^\omega$ is regular iff $[f(L)] \subseteq \Gamma^\omega$ is regular for any $L \subseteq \Sigma^\omega$.

These properties then allow to transfer the lower bound from Theorem 7 to the general case:

Theorem 8. *Let (Γ, I) be a non-diagonal independence alphabet. Then the problems* $\mathrm{DisjP}_\omega(\Gamma, I)$, $\mathrm{IncP}_\omega(\Gamma, I)$, *and* $\mathrm{UnivP}_\omega(\Gamma, I)$ *are hard for* Π^1_1.

7 Summary and Open Questions

Our results for ω-languages are collected in Table 2. The upper bounds for the inclusion problem transfer easily to the equality and the universality problem; the full version of this paper demonstrates that also the lower bounds of the inclusion problem hold for these two problems. It is open whether RegP_ω(TransA) and DisjP_ω(DiagA) are decidable or not; looking at the $*$-versions of these problems, decidability looks more likely to me. Also regarding IncP_ω(IndA) and RegP_ω(IndA), there is an open gap between the lower and the upper bound. Finally, we did not look at all at the problems whether $[L^\omega(\mathcal{A})] \cap [L^\omega(\mathcal{B})]$ or $\Gamma^\omega \setminus [L^\omega(\mathcal{A})]$ is the closure of some regular set – these will be the subject of a follow-up paper.

References

1. Aalbersberg, I., Hoogeboom, H.: Characterizations of the decidability of some problems for regular trace languages. Math. Syst. Theory **22**, 1–19 (1989)
2. Aalbersberg, I., Welzl, E.: Trace languages defined by regular string languages. RAIRO, Inf. Théor. Appl. **20**,pp. 103–119 (1986)
3. Breuers, S., Löding, C., Olschewski, J.: Improved Ramsey-Based Büchi Complementation. In: Birkedal, L. (ed.) FoSSaCS 2012. LNCS, vol. 7213, pp. 150–164. Springer, Heidelberg (2012). https://doi.org/10.1007/978-3-642-28729-9_10
4. Problèmes combinatoires de commutation et réarrangements. LNM, vol. 85. Springer, Heidelberg (1969). https://doi.org/10.1007/BFb0079468
5. Castro, J., Cucker, F.: Nondeterministic ω-computations and the analytical hierarchy. Zeitschr. f. math. Logik und Grundlagen d. Math. **35**,pp.333-342 (1989)
6. Cohen, R., Gold, A.: ω-computations on Turing machines. Theoret. Comput. Sci. **6**, 1–23 (1978)
7. Diekert, V., Muscholl, A.: Deterministic asynchronous automata for infinite traces. In: Enjalbert, P., Finkel, A., Wagner, K.W. (eds.) STACS 1993. LNCS, vol. 665, pp. 617–628. Springer, Heidelberg (1993). https://doi.org/10.1007/3-540-56503-5_61
8. Diekert, V., Rozenberg, G.: The Book of Traces. World Scientific Publ, Co (1995)
9. Ebinger, W., Muscholl, A.: Logical definability on infinite traces. In: Lingas, A., Karlsson, R., Carlsson, S. (eds.) ICALP 1993. LNCS, vol. 700, pp. 335–346. Springer, Heidelberg (1993). https://doi.org/10.1007/3-540-56939-1_84
10. Ganardi, M., Majumdar, R., Zetzsche, G.: The complexity of bidirected reachability in valence systems. In: LICS'22, pp. 26:1–26:15. ACM (2022)
11. Gastin, P.: Infinite traces. In: Guessarian, I. (ed.) LITP 1990. LNCS, vol. 469, pp. 277–308. Springer, Heidelberg (1990). https://doi.org/10.1007/3-540-53479-2_12
12. Gastin, P.: Recognizable and rational languages of finite and infinite traces. In: Choffrut, C., Jantzen, M. (eds.) STACS 1991. LNCS, vol. 480, pp. 89–104. Springer, Heidelberg (1991). https://doi.org/10.1007/BFb0020790
13. Gastin, P., Petit, A., Zielonka, W.: An extension of Kleene's and Ochmanski's theorems to infinite traces. Theoret. Comput. Sci. **125**, 167–204 (1994)
14. Hopcroft, J., Ullman, J.: Introduction to automata theory, languages, and computation. Addison Wesley (1979)
15. Hutagalung, M., Hundeshagen, N., Kuske, D., Lange, M., Lozes, É.: Multi-buffer simulations: decidability and complexity. Inf. Comput. **262**(2), 280–310 (2018)
16. Ibarra, O.: Reversal-bounded multicounter machines and their decision problems. J. ACM **25**(1), 116–133 (1978)
17. Köcher, C., Kuske, D.: Backwards-reachability for cooperating multi-pushdown systems. J. Comput. Syst. Sci. **148** (2025), article no. 103601
18. Kuske, D.: The theory of reachability in trace-pushdown systems. In: CiE 2025, pp. 283–298. Lecture Notes in Comp. Science. vol. 15764, Springer (2025)
19. Lohrey, M., Steinberg, B.: The submonoid and rational subset membership problems for graph groups. In: LATA'07, pp. 367–378. Report 35/07, Research Group on Mathematical Linguistics, Universitat Rovira i Virgili, Tarragona (2007)
20. Lohrey, M., Zetzsche, G.: The complexity of knapsack in graph groups. In: STACS'17. LIPIcs, vol. 66, pp. 52:1–52:14. Schloss Dagstuhl - Leibniz-Zentrum für Informatik (2017)
21. Matijasevich, Y.: Hilbert's Tenth Problem. MIT Press, Foundations of Computing Series (1993)

22. Mazurkiewicz, A.: Concurrent program schemes and their interpretation. Tech. rep., DAIMI Report PB-78, Aarhus University (1977)
23. Minsky, M.: Recursive unsolvability of Post's problem of "tag" and other topics in theory of Turing machines. Ann. Math. **74**(3),pp. 437–455 (1961)
24. Muscholl, A.: Decision and complexity issues on concurrent systems (1999), Habilitationsschrift, Universität Stuttgart
25. Odifreddi, P.: Classical recursion theory. The theory of functions and sets of natural numbers, Stud. Logic Found. Math., vol. 125. Amsterdam etc. North-Holland (1989)
26. Peled, D., Wilke, T., Wolper, P.: An algorithmic approach for checking closure properties of temporal logic specifications and omega-regular languages. Theoret. Comput. Sci. **195**(2), 183–203 (1998)
27. Perrin, D., Pin, J.E.: Infinite words. Pure and Applied Mathematics vol. 141, Elsevier (2004)
28. Sakarovitch, J.: On regular trace languages. Theor. Comput. Sci. **52**, 59–75 (1987)
29. Sakarovitch, J.: The "last" decision problem for rational trace languages. In: LATIN'92, pp. 460–473. Lecture Notes in Comp. Science vol. 583, Springer (1992)
30. Thomas, W.: Automata on infinite objects. In: van Leeuwen, J. (ed.) Handbook of Theoretical Computer Science, pp. 133–191. Elsevier Science Publ. B.V. (1990)
31. Wolk, E.: The comparability graph of a tree. Proc. AMS **13**, 789–795 (1962)
32. Zetzsche, G.: Silent transitions in automata with storage. In: Fomin, F.V., Freivalds, R., Kwiatkowska, M., Peleg, D. (eds.) ICALP 2013. LNCS, vol. 7966, pp. 434–445. Springer, Heidelberg (2013). https://doi.org/10.1007/978-3-642-39212-2_39
33. Zetzsche, G.: The emptiness problem for valence automata over graph monoids. Inf. Comput. **277** (2021), article no. 104583

Linear Time Algorithms for Colorful 3-Rainbow Domination in Block and Cactus Graphs

Tetiana Lavynska(✉)

Otto von Guericke University Magdeburg, Magdeburg, Germany
tetiana.lavynska@ovgu.de

Abstract. Dominance is a fundamental concept in graph theory. Motivated by applications in facility location problems, we study colorful k-rainbow domination. Given a graph $G = (V, E)$ and a set of k colors, we consider a function f that assigns a subset of colors to each vertex. This function is called the colorful k-rainbow dominating function of G if two conditions hold. First, for each vertex v, its closed neighborhood $N[v]$ contains all k different colors. Second, each vertex v gets a non-empty color set. The weight of f is the total number of colors assigned. The goal of the colorful k-rainbow domination problem is to find a colorful k-rainbow dominating function with the smallest weight, called the colorful k-rainbow domination number. Colorful 3-rainbow domination is NP-complete in general graphs, but solvable in linear time for trees. We give linear time algorithms that compute an optimal colorful 3-rainbow domination function if an input graph is a block or a cactus graph. Furthermore, we show that the corresponding problem remains NP-complete if we restrict the input graph to the class of maximum degree 3 graphs.

Keywords: domination · colorful rainbow domination · NP-complete · maximum degree 3 graph · block graph · cactus graph

1 Introduction

Dominance along with its variants is a fundamental concept in graph theory with various applications, for example, in network monitoring and facility location. The goal of the classical domination problem is to find a smallest set of vertices that dominates a given graph G, where a vertex v *dominates* itself and all its adjacent vertices, thus its closed neighborhood $N[v]$. Throughout this paper, a neighborhood is assumed to be closed. One of the variations of the domination problem is rainbow domination, introduced in 2007 by Brešar and Šumenjak [3]. Given a graph $G = (V, E)$ and a set of k colors $\{1, \dots, k\}$ we consider a function $f : V \to 2^{\{1,\dots,k\}}$ that assigns a subset of colors to each vertex. For a function to be a k-rainbow dominating function, any vertex without a color must have all k colors present in its neighborhood. More precisely, if for each vertex v with

A. Jeż and J. Otop (Eds.): FCT 2025, LNCS 16106, pp. 336–349, 2026.
https://doi.org/10.1007/978-3-032-04700-7_25

$f(v) = \emptyset$ we have $\bigcup_{x \in N[v]} f(x) = \{1, \ldots, k\}$, then f is a k-rainbow dominating function of G. The goal is to find a k-rainbow dominating function with the smallest total number of colors assigned.

Rainbow domination can be applied to scenarios like network monitoring, where different colors represent various types of monitoring devices, ensuring that every part of the network has access to all types of monitoring devices. Similarly, in facility location problems, different colors can represent various types of facilities. Imagine that we want to place different kinds of facilities, for example, pharmacies, gyms, and supermarkets, in various villages, such that each village has access to each type of facility, either directly or in its neighborhood. To make an allocation fair, each village should get at least one facility. Such a problem can be modeled using a variation of k-rainbow domination, which is called colorful k-rainbow domination [8]. Here, each vertex v gets a non-empty color set: $f(v) \neq \emptyset$. Furthermore, for each vertex v its neighborhood $N[v]$ contains all k different colors: $\bigcup_{x \in N[v]} f(x) = \{1, \ldots, k\}$.

Related Work. For $k = 1$, k-rainbow domination problem is equivalent to a classical domination problem, which is known to be NP-complete [5]. For $k = 2$, Brešar and Šumenjak [3] show its NP-completeness even if we restrict the input graph to be bipartite or chordal. At the same time, Brešar et al. show that it is solvable in linear time in trees [2]. Later, Chang et al. consider general k-rainbow domination for any k and show that the problem is NP-complete in bipartite and in chordal graphs, but solvable in linear time for trees [4].

Different variations of rainbow domination exist, such as total k-rainbow domination [1], k-rainbow total domination [12], independent k-rainbow domination [9], and colorful k-rainbow domination [8]. For colorful k-rainbow domination, the problem can be solved in linear time for $k = 2$ and is NP-complete for each $k \geq 3$ [8]. For $k = 3$, it remains NP-complete in split and in bipartite graphs. Split graphs are a subset of well-known chordal graphs. Hence, the problem is NP-complete in chordal graphs as well. For some special graphs, like paths, cycles, cliques, and bicliques, an optimal 3-rainbow dominating function together with its weight is given, as well as efficient linear time algorithms for trees and interval graphs [8]. Both trees and interval graphs are also subclasses of chordal graphs. Therefore, in this paper, we address the question, in which more general graph classes that lie between these classes and the class of chordal graphs colorful 3-rainbow domination problem can still be solved efficiently?

One of the well-known generalizations of trees is the class of block graphs, which was introduced by Harary [6] to analyze the structure of blocks and cut vertices in a connected graph. A block graph is a graph where every biconnected component is a clique [6]. Therefore, intuitively, a block graph is a tree-like structure of cliques where cliques are welded together at some vertices or connected by bridge edges. Another generalization of trees is a class of cactus graphs [7], that are graphs where every block is either an edge or a cycle. So here we have a tree-like structure of cycles and edges where cycles are welded together at some vertices or connected by bridge edges.

Our Contribution. We give linear time algorithms that compute an optimal colorful 3-rainbow domination function if an input graph is a block or a cactus graph and therefore answer an open question from [8]. Furthermore, we show that the colorful 3-rainbow domination problem remains NP-complete if we restrict the input graph to the class of maximum degree 3 graphs.

2 Preliminaries

Given a graph $G = (V, E)$ and a set of k colors $\{1, \ldots, k\}$ we consider a function f that assigns a subset of colors to each vertex, that is $f : V \to 2^{\{1,\ldots,k\}}$. If for each vertex v we have $\bigcup_{x \in N[v]} f(x) = \{1, \ldots, k\}$ and $f(v) \neq \emptyset$, then f is a *colorful k-rainbow dominating function* of G. The *weight* of f is $w(f) = \sum_{v \in V} |f(v)|$. The minimum weight is called the *colorful k-rainbow domination number* of G and is denoted by $\gamma_{crk}(G)$. Given a graph G and a number d, the goal of the *Colorful k-Rainbow Domination* problem is to decide whether $\gamma_{crk}(G) \leq d$.

In the following, if we say that vertex v gets color i and $i \notin \{1, \ldots, k\}$, then we mean the color ($i \mod k$) if $i \not\equiv 0 \mod k$ and color k if $i \equiv 0 \mod k$.

Given a graph $G = (V, E)$, a block is a maximal biconnected subgraph of G or a bridge edge [6]. A cut vertex of a graph G is a vertex that is a part of at least two blocks. Any connected graph can be decomposed into a block-cut tree [11] in time $O(|V|+|E|)$. The block-cut tree has a vertex for each block and each cut vertex. The vertex that corresponds to block b is connected to the vertex that corresponds to cut vertex c if block b contains vertex c. A *block graph* is a graph where every block is a clique. A *cactus graph* is a graph where every block is either an edge or a cycle.

3 Efficient Algorithms for Colorful 3-Rainbow Domination

In the following, we assume that graph G is connected. Otherwise, an optimal solution can be found separately for each connected component of G.

Lemma 1. *For each graph G with $n \geq 3$ vertices, an optimal solution exists that assigns at least two colors to each vertex that has neighbors with degree one.*

Proof. Let v be a vertex that has neighbors with degree one, and let u be one of its neighbors with $\deg(u) = 1$. Vertex u must have all three colors in its neighborhood. If u gets a single color, then v must get at least two colors. If u gets more than one color, then we can move all colors of u except one to v.

In the figures, we draw vertices that have neighbors with degree one as square nodes and all other vertices as circle nodes. We say that vertex v is a **box** vertex if v has neighbors with degree one. Lemma 1 implies that we can always assume that each box vertex gets at least two colors.

3.1 Block Graphs

Theorem 1. *Let $G = (V, E)$ be a block graph with $n \geq 3$ vertices. Let n_1 be the number of box vertices. Then $\gamma_{cr3}(G) = n + n_1$. We can compute an optimal colorful 3-rainbow dominating function in time $O(|V| + |E|)$.*

Proof. Lemma 1 implies $\gamma_{cr3}(G) \geq n + n_1$. We give an algorithm that computes a colorful 3-rainbow dominating function with weight $n + n_1$ in linear time, and thus show that $\gamma_{cr3}(G) = n + n_1$ for block graphs. We start with computing the block-cut tree T for G in time $O(|V| + |E|)$ [11]. We choose some leaf block b_r as a root and consider tree T as a rooted tree, as illustrated in Fig. 1. We traverse the tree and assign one color to each cut vertex using the following rules. The cut vertex corresponding to the only child of b_r gets color 1. If cut vertex c is a child of block b', which is a child of cut vertex c', and c' gets color i, then c gets color $i + 1$.

Now we traverse the tree a second time and assign colors to non-cut vertices of each block. For this, we distinguish if a block is a root, leaf, or non-leaf. If b is a root block b_r, then it contains only one cut vertex c, and c is a child of b with color 1. If b_r contains only one non-cut vertex, then it gets color 2. Otherwise, we assign color 2 to one vertex and color 3 to the remaining vertices. If b is a leaf block, then it contains only one cut vertex c, and c is the parent of b. Let c have color i. If b contains only one non-cut vertex, then it gets color $i + 1$. Otherwise, we assign color $i + 1$ to one vertex and color $i + 2$ to the remaining vertices. If b is a non-leaf block, then it contains its parent cut vertex c with color i and its children cut vertices $c_1, \ldots, c_m$ with color $i + 1$. We can assign color $i + 2$ to all remaining vertices. Observe that block b does not necessarily have any remaining vertices, since each vertex of b can be a cut vertex, as it happens for block $b_3 = \{8, 11, 12\}$ in Fig. 1.

Each vertex now has exactly one color. Each box vertex needs one additional color. Observe that a box vertex is always a cut vertex. Let c be a box vertex with color i. All its neighbors with degree one lie in blocks that are connected to c and consist of a single non-cut vertex, and thus, have color $i + 1$. Hence, we add additional color $i + 2$ to vertex c.

We prove that this color assignment is a colorful 3-rainbow dominating function. Recall that since G is a block graph, each block of G is a clique. Hence, all vertices of the block are connected. Each non-cut vertex belongs to exactly one block, and each cut vertex belongs to at least two blocks. Let us consider an arbitrary vertex v and the following possible cases.

Case a: Vertex v is a non-cut vertex and belongs to root block b_r. If b_r has only two vertices, then by construction, the cut vertex of b_r has color set $\{1, 3\}$ and the non-cut vertex has color 2. Otherwise, b_r has the cut vertex with color 1, one non-cut vertex with color 2, and at least one non-cut vertex with color 3. Therefore, vertex v has all three colors in its neighborhood.

Case b: Vertex v is a non-cut vertex and belongs to leaf block b. Let c be the only cut vertex of b that during our first traversal gets color i. If block b contains only one non-cut vertex, then its only cut vertex is a box vertex and gets color set $\{i, i + 2\}$, and its non-cut vertex has color $i + 1$. Otherwise, one

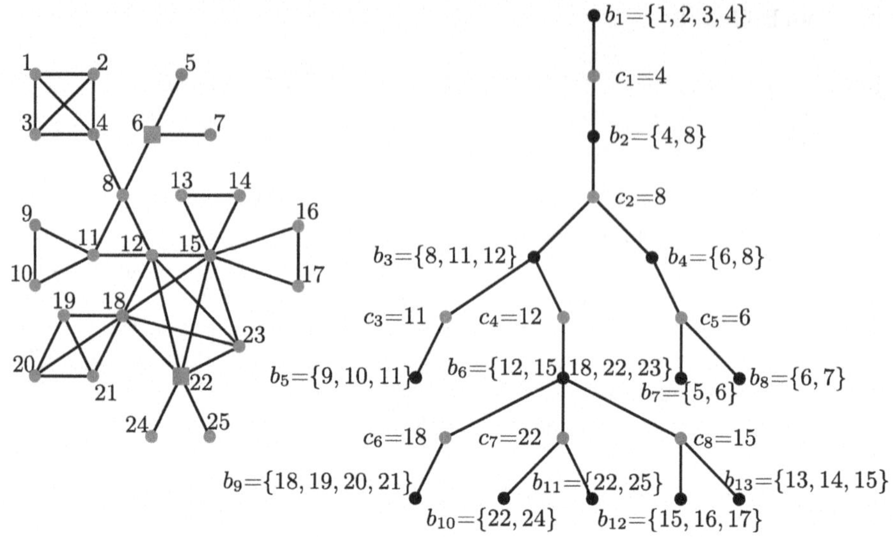

Fig. 1. An example of a block graph G and its block-cut tree T.

non-cut vertex of b gets color $i+1$, and at least one non-cut vertex gets color $i+2$. Therefore, v has all three colors in its neighborhood.

Case c: Vertex v is a non-cut vertex and belongs to non-leaf block b. If b is a non-leaf block, then it contains its parent cut vertex c and at least one child cut vertex c'. If c gets color i during our first traversal, then c' gets color $i+1$. All non-cut vertices get color $i+2$. Therefore, v has all three colors in its neighborhood.

Case d: Vertex v is a cut vertex c with parent block b_p and at least one child block b_{ch}. If b_p is a root block b_r, then the proof is basically identical to the first case. Otherwise, b_p has parent cut vertex c'. If c has color i, then c' has color $i-1$. Furthermore, at least one vertex in b_{ch} has color $i+1$. It is either a child cut vertex of b_{ch} if b_{ch} is non-leaf, or a non-cut vertex of b_{ch} if it is a leaf. Therefore, v has all three colors in its neighborhood.

In summary, in each case, vertex v has all three colors in its neighborhood. The weight of this solution is $n+n_1$. Since the computation of the block-cut tree takes $O(|V|+|E|)$ time, and afterward we only traverse the tree twice, the runtime of the algorithm is also $O(|V|+|E|)$.

3.2 Cactus Graphs

For cactus graphs, we also use the corresponding block-cut tree. If a cactus graph G does not have any cut vertices, then it is either a cycle or an edge. In this trivial case, an optimal solution can be found in linear time [8]. Thus, in the following, we assume that G has at least one cut vertex. Now, we analyze which cycles can be colored such that each non-box vertex gets exactly one color.

Lemma 2. *Let C be a leaf cycle with m non-cut vertices and no box vertex. Then a colorful 3-rainbow dominating function that assigns a single color to each vertex of this cycle exists if and only if $m \equiv 2 \mod 3$.*

Proof. Let C be $(c, u_1, \ldots, u_m)$ where c is a single cut vertex. W.l.o.g. let the color of c be 1 and the color of u_1 be 2. Vertex u_1 must have all three colors in its neighborhood $\{u_1, c, u_2\}$, therefore the color of u_2 must be 3. Applying this argument inductively, we get that the color of vertices $u_1, u_4, \ldots$ must be 2, the color of $u_2, u_5, \ldots$ must be 3, and the color of $u_3, u_6, \ldots$ must be 1. Therefore, vertex u_m has all three colors in its neighborhood $\{u_m, u_{m-1}, c\}$ if and only if $m \equiv 2 \mod 3$ since then the color of u_{m-1} is 2 and the color of u_m is 3.

In the following, we call a leaf cycle **demanding**, if it has no box vertex and for the number m of its non-cut vertices, we have $m \equiv 0 \mod 3$ or $m \equiv 1 \mod 3$. Lemma 2 implies that in each demanding leaf cycle, at least one non-box vertex gets at least two colors.

Next, we consider non-leaf cycles. Let us consider two consecutive cut vertices c_1 and c_2 of the cycle and the corresponding path $(c_1, u_1, \ldots, u_m, c_2)$ consisting of non-cut vertices $\{u_1, \ldots, u_m\}$ that connects them.

Lemma 3. *Let f be a colorful 3-rainbow dominating function that assigns a single color to each vertex from the path $(c_1, u_1, \ldots, u_m, c_2)$, whereby c_1, c_2 are cut vertices and $\{u_1, \ldots, u_m\}$ are non-cut vertices. Then:*

(i) If $m \equiv 0 \mod 3$ and $m > 0$, then c_1 and c_2 have different colors.
(ii) If $m \equiv 1 \mod 3$, then c_1 and c_2 have different colors.
(iii) If $m \equiv 2 \mod 3$, then c_1 and c_2 have the same color.
(iv) If $m = 0$, then c_1 and c_2 may have different or the same colors.

Proof. Each u_i has only two adjacent vertices, therefore u_i and both its adjacent vertices have different colors. This implies that c_1, u_1, u_2 have different colors. The same holds for u_1, u_2, u_3. Thus, c_1 and u_3 have the same color, which is different from the color of u_1 and the color of u_2. Applying this argument inductively we get that vertices $\{c_1, u_3, u_6, \ldots\}$ have the same color, as well as vertices $\{u_1, u_4, \ldots\}$ and vertices $\{u_2, u_5, \ldots\}$.

In the following, we call a path $(c_1, u_1, \ldots, u_m, c_2)$ between two consecutive cut vertices c_1 and c_2 of the cycle **changing**, if either $m \equiv 0 \mod 3$ and $m > 0$ or $m \equiv 1 \mod 3$, and **propagating** if $m \equiv 2 \mod 3$. Lemma 3 implies that if a colorful 3-rainbow dominating function assigning a single color to each vertex on the path exists and the path is changing, then the colors of its two cut vertices are different. If the path is propagating, the colors are the same.

Lemma 4. *Let C be a cycle with at least two cut vertices and no box vertex. Let each path connecting two consecutive cut vertices be propagating except one, which is changing. Then there is no colorful 3-rainbow dominating function that assigns a single color to each vertex of this cycle.*

Proof. Let $c_1, c_2, \ldots, c_m$ be cut vertices in the order of their occurrence along the cycle. Let the only changing path be the path that connects c_l and c_{l+1}, whereby $c_{m+1} = c_1$. Let f be a colorful 3-rainbow dominating function that assigns a single color to each vertex of this cycle. By Lemma 3, the colors of $c_1, c_2, \ldots, c_l$ are equal since the paths between these cut vertices are propagating. Futhermore, the colors of $c_{l+1}, c_{l+2}, \ldots, c_m, c_1$ are equal. Thus, the colors of c_l and c_{l+1} are equal, which contradicts Lemma 3, since the corresponding path is changing.

In the following, we call a non-leaf cycle **demanding** if it has no box vertex and each path connecting two consecutive cut vertices is propagating except one, which is changing.

Putting it together, Lemmas 2 and 4 imply that each demanding cycle (leaf and non-leaf) contains at least one vertex that gets at least one additional color. We call a subset V' of vertices of demanding cycles **valid** if each demanding leaf and demanding non-leaf cycle contains at least one vertex from V'. In each solution, the subset of vertices of demanding cycles that get at least two colors must be valid. For an optimal solution, the cardinality of this valid set should be minimal. Since some demanding cycles may share cut vertices, this number could be less than the number of demanding cycles. In the first phase of the algorithm, we find a valid set with the smallest cardinality n' and call the vertices from this subset **chosen**. Demanding cycles do not contain box vertices. Thus, $\gamma_{cr3}(G) \geq |V| + n_1 + n'$, whereby n_1 is the number of box vertices. In the second phase, we assign colors to the vertices of each block such that chosen vertices from the first phase and box vertices get two colors, and all other vertices get one color. There exists a particular case that must be considered separately.

In the following, we call a demanding leaf cycle b **special** if for the number m of its non-cut vertices holds $m \equiv 0 \mod 3$. A demanding non-leaf cycle b is called **special** if for the number m of non-cut vertices of its single changing path holds $m \equiv 0 \mod 3$.

Theorem 2. *Let $G = (V, E)$ be a cactus graph with $n \geq 3$ vertices and all blocks being special demanding leaf cycles. Then $\gamma_{cr3}(G) = n + 2$.*

Proof. If all blocks are demanding leaf cycles, then the block-cut tree of the graph consists of one cut vertex c and blocks $b_1, b_2, \ldots, b_l$ that share this cut vertex. Since $n' = 1$ and $n_1 = 0$ we get $\gamma_{cr3}(G) \geq n + 1$. If a solution with one additional color exists, then the only cut vertex should get two colors i, j. Each cycle b_t is special demanding, therefore for the number m_t of its non-cut vertices holds $m_t \equiv 0 \mod 3$. We need to color each non-cut vertex of each cycle such that it has all three colors in its neighbourhood. Thus, we should move along each cycle and assign colors alternatingly. The only possible coloring will give colors i and j to both cycle neighbors of c in each cycle, and therefore c would not have color $k \in \{1, 2, 3\} - \{i, j\}$ in its neighborhood. To construct a solution with two additional colors, we assign color set $\{1, 2, 3\}$ to the cut vertex c. In each cycle, to the non-cut vertices colors $1, 2, 3$ are assigned alternatingly.

Theorem 3. *Let $G = (V, E)$ be a cactus graph with $n \geq 3$ vertices, at least one cut vertex, and at least one block that is not a special demanding leaf cycle. Let*

n_1 be the number of box vertices and n' be the smallest cardinality of a valid set. Then $\gamma_{cr3}(G) = n + n_1 + n'$. We can compute an optimal colorful 3-rainbow dominating function for G in time $O(|V| + |E|)$.

Proof. We compute the block-cut tree T for G [11]. First, we find a valid set with the smallest cardinality n'. If we consider only the demanding cycles of the graph, then the block-cut tree falls apart into several connected components, whereby each component is a tree; thus demanding cycles form a forest. We start with a demanding cycle that is a leaf in one of the trees in the forest. If this leaf is not isolated, then it is connected to its tree by some cut vertex. An optimal valid set always exists that contains this cut vertex. We mark this cut vertex as chosen and remove all demanding cycles that contain this chosen vertex from the forest. If this leaf is isolated, we mark some cut vertex of it as chosen and remove this isolated leaf from the forest. The remaining part is still a forest. Thus we can proceed by taking again some demanding cycle that is now a leaf until no demanding cycles are left. This greedy approach provides a valid set of chosen vertices with the smallest cardinality.

In the second phase, we choose a block b_r as the root and consider tree T as a rooted tree. We traverse the tree and assign colors to the vertices of each block. Each box vertex and each chosen vertex gets two colors. The remaining vertices get one color. To ensure correctness, we maintain the following invariant:

Invariant 4 *After block b is processed, each non-cut vertex in b has all three colors in its neighborhood. The cut vertex c, which is the parent of b, also has all three colors in its neighbourhood. Each cut vertex c', which is a child of b, has at least two different colors in its neighborhood. If c' has already gotten two colors, then c' has all three colors in its neighborhood.*

To ensure this invariant for the root block, we choose the root by using the following priority list and take an existing block with the highest priority:

1. A leaf block that is an edge and hence consists of a box vertex and its neighbor with degree one.
2. A non-leaf block that is an edge.
3. A leaf cycle that is not a special demanding leaf cycle.
4. A non-leaf cycle that is not a special demanding non-leaf cycle.
5. A special demanding non-leaf cycle that is connected to some special demanding leaf cycle.

Since G has at least one block that is not a special demanding leaf cycle, at least one candidate for the root exists. Now, first we assign colors to the vertices of the root block b_r, whereby a procedure depends on its priority. Observe that if the root block has a priority greater than one, the graph has no box vertices.

If b_r has **priority one**, then its cut vertex c is a box vertex and we assign colors 1 and 2 to it and color 3 to the non-cut vertex, as illustrated in Fig. 2.

If b_r has **priority two**, then both its vertices are cut vertices and we assign color 1 to one of them and 2 to the other one.

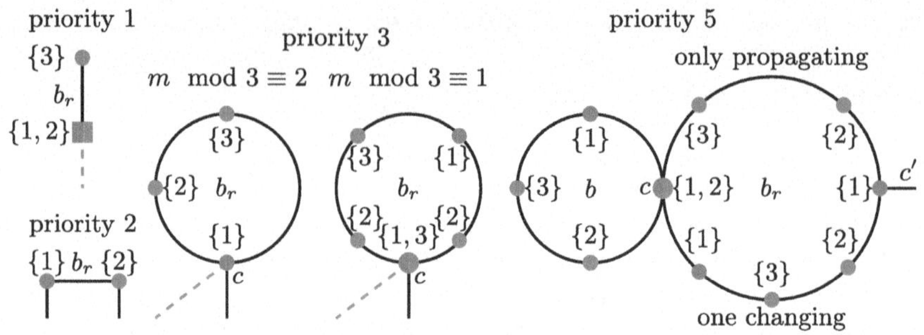

Fig. 2. Color assignments to vertices of root block b_r with different priorities.

If b_r has **priority three**, then let c be its only cut vertex and m be the number of non-cut vertices. Cut vertex c cannot be a box vertex, otherwise, we would not choose this cycle as a root. If b_r is not demanding, then $m \equiv 2 \mod 3$ and we travel along the cycle and assign colors 1, 2, 3 alternatingly. If b_r is demanding, then it is not special and therefore $m \equiv 1 \mod 3$. Its only cut vertex is chosen. We travel along the cycle and assign colors 1, 2, 3 alternatingly. When we arrive at c again, it gets an additional color 3, and therefore c has all three colors in its neighborhood.

If b_r has **priority four**, we distinguish different cases.

Case a: b_r is demanding, but not special. Since b_r is demanding, at least one cut vertex is chosen. Let c be some chosen vertex. Each path connecting two consecutive cut vertices is propagating except one, which is changing. Since b_r is not special, for the number m of non-cut vertices of this changing path holds $m \equiv 1 \mod 3$. We travel along the cycle starting in c and assign colors 1, 2, 3 alternatingly. As we arrive at c again, it should get color 3 since we had exactly one changing path on our way with $m \equiv 1 \mod 3$. We add this additional color to the color set of c. This procedure assigns two colors to c and one color to the remaining vertices. Each vertex has all three colors in its neighborhood.

Case b: b_r is not demanding and has at least two consecutive cut vertices connected by an edge. Let them be c and c'. We travel along the cycle from c' to c, starting in c' and ending in c, and assign colors 1, 2, 3 alternatingly. This procedure assigns one color to each vertex of b_r. Each vertex except c and c' has all three colors in its neighborhood. The colors of cut vertex c and its adjacent vertex along the cycle that is not c' are different, therefore c has two colors in $N[c]$, and at most one color is missing. The same holds for c'.

Case c: b_r is not demanding, and the number of changing paths connecting two consecutive vertices is not equal to one. Let $c_1, \ldots, c_l$ be cut vertices of b_r. We travel along the cycle starting in c_1 and assign colors 1, 2, 3 alternatingly. If the number of changing paths is zero, then each cut vertex gets color 1. Otherwise, the number of changing paths is at least two. If it is even, we can change the color of cut vertices back and forth such that as we arrive at c_1 again, we get the proper color 1 of it. Depending on the length of the first changing path we meet,

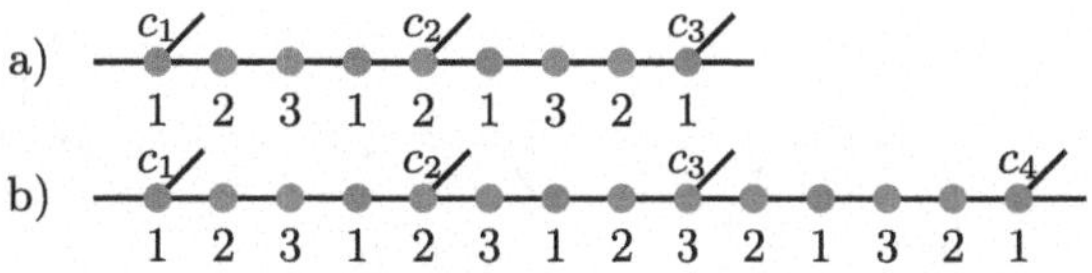

Fig. 3. Examples of changing the cut vertex color in a) two and in b) three steps.

we change the color of the cut vertex from 1 to either 2 or 3. Along the next changing path, we change back to 1 by choosing an appropriate order of colors $1, 2, 3$ that we use for this path. And so on till we arrive at c_1 again. In Fig. 3 a), an example of the color change in two steps is given. Because of the possible change of the order, some cut vertices from $c_1, \ldots, c_l$ may have only two colors in their neighborhoods. If the number of changing paths is odd, we can first change the color of cut vertices in three steps, either from 1 to 2, then to 3, and then back to 1, or from 1 to 3, then to 2, and then back to 1. This uses three changing paths, as in an example in Fig. 3 b). Thus, the remaining number of changing paths is now even, and we can change the color back and forth such that as we arrive at c_1 again, we get the proper color of c_1. This procedure assigns one color to each vertex of b_r. Each non-cut vertex has all three colors in its neighborhood. Cut vertices $c_1, \ldots, c_l$ have at least two colors in their neighborhoods.

If b_r has **priority five**, let b be a special demanding leaf block to which b_r is connected. W.l.o.g. let c be a cut vertex of b. Since b_r is not a leaf, it has at least two cut vertices. Let $c' \neq c$ be some other cut vertex of b_r Here, to assure Invariant 4, we need to assign colors to both b_r and b at once. Since c is a single cut vertex of b, it is chosen and gets two colors $1, 2$. We assign colors $1, 3, 2$ alternatingly to non-cut vertices of b as illustrated in Fig. 2. Since b is special after such an assignment, color 3 is missing in $N[c]$. There are two ways along the cycle from c to c'. Since b_r is special demanding all the paths on one way are propagating. On the other way, one path is changing with m non-cut vertices, whereby $m \equiv 0 \mod 3$ and all other paths are propagating. We travel along the way with only propagating paths and assign colors $3, 2, 1$ alternatingly. This assures that c now has color 3 in its neighborhood. We arrive at c' with color 1. Now, we travel along the way with one changing path and assign $1, 3, 2$ alternatingly. Then we arrive at c' again with color 1.

Now, after the root block is processed, we traverse the tree and distinguish the following cases depending on the role of block b.

Case 1: Block b is a leaf block. Then it contains a single cut vertex c, and c is the parent of b. Vertex c has already been given either one or two colors. If c has one color, let it be color i. By Invariant 4, vertex c already has at least two colors in its neighborhood. Let the second color be k. This implies that color $j \in \{1, 2, 3\} - \{i, k\}$ can be missing in $N[c]$. If b is an edge, then c is a box vertex, and we assign color set $\{i, j\}$ to it and color k to its non-cut vertex. If b is a cycle, then we travel along the cycle and assign colors j, k, i alternatingly such that c gets color j in its neighborhood: the first neighbor of c gets color j, the next vertex gets color k, and so on. Finally, we arrive at c again. If it gets a color

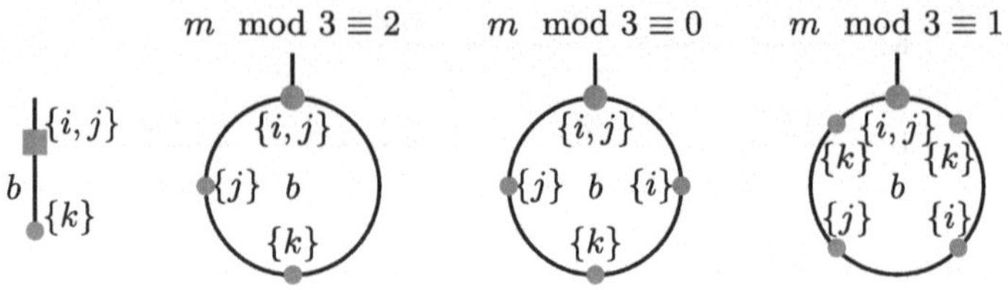

Fig. 4. Leaf block b with m non-cut vertices and cut vertex with colors i and j.

that is different from i, then it is either a box vertex or a chosen vertex, and we add this additional color to it. If c already has two colors i and j, then due to Invariant 4, it has all three colors in its neighborhood. So, we need to choose an appropriate order of colors for traveling along the cycle, such that we arrive at the cut vertex with color i or j, as shown in Fig. 4.

Case 2: Block b is a non-leaf cycle block. Then it contains its parent cut vertex c and its children cut vertices $c_1, \ldots, c_l$. Let $c_0 = c = c_{l+1}$. If c already has two colors i and j, then we can proceed in the same way as for a leaf cycle block. If c has one color i, then by Invariant 4, let k be the second color from the neighborhood of c. Thus, color $j \in \{1, 2, 3\} - \{i, k\}$ can be missing in $N[c]$. We assign colors to the vertices of b such that c gets color j in its neighborhood, and Invariant 4 holds. Here, we distinguish different subcases.

Case 2.1: Block b is demanding. No cut vertex in b is a box vertex by definition. At least one cut vertex in b is chosen. Let it be c'. Each path connecting two consecutive cut vertices is propagating except one, which is changing. If $c \neq c'$, we travel along the cycle to c' using the way where each path is propagating and assign colors i, j, k alternatingly, as shown in Fig. 5. We arrive at c' with color i. Now we travel in the opposite direction, where one path is changing with m non-cut vertices. If $m \mod 3 \equiv 1$ we use the order i, k, j. If $m \mod 3 \equiv 0$ we use the order j, k, i. In both cases, we arrive at c' with color j and add it to vertex c'. It now has a color set $\{i, j\}$ and at least one neighbor with color k. If $c = c'$, we travel along the cycle and assign colors i, j, k alternatingly. Finally, we arrive at c with the color not equal to i and add it to the color set of c, which now has two colors in its set and all three colors in its neighborhood.

Case 2.2: Block b is not demanding and has at least one box vertex. The construction is similar to the case of a demanding cycle, whereby the box vertex gets two colors. To ensure that this box vertex has the third color in its neighborhood, we assign this third color to its neighbor with degree one.

Case 2.3: Block b is not demanding, has no box vertex, and has at least two consecutive cut vertices connected by an edge. Let them be c_t and c_{t+1}. We travel along the cycle from c to c_t and assign colors i, j, k alternatingly. Afterward, we travel along the cycle in the other direction from c to c_{t+1} and assign colors k, j, i alternatingly. Each vertex except c_t and c_{t+1} has all three colors in its neighborhood. The colors of cut vertex c_t and its adjacent vertex along the cycle that is not c_{t+1} are different, therefore c_t has at least two colors in its neighborhood. The same holds for c_{t+1}.

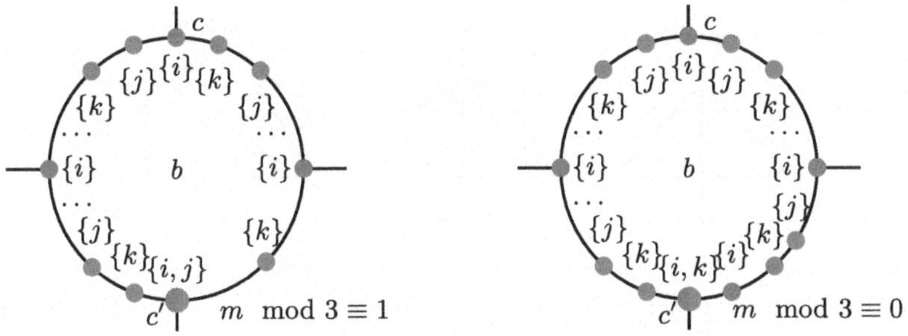

Fig. 5. Color assignments to vertices of demanding non-leaf block b with one changing path consisting of m non-cut vertices.

Case 2.4: Block b is not demanding, has no box vertex, and the number of changing paths connecting two consecutive vertices is not equal to one. Then the procedure is similar to Case c for the root block with priority four. We travel along the cycle starting from c and assign colors i, j, k alternatingly. Vertex c already has color i, the next vertex gets color j, and so on. Taking into account the parity of the number of changing paths and their lengths, we use an appropriate order of colors such that we arrive at c again with color i.

Case 3: Block b is a non-leaf edge block. Then it contains its parent cut vertex c, its only child cut vertex c', and no other vertices. If c has one color i, then let k be the second color from $N[c]$. Thus, color $j \in \{1, 2, 3\} - \{i, k\}$ can be missing in $N[c]$. We assign color j to c'. Therefore, vertex c gets color j in $N[c]$. Vertex c' has color j and a different color i in $N[c']$. If c has already two colors i, j we assign color k to c'.

In summary, in each case, Invariant 4 holds. Each non-cut vertex v is a part of exactly one block b. Thus, after b is processed, v has all three colors in its neighborhood, by Invariant 4. Each cut vertex c is always the parent of some block b. Therefore, after b is processed, due to Invariant 4, vertex c has all three colors in its neighborhood. Hence, we have found a colorful 3-rainbow dominating function with weight $n + n_1 + n'$, which is equal to our lower bound. Both the computation of the block-cut tree and the greedy computation of the valid set of chosen vertices take $O(|V| + |E|)$ time. Afterward, in the second phase, we only traverse the tree once. Therefore, the runtime of the algorithm is $O(|V| + |E|)$.

4 NP-Completeness of Colorful 3-Rainbow Domination

Theorem 5. *Colorful 3-Rainbow Domination is NP-complete in maximum degree 3 graphs.*

Proof. We reduce the 3-Colorability problem, which is NP-complete [10], to the Colorful 3-Rainbow Domination problem in maximum degree 3 graphs. First, we delete all isolated vertices, since they do not influence the 3-colorability of the graph. Now, we construct a new graph G' where we add a new vertex v_{xy}

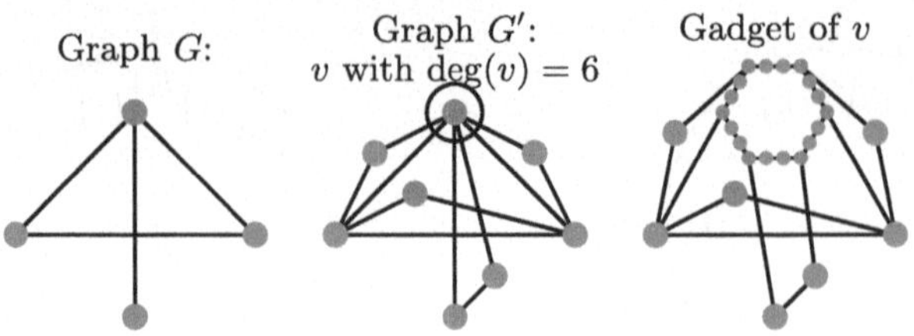

Fig. 6. An illustration of the reduction from 3-Colorability to Colorful 3-Rainbow Domination.

for each edge $\{x, y\} \in E$ and connect it to both old vertices x and y, as shown in Fig. 6. Each new vertex has degree 2. Old vertices can have a larger degree, which we want to eliminate, and construct a new graph G''. We exchange each vertex v of degree greater than 3 by the following gadget. Let v have neighbours $u_0, \ldots, u_{\deg(v)-1}$. The gadget is a cycle that contains $3\deg(v)$ vertices $v_0, v_1, \ldots v_{3\deg(v)-1}$. We connect each third vertex to the corresponding neighbour of v by an edge, namely v_{3i} is connected to u_i. Observe, that the path between v_{3i} and v_{3i+3} is propagating. We choose d to be equal to the number of vertices in G''. Thus, $\gamma_{cr3}(G'') \leq d$ only if a colorful 3-rainbow dominating function exists where exactly one color is assigned to each vertex.

We prove that G is 3-colorable if and only if $\gamma_{cr3}(G'') \leq d$. If G is 3-colorable, we choose a 3-coloring of G and assign the same colors to the old vertices in G'. Each new vertex v_{xy} corresponds to an edge $\{x, y\} \in E$. Therefore, x and y have different colors and we assign the third remaining color to a vertex v_{xy}. Hence, all new vertices have all three colors in their neighborhood. Since each old vertex is not isolated, it has at least one adjacent edge and therefore has all three colors in its neighborhood. Now for each old vertex v with $\deg(v) > 3$ and color i, a gadget with vertices $v_0, v_1, \ldots v_{3\deg(v)-1}$ can be colored in such a way, that vertices $v_0, v_3, \ldots$ get color i, vertices $v_1, v_4, \ldots$ get color $i+1$, and vertices $v_2, v_5, \ldots$ get color $i+2$. Then each vertex of the gadget has all three colors in its neighborhood. If for G'' a colorful 3-rainbow domination function with weight at most d exists, then each vertex gets exactly one color. Let v be an old vertex that is exchanged by a gadget with vertices $v_0, v_1, \ldots v_{3\deg(v)-1}$. Since each path between v_{3i} and v_{3i+3} is propagating vertices $v_{3i}, 0 \leq i \leq \deg(v) - 1$ have the same color. Therefore, we get a color assignment of the vertices of the original graph G. For each edge $\{x, y\} \in E$, the corresponding vertex v_{xy} must have all three colors in its neighborhood. Hence, vertices x and y have different colors. Thus, this coloring is a proper 3-coloring of the original graph G, such that no two vertices sharing the same edge have the same color.

5 Conclusion

We consider the colorful 3-rainbow domination problem and give linear time algorithms that compute an optimal colorful 3-rainbow domination function if an input graph is a block or a cactus graph, and therefore answer an open question from [8]. Our efficient algorithms use the block-cut tree to find an optimal solution. For cactus graphs, an important part of the algorithm is to find a block that plays the role of an entry point in the graph, starting from which colors are assigned. Moreover, we show that the problem remains NP-complete if we restrict the input graph to the class of maximum degree 3 graphs.

References

1. Ahangar, H.A., Amjadi, J., Rad, N.J., Samodivkin, V.: Total k-rainbow domination numbers in graphs. Commun. Comb. Optim **3**(1), 37–50 (2018). https://doi.org/10.22049/cco.2018.25719.1021
2. Brešar, B., Henning, M.A., Rall, D.F.: Rainbow domination in graphs. Taiwanese J. Math. **12**(1), 213–225 (2008). https://doi.org/10.11650/twjm/1500602498
3. Brešar, B., Šumenjak, T.K.: On the 2-rainbow domination in graphs. Discret. Appl. Math. **155**(17), 2394–2400 (2007). https://doi.org/10.1016/J.DAM.2007.07.018
4. Chang, G.J., Wu, J., Zhu, X.: Rainbow domination on trees. Discret. Appl. Math. **158**(1), 8–12 (2010). https://doi.org/10.1016/J.DAM.2009.08.010
5. Garey, M.R., Johnson, D.S.: Computers and Intractability: A Guide to the Theory of NP-Completeness. Freeman, W. H (1979)
6. Harary, F.: A characterization of block-graphs. Can. Math. Bull. **6**(1), 1–6 (1963). https://doi.org/10.4153/CMB-1963-001-x
7. Harary, F., Uhlenbeck, G.E.: On the number of Husimi trees: I. PNAS **39**(4), 315–322 (1953). https://doi.org/10.1073/pnas.39.4.315
8. Lavynska, T.: Colorful 3-rainbow domination. In: SOFSEM 2025, LNCS, vol. 15539, pp. 99–111. Springer (2025). https://doi.org/10.1007/978-3-031-82697-9_8
9. Shao, Z., Li, Z., Peperko, A., Wan, J., Žerovnik, J.: Independent rainbow domination of graphs. Bullet. Malays. Math. Sci. Soc. **42**(2), 417–435 (2017). https://doi.org/10.1007/s40840-017-0488-6
10. Stockmeyer, L.J.: Planar 3-colorability is polynomial complete. SIGACT News **5**(3), 19–25 (1973). https://doi.org/10.1145/1008293.1008294
11. Tarjan, R.E., Vishkin, U.: Finding biconnected components and computing tree functions in logarithmic parallel time. In: FOCS 1984, pp. 12–20. IEEE (1984). https://doi.org/10.1109/SFCS.1984.715896
12. Tepeh, A.: Total domination in generalized prisms and a new domination invariant. Discuss. Math. Graph Theory **41**(4), 1165–1178 (2021). https://doi.org/10.7151/DMGT.2256

Complexity of Minimal Faithful Permutation Degree for Fitting-Free Groups

Michael Levet[1(✉)], Pranjal Srivastava[2], and Dhara Thakkar[3]

[1] College of Charleston, Charleston, SC 29492, USA
levetm@cofc.edu

[2] Indian Institute of Information Technology Vadodara-ICD, Diu, India
pranjal.srivastava194@gmail.com

[3] Graduate School of Mathematics, Nagoya University, Nagoya, Japan
thakkar_dhara@math.nagoya-u.ac.jp

Abstract. In this paper, we investigate the complexity of computing the minimal faithful permutation degree for groups without abelian normal subgroups (a.k.a. Fitting-free groups). When our groups are given as quotients of permutation groups, we establish that this problem is in P. Furthermore, in the setting of permutation groups, we obtain an upper bound of NC for this problem. This improves upon the work of Das and Thakkar (STOC 2024), who established a Las Vegas polynomial-time algorithm for this class in the setting of permutation groups.

Keywords: Permutation Group Algorithms · Minimum Faithful Permutation Degree · Circuit Complexity

1 Introduction

Cayley's theorem states that for every finite group G, there exists $m \in \mathbb{N}$ such that G can be embedded into $\mathrm{Sym}(m)$. The *minimal faithful permutation degree*, denoted $\mu(G)$, is the smallest such m for which G can be embedded into $\mathrm{Sym}(m)$. A permutation representation of G on $\mu(G)$ points is called a *minimal faithful permutation representation*. Note that $\mu(G) \leq |G| \leq \mu(G)!$. By convention, $\mu(G) = 0$ if and only if $G = 1$. Johnson [17] exhibited the following characterization of $\mu(G)$.

Proposition 1 ([17]). *Let G be a finite group, and let $\mathcal{G}$ be the collection of all subgroups of G. We have that:*

$$\mu(G) := \min_{\emptyset \neq \mathcal{H} \subseteq \mathcal{G}} \left\{ \sum_{H \in \mathcal{H}} [G : H] \;\middle|\; \bigcap_{H \in \mathcal{H}} Core_G(H) = 1 \right\},$$

where $Core_G(H) = \bigcap_{g \in G} gHg^{-1}$ and $|\mathcal{H}| \leq \log |G|$.

A. Jeż and J. Otop (Eds.): FCT 2025, LNCS 16106, pp. 350–361, 2026.
https://doi.org/10.1007/978-3-032-04700-7_26

In this paper, we investigate the computational complexity of computing $\mu(G)$. Precisely, we will consider the decision variant, MIN-PER-DEG, which takes as input a finite group G and $k \in \mathbb{N}$, and asks if $\mu(G) \leq k$. Key motivation for this comes from the setting of permutation group algorithms. Given a group $G \leq \text{Sym}(n)$, there is a natural divide-and-conquer strategy where we decompose G in terms of a normal subgroup $N \trianglelefteq G$ and the corresponding quotient G/N. We then seek to solve the algorithmic problem on N and G/N, before recombining the solutions on the subproblems to a solution for G. The techniques involved when reasoning about quotients of two permutation groups are considerably more involved than when dealing with a single permutation group [19]. Thus, if G/N has a *small* faithful permutation representation that can be efficiently constructed, then we can take advantage of the extensive suite of permutation group algorithms (e.g., [1,5,27]). This approach is not always possible, as there exist quotients with exponentially large degree. Let D_8 be the dihedral group of order 8. Neumann observed that D_8^ℓ admits a faithful permutation representation on 4ℓ points, but admits a quotient Q that is an extraspecial 2-group with $\mu(Q) \geq 2^{\ell+1}$ [25]. For a survey of algorithms to deal with quotient groups, we refer to [19,27].

The parameter $\mu(G)$ has received considerable attention, primarily from a mathematical perspective. We refer to [10,22] for a survey. To the best of our knowledge, Kantor and Luks [19] were the first to propose investigating the computational complexity of MIN-PER-DEG. In the setting of permutation groups and their quotients, MIN-PER-DEG belongs to NP [10,19]. Proposition 1 yields a trivial upper-bound of $|G|^{O(\log^2 |G|)}$-time for this problem, in the Cayley table model. This is essentially the generator-enumeration strategy. We enumerate over all such collections of $\leq \log |G|$ subgroups, and each subgroup is specified by $\leq \log |G|$ many generators.

Das and Thakkar [10] recently exhibited significant progress on the computational complexity of MIN-PER-DEG. In the setting of permutation groups, they exhibited a Las Vegas polynomial-time algorithm for groups without abelian normal subgroups, and a quasipolynomial-time algorithm for primitive permutation groups. When the groups are given by their multiplication (Cayley) tables, Das and Thakkar exhibited an upper bound of $\mathsf{DSPACE}(\log^3 |G|)$, which improves upon the trivial $|G|^{O(\log^2 |G|)}$ bound.

Our Contributions. In this paper, we further investigate the MIN-PER-DEG problem for groups without abelian normal subgroups, addressing one of the open questions posed by Das and Thakkar [10, Section 9]. Our main result is the following.

Theorem 1. *Let* $\mathbf{G}$ *be a Fitting-free group (a.k.a. group without abelian normal subgroups).*

(a) Suppose $\mathbf{G}$ *is given as the quotient* G/K *of two permutation groups* $K \trianglelefteq G \leq \mathit{Sym}(n)$*. Then we can compute* $\mu(\mathbf{G})$ *in polynomial-time.*

(b) If $\mathbf{G}$ *is given as a permutation group* ($K = 1$)*, then we can compute* $\mu(\mathbf{G})$ *in* NC*.*

Given a finite group G by its Cayley table, we can construct the regular representation in NC. Thus, Theorem 1 yields the following immediate corollary.

Corollary 1. *Let G be a Fitting-free group given by its Cayley table. We can compute $\mu(G)$ in NC.*

We compare Theorem 1 to the previous result of Das and Thakkar [10]. Here, Das and Thakkar established a Las Vegas polynomial-time analogue of Theorem 1 in the setting of *permutation groups*. They explicitly did not deal with quotients [10]. Das and Thakkar crucially leveraged randomness by utilizing the suite of Las Vegas algorithms for constructive recognition of finite simple groups in the black-box model [18]. Note that the black-box model also includes matrix groups, where order-finding is already at least as hard as DISCRETE LOGARITHM, and so Las Vegas polynomial-time is essentially optimal in the black-box setting. However, in the setting of permutation groups and their quotients, constructive recognition of finite simple groups is *deterministically* computable in polynomial-time [19]. In order to establish Theorem 1(a), we leverage this deterministic procedure for constructive recognition of finite simple groups, in tandem with the extensive library of quotient group algorithms due to Kantor and Luks [19].

In order to prove Theorem 1(b), we similarly take advantage of the NC-framework for constructive recognition of finite simple groups [20], in tandem with a large suite of NC algorithms for permutation groups [1] and standard parallel algorithms for linear algebra (e.g., [24]). Still, we require additional ingredients; namely, computing the socle in NC. While such polynomial-time algorithms have long been known for both permutation groups (see e.g., [16,27]) and their quotients [19], it is open whether we can compute the socle in NC. Prior to this work, it was open, even in the restricted setting where we have a permutation group that is Fitting-free. To this end, we establish the following.

Proposition 2. *Let G be a Fitting-free group specified by a generating sequence of permutations from $\mathrm{Sym}(n)$. We can compute $\mathrm{Soc}(G)$ in NC.*

In order to prove Proposition 2, we take advantage of an NC-computable series of length $O(\log^2 n)$, where each term is normal in G and the successive quotients are semisimple (direct products of simple groups) [1]. For an arbitrary group, $\mathrm{Soc}(G)$ may contain abelian normal subgroups. Constructing generators for both the abelian and non-abelian parts of the socle together, in NC, remains open.

Extending Proposition 2 to the setting of quotients is also non-trivial. To prove Proposition 2, we take advantage of the fact that we can, in NC, compute the centralizer of a normal subgroup [1]. While such a polynomial-time algorithm exists for quotients, the machinery is considerably more complicated than for permutation groups and appears resistant to parallelization [19, Section 6]. This, in turn, presents a significant obstacle to improving the bound in Theorem 1(a) from P to NC.

Remark 1. We also compare Theorem 1 to the work of Cannon and Holt [5], for practical isomorphism testing of permutation groups. Cannon and Holt start with a permutation group G, and compute the solvable radical $\text{Rad}(G)$. They then deal with the Fitting-free quotient $G/\text{Rad}(G)$ by explicitly constructing a faithful permutation representation. Given a quotient group $\mathbf{G} := G/K$, with $K \trianglelefteq G \leq \text{Sym}(n)$, we can compute $\text{Rad}(\mathbf{G})$ in polynomial-time [19]. Combining this with Theorem 1(a) formalizes that much of the framework established by Cannon and Holt [5] is in fact polynomial-time computable. It remains open whether the solvable radical can be computed in NC, even in the setting of permutation groups.

Further Related Work. Fitting-free groups have received considerable attention from the computational complexity community. In the multiplication Cayley table model, isomorphism testing of Fitting-free groups is known to be in P, through a series of two papers [2,3]. This class of groups, as well as important sub-classes such as direct products of non-abelian simple groups and almost simple groups, have recently received attention from the perspective of the Weisfeiler–Leman algorithm [4,13,14].

2 Preliminaries

Groups. All groups are assumed to be finite. The *normal closure* of a subset $S \subseteq G$, denoted $\text{ncl}_G(S)$, is the smallest normal subgroup of G that contains S. The *socle* of a group G, denoted $\text{Soc}(G)$, is the subgroup generated by the minimal normal subgroups of G. If G has no abelian normal subgroups, then $\text{Soc}(G)$ decomposes uniquely as the direct product of non-abelian simple factors. The *solvable radical* $\text{Rad}(G)$ is the unique maximal solvable normal subgroup of G. Every finite group can be written as an extension of $\text{Rad}(G)$ by the quotient $G/\text{Rad}(G)$, the latter of which has no abelian normal subgroups. We refer to the class of groups without abelian normal subgroups as *Fitting-free*.

For background on finite simple groups, we refer to [28, Section 2.2] and [7,21,33]. There are four types of automorphism of a finite simple groups known as the inner automorphism, diagonal automorphism, field automorphism, and graph automorphism. Any automorphism of a classical simple group or exceptional group of Lie type can be decomposed as a product of these four automorphisms [7]. We will not need the technical details about these automorphisms, and instead refer to [22, §2.4] for additional background.

Permutation Group Algorithms. We will consider groups specified succinctly by generating sequences. The first model we will consider is that of permutation groups. Here, a group G is specified by a sequence S of permutations from $\text{Sym}(n)$. Alternatively, we will consider the more general quotients of permutation groups model, which we will refer to as the *quotient model.* To the best of our knowledge, this model was first introduced by Kantor and Luks [19]. Here, we will be given $K \trianglelefteq G \leq \text{Sym}(n)$, represented in the following way: (i) G, K are both specified by a generating sequence of permutations in $\text{Sym}(n)$, and (ii) each element of $\mathbf{G} := G/K$ is specified by a single coset representative. The computational complexity for both the permutation group and quotient model will be measured in terms of n. We refer to [1] for a suite of NC algorithms for permutation groups.

Lemma 1. *Let $c > 0$. Let $K \trianglelefteq G \leq Sym(n)$, and let $\mathbf{G} := G/K$. If $|\mathbf{G}| \leq n^c$, then we can list $\mathbf{G}$ in NC.*

Constructive Recognition for Finite Simple Groups. We recall the framework for constructive recognition of finite simple groups given as either permutation groups or their quotients [19,20].

Theorem 2 ([19,20]). *Let $K \trianglelefteq G \leq Sym(n)$, and let $\mathbf{G} := G/K$. If $\mathbf{G}$ is a non-abelian simple group, then we can, in polynomial-time, determine the name S of $\mathbf{G}$, as well as construct an isomorphism φ from $\mathbf{G}$ to an isomorphic copy $\mathbf{G}_1$ of $\mathbf{G}$. Furthermore, if $\mathbf{G}$ is a classical simple group, then the following hold:*

1. *The elements of $\mathbf{G}_1$ are matrices acting on an $\mathbb{F}_p$-vector space V of dimension $O(\log n)$,*
2. *The isomorphism φ comes with a basis β for V and a data structure that:*
   (a) Given a permutation $g \in \mathbf{G}$, we can, in polynomial-time, construct the corresponding matrix representation T with respect to β, and
   (b) Given a matrix $T \in S$, we can, in polynomial-time, construct the corresponding permutation $g \in \mathbf{G}$.

   If $\mathbf{G}$ is given as a permutation group ($K = 1$), then this correspondence is NC-computable.

The Minimal Faithful Permutation Degree of Simple and Almost Simple Groups.

Lemma 2. *Let $G, K \leq Sym(n)$ with $K \trianglelefteq G$. Let $\mathbf{S} = G/K$ be a simple group then we can compute $\mu(\mathbf{S})$ in polynomial-time. Furthermore, if $K = 1$, then we can compute $\mu(\mathbf{S})$ in NC.*

Proof. We first use Theorem 2 to identify the type of $\mathbf{S}$. Now if $\mathbf{S} \cong A_n$ then clearly, $\mu(\mathbf{S}) = n$. If $\mathbf{S}$ is either a classical simple group or an exceptional group of Lie type, then we use [15, Table 4]. If $\mathbf{S}$ is a sporadic simple group, we refer to [9]. For the explicit value, refer to [8, Page 10].

For the complexity bounds, we note that Theorem 2 yields polynomial-time bounds in the general case (quotients of permutation groups), and NC bounds when $K = 1$ (permutation groups). The result now follows.

Let S be a simple group, then a group A with $S \leq A \leq \text{Aut}(S)$ is called an *almost simple group*. A normal subgroup $H \neq 1$ of G is called *minimal normal* if the only normal subgroups of G contained in H are 1 and H. A minimal normal subgroup H of a finite group G is either simple or a direct product of isomorphic simple groups (see e.g., [26, Page 106]).

Proposition 1. *[6] Let $S \leq A \leq \text{Aut(S)}$ for a finite non-abelian simple group S. Then $\mu(A) = \mu(S)$, except for the following cases described in the Table 1.*

Table 1. Table containing μ value of certain simple groups and almost simple groups

	S	A	$\mu(S)$	$\mu(A)$
1	A_6	$A \not\leq S_6$	6	10
2	$\mathrm{PSL}(2,7)$	$\mathrm{PGL}(2,7)$	7	8
3	M_{12}	$\mathrm{Aut}(M_{12}) = M_{12}.2$	12	$2\mu(S)$
4	O′N	$\mathrm{Aut}(O'N) = O'N.2$	122760	$2\mu(S)$
5	$\mathrm{PSU}(3,5)$	$A \not\leq \mathrm{P\Sigma U}(3,5)$	50	126
6	$\mathrm{P\Omega}^+(8,2)$	$3 \mid \lvert A/S\rvert$	120	$3\mu(S)$
7	$\mathrm{P\Omega}^+(8,3)$	$3 \mid \lvert A/S\rvert, 12 \nmid \lvert A/S\rvert$	1080	$3\mu(S)$
8	$\mathrm{P\Omega}^+(8,3)$	$12 \mid \lvert A/S\rvert$	1080	3360
9	$G_2(3)$	$\mathrm{Aut}(\mathrm{G}_2(3))$	351	$2\mu(S)$
10	$\mathrm{P}\Omega^+(8,q)$, $q \geq 4$	$3 \mid \lvert A/S\rvert$	$\frac{(q^4-1)(q^3+1)}{q-1}$	$3\mu(S)$
11	$\mathrm{PSL}(d,q)$, $d \geq 3$ $(d,q) \neq (3,2),(4,2)$	$A \not\leq \mathrm{P\Gamma L}(d,q)$	$\frac{q^d-1}{q-1}$	$2\mu(S)$
12	$\mathrm{PSp}(4,2^e)$, $e \geq 2$	$A \not\leq \mathrm{P\Gamma Sp}(4,2^e)$	$\frac{2^{4e}-1}{2^e-1}$	$2\mu(S)$
13	$\mathrm{P}\Omega^+(2d,3)$, $d \geq 4$	$3 \nmid \lvert A/S\rvert, (*)$	$\frac{3^{d-1}(3^d-1)}{2}$	$\frac{(3^d-1)(3^{d-1}+1)}{2}$
14	$G_2(3^e)$, $e > 1$	$A \not\leq \Gamma\mathrm{G}_2(3^e)$	$\frac{(q^6-1)}{q-1}$	$2\mu(S)$
15	$F_4(2^e)$	$A \not\leq \Gamma\mathrm{F}_4(2^e)$	$\frac{(q^{12}-1)(q^4+1)}{q-1}$	$2\mu(S)$
16	$E_6(q)$	$A \not\leq \Gamma\mathrm{E}_6(q)$	$\frac{(q^9-1)(q^8+q^4+1)}{q-1}$	$2\mu(S)$

The condition $(*)$ in the row 13 is $A \not\leq \mathrm{PO}^+(2d,3)$ when $d > 4$, and A is not contained in any conjugate of $\mathrm{PO}^+(8,3)$ in $\mathrm{Aut}(\mathrm{P\Omega}^+(8,3))$ when $d = 4$.

3 Computing Socle of Fitting-Free Group

In this section, we show that we can compute the socle for Fitting-free groups in NC. We require the following key lemmas. The authors thank Joshua A. Grochow [12] for sharing these results with us.

Lemma 3. *[12] If G is Fitting-free and $N \trianglelefteq G$, then N is Fitting-free and $Soc(N) = Soc(G) \cap N$.*

Lemma 4. *[12] Suppose G is Fitting-free and $1 \leq N_1 \leq \cdots \leq N_r = G$ is a normal series (all N_i are normal in G) such that each factor N_{i+1}/N_i is a direct product of simple groups, and is either abelian, or contains no abelian direct factors. Define inductively $M_1 := N_1$, $M_i := M_{i-1} \times (N_i \cap C_G(M_{i-1}))$. Then for all i, we have $M_i = Soc(N_i)$, and in particular, $M_r = Soc(G)$.*

Now we are ready to prove the main result of this section.

Proposition 2. *Let $G \leq Sym(n)$ be a Fitting-free permutation group. We can compute generators for $Soc(G)$ in NC. Furthermore, we can in NC, decompose $Soc(G)$ into a direct product of non-abelian simple groups.*

Proof. Babai, Luks, and Seress [1] established that there exists an NC-computable series $1 = N_0 \leq N_1 \leq \ldots \leq N_r = G$ of length $r \in O(\log^2 n)$ such that for all $i \in [r]$, $N_i \trianglelefteq G$ and each N_i/N_{i-1} is a direct product of simple groups. In particular, N_i/N_{i-1} is either abelian, or has no abelian direct factor. We crucially take advantage of both the length of the series, as well as the fact that the successive quotients are semisimple to compute $\text{Soc}(G)$ in NC.

We proceed inductively along the series. We begin with N_1. As G is Fitting-free, N_1 is the direct product of non-abelian simple groups. By Lemma 3, $N_1 = \text{Soc}(N_1) \trianglelefteq \text{Soc}(G)$. Then by [16, Lemma 10.2] have $\text{Soc}(G) = N_1 \times \text{Soc}(C_G(N_1))$.

Define $M_1 := N_1 (= \text{Soc}(N_1))$. And for $i > 1$, define $M_i := M_{i-1} \times (N_i \cap C_G(M_{i-1}))$. As each N_i is normal in G, we have that each M_i is normal in G. Now by [1], we may compute $C_G(M_{i-1})$ in NC, provided we are given generators for M_{i-1}. By [1], we may then compute $C_G(M_{i-1}) \cap N_i$ in NC. Thus, we may compute M_i in NC, provided we are given generators for M_{i-1}. As $r \in O(\log^2 n)$, we may compute $M_1, \ldots, M_r$ in NC. By Lemma 4, we have that $M_r := \text{Soc}(G)$.

Once we have generators for $\text{Soc}(G)$, we may decompose $\text{Soc}(G)$ into a direct product of non-abelian simple groups in NC using [1]. The result now follows.

Remark 2. In the setting of both permutation groups and their quotients, the socle is polynomial-time computable [19]. However, obtaining NC bounds is a long-standing open problem. Extending Proposition 2 to the setting of quotients is also non-trivial. We rely crucially on the NC algorithm of Babai, Luks, and Seress [1] to compute the centralizer of a normal subgroup H in NC. Kantor and Luks [19, Section 6] note that the techniques for computing $C_G(H)$ in the setting of permutation groups appear not to work for quotients. Instead, Kantor and Luks bring to bear the extensive machinery for computing Sylow subgroups of permutation groups in order to compute cores of quotients. Kantor and Luks then reduce computing $C_G(H)$ to computing an appropriate core. It is unclear how to compute cores of quotients in NC.

Proposition 3. *Let $K \trianglelefteq G \leq Sym(n)$, and let $\mathbf{G} := G/K$ be Fitting-free. Suppose we are given the non-abelian simple factors of Soc($\mathbf{G}$). Then in NC, we can compute the set of minimal normal subgroups of $\mathbf{G}$.*

Proposition 4. *Let $G, K \leq Sym(n)$ with $K \trianglelefteq G$. Let $\mathbf{G} = G/K$. Suppose that $\mathbf{G}$ is Fitting-free, and we are given the non-abelian simple factors of Soc($\mathbf{G}$). Let S be such a simple factor. Computing $N_{\mathbf{G}}(S)$ is NC-reducible to* Pointwise Stabilizer*. In particular, we can compute $N_{\mathbf{G}}(S)$ in NC.*

4 The Minimal Faithful Permutation Degree of Fitting-Free Groups

Definition 1. *Let G be a group and let N be a minimal normal subgroup of G such that $N = S_1 \times \cdots \times S_\ell$, where S_i's are isomorphic simple groups. Let $\phi : N_G(S_1) \to \text{Aut}(S_1)$ be the conjugation action of $N_G(S_1)$ on S_1. Let A be the subgroup of $\text{Aut}(S_1)$ induced by this conjugation action ϕ. We define $\mu(G, N) := \mu(A)$.*

We note that $S_1 \leq A \leq \mathrm{Aut}(S_1)$ [11]. The well-definedness of $\mu(G, N)$ follows form the fact that S_i's are isomorphic under conjugation action by elements of G.

Theorem 3. *Let $G, K \leq Sym(n)$ with $K \trianglelefteq G$. Let $\mathbf{G} = G/K$. Suppose that $\mathbf{G}$ is Fitting-free, and let N be a minimal normal subgroup of $\mathbf{G}$. Suppose that we are given generators for each of the (unique) non-abelian simple direct factors of N. We can compute $\mu(\mathbf{G}, N)$ in polynomial-time. Furthermore, If $\mathbf{G}$ is given as a permutation group ($K = 1$), then we can compute $\mu(\mathbf{G}, N)$ in NC.*

Before we go into the proof of this theorem we first show how the above theorem can be used to compute $\mu(G)$ of a given Fitting-free G.

Lemma 5 (*[6,* ***Proposition 2.5****]*)**.** *Let G be a Fitting-free group. Let $\mathrm{Soc}(G) = N_1 \times \cdots \times N_r$, where N_i's are the minimal normal subgroups of G with $N_i = S_{i1} \times \cdots \times S_{i\ell_i}$. Then $\mu(G) = \sum_{i=1}^{r} \ell_i \mu(G, N_i)$.*

Theorem 4. *Let $G, K \leq Sym(n)$ with $K \trianglelefteq G$. Let $\mathbf{G} = G/K$. Suppose that $\mathbf{G}$ is Fitting-free. Then in polynomial-time, we can compute $\mu(\mathbf{G})$. Furthermore, If $\mathbf{G}$ is given as a permutation group ($K = 1$), then we can compute $\mu(\mathbf{G})$ in NC.*

Proof. Note that $\mathrm{Rad}(\mathbf{G}) = 1$. We first compute $\mathrm{Soc}(\mathbf{G})$ and its direct product decomposition into non-abelian simple groups $\mathrm{Soc}(\mathbf{G}) = S_1 \times \cdots \times S_k$. This step is polynomial-time computable in the quotients model [19], and NC computable when $K = 1$ (Proposition 2).

Given such a decomposition, we may easily recover in NC the minimal normal subgroups of $\mathbf{G}$ by considering the conjugation action of $\mathbf{G}$ on $S_1, \ldots, S_k$.

Let $\mathrm{Soc}(\mathbf{G}) = N_1 \times \cdots \times N_r$, where N_i's are minimal normal subgroups of $\mathbf{G}$. Suppose that $N_i = S_{i1} \times \cdots \times S_{i\ell_i}$. From Lemma 5 we have $\mu(\mathbf{G}) = \sum_{i=1}^{r} \ell_i \mu(\mathbf{G}, N_i)$. Note that given a direct product decomposition of $\mathrm{Soc}(\mathbf{G})$ and the N_i's, we can find r and ℓ_i's in NC (Proposition 3). As the N_i's are non-abelian, we have by Theorem 3 that we can compute $\mu(\mathbf{G}, N_i)$ in polynomial-time for quotients, and in NC if $K = 1$. Thus, the theorem follows.

Now we prove Theorem 3 in the remainder of this section.

Proof of Theorem 3. We follow the strategy of [10, Theorem 4.2]. Let $N = S_1 \times \cdots \times S_\ell$. The next task is to find the group A mentioned in Definition 1. Since $\mu(\mathbf{G}, N) = \mu(A)$, our task is to find $\mu(A)$.

As we are given the direct decomposition of N into $S_1 \times \cdots \times S_\ell$, we have by Proposition 4, that we can in NC (for both the quotients and permutation groups models), compute $N_{\mathbf{G}}(S_1) = \langle g_1, \ldots, g_t \rangle$. Each element $g \in N_{\mathbf{G}}(S_1)$ induces an automorphism $C_g : S_1 \to S_1$ which maps an element s to $g^{-1}sg$ i.e., $C_g(s) = g^{-1}sg$. Since $N_{\mathbf{G}}(S_1) = \langle g_1, \ldots, g_t \rangle$, we have $A = \langle C_{g_1}, \ldots, C_{g_t} \rangle$ and $N_{\mathbf{G}}(S_1)/C_{\mathbf{G}}(S_1) \cong A$. The latter equality can be used to compute $|A|$. As $C_{N_{\mathbf{G}}(S_1)}(S_1) = C_{\mathbf{G}}(S_1) \trianglelefteq N_{\mathbf{G}}(S_1)$, we can compute $C_{\mathbf{G}}(S_1)$ in polynomial-time in the setting of quotients [19], and in NC for permutation groups [1].

To compute $\mu(A)$, we use [6, Proposition 2.2], which says that $\mu(A)$ equals $\mu(S_1)$ except for the cases mentioned in Table 1. If S_1 and A do not correspond to any of the cases mentioned in the table, then $\mu(A) = \mu(S_1)$ can be computed using Lemma 2. Using Theorem 2, we first determine the name of S_1 in polynomial-time for quotients and NC for permutation groups. Once we have the name of S_1, we are able to determine if S_1 is one of the finite simple groups on Table 1. Next, we need to test if A satisfies the condition mentioned in the 3rd column of the same row in the table. If the conditions mentioned in the Table 1 are satisfied, then the table itself gives the value of $\mu(A)$. Otherwise, by Proposition 1, we know that $\mu(A) = \mu(S_1)$, and we can compute $\mu(S_1)$ using Lemma 2.

The groups mentioned rows 1–9 of the Table 1 are of constant size and can be handled in NC. We handle other rows separately. First, we need to check if S_1 is isomorphic to one of the simple groups on rows 10–16 of Table 1. For classical simple groups $\mathrm{P}\Omega^+(8,q), \mathrm{PSL}(d,q), \mathrm{PSp}(4,2^e), \mathrm{P}\Omega^+(2d,3)$ this check is computable polynomial-time in the setting of quotients, and in NC for permutation groups (Theorem 2). If S_1 is isomorphic to a simple group of Lie type, then its order is bounded by n^9 and so can be identified in NC. We can then compute $\mu(S_1)$ using [29–32]. Therefore, we can check in NC if S_1 isomorphic to one of $G_2(3^e)$, $F_4(2^e)$, $E_6(q)$. We now need to check if A satisfies the corresponding conditions in the 3rd column of Table 1. Consider the following cases.

Case 1: $S_1 \cong \mathrm{P}\Omega^+(8,q)$. See [22, §5.1] for details.

Case 2: Consider the case when $S_1 \cong \mathrm{PSL}(d,q)$, for $d \geq 3$ and $(d,q) \neq (3,2),(4,2)$. Our goal is to determine whether $A \not\leq \mathrm{P}\Gamma\mathrm{L}(d,q)$, as indicated in the third column. It is known that $\mathrm{P}\Gamma\mathrm{L}(d,q) \leq \mathrm{Aut}(\mathrm{PSL}(d,q)) \cong \mathrm{Aut}(\mathrm{SL}(d,q))$. Given that $A \leq \mathrm{Aut}(S_1)$ and $S_1 \cong \mathrm{PSL}(d,q)$, we can represent A as a subgroup of $\mathrm{Aut}(\mathrm{PSL}(d,q))$ via the isomorphism Iso. Furthermore, we have that $\mathrm{Aut}(\mathrm{PSL}(d,q)) \cong \mathrm{Aut}(\mathrm{SL}(d,q))$, allowing us to also view A as a subgroup of $\mathrm{Aut}(\mathrm{SL}(d,q))$.

Now, we have both A and $\mathrm{P}\Gamma\mathrm{L}(d,q)$ as subgroups of $\mathrm{Aut}(\mathrm{SL}(d,q))$. Furthermore, the automorphisms that contain the graph automorphism $\mathfrak{g}$ in their decompositions are exactly the elements in $\mathrm{Aut}(\mathrm{SL}(d,q)) \setminus \mathrm{P}\Gamma\mathrm{L}(d,q)$ (see [22, §2.4]). Therefore, to verify whether A is a subgroup of $\mathrm{P}\Gamma\mathrm{L}(d,q)$, it suffices to check that each generator of A is free from the graph isomorphism.

Let us now see the computational details of the process described above. Let $\mathrm{Iso} : S_1 \to \mathrm{PSL}(d,q)$ be the isomorphism from Theorem 2. Let $N_{\mathbf{G}}(S_1) = \langle g_1,\dots,g_t\rangle$ and $A = \langle C_{g_1},\dots,C_{g_t}\rangle$, where $C_{g_i} : S_1 \to S_1$ is the map sending $x \mapsto g_i x g_i^{-1}$. Let $\mathrm{SL}(d,q) = \langle L\rangle$, where $L = \{I + \beta e_{i,j} \mid \beta \in \mathbb{F}_q^*, i,j \in [d], i \neq j\}$, $e_{i,j}$ is the matrix which has 1 in the (i,j)th entry and 0 in the remaining entries (see e.g., [7, Page 185]).

Define $\lambda_{g_i} \in \mathrm{Aut}(\mathrm{PSL}(d,q))$ as $\lambda_{g_i} = \mathrm{Iso} \circ C_{g_i} \circ \mathrm{Iso}^{-1}$. Thus, A is embedded as the subgroup $\langle \lambda_{g_1},\dots,\lambda_{g_t}\rangle$ in $\mathrm{Aut}(\mathrm{PSL}(d,q))$. We apply [22, Lem. 4.4] on each

λ_{g_i} to get $\alpha_{g_i} \in \mathrm{Aut}(\mathrm{SL}(d,q))$ such that $\lambda_{g_i} = \bar{\alpha}_{g_i}$. The subgroup $\langle \alpha_{g_1}, \ldots, \alpha_{g_t} \rangle$ is the natural embedding of A in $\mathrm{Aut}(\mathrm{SL}(d,q))$.

Let $q = p^e$. Consider the Frobenius automorphism σ_0 of $\mathrm{Aut}(\mathbb{F}_q)$ given by $a \mapsto a^p$. For each α_{g_i}, define $\alpha_{g_i}^{t',t''} = \alpha_{g_i} \mathfrak{f}_{\sigma_0}^{-t'} \mathfrak{g}^{-t''}$, where $\mathfrak{f}_{\sigma_0}$ is a field automorphism, $\mathfrak{g}$ is a graph automorphism, and $1 \leq t' \leq e$, $t'' \in \{0,1\}$. The maps $\alpha_{g_i}^{t',t''}$ help us to identify the type of the automorphism α_{g_i}.

For each i, and each t', t'', the automorphism $\alpha_{g_i}^{t',t''}$ is specified on a generating set L. Observe that for a fixed i, there is exactly one choice of t' and t'' such that $\alpha_{g_i}^{t',t''}$ is a diagonal automorphism or an inner automorphism. For each t', t'', we use [22, Obs. 4.6, Lem. 4.5] (with $B = L$) to check, in NC^2, if $\alpha_{g_i}^{t',t''}$ is an inner or a diagonal automorphism. Let t', t'' be such that $\alpha_{g_i}^{t',t''}$ is either an inner automorphism or a diagonal automorphism. Note that $t'' = 0$ if and only if $\alpha_{g_i} = \alpha_{g_i}^{t',t''} \circ \mathfrak{f}_{\sigma_0}^{t'}$ and α_{g_i} do not have a graph automorphism in its decomposition, i.e., $\alpha_{g_i} \in \mathrm{P\Gamma L}(d,q)$. If for each i, α_{g_i} turns out to be in $\mathrm{P\Gamma L}(d,q)$ we can conclude that $A \leq \mathrm{P\Gamma L}(d,q)$. Thus, we can test if $A \not\leq \mathrm{P\Gamma L}(d,q)$ in polynomial-time in the setting of quotients and NC in the setting of permutation groups. If $A \leq \mathrm{P\Gamma L}(d,q)$ then $\mu(\mathbf{G}, N) = \mu(S_1)$. Otherwise, $\mu(\mathbf{G}, N) = 2\mu(S_1)$ (see e.g., [6]). This completes the case when $S_1 \cong \mathrm{PSL}(d,q)$.

Case 3: The case when $S_1 \cong \mathrm{PSp}(4, 2^e), e \geq 2$, is handled similarly as Case 2. See [22, §5.3] for details.

Case 4: The case when $S_1 \cong \mathrm{P}\Omega^+(2d, 3), d \geq 4$ is handled similarly as Case 2. See [22, §5.4] for details.

Case 5: Suppose that $S_1 \cong \mathcal{L}(\mathbb{F}_q)$, $\mathcal{L}(\mathbb{F}_q) \in \{G_2(3^e), F_4(2^e), E_6(q)\}$ (rows 14–16 in Table 1). Our approach deviates slightly than what we have done in the previous sections. The reason is that the exceptional groups have small orders, allowing us to compute their Cayley tables. However, we also do not have the matrix group structure as in earlier cases, which makes the analysis more intricate. We now provide an overview of our approach.

We start with the identification of S_1 with $\mathcal{L}(\mathbb{F}_q)$, i.e. $S_1 \cong \mathcal{L}(\mathbb{F}_q)$. These groups have order at most n^9 [23], so we may write down the multiplication table for S_1 in NC (Lemma 1). To find a generating set of the group $\mathcal{L}(\mathbb{F}_q)$, we will leverage the Chevally presentation of the group $\mathcal{L}(\mathbb{F}_q)$ (cf. [22, §2.4]). We then try all possible 2-element subsets of $\mathcal{L}(\mathbb{F}_q)$ to find a generating set of the group $\mathcal{L}$ satisfying the Chevally presentation. Using generator-enumeration, we find an isomorphism Iso : $S_1 \to \mathcal{L}(\mathbb{F}_q)$, if $S_1 \cong \mathcal{L}(\mathbb{F}_q)$. Note that, our aim in this case is to determine whether $A \not\leq \Gamma G_2(3^e)$ in the 3rd column of Table 1. We have $A, \Gamma G_2(3^e) \leq \mathrm{Aut}(G_2(3^e))$, and $\mathrm{Aut}(G_2(3^e)) \setminus \Gamma G_2(3^e)$ are exactly those automorphisms which contain a graph automorphism in their decompositions. Thus it is enough to check if each generator of A is free of graph automorphism. See [22, §5.5] for the remaining details.

References

1. Babai, L., Luks, E., Seress, A.: Permutation groups in NC. In: STOC 1987, STOC '87, pp. 409–420. Association for Computing Machinery, New York (1987). https://doi.org/10.1145/28395.28439
2. Babai, L., Codenotti, P., Grochow, J.A., Qiao, Y.: Code equivalence and group isomorphism. In: Proceedings of the Twenty-Second Annual ACM–SIAM Symposium on Discrete Algorithms (SODA11), pp. 1395–1408. SIAM, Philadelphia (2011). https://doi.org/10.1137/1.9781611973082.107
3. Babai, L., Codenotti, P., Qiao, Y.: Polynomial-time isomorphism test for groups with no abelian normal subgroups. In: Czumaj, A., Mehlhorn, K., Pitts, A., Wattenhofer, R. (eds.) ICALP 2012. LNCS, vol. 7391, pp. 51–62. Springer, Heidelberg (2012). https://doi.org/10.1007/978-3-642-31594-7_5
4. Brachter, J.: Combinatorial approaches to the group isomorphism problem. Ph.D. thesis, TU Darmstadt (2023). https://tuprints.ulb.tu-darmstadt.de/26387/1/thesis_brachter.pdf
5. Cannon, J.J., Holt, D.F.: Automorphism group computation and isomorphism testing in finite groups. J. Symb. Comput. **35**, 241–267 (2003). https://doi.org/10.1016/S0747-7171(02)00133-5
6. Cannon, J.J., Holt, D.F., Unger, W.R.: The use of permutation representations in structural computations in large finite matrix groups. J. Symbolic Comput. **95**, 26–38 (2019). https://doi.org/10.1016/j.jsc.2018.09.001
7. Carter, R.W.: Simple Groups of Lie Type, vol. 22. John Wiley & Sons, New York (1989)
8. Chamberlain, R.: Permutation Representations of Group Quotients and of Quasisimple Groups. Ph.D. thesis, University of Warwick (2020). https://wrap.warwick.ac.uk/id/eprint/149881/
9. Conway, J.H., Curtis, R.T., Norton, S.P., Parker, R.A., Wilson, R.A.: ATLAS of finite groups. Oxford University Press, Eynsham, Eynsham, UK (1985). https://doi.org/10.1017/S0025557200139440. maximal subgroups and ordinary characters for simple groups, With computational assistance from J. G. Thackray
10. Das, B., Thakkar, D.: The minimal faithful permutation degree of groups without abelian normal subgroups. In: Proceedings of the 56th Annual ACM Symposium on Theory of Computing, STOC 2024, pp. 118–129. Association for Computing Machinery, New York (2024). https://doi.org/10.1145/3618260.3649641
11. Dixon, J.D., Mortimer, B.: Permutation groups, Graduate Texts in Mathematics, vol. 163. Springer-Verlag, New York (1996). https://doi.org/10.1007/978-1-4612-0731-3,https://doi.org/10.1007/978-1-4612-0731-3
12. Grochow, J.A.: Lemmas about Socle. Personal Communication (2025)
13. Grochow, J.A., Levet, M.: On the descriptive complexity of groups without Abelian normal subgroups (extended abstract). In: Achilleos, A., Monica, D.D. (eds.) Proceedings of the Fourteenth International Symposium on Games, Automata, Logics, and Formal Verification, GandALF 2023, Udine, Italy, 18-20th September 2023. EPTCS, vol. 390, pp. 185–202 (2023). https://doi.org/10.4204/EPTCS.390.12, preprint of full version at arXiv:2209.13725 [cs.LO]
14. Grochow, J.A., Levet, M.: On the parallel complexity of group isomorphism via Weisfeiler-Leman. In: Fernau, H., Jansen, K. (eds.) Fundamentals of Computation Theory - 24th International Symposium, FCT 2023, Trier, Germany, 18–21 September 2023, Proceedings. Lecture Notes in Computer Science, vol. 14292, pp. 234–247. Springer (2023).https://doi.org/10.1007/978-3-031-43587-4_17. preprint of full version at arXiv: 2112.11487 [cs.DS]

15. Guest, S., Morris, J., Praeger, C., Spiga, P.: On the maximum orders of elements of finite almost simple groups and primitive permutation groups. Trans. Am. Math. Soc. **367**(11), 7665–7694 (2015). https://doi.org/10.1090/S0002-9947-2015-06293-X
16. Holt, D.F., Eick, B., O'Brien, E.A.: Handbook of Computational Group Theory. Chapman and Hall/CRC (2005)
17. Johnson, D.L.: Minimal permutation representations of finite groups. Am. J. Math. **93**(4), 857–866 (1971). https://doi.org/10.2307/2373739
18. Kantor, W., Seress, A.: Black box classical groups. Mem. Am. Math. Soc. **708** (2001). https://doi.org/10.1090/memo/0708
19. Kantor, W.M., Luks, E.M.: Computing in quotient groups. In: Ortiz, H. (ed.) Proceedings of the 22nd Annual ACM Symposium on Theory of Computing, May 13-17, 1990, Baltimore, Maryland, USA, pp. 524–534. ACM (1990). https://doi.org/10.1145/100216.100290
20. Kantor, W.M., Luks, E.M., Mark, P.D.: Sylow subgroups in parallel. J. Algorithms **31**(1), 132–195 (1999). https://doi.org/10.1006/JAGM.1999.0982
21. Kleidman, P.B., Liebeck, M.W.: The Subgroup Structure of The Finite Classical Groups, vol. 129. Cambridge University Press, Cambridge (1990). https://doi.org/10.1017/CBO9780511629235
22. Levet, M., Srivastava, P., Thakkar, D.: Complexity of minimal faithful permutation degree for fitting-free groups (2025). https://arxiv.org/abs/2501.16039
23. Liebeck, M.W.: On minimal degrees and base sizes of primitive permutation groups. Arch. Math. **43**(1), 11–15 (1984). https://doi.org/10.1007/BF01193603
24. Mulmuley, K.: A fast parallel algorithm to compute the rank of a matrix over an arbitrary field. Comb. **7**(1), 101–104 (1987). https://doi.org/10.1007/BF02579205
25. Neumann, P.M.: Some algorithms for computing with finite permutation groups. In: Proceedings of Groups St. Andrews 1985. vol. 121, pp. 59–92. Cambridge Univ. Press, Cambridge, Cambridge (1985). https://doi.org/10.1017/CBO9780511600647.006
26. Rotman, J.J.: An Introduction to The Theory of Groups, Graduate Texts in Mathematics, fourth edn., vol. 148. Springer-Verlag, New York (1995). https://doi.org/10.1007/978-1-4612-4176-8
27. Seress, A.: Permutation Group Algorithms. Cambridge Tracts in Mathematics, Cambridge University Press (2003). https://doi.org/10.1017/CBO9780511546549
28. Thakkar, D.: On Computing Optimal Representations of Finite Groups. Ph.D. thesis, Indian Institute of Technology Gandhinagar (2024). https://drive.google.com/file/d/1e2W5NvGr6DTHj6UHYxudw3xBQReHUqsN/view
29. Vasilěv, A.V.: Minimal permutation representations of finite simple exceptional groups of types G_2 and F_4. Algebra Logic **35**(6), 371–383 (1996). https://doi.org/10.1007/BF02366397
30. Vasilěv, A.V.: Minimal permutation representations of finite simple exceptional groups of types E_6, E_7, and E_8. Algebra Logic **36**(5), 302–310 (1997). https://doi.org/10.1007/BF02671607
31. Vasilěv, A.V.: Minimal permutation representations of finite simple exceptional twisted groups. Algebra Logic **37**(1), 9–20 (1998). https://doi.org/10.1007/BF02684081
32. Vasilěv, A.V., Mazurov, V.D.: Minimal permutation representations of finite simple orthogonal groups. Algebra Logic **33**(6), 337–350 (1995). https://doi.org/10.1007/BF00756348
33. Wilson, R.: The finite simple groups, Graduate Texts in Mathematics, vol. 251. Springer, London (2009). https://doi.org/10.1007/978-1-84800-988-2

A Study of the Binary and Boolean Rank of Matrices with Small Constant Real Rank

Michal Parnas(✉) and Adi Shraibman

School of Computer Science, Academic College of Tel Aviv-Yaffo, Tel Aviv-Yaffo, Israel
{michalp,adish}@mta.ac.il

Abstract. We initiate the study of the binary and Boolean rank of 0, 1 matrices that have a small rank over the reals. The relationship between these three rank functions is an important open question, and here we prove that when the real rank d is a small constant, the gap between the real and the binary and Boolean rank is a small constant. We give tight upper and lower bounds on the Boolean and binary rank of matrices with real rank $1 \leq d \leq 4$, as well as determine the size of the largest isolation set in each case. Furthermore, we prove that for $d = 3, 4$, the circulant matrix defined by a row with $d-1$ consecutive ones followed by $d-1$ zeros, is the only matrix of size $(2d-2) \times (2d-2)$ with real rank d and Boolean and binary rank and isolation set of size $2d-2$, and this matrix achieves the maximal gap possible between the real and the binary and Boolean rank for these values of d.

Our results can also be interpreted in other equivalent terms, such as finding the minimum number of bicliques needed to partition or cover the edges of a bipartite graph whose reduced adjacency matrix has real rank $1 \leq d \leq 4$. We use a combination of combinatorial and algebraic techniques combined with the assistance of a computer program.

Keywords: Real rank · Boolean rank · Binary rank · Biclique cover number · Biclique partition number

1 Introduction

The real (standard) rank of an $n \times m$ matrix M over $\mathbb{R}$, denoted by $\text{rank}_{\mathbb{R}}(M)$, is an important concept in many applications and fields of mathematics and computer science and has several equivalent definitions. The following definition is used here: the real rank of M is the minimal d for which there exist real matrices A and B of size $n \times d$ and $d \times m$ respectively, such that $M = A \cdot B$, where the operations are over $\mathbb{R}$. In a similar way, consider the following rank functions defined over a semi-ring (see e.g. Gregory and Pullman [10]). The *binary rank* of a 0, 1 matrix M of size $n \times m$, denoted by $\text{rank}_{\text{bin}}(M)$, is the minimal d for which there exist 0, 1 matrices A and B of size $n \times d$ and $d \times m$

A. Jeż and J. Otop (Eds.): FCT 2025, LNCS 16106, pp. 362–375, 2026.
https://doi.org/10.1007/978-3-032-04700-7_27

respectively, such that $M = A \cdot B$, where the operations are over the integers. The *Boolean rank* of a 0,1 matrix M, denoted by $\text{rank}_{\mathbb{B}}(M)$, is defined similarly to the binary rank, but here the operations are under Boolean arithmetic (namely, $0 + x = x + 0 = x$, $1 + 1 = 1 \cdot 1 = 1$, and $x \cdot 0 = 0 \cdot x = 0$). Such decompositions $M = A \cdot B$ are called *optimal* for the given rank function. The broad interest in the Boolean and binary rank stems also from the following equivalent definitions, and we will make use of these alternative definitions throughout the paper.

A *monochromatic combinatorial rectangle* in a 0,1 matrix M is a sub-matrix of M, all of whose entries have the same value. The *partition number* of M is the minimum number of monochromatic rectangles required to partition the ones of M, and is equal to $\text{rank}_{\text{bin}}(M)$. The *cover number* of M is the minimum number of monochromatic rectangles required to cover the ones of M, where rectangles may overlap, and is equal to $\text{rank}_{\mathbb{B}}(M)$ (see Gregory, Pullman, Jones, and Lundgren [11]).

The binary and Boolean rank are also related to concepts in graph theory. Given a bipartite graph G, the *biclique partition number*, $bp(G)$, and *biclique cover number*, $bc(G)$, are equal to the minimum number of bicliques needed to partition or cover, respectively, all edges of G. Let M be the reduced adjacency matrix of G, that is, if G has n and m vertices on each side, then M has n rows and m columns and $M_{i,j} = 1$ if and only if (i, j) is an edge of G, where $M_{i,j}$ is the element on the i'th row and j'th column of M. A monochromatic combinatorial rectangle of ones in M corresponds to a biclique in G, and, thus, $bp(G) = \text{rank}_{\text{bin}}(M)$ and $bc(G) = \text{rank}_{\mathbb{B}}(M)$ (see [11]).

The binary and Boolean rank are also closely related to the field of communication complexity. Here two players are given a joint matrix M, where one player receives a row index i and the second a column index j, and their goal is to determine $M_{i,j}$ while minimizing the number of bits communicated between them in the worst case. It was shown that $\log \text{rank}_{\text{bin}}(M)$ approximates the deterministic communication complexity of M up to a polynomial, and determines exactly the unambiguous nondeterministic communication complexity of M, whereas $\log \text{rank}_{\mathbb{B}}(M)$ determines exactly the non-deterministic communication complexity of M. Furthermore, the famous log-rank conjecture of Lovász and Saks [15] suggests that the deterministic communication complexity is polynomially related to $\log \text{rank}_{\mathbb{R}}(M)$ (see, e.g., Lovett [16]). Since the log-rank conjecture seems so far unreachable, any further understanding of the properties of the Boolean and binary rank in comparison with the real rank can assist.

Computing the Boolean and binary rank is NP-complete (See Orlin [17] and Jiang and Ravikumar [14]). Chalermsook, Heydrich, Holm, and Karrenbauer showed in [4] that it is hard in general to approximate the Boolean rank of a matrix M of size $n \times n$ to within a factor of $n^{1-\epsilon}$ for any given $\epsilon > 0$. Chandran, Issac and Karrenbauer [5] give a polynomial time algorithm with an approximation ratio of $n/\log n$ for the Boolean and binary rank.

In the framework of parameterized algorithms, Fleischner, Mujuni, Paulusma, and Szeider showed in [7] that the Boolean and binary rank are fixed-parameter tractable with running time $f(d)poly(n, m)$, where d is the rank in question. Gramm, Guo, Hüffner, and Niedermeier [9] give an exact solution for the Boolean rank with $f(d) = 2^{2^{O(d)}}$, and Chandran, Issac and Karrenbauer provide in [5] an algorithm for the binary rank d with $f(d) = 2^{O(2^d)}$. Moreover, [5] prove that there is no algorithm for the Boolean rank with less than a double exponential complexity in d unless the Exponential Time Hypothesis (ETH) is false, thus matching up to polynomial factors the algorithm of [9].

There are simple bounds on the relationship between the real rank and the binary and Boolean rank. It always holds that $\text{rank}_{\mathbb{R}}(M) \leq \text{rank}_{\text{bin}}(M)$, whereas the Boolean rank can be smaller or larger than the real rank and is always bounded above by the binary rank. The identity matrix is a trivial example where all three ranks are equal, and the binary and Boolean rank are always equal to the real rank for real rank 1 or 2. When $\text{rank}_{\mathbb{R}}(M) \geq 3$ there are examples of matrices for which all three rank functions differ (see Sect. 2).

A useful tool for proving lower bounds on the binary and Boolean rank of M is finding a large *isolation set* in M (or a *fooling set* as it is called in communication complexity), which is a subset of 1 entries of M, such that no two of these ones belong to an all-one 2×2 sub-matrix of M, and no two of these ones are in the same row or column of M. Thus, no two ones of an isolation set can belong to the same monochromatic rectangle, and the size of a maximum isolation set of M, denoted by $i(M)$, provides a lower bound on $\text{rank}_{\text{bin}}(M)$ and $\text{rank}_{\mathbb{B}}(M)$ (see, e.g., Beasley [1]). However, there can be a large gap between $i(M)$ and $\text{rank}_{\mathbb{B}}(M), \text{rank}_{\text{bin}}(M)$. For example, the matrix D_n of size $n \times n$, $n \geq 3$, which has an all-zero main diagonal and ones elsewhere, has $i(D_n) = 3$ whereas $\text{rank}_{\mathbb{B}}(D_n) = \Theta(\log_2 n)$ and $\text{rank}_{\mathbb{R}}(D_n) = \text{rank}_{\text{bin}}(D_n) = n$ (see de Caen, Gregory, and Pullman [3] and Haviv and Parnas [12]).

1.1 Our Results

Our goal is to find the largest gap possible between the real rank d and the Boolean and binary rank of a 0,1 matrix M, for small constant values of d, and to characterize these matrices from a combinatorial and algebraic point of view, as well as understand the size of the largest isolation set possible. It is enough to consider only *basic* matrices, that is, matrices that do not have all-zero rows and columns, and do not have identical rows or columns.

Using a result of de Caen, Gregory, Henson, Lundgren, and Maybee [2] it is possible to prove the following theorem and get a multiplicative gap of almost 2 between the real and the binary and Boolean rank as the size of the matrix grows. The matrices used for the proof of this theorem also have large isolation sets which are equal to the binary and Boolean rank of the matrix.

Theorem 1. *For all $n \geq 4$, $n \neq 5$, there exists a basic 0,1 matrix M of size $n \times n$, with $\text{rank}_{\mathbb{R}}(M) = \lfloor n/2 \rfloor + 1$, $\text{rank}_{\text{bin}}(M) = \text{rank}_{\mathbb{B}}(M) = i(M) = 2\lfloor n/2 \rfloor$.*

The even sized matrices used in the proof of Theorem 1 are the circulant matrices C_n of size $n \times n$ defined by a row with $n/2$ consecutive ones followed by $n/2$ zeros. The odd sized matrices are constructed by taking an even sized matrix C_n and adding to it a carefully selected row and column which are a linear combination of rows and columns of C_n. Note that the theorem does not hold for $n = 5$, and as we prove, there does not exist a basic 0,1 matrix of size 5×5 and real rank $\lfloor 5/2 \rfloor + 1 = 3$ (see Lemma 11, Sect. 3).

By taking any of the matrices M given in Theorem 1 and computing the Kronecker product of M with itself s times, it is possible to amplify the gap and get a matrix $M^{\otimes s}$ with real rank $(\lfloor n/2 \rfloor + 1)^s$ and binary rank $(2\lfloor n/2 \rfloor)^s$ since $i(M) = 2\lfloor n/2 \rfloor$. See also Dietzfelbinger, Hromkovič and Schnitger [6] who proved a similar result for $n = 4$. Friesen, Hamed, Lee and Theis show in [8] that for any $d \geq 1$ there exists a matrix M_d over $\mathbb{Q}$, such that $\text{rank}_{\mathbb{R}}(M_d) = d$ and M_d has an isolation set of size $\binom{d+1}{2}$. Shigeta and Amano [19] construct 0,1 matrices of size $n \times n$ with real rank $n^{1/2+o(1)}$ and an isolation set of size n.

It is well known that the number of distinct rows and columns of a 0,1 matrix M is bounded above by $2^{\text{rank}_{\mathbb{R}}(M)}$ (see, for example, Hrubeš [13]). This gives a trivial upper bound of $2^{\text{rank}_{\mathbb{R}}(M)}$ on the Boolean and binary rank of a matrix as a function of the real rank. But for matrices with small constant real rank we prove that the gap between the real and the binary and Boolean rank is considerably smaller, and the result in Theorem 1 is tight.

Theorem 2. *Let M be a 0,1 matrix with $\text{rank}_{\mathbb{R}}(M) = d$. For $d = 3, 4$ it holds that $d \leq \text{rank}_{\text{bin}}(M) \leq 2d - 2$ and $d - 1 \leq \text{rank}_{\mathbb{B}}(M), i(M) \leq 2d - 2$.*

The bounds given in Theorem 2 are tight: there exist matrices whose binary and Boolean rank match the lower bounds given, and others for which the upper bound is the correct value. The proof of Theorem 2 reveals other interesting facts about matrices with real rank $d = 3, 4$. Specifically, we show that there does not exist a basic 0,1 matrix of size 5×5 and real rank 3. For $d = 4$ we prove that any basic 0,1 matrix of size 7×7 and real rank 4 has binary rank at most 6, and there does not exist a basic 0,1 matrix of size 8×8 with real rank 4.

It is interesting to note that Shitov [20] proved that if M is a non-negative matrix of size $n \times n$, $n > 6$, and $\text{rank}_{\mathbb{R}}(M) = 3$, then the non-negative rank of M is at most $6n/7$. Shitov also showed that there exist non-negative matrices of size $n \times n$ with real rank 3 and non-negative rank n, for $n = 3, 4, 5, 6$.

A useful concept in our proofs and analysis is the *kernel* of M, which we define as a sub-matrix of M achieved by deleting all zero rows and columns and all duplicate rows and columns of M. Studying the structure of the kernel of a matrix M of real rank 3 allows us to prove the following.

Theorem 3. *Let M be a 0,1 matrix with $\text{rank}_{\mathbb{R}}(M) = 3$. Then:*

- *The kernel of M has at most 4 rows or columns.*
- $\text{rank}_{\text{bin}}(M) = \text{rank}_{\mathbb{B}}(M) = 4$ *if and only if the kernel of M includes C_4 as a sub-matrix.*
- $i(M) = \text{rank}_{\text{bin}}(M) = \text{rank}_{\mathbb{B}}(M)$ *unless the kernel of M is the following sub-matrix, in which case* $\text{rank}_{\text{bin}}(M) = 3, \text{rank}_{\mathbb{B}}(M) = i(M) = 2$:

$$\begin{pmatrix} 1 & 1 & 0 \\ 0 & 1 & 1 \\ 1 & 1 & 1 \end{pmatrix}$$

Our proofs of Theorems 2 and 3, as mentioned earlier, give additional interesting insights as to the structure of basic 0, 1 matrices with real rank $d = 3, 4$. We characterize the binary and Boolean rank and the size of the largest isolation sets of basic 0, 1 matrices of size 6×6 and real rank 4. Moreover, we prove that the only 0, 1 matrix of size $n \times n$, for $n = 4, 6$, with $\text{rank}_{\mathbb{R}}(M) = 1 + n/2$, $\text{rank}_{\text{bin}}(M) = \text{rank}_{\mathbb{B}}(M) = i(M) = n$, is the matrix C_n. The proof of Theorem 2 shows that the bounds in Theorem 1 are tight also for 0, 1 matrices of size 7×7, and that the kernel of a 0, 1 matrix M with $\text{rank}_{\mathbb{R}}(M) = 4$ has at most 7 rows or columns.

Our results give tight bounds on $bp(G)$ and $bc(G)$ for a bipartite graph G whose reduced adjacency matrix M has real rank 3 or 4. The kernel of the matrix M corresponds to the kernel of the graph G as used in [7] in the context of parameterized algorithms. There the kernel is a sub-graph of G obtained by removing all vertices with no neighbors (which correspond to all-zero rows or columns in M), and removing all vertices but one from any set of vertices with the same neighbor set (which corresponds to removing duplicate rows and columns in M). Thus, the number of rows and columns of the kernel of M is the same as the number of vertices in the kernel of G. As stated in [7], the number of vertices on each side of the kernel is at most $2^{bp(G)}$, thus achieving a kernel with at most $2^{bp(G)+1}$ vertices in total. As our results show, for a reduced adjacency matrix M with real rank $d = 3, 4$, the number of vertices in the kernel of G is significantly smaller. Translating Theorems 2 and 3 into the language of graphs, and noticing that the circulant matrix C_4 is the reduced adjacency matrix of a cycle of length 8, we get the following for real rank 3:

Corollary 1. *Let G be a bipartite graph whose reduced adjacency matrix M has real rank 3. Then:*

- $3 \leq bp(G) \leq 4$ *and* $2 \leq bc(G) \leq 4$.
- $bp(G) = bc(G) = 4$ *if and only if the kernel of G has a cycle of length 8 as a subgraph.*
- $bp(G) = bc(G)$ *unless the kernel of G is a cycle of length 6 with a chord bisecting the cycle.*
- *The kernel of G has at most 4 vertices on one of its sides.*

Finally, our results can also give efficient approximation algorithms up to a small additive constant for the binary and Boolean rank when $1 \leq \text{rank}_{\mathbb{R}}(M) \leq 4$. The approximation algorithm can simply compute the real rank of the kernel of the matrix and output it as an approximation to the binary and Boolean rank, where for $\text{rank}_{\mathbb{R}}(M) = 3$ it can even determine the binary and Boolean rank exactly by determining the structure of the kernel as described in Theorem 3.

We conclude with a few open problems. First, what is the maximal real rank d for which the bound in Theorem 2 holds, that is, the binary rank is at most $2d - 2$? We also wonder how far can a computer search assist us here. We have preliminary results that there exist basic 0, 1 matrices of size 11×11 with real rank $d = 5$. However, the binary rank of the matrices we found so far is at most

$8 = 2d - 2$. Is 8 the largest binary rank possible for real rank 5 as exists in the matrix C_8, and is there a 12×12 basic 0, 1 matrix with real rank 5? Another interesting problem is to determine for what n is C_n the only matrix of size $n \times n$ with $\text{rank}_{\mathbb{R}}(M) = 1 + n/2$, $\text{rank}_{\text{bin}}(M) = \text{rank}_{\mathbb{B}}(M) = i(M) = n$.

1.2 Our Techniques

We prove our results using both algebraic and combinatorial techniques combined with a computer program which assisted us in achieving some of the theoretical results. Since the number of possible matrices we had to examine grows exponentially with the size of the matrix, we use the following algorithmic ideas to prune the number of options considered by the computer program:

1. It is enough to consider basic 0, 1 matrices. This reduces significantly the number of matrices we had to consider.
2. A 0, 1 matrix has the *2-sum property* if it has two rows (columns) whose sum equals a third row (column). The binary rank of such a matrix cannot be full (see Lemma 3), and thus, we can remove all these matrices when looking for a matrix with full binary rank.
3. Two matrices will be called *equivalent* if they are identical up to a permutation of the rows and columns or the transpose operation. Our algorithm will consider only one matrix from each class of equivalent matrices, and this matrix will be a *representative* for this class.
4. Each row of a matrix was labeled by an integer which corresponds to the binary number represented by this row vector. This allowed us to sort the rows of each matrix, and then remove efficiently matrices which are identical up to a permutation of the rows. The same can be done for the columns.

Our proof of Theorem 2 for matrices with real rank $d = 3$ uses solely algebraic and combinatorial ideas. The main ingredient is proving that there does not exist a basic 0, 1 matrix of size 5×5 with real rank 3. This can be done by using the fact that each matrix with real rank 3 must include a 3×3 sub-matrix of real rank 3, and as the number of possible representatives of matrices of size 3×3 and real rank 3 is relatively small, it is possible to go over each such representative and prove that it cannot be extended to a 5×5 basic 0, 1 matrix with real rank 3.

However, when $d = 4$ the number of representatives of matrices of size 4×4 and real rank 4 becomes significantly larger. Therefore, we used the algorithmic ideas described above to write a computer program that checks what is the largest k for which there exists a basic 0, 1 matrix of size $k \times k$ with real rank d, and for such matrices get an upper bound on their binary rank, using combinatorial and algebraic arguments. The program starts by considering all basic 0, 1 matrices of size $d \times d$ and real rank d, and then prunes this set of matrices as explained above to get just one representative of each class of equivalent matrices. Then the program tries to add additional rows and columns to each representative, so as to get larger matrices while not increasing the real rank. In each step when a new class of larger matrices is found, the matrices are pruned again so that only the new representatives remain and carried over to the next step. For $d = 4$ we show that there are no basic 0, 1 matrices of size 8×8 with real rank 4, and prove that any basic matrix with at most 7 rows or columns and real rank 4 has binary rank at most 6. All missing proofs can be found in the full version of the paper ([18]).

2 Preliminaries

We now state a few simple lemmas used throughout the paper, and also prove Theorem 1. Since identical rows and columns and all-zero rows and columns do not change the rank, it is enough to consider *basic* matrices, and for any matrix which is not basic, consider its *kernel* as defined in the introduction.

Lemma 1. *Let M be a basic 0, 1 matrix of size $n \times m$ with* $\text{rank}_{\mathbb{R}}(M) = d$. *Then for any s, t such that $d \leq s \leq n, d \leq t \leq m$, there exists a basic sub-matrix A of M of size $s \times t$ such that* $\text{rank}_{\mathbb{R}}(A) = d$.

Lemma 2. *Let M be a basic 0, 1 matrix and let M' be a matrix achieved by augmenting M with a row which is identical to one of the rows of M or with an all-zero row, and then adding a column to the augmented matrix. If M' is basic then* $\text{rank}_{\mathbb{R}}(M') > \text{rank}_{\mathbb{R}}(M)$.

Definition 1. *A matrix M has the k-*sum property *if it has k rows (columns) whose sum equals another row (column) in M.*

Lemma 3. *Let M be a 0, 1 matrix which has the k-sum property. Then it is possible to partition all ones in the $k + 1$ rows (columns) involved in this sum into at most k monochromatic rectangles of ones.*

Proof. Let X be a row in M which is the sum of rows $R_1, ..., R_k$. Since M is a 0, 1 matrix, the ones in $R_1, ..., R_k$ are in distinct columns, and row X contains exactly the ones in these k rows. Thus, the ones in R_1 and the ones in X that belong to the same columns define one monochromatic rectangle, the ones in R_2 and the ones in X in the same columns define a second rectangle and so on, and we get a total of k rectangles as required. □

Lemma 4. *Let M be a real matrix, and let $M' = (M|X)$ be a matrix achieved by augmenting M with a column X. If* $\text{rank}_{\mathbb{R}}(M) = \text{rank}_{\mathbb{R}}(M')$ *then any dependence between the rows of M is preserved in M'. A similar claim holds when the role of the columns and rows is reversed.*

Proof. A linear dependence between the rows of M can be expressed as $\alpha^t M = 0$, for some vector of coefficients α. To prove the claim it is enough the show that also $\alpha^t X = 0$. Indeed, if $\text{rank}_{\mathbb{R}}(M) = \text{rank}_{\mathbb{R}}(M')$ then $X = M\beta$ for some other vector β, and therefore $\alpha^t X = \alpha^t(M\beta) = (\alpha^t M)\beta = 0$. □

The following lemmas determine the Boolean and binary rank of a matrix with real rank $d = 1, 2$, and prove that for $d \geq 3$ the Boolean rank can be smaller than the real rank, however, for $d = 3, 4$ it is at least $d - 1$.

Lemma 5. *For any* $0, 1$ *matrix* M *it holds:*

- $\text{rank}_{\mathbb{R}}(M) = 1$ *if and only if* $\text{rank}_{\mathbb{B}}(M) = 1$ *if and only if* $\text{rank}_{\text{bin}}(M) = 1$.
- $\text{rank}_{\mathbb{R}}(M) = 2$ *if and only if* $\text{rank}_{\text{bin}}(M) = 2$, *and if* $\text{rank}_{\mathbb{R}}(M) = 2$ *then* $\text{rank}_{\mathbb{B}}(M) = 2$.

Proof. The first item is easy to verify from the definition. We now prove the second item. If $\text{rank}_{\text{bin}}(M) = 2$ then $\text{rank}_{\mathbb{R}}(M) = 2$ since $\text{rank}_{\mathbb{R}}(M) \leq \text{rank}_{\text{bin}}(M)$ and if $\text{rank}_{\mathbb{R}}(M) = 1$ then $\text{rank}_{\text{bin}}(M) = 1$ by the first item.

Assume now that $\text{rank}_{\mathbb{R}}(M) = 2$. Thus, M has two distinct non-zero rows A, B such that any other row X is a linear combination of A and B. Let $X = \alpha A + \beta B$. The possible values for α and β are $1, -1$ or 0, as we now show. Since A, B are distinct, there exists a position in which A and B differ, where one has a zero in this position and the other has a one. Assume without loss of generality that B has a 1. Thus, β is either 0 or 1. It is easy to verify that α is restricted to $1, -1, 0$. Therefore, all other rows of M are one of $0, A, B, A + B, B - A$. Also note, that rows $A + B$ and $B - A$ cannot coincide in M. Thus, we can choose A and B so that all other rows are one of $0, A, B, A + B$. Hence, by Lemma 3 we can cover and partition M with at most 2 rectangles. By the first item of this lemma, we can again rule out the option that $\text{rank}_{\mathbb{B}}(M) = 1$. □

Lemma 6 (Beasley [1]). *For* $k = 1, 2$, $i(M) = k$ *if and only if* $\text{rank}_{\mathbb{B}}(M) = k$.

Lemma 7. *For any* $d \geq 3$ *there exists a basic* $0, 1$ *matrix* M *with* $\text{rank}_{\mathbb{R}}(M) = d$ *and* $\text{rank}_{\mathbb{B}}(M) = i(M) = d - 1$.

Lemma 8. *If* $\text{rank}_{\mathbb{R}}(M) = d$, $d = 3, 4$, *then* $\text{rank}_{\mathbb{B}}(M), i(M) \geq d - 1$.

Proof. Let $d = 3$. Assume by contradiction that $\text{rank}_{\mathbb{B}}(M) = 1$. Then by Lemma 5, $\text{rank}_{\mathbb{R}}(M) = 1$. Thus, $\text{rank}_{\mathbb{B}}(M) \geq 2$, and by Lemma 6, $i(M) \geq 2$.

Now let $d = 4$ and assume by contradiction that $\text{rank}_{\mathbb{B}}(M) \leq 2$. Therefore, the ones of M can be covered by at most 2 monochromatic rectangles, and so M has at most 3 different rows that are not all zeros (two rows, each including exactly one of the 2 rectangles, and a row including both rectangles). But then the real rank of M is at most 3 and we get a contradiction. Thus, $\text{rank}_{\mathbb{B}}(M) \geq 3$, and using Lemma 6 we get that $i(M) \geq 3$ as well. □

The following lemma was proved by de Caen, Gregory, Henson, Lundgren, and Maybee [2], and deals with even sized matrices. As we prove in Lemma 10 their result can be generalized to odd sized matrices. Theorem 1 follows.

Lemma 9. *For* $k \geq 2$, *let* C_{2k} *be the* $2k \times 2k$ *basic circulant* $0, 1$ *matrix defined by a row with* k *consecutive ones followed by* k *zeros. Then* $\text{rank}_{\mathbb{R}}(C_{2k}) = k + 1$, $\text{rank}_{\text{bin}}(C_{2k}) = \text{rank}_{\mathbb{B}}(C_{2k}) = i(C_{2k}) = 2k$.

$$
\left(\begin{array}{cccccccccc|c}
1&1&1&1&1&0&0&0&0&0&0\\
0&1&1&1&1&1&0&0&0&0&0\\
0&0&1&1&1&1&1&0&0&0&0\\
0&0&0&1&1&1&1&1&0&0&1\\
0&0&0&0&1&1&1&1&1&0&0\\
0&0&0&0&0&1&1&1&1&1&1\\
1&0&0&0&0&0&1&1&1&1&1\\
1&1&0&0&0&0&0&1&1&1&1\\
1&1&1&0&0&0&0&0&1&1&0\\
1&1&1&1&0&0&0&0&0&1&1\\
1&1&1&1&1&1&1&1&1&1&1
\end{array}\right)
$$

Fig. 1. A matrix of size $(2k + 1) \times (2k + 1)$ built by adding a row and a column to the circulant matrix C_{2k}, where $k = 5$. A partition of the ones into $2k$ rectangles is described, where each rectangle is described by a different color, and since the first $2k$ ones on the main diagonal are an isolation set of size $2k$, the binary rank is exactly $2k$.

Lemma 10. *For $k \geq 3$, there exists a basic 0,1 matrix M of size $(2k + 1) \times (2k + 1)$, such that* $\text{rank}_{\mathbb{R}}(M) = k + 1$, $\text{rank}_{\text{bin}}(M) = \text{rank}_{\mathbb{B}}(M) = i(M) = 2k$.

Proof. Let C_{2k} be the circulant matrix described in Lemma 9. We show how to add a row and a column to C_{2k} and get a matrix M of size $(2k + 1) \times (2k + 1)$ with the claimed properties.

First add to C_{2k} a row which is the sum of rows 1 and $k + 1$ of C_{2k}. The resulting new row is the all-one vector. Next add to the augmented matrix a new column $Y_{2k+1} = Y_{2k} - Y_{2k-1} + Y_{2k-2}$, which is a linear combination of the last three columns $Y_{2k-2}, Y_{2k-1}, Y_{2k}$ of the augmented matrix. The new column is $(0, 0, ..., 0, 1, 0, 1, 1, ..., 1, 0, 1, 1)^T$, where the first block of zeros and the middle block of ones are each of length $k-2$. Column Y_{2k+1} is well defined since $k-2 \geq 1$ as $k \geq 3$, and it is easy to verify from the structure of the circulant matrix C_{2k} that Y_{2k+1} contains only zeros and ones (see Fig. 1).

The matrix M is basic since C_{2k} is basic and the new row and column are different from all rows and columns of C_{2k} (again since $k - 2 \geq 1$). Furthermore, $\text{rank}_{\mathbb{R}}(M) = \text{rank}_{\mathbb{R}}(C_{2k}) = k + 1$, since the new row and column are a linear combination of rows and columns of C_{2k}. Finally, $\text{rank}_{\text{bin}}(M) = \text{rank}_{\mathbb{B}}(M) = 2k$ since the first $2k$ ones on the main diagonal of M are an isolation set of size $2k$, and using Lemma 3 it is possible to partition the ones of M into $2k$ monochromatic rectangles since the last row of M is also a sum of two rows (see Lemma 4). That is, the ones on rows $1, k + 1, 2k + 1$ can be partitioned into 2 rectangles, and the ones of the remaining $2k - 2$ rows can be partitioned into $2k - 2$ additional rectangles (see Fig. 1).

3 Real Rank 3

In this section we consider 0,1 matrices with real rank 3. The main observation is that any 0,1 matrix M with $\text{rank}_{\mathbb{R}}(M) = 3$ must contain a 3×3 sub-matrix of real rank 3 which is equivalent to one of the 7 representatives in Fig. 2. This allows us to partition the basic matrices of real rank 3 into 7 classes, analyze each class separately by considering only its representative, and prove that it is not possible to augment any of these representatives into a 5×5 basic 0, 1 matrix with real rank 3 (Lemma 11). From this will follow Theorem 2 for $d = 3$. These representatives also help us characterize the maximal size of an isolation set according to the structure of the kernel of M, thus proving Theorem 3.

$$A_1 = \begin{pmatrix} 1&0&0 \\ 0&1&0 \\ 0&0&1 \end{pmatrix} A_2 = \begin{pmatrix} 1&0&0 \\ 0&1&0 \\ 0&1&1 \end{pmatrix} A_3 = \begin{pmatrix} 1&0&0 \\ 0&1&0 \\ 1&1&1 \end{pmatrix} A_4 = \begin{pmatrix} 1&0&0 \\ 1&1&0 \\ 0&1&1 \end{pmatrix}$$

$$A_5 = \begin{pmatrix} 1&0&0 \\ 1&1&0 \\ 1&1&1 \end{pmatrix} A_6 = \begin{pmatrix} 1&1&0 \\ 0&1&1 \\ 1&0&1 \end{pmatrix} A_7 = \begin{pmatrix} 1&1&0 \\ 0&1&1 \\ 1&1&1 \end{pmatrix}$$

Fig. 2. The seven possible representatives of matrices of real rank 3 and size 3×3.

The following lemma can be verified by a computer program, but it can also be proved using only algebraic and combinatorial arguments. We include the full theoretical proof since the analysis is required for the proof of Theorem 3.

Lemma 11. *There does not exist a* 5×5 *basic* 0, 1 *matrix* M *with* $\text{rank}_{\mathbb{R}}(M) = 3$.

Proof. Assume by contradiction that there exists such a matrix M. Then M has a sub-matrix of size 3×3 and real rank 3, which is equivalent to one of the seven sub-matrices A_i in Fig. 2. It is easy to verify that the matrices A_i, $1 \le i \le 7$, are the only 7 possible representatives of size 3×3 and real rank 3, by considering all options of the possible number of ones in each row and column and making sure that the three rows and columns are independent.

We now show that it is not possible to extend any of these seven representatives into a basic 0, 1 matrix M of size 5×5 with $\text{rank}_{\mathbb{R}}(M) = 3$. We first augment each matrix A_i with two additional rows which are a linear combination of the existing rows of A_i, and then show that it is not possible to add two additional distinct non-zero columns to the augmented sub-matrix without increasing the real rank. By Lemma 2 if we augment A_i with a row which is identical to one of the rows of A_i or augment A_i with an all zero row, and then add columns to this augmented matrix, the resulting matrix, if basic, will have a strictly larger real rank than A_i. Thus, any row added to A_i should be a non-trivial linear combination of at least two rows of A_i. Since the rows of A_i are a base, they span all 0, 1 row vectors, but we explicitly find the possible non-trivial linear combinations of the rows of A_i, since the coefficients in each combination determine the columns that can be added to the augmented matrix (see Lemma 4).

Case 1: The only possible linear combinations of two or more rows of A_1 are a simple sum and give the following row vectors: $(1,1,0), (0,1,1), (1,0,1), (1,1,1)$.

Consider first the matrix A_1' achieved by extending A_1 with the two rows: $(1,1,0),(0,1,1)$, and assume by contradiction that A_1' can be augmented with two additional columns X_1, X_2, so that the resulting 5×5 matrix is basic with real rank 3.

First note, that for $i = 1, 2$, there is at least one 1 in the first three positions of X_i. Otherwise, X_i is the all zero vector as the last two positions will also be a zero by Lemma 4, and thus the resulting matrix is not basic. Furthermore, X_i has at least 2 ones in the first three positions, otherwise, these three positions are identical to one of the columns of A_1, and then X_i is identical to one of the columns of A_1' (again using Lemma 4). Next, recall that the non-zero coefficients of the linear combinations are 1, that is $(1,1,0) = 1 \cdot (1,0,0) + 1 \cdot (0,1,0)$ and $(0,1,1) = 1 \cdot (0,1,0) + 1 \cdot (0,0,1)$. Thus, if for example the first row of A_1 is extended with a pair of elements x, y, such that $x = 1$ or $y = 1$, then the extension of the second row of A_1 cannot contain a 1 in the same position, and similarly, for the second and third row of A_1. But, this means that we can only extend A_1' with one column, the column $(1,0,1,1,1)^T$, and we get a contradiction.

A similar contradiction is achieved if we extend A_1 with the remaining possibilities of two rows, each with two ones. Finally, consider the case that A_1 is extended with a row with two ones $(1,1,0)$ and a row with three ones. Again, the coefficients of the linear combinations are 1, and here since row $(1,1,1)$ is a linear combination of all three rows of A_1, then it is possible to have only one 1 in the first three positions of any new column added. But then there are identical columns in the augmented matrix and we get a contradiction as required.

Case 2: The possible linear combinations of two or more rows of A_2 are:

$$
\begin{aligned}
(0,0,1) &= -1 \cdot (0,1,0) + 1 \cdot (0,1,1),\\
(1,1,0) &= 1 \cdot (1,0,0) + 1 \cdot (0,1,0),\\
(1,0,1) &= 1 \cdot (1,0,0) - 1 \cdot (0,1,0) + 1 \cdot (0,1,1),\\
(1,1,1) &= 1 \cdot (1,0,0) + 1 \cdot (0,1,1).
\end{aligned}
$$

If we extend A_2 with row $(0,0,1)$ we get a matrix that contains A_1 as a submatrix and we can continue as in Case 1. If we extend A_2 with both rows $(1,1,0)$ and $(1,0,1)$, then again, we cannot add a column which is identical in the first three positions to one of the columns of A_2. We also cannot add a column that has two ones in the first two positions, since row $(1,1,0)$ is the sum of the first two rows of A_2. Similarly, we cannot add a column whose first three positions are $(1,0,1)^T$, since the coefficients in the linear combination which results in row $(1,0,1)$ are $1,-1,1$, and thus we will get the value 2. Finally, we cannot add a column whose first three positions are $(0,1,0)^T$ since then the matrix M will contain the matrix A_1 and we can continue as in Case 1. Thus, there is no way to complete A_2, when extended with rows $(1,1,0),(1,0,1)$, into a basic 5×5 matrix with rank 3 as required.

Next we try to extend A_2 with rows $(1,1,0),(1,1,1)$. A similar argument shows that there is no way to add columns to the augmented matrix in this case without getting two identical columns or values that are not 0, 1. Extending A_2 with rows $(1,0,1),(1,1,1)$ results in a similar contradiction.

Case 3: The possible linear combinations of two or more rows of A_3 are:

$$\begin{aligned}(1,1,0) &= 1\cdot(1,0,0)+1\cdot(0,1,0),\\ (0,1,1) &= -1\cdot(1,0,0)+1\cdot(1,1,1),\\ (1,0,1) &= -1\cdot(0,1,0)+1\cdot(1,1,1)\\ (0,0,1) &= -1\cdot(1,0,0)-1\cdot(0,1,0)+1\cdot(1,1,1).\end{aligned}$$

Extending A_3 with $(0,0,1)$ gives us A_1 and we are done. Extending A_3 with $(1,0,1)$ or with $(0,1,1)$ gives us A_2 up to a permutation of rows and columns, and we are done. Thus, we can only extend A_3 with the row $(1,1,0)$. But then we do not get a 5×5 matrix and again we are done.

Case 4: The possible linear combinations of two or more rows of A_4 are:

$$\begin{aligned}(0,1,0) &= -1\cdot(1,0,0)+1\cdot(1,1,0),\\ (1,1,1) &= 1\cdot(1,0,0)+1\cdot(0,1,1),\\ (0,0,1) &= 1\cdot(1,0,0)-1\cdot(1,1,0)+1\cdot(0,1,1),\\ (1,0,1) &= 2\cdot(1,0,0)-1\cdot(1,1,0)+1\cdot(0,1,1).\end{aligned}$$

We cannot extend A_4 with $(0,1,0)$ or with $(0,0,1)$ since then we get A_2 as a sub-matrix. If we extend A_4 with $(1,0,1),(1,1,1)$, then, as in the previous cases we cannot add two distinct non-zero columns.

Case 5: The possible linear combinations of two or more rows of A_5 are:

$$\begin{aligned}(0,1,0) &= -1\cdot(1,0,0)+1\cdot(1,1,0),\\ (0,1,1) &= -1\cdot(1,0,0)+1\cdot(1,1,1),\\ (0,0,1) &= -1\cdot(1,1,0)+1\cdot(1,1,1),\\ (1,0,1) &= 1\cdot(1,0,0)-1\cdot(1,1,0)+1\cdot(1,1,1).\end{aligned}$$

Extending A_5 with $(0,1,0)$ gives A_3 as a sub-matrix; extending A_5 with $(0,0,1)$ gives A_2 as a sub-matrix, and extending A_5 with $(0,1,1)$ gives A_4 as a sub-matrix. Thus, we can only extend A_5 with $(1,0,1)$ and again we are done.

Case 6: The possible linear combinations of two or more rows of A_6 are:

$$\begin{aligned}(1,1,1) &= \frac{1}{2}\cdot(1,1,0)+\frac{1}{2}\cdot(0,1,1)+\frac{1}{2}\cdot(1,0,1),\\ (1,0,0) &= \frac{1}{2}\cdot(1,1,0)-\frac{1}{2}\cdot(0,1,1)+\frac{1}{2}\cdot(1,0,1),\\ (0,0,1) &= -\frac{1}{2}\cdot(1,1,0)+\frac{1}{2}\cdot(0,1,1)+\frac{1}{2}\cdot(1,0,1),\\ (0,1,0) &= \frac{1}{2}\cdot(1,1,0)+\frac{1}{2}\cdot(0,1,1)-\frac{1}{2}\cdot(1,0,1).\end{aligned}$$

If we extend A_6 with one of $(1,0,0),(0,1,0),(0,0,1)$ we get a matrix that contains A_4 we are done. Thus, we can only extend A_6 with $(1,1,1)$ and again we are done.

Case 7: The possible linear combinations of two or more rows of A_7 are:

$$\begin{aligned}(0,0,1) &= -1 \cdot (1,1,0) + 1 \cdot (1,1,1),\\ (1,0,0) &= -1 \cdot (0,1,1) + 1 \cdot (1,1,1),\\ (0,1,0) &= 1 \cdot (1,1,0) + 1 \cdot (0,1,1) - 1 \cdot (1,1,1),\\ (1,0,1) &= -1 \cdot (1,1,0) - 1 \cdot (0,1,1) + 2 \cdot (1,1,1).\end{aligned}$$

Extending A_7 with $(0,0,1)$ or with $(1,0,0)$ gives us a matrix that contains A_4 as a sub-matrix and we can continue as in case 4. Extending A_7 with $(0,1,0)$ gives us a matrix that contains A_5 as a sub-matrix and we can continue as in case 5, and thus we can extend A_7 only with $(1,0,1)$ and we are done. □

Theorem 4. *For any* $0,1$ *matrix* M *with* $\text{rank}_{\mathbb{R}}(M) = 3$ *it holds that* $2 \leq \text{rank}_{\mathbb{B}}(M), i(M) \leq 4$ *and* $3 \leq \text{rank}_{\text{bin}}(M) \leq 4$.

Proof. It is enough to consider basic matrices. By Lemma 1 if there exists a basic $0,1$ matrix M of real rank 3 and size $n \times m$, where $n, m \geq 5$, then there exists a basic sub-matrix of size 5×5 and real rank 3, in contradiction to Lemma 11. Therefore, M has at most 4 rows or at most 4 columns, and thus $\text{rank}_{\mathbb{B}}(M), \text{rank}_{\text{bin}}(M) \leq 4$. The theorem follows by combining this upper bound with Lemma 8 and the fact that $\text{rank}_{\mathbb{R}}(M) \leq \text{rank}_{\text{bin}}(M)$. □

Corollary 2. *The only* $0,1$ *matrix* M *of size* 4×4 *with* $\text{rank}_{\mathbb{R}}(M) = 3$ *and* $\text{rank}_{\text{bin}}(M) = \text{rank}_{\mathbb{B}}(M) = i(M) = 4$ *is the circulant matrix* C_4.

Proof. If M is basic, the corollary follows from Theorem 3. Otherwise, M has an all-zero row or column or at least two identical rows or columns and thus, $\text{rank}_{\text{bin}}(M) \leq 3$. □

References

1. Beasley, L.B.: Isolation number versus Boolean rank. Linear Algebra Appl. **436**(9), 3469–3474 (2012)
2. de Caen, D., Gregory, D., Henson, T., Lundgren, R., John, M.: Real rank as a bound for nonnegative integer rank. J. Comb. Math. Comb. Comput. **011**, 47–53 (1992)
3. de Caen, D., Gregory, D.A., Pullman, N.J.: The Boolean rank of zero-one matrices. In: Proceedings of the 3rd Caribbean Conference on Combinatorics and Computing, pp. 169–173 (1981)
4. Chalermsook, P., Heydrich, S., Holm, E., Karrenbauer, A.: Nearly tight approximability results for minimum biclique cover and partition. In: European Symposium on Algorithms, pp. 235–246. Springer (2014)
5. Chandran, S., Issac, D., Karrenbauer, A.: On the parameterized complexity of biclique cover and partition. In: 11th International Symposium on Parameterized and Exact Computation (IPEC 2016). Schloss Dagstuhl-Leibniz-Zentrum fuer Informatik (2017)

6. Dietzfelbinger, M., Hromkovič, J., Schnitger, G.: A comparison of two lower-bound methods for communication complexity. Theoret. Comput. Sci. **168**(1), 39–51 (1996)
7. Fleischner, H., Mujuni, E., Paulusma, D., Szeider, S.: Covering graphs with few complete bipartite subgraphs. Theoret. Comput. Sci. **410**(21–23), 2045–2053 (2009)
8. Friesen, M., Hamed, A., Lee, T., Theis, D.O.: Fooling-sets and rank. Eur. J. Comb. **48**, 143–153 (2015)
9. Gramm, J., Guo, J., Hüffner, F., Niedermeier, R.: Data reduction and exact algorithms for clique cover. J. Exp. Algorithmics (JEA) **13**, 2–2 (2009)
10. Gregory, D.A., Pullman, N.J.: Semiring rank: Boolean rank and nonnegative rank factorization. J. Combin. Inform. System Sci. **8**(3), 223–233 (1983)
11. Gregory, D.A., Pullman, N.J., Jones, K.F., Lundgren, J.R.: Biclique coverings of regular bigraphs and minimum semiring ranks of regular matrices. J. Comb. Theory, Ser. B **51**(1), 73–89 (1991)
12. Haviv, I., Parnas, M.: The binary rank of circulant block matrices. Linear Algebra Appl. **656**, 277–303 (2023)
13. Hrubeš, P.: Hard submatrices for non-negative rank and communication complexity. In: 39th Computational Complexity Conference (CCC). Leibniz International Proceedings in Informatics (LIPIcs), vol. 300, pp. 1–12. Dagstuhl, Germany (2024)
14. Jiang, T., Ravikumar, B.: Minimal NFA problems are hard. SIAM J. Comput. **22**(6), 1117–1141 (1993)
15. Lovász, L., Saks, M.E.: Lattices, Möbius functions and communication complexity. In: IEEE 29th Annual Symposium on Foundations of Computer Science (FOCS), pp. 81–90 (1988)
16. Lovett, S.: Recent advances on the log-rank conjecture in communication complexity. Bull. EATCS **112** (2014)
17. Orlin, J.: Contentment in graph theory: covering graphs with cliques. Indag. Math. **80**(5), 406–424 (1977)
18. Parnas, M., Shraibman, A.: A study of the binary and Boolean rank of matrices with small constant real rank (2025). https://arxiv.org/abs/2507.05824
19. Shigeta, M., Amano, K.: Ordered biclique partitions and communication complexity problems. Discret. Appl. Math. **184**, 248–252 (2015)
20. Shitov, Y.: An upper bound for nonnegative rank. J. Comb. Theory, Ser. A **122**, 126–132 (2014)

Shallow-Rotation Distance via Forest Representations

Anoop S. K. M. (✉) and Jayalal Sarma

Indian Institute of Technology Madras, Chennai, India
{skmanoop,jayalal}@cse.iitm.ac.in

Abstract. Rotation distance is a fundamental parameter in the study of tree balancing and data structures. One of the fundamental open problem in the area is to compute the minimum number of rotations required to transform one tree to the other, in which both trees have same number of nodes. We define *shallow rotation distance* with parameter k (which we call k-shallow rotation distance, denoted by $d_k(T_1, T_2)$) to be the rotation distance between any two full binary trees when the rotation is allowed only at nodes within depth k. Indeed, 2-shallow rotation distance is well defined and is already the frontier of unknown territory in algorithmic approaches to rotation distance problem.

In this paper, we identify a special case of rotation distance problem, known as the left-λ-RotDist and right-λ-RotDist and prove that they are as hard as the general version of the problem. Complementing this, we give a polynomial time algorithms for computing the $d_2(T_1, T_2)$ in three scenarios (a) when T_1 is a right-λ-tree and T_2 is a right-comb tree, (b) when T_1 is a left-λ tree and T_2 is a left-comb-tree and (c) when T_1 is a left-right-λ-tree and T_2 is a double comb with matching subtree size, where a *double-comb* is a tree in which all the nodes are present only on the right-most or left-most path from the root.

We obtain our results by using a known connection between the rotation distance and the well-studied algebraic object called Thompson's group $\mathcal{F}$ [8]. We use the forest representation for a pair of binary trees introduced by Belk and Brown (2005), where we design a new measure for pairs of spaces in the representation, and demonstrate that this new measure indeed captures the shortest representation of the group element in the above three scenarios.

1 Introduction

Binary trees are ubiquitous in computer science. A binary tree is a tree in which each node can have at most 2 children. A binary tree in which each node has either 0 or 2 children is called a full binary tree. One of the fundamental tasks about *binary trees* that is critical in data structures is that of balancing binary trees using the left and right rotations (See [9]). A right (resp. left) rotation of a full binary tree at an internal node consists of moving the right (resp. left) subtree of the left (resp. right) child of the node as the left (resp. right) child of

A. Jeż and J. Otop (Eds.): FCT 2025, LNCS 16106, pp. 376–390, 2026.
https://doi.org/10.1007/978-3-032-04700-7_28

the node. The peculiarity of these operations is that they preserve the in-order traversal of the tree, and help reducing the height of the tree - thus leading to, for instance, the amortized cost of operations in binary trees in certain situations.

One of the fundamental open problems related to rotations in binary tree is that of the rotation distance. The *rotation distance* between two full binary trees $d(T_1, T_2)$ with n internal nodes each is the minimum number of rotations required to transform T_1 to T_2. There are two natural questions related to this parameter - one as a combinatorial problem and the other as a computational problem. As a combinatorial problem, we are interested in finding how far can two n-node full binary trees be in terms of rotation distance. A rotation graph on n nodes, denoted as RG_n, consists of all n node trees as vertices (Catalan number many of them) with an edge connecting two if trees represented by them differ by a single rotation. Indeed, the rotation distance between two full binary trees is the length of the shortest path between two given trees in the corresponding rotation graph. Culik and Wood [7] showed an upper bound $2n-2$ distance for any pair of trees by using the fact that any tree can be turned into a right comb tree (all nodes on right path from root) of height n using at most $n-1$ rotations. Later Sleator, Tarjan and Thurston improved this upper bound to $2n-10$ for $n \geq 9$ [15]. Using the relationship between binary tree rotations and the flip distance between triangulations of a polygon, Dehornoy [8] showed the existence of n node trees that are at distance $2n - O(\sqrt{n})$ and conjectured a lower bound of $2n - 10$. Pournin [13] proved this bound and thus settled the problem.

The computation version of the decision problem asks for a given pair of trees T_1, T_2 and a value k, is $\textsc{RotDist}(T_1, T_2) \leq k$? Despite many efforts, this problem has remained surprisingly evasive. Till date, we neither know a deterministic polynomial time algorithm to solve the rotation distance problem, nor we know that it is NP-complete.

Cleary and John [6] has given a 2-approximation algorithm and they have also given a fixed parameter tractable algorithm parametrized by the rotation distance [5]. Restricting the inputs, Lucas [10] presented a polynomial time algorithm in finding rotation distance between two degenerate trees, that is trees where every node has at most one non-leaf child. Restricting the rotations, if we only allow rotations either all along the right arm of the tree or only at the root and right child of the root [4], then there are linear-time algorithms for computing the resulting restricted rotation distance and right-arm rotation distances. Various heuristics [12] are known describing polynomial-time algorithms which estimate rotation distance but with no provable guarantees. When the intermediate trees are restricted skew-trees, [14] gave a polynomial time algorithm for computing the skew-rotation-distance between two given skew trees.

Our Results: For a tree, let us denote that root node as A, the left child as B and the right child as C. We define a *left-λ-tree* to be a full binary tree where the node C has two leaves as its children. Similarly, a *right-λ-tree* is a full binary tree where node B and the left child of node C are leaves. λ-$\textsc{RotDist}$ is the problem of given two λ-trees computing the rotation distance between them. Indeed, while it may look like a restricted variant of the problem, we first show (we defer the proof to the full version) that these are as hard as the general problem.

Proposition 1. RotDist *reduces to* λ-RotDist *via deterministic polynomial time reductions.*

We now consider a restriction on the rotations allowed on binary trees. Given two full binary trees on n internal nodes, we define *shallow-rotation distance* with parameter k (which we denote by k-shallow rotation distance), denoted by $d_k(T_1, T_2)$, to be the rotation distance between the trees T_1 and T_2 when the rotation is allowed to be at the root or any of the nodes at distance (strictly) less than k from the root. Note that 1-shallow rotation distance is not well-defined since it is easy to construct a pair of trees that cannot be transformed to one another if we restrict the rotations to the root node alone. We study the special case when $k = 2$. Note that this case is already beyond the current frontier for the rotation distance problem where we know efficient algorithms.

One of the noteworthy approaches of the rotation distance problem is in terms of the Thompson's Group $\mathcal{F}$, an infinite group with many remarkable properties, which has been extensively studied in the fields of combinatorial group theory, measure theory and logic (See [3] for a detailed introduction and a survey of properties of Thompson's group). In the context of rotation distance, a pair of trees (T_1, T_2) is associated with one of the elements of the group F, and the pair of identical trees corresponds to identity element of F. Each rotation on a tree in the pair (T_1, T_2) can be interpreted as a group operation on the pair of trees, and hence rotation distance has a natural interpretation in the Cayley graph of the Thompson's group with respect to the generating set of elements which represents the set of rotations that we are interested. The most general version of the formulation allows any pair of trees that are single rotation away from each other to be in the generating set. Thus, computing the rotation distance between two trees is equivalent to computing the length of the shortest representation of the group element (T_1, T_2) in F.

This suggests natural techniques for understanding restricted versions of rotation distances in terms of generators of Thompson's group that is used in the above formulation. In fact, it is well known that Thompson's group has a two element generating set (A and B). In a surprising result, Fordham [2] designed a linear time algorithm which given any element of F, gives out the length of the shortest representation in terms of the generators A and B (and their inverses). In fact, the two elements A and B can be interpreted as the pairs corresponding to the left-rotation at the root of the first tree, and the left-rotation at the right child of the root of the first tree, respectively. Using this interpretation, [4] showed a linear time algorithm for computing the restricted rotation distance where rotation is allowed only at the root or right child of the root. We call such sequence of rotations between the two trees as *AB-paths*.

To state the exact connection between the Fordham's approach and our setting, we define C to be the Thompson's group element corresponding to the (reduced) pair of trees where the rotation operation is performed in the left child of the root node. Thus, A, B and C together forms all rotations up to height 2. The *ABC-path* between two trees T_1 and T_2 is defined as the minimum number of rotations when rotations are restricted to elements corresponding to elements A, B, and C.

As our first result, we demonstrate the applicability of Fordham's framework to compute ABC-path (which is exactly the 2-shallow rotation distance) under some restricted setting about the second tree.

Theorem 1. *There is a polynomial time algorithm for computing the 2-shallow rotation distance (or equivalently the shortest ABC path) between any two trees* (T_1, T_2) *where* T_1 *is a right-*λ*-tree and* T_2 *is a right-comb.*

It is already known (cf. [14]) that rotation distance of any given tree to left-comb tree or the right comb tree can be exactly computed in polynomial time. However, the novelty in the above theorem is in the technique - the generalization of Fordham's method based on Thompson's group - in the context of height-restricted rotation distances (namely ABC-path).

We further extend the above methods to show a polynomial time when the first tree is a left-λ-tree as well, which we state below.

Theorem 2. *There is a polynomial time algorithm for computing the 2-shallow rotation distance (or equivalently the shortest ABC path) between any two trees* (T_1, T_2) *where* T_1 *is a left-*λ*-tree and* T_2 *is a left-comb.*

To generalize the above result further, we define the *left-right-*λ*-tree* to be a full binary tree where the node B has a leaf as the left child. Now we can generalize the above techniques to show methods to compute rotation distance when the second tree is a double comb - a *double-comb* is a tree in which all the nodes are present only on the left-most or right-most path from the root. A double-comb tree T_2 is said to match in size with a tree T_1 if the number of leaf nodes to the left and the right of the root of the tree is exactly the same in both the trees.

Theorem 3. *There is a polynomial time algorithm for finding the 2-shallow rotation distance (or equivalently the shortest ABC path) between any two* (T_1, T_2) *where* T_1 *is a left-right-*λ*-tree and* T_2 *is a double-comb matching in size with* T_1.

We remark that the above results can be further generalized to convert an arbitrary double comb to the one which matches in size with the tree T_1 by applying rotations at the root node for the tree T_2.

Our Techniques: Our main tool is the forest representation for a pair of binary trees introduced by Belk and Brown [1]. They have given a reinterpretation of Fordham's algorithm using forest representations to design a linear-time algorithm for finding a minimum AB-path. The argument by Belk and Brown proceeds as follows - they identify the type of spaces in the tree-diagram and assigns a measure for each pair of spaces inspired by the methods of Fordham [2]. Michaud [11] extended this idea and gave method for finding a shortest word representation of a Thompson's group $\mathcal{F}$ element using a three generator set which introduced another new group element in addition to A and B. However, the third generator used by them does not correspond to a single rotation, and hence this renders to be not directly useful for any restricted version of rotation distance problem.

Our main technical contribution in the paper is a novel weight assignment between spaces in the forest diagram of the pair of trees. The weight assignment is inspired by some new insights regarding the effect of the group element C on the forest diagram. We prove that the minimum length representation of any group element corresponding to the pair of trees (T_1, T_2) in the three theorems can be obtained by a direct formula for the weights, similar to Fordham's algorithm, and hence this gives a linear-time algorithm for the problem.

2 Preliminaries

We now introduce Thompson's group F. It can be defined in combinatorial, algebraic and geometric methods (see [3] for a survey). Algebraically, this group has a presentation with an infinite number of generators in the following form $\{x_0, x_1, \dots \mid x_i^{-1} x_n x_i = x_{n+1} \forall i < n\}$. It also has a finite presentation with only two generators x_0 and x_1 as well (sometimes denoted by A and B respectively), where all the above relations can be expressed in terms of these two elements (see [3]).

Geometrically, each element in the Thompson's group $\mathcal{F}$ can be described using tree pair diagrams, each of which is a pair of rooted binary trees with the same number of leaves. We say that a tree pair diagram (T_1, T_2) is *not reduced* if both T_1 and T_2 contain a node with two leaves numbered consecutively when the leaves are numbered from left to right in both trees. Any tree pair diagram which does not have the above property is said to be *reduced.* Every element of F can be shown to correspond to a reduced tree pair diagram, and the operation of the group is composition. However, in order to compose two group elements, the tree pairs may have to be expanded, introducing more leaf nodes to get unreduced representatives of the same elements of F, in order to perform this composition operation.

In this view, the generators A and B can be shown to exactly match with the tree pair where the second tree is obtained from the first by right rotating at the root and the right child of the root respectively. With a slight abuse of notation, we use A for the operation of left rotation at root, B to be the left rotation at right child of root, and C to be the left rotation at left child of root. Viewing these as group elements in Thompson's group, the corresponding inverse rotations are labeled as A^{-1}, B^{-1} and C^{-1}, and they match exactly with the corresponding reverse rotations as well.

Forest Diagrams: One of the important ingredients in our arguments is the Forest diagrams studied by [1]: Given a tree, we get its corresponding forest diagram by peeling off the outer-most vertices and its corresponding edges. Figure 1 shows an example of finding out the corresponding forest diagram from a given binary tree. This operation converts a given tree to its corresponding forest representation. A trivial tree in a forest is represented by a single dot. All non-trivial trees contains at least one internal node (which we call a *caret node*).

We use the forest diagram introduced by Belk and Brown [1] to represent a pair of binary trees. Given a pair of full binary trees on n nodes, the internal nodes are labeled in in-order. A forest diagram is obtained by placing one tree on top and the other on bottom in such a way that their leaf labels match each other as shown in Fig. 1 below. An identity element in Thompson's Group $\mathcal{F}$ has both trees identical. Identity element consists of only trivial trees on both top and bottom with both pointers being opposite to each other in the forest diagram representation.

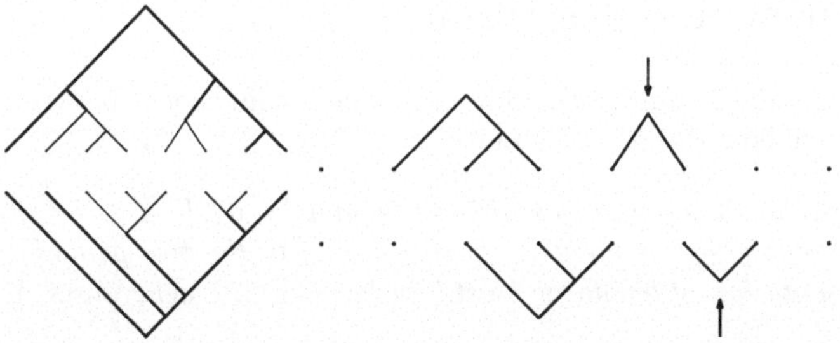

Fig. 1. A tree pair and its corresponding forest diagram

Our aim is to find a minimal length representation for a pair of binary trees in terms of the above three rotations A, B and C. It can be noted that A and B correspond to the generating elements x_0 and x_1 respectively in Thompson's Group $\mathcal{F}$. C corresponds to the group element $x_0 x_1 x_0^{-2}$.

Interpreting the A, B, and C **Operations in Terms of Forest Diagrams:** Let f be a group element representing a pair of reduced trees. We represent the action of C and C^{-1} on the forest diagram in Fig. 2 (for the action by elements A and B, we refer the reader to [1]).

A : Moves the top pointer one tree to the right.
A^{-1} : Moves the top pointer one tree to the left.
B : Adds a caret connecting the tree pointed by the pointer with the one on its right. The pointer points to the root of this combined tree.
B^{-1} : Removes a caret from the top of the tree at the pointer. The pointer points to the root of the tree on the left.

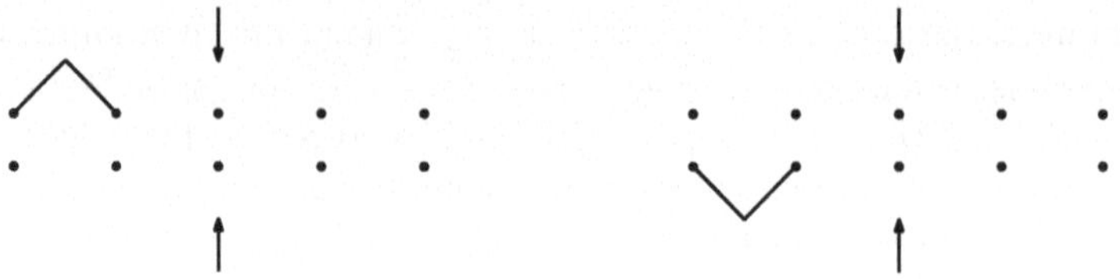

Fig. 2. Forest representation of C and C^{-1}

C: Adds a caret connecting the trees to the double left of the pointer with the one on the left of the pointer. The position of the pointer is unaltered.

C^{-1}: Removes a caret from the top of tree on the left of the pointer. If that tree is trivial, a caret gets added on the bottom tree, exactly opposite where a C operation would have created a node. The position of the pointer is unaltered.

3 2-Shallow Rotation Distance

We first establish (see full version for a proof) the equivalence between the ABC-path and 2-shallow Rotation Distance.

Proposition 2. *Given any pair of trees T_1 and T_2, $d_2(T_1, T_2) \leq \ell$ if and only if the Thompson's group element X corresponding to the tree pair (T_1, T_2), has a word representation of length at most ℓ in terms of the generators A, B and C.*

A *space* is the portion between two leaves in a forest diagram. A space is said to be in *support* if there exists at least one non-trivial tree either to the left or right of it. So a *support* is the set of contiguous spaces in a forest diagram where trees exist. That portion of the set of non-trivial trees lying on the left of left-most non-trivial tree and right of right-most non-trivial tree is not in support. Notice that if we apply A, B, C or their inverses to a tree pair corresponding to a forest diagram, the support of the diagram may increase or decrease by at most one.

Main Idea: Our algorithm is going to directly compute the shortest word representation of the Thompson's group element for a given tree diagram, by considering the task of repeatedly applying the group operations to clean up the top and bottom part of the diagram to reach the identity representation. The idea is to assign weights to the spaces for a pair of top and bottom trees based on the number of movements of the top pointer (which corresponds to rotation) across the spaces in order to clean the space. Once these weights are defined, we formally prove that these weights indeed works and gives the optimal length representation of the group element represented by the pair of trees, in terms of the generating elements.

3.1 The Space Types and the Weight

We define the following types of spaces between two trees in a forest diagram representing a pair of n node reduced binary trees as follows.

1. E_{null} : Mark the external space to immediate right of the tree on the extreme left of the pointer as E_{null}, if the tree is a single caret or it is a right comb.
2. E_L : Mark all spaces except E_{null} that are exterior on the left of the pointer as E_L.
3. E_R : Mark all spaces that are exterior on the right of the pointer as E_R.
4. I_0 : Mark all internal node spaces formed by connecting two trivial trees as I_0.
5. I_1 : Mark all node spaces immediately to the left of I_0 as I_1 if it's immediate left is not a part of right comb.
6. I_2 : Mark all node spaces immediately to the left of I_1 as I_2 if its immediate left is not a part of right comb.
7. I^L_{rc} : Mark all internal node spaces on the left of the pointer that are part of a right comb tree and are not already marked I_0 or I_1 as I^L_{rc}.
8. I^R_{rc} : Mark all internal node spaces on the right of the pointer that are part of a right comb and are not already marked I_0 or I_1 as I^R_{rc}.
9. I_{rrc} : Mark all internal nodes that are on immediate right of a right comb as I_{rrc}
10. I^L_{lc} : Mark all internal node spaces on the left of the pointer that are part of a left comb and are not already marked I_0 or I_{rrc} as I^L_{lc}.
11. I^R_{lc} : Mark all internal node spaces on the right of the pointer that is part of a left comb and are not already marked I_0 as I^R_{lc}.
12. I_{LR} : Mark all internal node spaces not already marked that contain carets on the immediate left and right as I_{LR}.
13. E_{er} : Mark the space on the extreme right as E_{er}, if its left-to-left-to-left space is I_{rc}.
14. E_{rrc} : Mark the space on the immediate right of a tree as E_{rrc}, if only the top part of the tree is a right-comb with more than 2 carets.

We use a weight function to calculate the minimum word length using A, B, C and its inverses that represents a given reduced tree pair. This is presented in the following table. In the context of our results, in the given pair of trees (T_1, T_2) the second tree is always the left or the right comb, and hence the type of the space can only be E_L or E_R. Due to this, we show only two rows of the weight chart (Table 1).

Table 1. Weight Chart

	E_L	E_R	I_0	I_1	I_2	I^L_{rc}	I^R_{rc}	I^L_{lc}	I^R_{lc}	I_{LR}	E_{er}	I_{rrc}	E_{rrc}	E_{null}
E_L	2	1	0	0	1	0	0	1	0	1	1	3	3	0
E_R	1	2	0	1	1	1	1	1	1	1	2	2	2	0

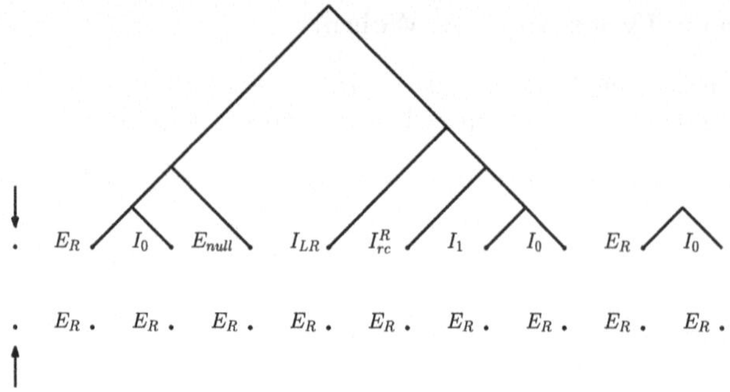

Fig. 3. Example: Top and bottom pointers at the extreme left

3.2 Length of Shortest Representation

We also adapt the formula by Belk and Brown [1] to calculate the minimum word length in our context. Let f be the group element that represents a reduced pair of trees. We find the length of f using the formula.

$$\ell(f) = \ell_0(f) + \ell_1(f) \tag{1}$$

where: $\ell_0(f)$ is the sum of the weights of space pairs in the forest diagram of f and $\ell_1(f)$ is the number of carets in the forest diagram of f.

By careful examination, we can see that $\ell_0(f)$ counts the number of A or its inverses in f and $\ell_1(f)$ counts the number of B, C or its inverses in f. Intuitively, with respect to the forest diagram, ℓ_0 counts the number of times the pointer moves over spaces and ℓ_1 counts the number of carets to be removed, in the shortest number of moves to reach the identity.

We use the technique used by Fordham [2], which in turn was used by Belk and Brown [1] to prove that this length formula indeed calculates the minimum word length of a Thompson's group $\mathcal{F}$ element.

Lemma 1. *Let G be a group with $S = \{A, B, C\}$ as its generating set and let $\ell : G \to \mathbb{N}$ be a function. Then ℓ is the length function for G with respect to S, if and only if*

1. *$\ell(e) = 0$, where e is the identity element in G.*
2. *$\mid \ell(gs) - \ell(g) \mid \leq 1$, for all $g \in G$ and $s \in S$.*
3. *$\forall g \in G \setminus \{e\}$, $\exists s \in S \cup S^{-1}$, such that $\ell(gs) < \ell(g)$.*

Clearly, the above implies that for any $g \in G$, the length function $\ell(g)$ must indeed be equal to the length of the minimum length representation of the group element g in terms of the generators S. By using Theorem 1, we now show how a minimum length ABC-representation (equivalently 2-shallow rotation sequence) can be found.

Let $f \in G$ represents a pair of n-node trees represented using its infinite length generators. Let $S = \{A, B, C\}$ and $\ell(f) = k, k \in \mathbb{N}$ be the minimum length word representation of f in terms of $S \cup S^{-1}$, Y represents a zero length word and $f' = f$. We repeatedly do the following till f' becomes identity.

1. Select an $s \in S \cup S^{-1}$ for which $\ell(f's) < \ell(f')$.
2. Update Y by concatenating it with s.
3. Update f' to $f's$.

We see that Y is a word that converts f to identity in $\ell(f)$ steps. Hence $X = Y^{-1}$ is a minimum length word representing f in terms of $S \cup S^{-1}$.

Hence as the main technical contribution of our paper, we prove Theorem 1 for the function ℓ defined by Eq. 1 for the cases to prove Theorem 2 and Theorem 1 from the introduction.

4 Taming the Right-λ-Trees : Proof of Theorem 1

We prove Theorem 1 in this subsection. Notice that since the tree T_2 is a right-comb tree, all the trees in the bottom side of the forest diagram are trivial and hence all the types of the spaces in the bottom of the diagram are E_L or E_R only. Since the tree T_1 is a right-λ-tree, and T_2 is a right-comb tree, in the tree representation, the top and bottom pointers are on the extreme left.

The intuition in the remaining part of the argument is as follows. Recall that the identity element of the Thompson's group has a forest representation that has trivial trees on the top and the bottom parts of the diagram and the arrows point to each other. In order to convert any given element represented by the pair of trees (T_1, T_2), the aim is to clean the trees in the top part of the forest diagram (by using group operations A, B, C) and the bottom part (by using reductions after the matching structure of tree is built on the top part by using group operations A, B, and C) of the diagram and bring the arrows to the original position. We show such an optimal clean-up strategy for the case when T_1 is the right-λ-tree and T_2 is the right-comb, with the pointers being at the extreme left pointing to each other.

Firstly, observe that, in the minimum length word representation of the tree pair need not contain B^{-1} or C^{-1}. This is because no carets are to be created on the bottom forest. So, neither B nor C operation needs to be performed while transforming an element on a shortest path to identity .

We use B^{-1} or C^{-1} to remove all the carets. A and A^{-1} is used to move the pointer one tree right and left respectively. The optimal way to do the removal is from right to left as it minimizes the number of skips over each space. We propose the following rules to compute a minimum length path to identity.

1. Apply A to move the top pointer to the right-most non trivial tree if it is not a right comb, otherwise move it to the trivial tree right of the right-most right comb.
2. Apply C^{-1} to remove all carets above I_{rc}.
3. Apply B^{-1} to remove all other caret types.
4. Apply A^{-1} to move the top pointer backwards, to reach the current right-most non-trivial tree after caret removals or till it matches with the bottom pointer.

All spaces in the bottom forest are of type E_R. So the relevant row for calculating the weight is the one for E_R in the Table 1.

We now translate the above intuition into a proof by proving Lemma 1 explicitly. This will be done via the following three propositions. A group element is said to be *strongly positive* [1] if the forest diagram has both pointers on the extreme left and the bottom forest contains only trivial trees.

Proposition 3. *If $g \in G$ and g is strongly positive, then for all f in a minimum length path to identity $| \ell(fB) - \ell(f) | \leq 1$ and $| \ell(fC) - \ell(f) \leq 1$.*

Proof. Let us first consider the case when B is applied when the pointer is on a trivial tree and the tree to the right is also trivial. The space on the right is E_R and the weight of the space $\begin{bmatrix} E_r \\ E_r \end{bmatrix}$ is 2. Upon applying B, this right space changes to I_0 and the weight of $\begin{bmatrix} I_0 \\ E_r \end{bmatrix}$ is 0. Hence the ℓ_0 value of the new element gets reduced by 2. However, since a new caret got created, its ℓ_1 value gets incremented by 1, causing a reduction of length value by 1.

Consider another case when B is applied when the pointer is on a non-trivial tree and the tree to the right is also non-trivial. The space on the right is E_R and the weight of the space $\begin{bmatrix} E_R \\ E_R \end{bmatrix}$ is 2. Upon applying B, this space changes to I_{LR}. No other space gets changed. Since $\begin{bmatrix} I_{LR} \\ E_R \end{bmatrix}$ is of weight 1, the ℓ_0 value of the new element decreases by 1 and the ℓ_1 value gets incremented by 1 due to the creation of a new caret. Here also the change in length value is ≤ 1.

Now let us consider the case when B is applied when the pointer is on trivial tree and the tree to the right is non-trivial. The space on the right is E_R and the weight of the space is 2. Upon applying B, this space changes to I_{rc}. No other space gets changed. Since $\begin{bmatrix} I_{rc} \\ E_R \end{bmatrix}$ is of weight 1, the ℓ_0 value of the new element get reduced by 1, but the addition of a new caret increments its ℓ_1 value by 1. Thus the total length of the group element remains unaltered.

Let us consider the effect of C on a strongly positive element. By similar arguments as above, it can be verified that caret creation by one C operation does not change the length value by more than 1. Here, since the position of the pointer remains unaltered by applying a C operation, all other spaces remain unaltered. □

We prove the following two important propositions. These are proved by careful case analysis, where, due to space constraints, we only present a few cases and leave the rest to the full version of the paper.

Proposition 4. *If $g \in G$ and g is strongly positive, then for all f in a minimum length path to identity, $| \ell(fA^{-1}) - \ell(f) | \leq 1$. In particular, $\ell(fA^{-1}) = \ell(f) - 1$ unless one of the conditions holds:*

1. *fA^{-1} has larger support than f.*
2. *The left space of f has a top label E_L and there exists some non-trivial tree on the right of the pointer.*

Proposition 5. *If $g \in G$, g is strongly positive, $g \neq e$ and $S = \{A, B, C\}$, there exists $s \in S \cup S^{-1}$, such that $\ell(fs) < \ell(f)$, $\forall f$ in the minimum length path from g to identity.*

Proof. We give a sketch of the proof. To begin with, it can be noticed that the weight remains same for $I^L_{rc}, I^R_{rc}, I^L_{lc}$ and I^R_{lc} with respect to E_R. The remaining argument is a careful analysis of the possible cases. Let f denote the current element under consideration in all the following.

Case 1: Right of pointer is E_R or E_{er}:
Initially when the top pointer is at the left-most end, its right space is E_R. Applying A moves the pointer one tree to its right. Since no extra carets are being added or removed, ℓ_1 value remains the same. But this move changes one space from E_R to E_L. Since $\begin{bmatrix} E_R \\ E_R \end{bmatrix}$ is of weight 2 and $\begin{bmatrix} E_L \\ E_R \end{bmatrix}$ is of weight 1, this reduces ℓ_0 by 1 and so $\ell(fA) = \ell(f) - 1$.

Case 2: Pointer on top of a tree with right-comb structure on top:
Here, the right space is either E_R or E_{er}. This is because on moving the pointer from right to left after removing carets on its right, C^{-1} could be used to remove carets that are part of right comb structure. Hence as in the previous case A can be applied to move the pointer one tree to its right. Since no extra carets are being added or removed, ℓ_1 value remains the same and this move changes one space from E_R to E_L or from E_{er} to E_L. Since $\begin{bmatrix} E_R \\ E_R \end{bmatrix}$ is of weight 2 and $\begin{bmatrix} E_L \\ E_R \end{bmatrix}$ is of weight 1, this reduces ℓ_0 by 1 and so $\ell(fA) = \ell(f) - 1$.

Case 3: Pointer on right of a right-comb :
One C^{-1} operation removes a caret from top of the tree to the left of the pointer. So a space I^L_{rc} changes to E_L. As the weight for both $\begin{bmatrix} I^L_{rc} \\ E_R \end{bmatrix}$ and $\begin{bmatrix} E_L \\ E_R \end{bmatrix}$ equals 1, its ℓ_0 remains the same. Since one caret gets removed, its ℓ_1 value decreases by 1 and so $\ell(fC^{-1}) = \ell(f) - 1$. Note that no other space changes its type as the pointer does not change its position.

Case 4: Pointer on right of a tree whose left and right subtrees are non-trivial:
As in the above case, a C^{-1} operation removes a caret from top of the tree to the left of the pointer. So the space under the caret of type I_{LR} changes to E_L. As the weight for both $\begin{bmatrix} I^L_{LR} \\ E_R \end{bmatrix}$ and $\begin{bmatrix} E_L \\ E_R \end{bmatrix}$ equals 1, its ℓ_0 remains the same. Since one caret gets removed, its ℓ_1 value decreases by 1 and so $\ell(fC^{-1}) = \ell(f) - 1$. Here also no other space changes its type as the pointer does not change its position.

Remaining Cases: The remaining possibilities include (5) the pointer on top of a tree with left-comb structure on its top, (6) the Pointer on top of tree whose right-most interior space is I_{rrc} and (7) the left of pointer is E_L with no space on the right in support. We refer the reader to the full version for the proof in these cases.

5 Taming the Left-λ-Trees: Proof of Theorem 2

We prove Theorem 2 in this subsection. Notice that since the tree T_2 is a right-comb tree, all the trees in the bottom side of the forest diagram are trivial and hence all the types of the spaces in the bottom of the diagram are E_L or E_R only. Since the tree T_1 is a right-λ-tree, and T_2 is a right-comb tree, in the tree representation, the top and bottom pointers are on the extreme left.

Similar to the previous section, we first propose the following rules to compute a minimum length path to identity, and prove that it is indeed optimal by the strategy outlined earlier in Lemma 1.

1. Apply A^{-1} to move the top pointer to the left-most non trivial tree if the top part of it is not a right comb, otherwise move to the tree immediately right of the left-most tree.
2. Apply C^{-1} to remove all carets in a right-comb by moving one tree to the right and move right using A if there are no more non-trivial trees on the left. If there are non-trivial trees on the left, then move left using A^{-1}.
3. Apply B^{-1} to remove all other caret types that are above I_1, I_2 and move right using A.
4. Apply A to move the top pointer towards right till it matches with the bottom pointer.

The spaces in the bottom forest are all of type E_L. Hence the relevant row for calculating the weight is the one for E_L on Table 1.

To further develop the intuition, we comment on why $\begin{bmatrix} I_{rrc} \\ E_L \end{bmatrix}$ is of weight 3. Consider an element f whose two pointers are on the extreme right. Intuitively, the minimum path to identity can be reached by moving the top pointer to the extreme left and returning to the original position by cleaning up or removing trees one by one. When the top of a tree is a left-comb, we apply B^{-1} and A repeatedly to remove it. After that, if there is a right-comb with more than 2 carets appearing, it is better to go to the right of the tree by applying A by skipping I_{rrc} space once and applying C^{-1} multiple times to remove the right-comb part. But suppose after that removal, if there still exists a non-right-comb tree, we need to apply A^{-1} to reach back to this tree, thereby skipping I_{rrc} again. Since at the end the pointer needs to be brought back to the initial position, I_{rrc} needs to be skipped again by applying A. Thus we see that the space $\begin{bmatrix} I_{rrc} \\ E_L \end{bmatrix}$ is of weight 3. It may be noted that after the first skip I_{rrc} space gets changed to E_L. From the table $\begin{bmatrix} E_L \\ E_L \end{bmatrix}$ is having weight 2. After the second skip, this space gets changed to E_R. The weight of $\begin{bmatrix} E_R \\ E_L \end{bmatrix}$ is 1. After the third skip, this space will not be there in the support. See full version for the details.

6 Taming the Left-Right-λ-Trees : Proof of Theorem 3

Before proving the main theorem, we first show the following proposition which is critical in proving the theorem. We defer the proofs to the full version.

Proposition 6. *Let $g \in G$ represents a pair of n nodes trees, has a forest representation with both the pointers pointing opposite to each other, then there exists g', g'' such that $\ell(g) = \ell(g') + \ell(g'')$, where $g' \in G$ is an element whose forest representation has both the pointers on the extreme right representing a pair of trees each with some $k < n$ nodes and $g'' \in G$ is an element whose forest representation has both the pointers on the extreme left representing a pair of trees each with $n - k$ nodes.*

Proposition 7. *If $g \in G$, g is an element whose forest representation has both the pointers pointing opposite to each other, $g \neq e$ and $S = \{A, B, C\}$, there exists $s \in S \cup S^{-1}$, such that $\ell(fs) < \ell(f)$, $\forall f$ in the minimum length path from g to identity.*

Combining the above two propositions, we get the proof of Theorem 3.

Acknowledgments. We thank the anonymous reviewers for their insightful comments which improved the readability of the paper.

References

1. Belk, J.M., Brown, K.S.: Forest diagrams for elements of Thompson s group F. Int. J. Algebra Comput. **15**(05n06), 815–850 (2005)
2. Blake Fordham, S.: Minimal length elements of Thompson's group F. Geom. Dedicata. **99**(1), 179–220 (2003)
3. Cannon, J., Floyd, W., Parry, W.: Introductory notes on Richard Thompson s groups. Enseign. Math. **42**, 215–256 (1996)
4. Cleary, S.: Restricted rotation distance between binary trees. Inf. Process. Lett. **84**(6), 333–338 (2002)
5. Cleary, S., John, K.S.: Rotation distance is fixed-parameter tractable. Inf. Process. Lett. **109**, 918–922 (2009)
6. Cleary, S., St. John, K.: A linear-time approximation algorithm for rotation distance. J. Graph Algorithms Appl. **14**(2), 385–390 (2010). https://doi.org/10.7155/jgaa.00212
7. Culik, K., Wood, D.: A note on some tree similarity measures. Inf. Process. Lett. **15**(1), 39–42 (1982)
8. Dehornoy, P.: On the rotation distance between binary trees. Adv. Math. **223**(4), 1316–1355 (2010)
9. Knuth, D.E.: The art of Computer Programming, volume 3: (2nd ed.) Sorting and Searching. Addison Wesley Longman Publishing Inc., USA (1998)
10. Lucas, J.M.: Untangling binary trees via rotations. Comput. J. **47**, 259–269 (2004)
11. Michaud, N.: A three-element generating set for Thompson's group F. Senior Projects 2007-2010 Archival Collection. 490 (2009)
12. Pallo, J.: An efficient upper bound of the rotation distance of binary trees. Inf. Process. Lett. **73**(3), 87–92 (2000)
13. Pournin, L.: The diameter of associahedra. Adv. Math. **259**, 13–42 (2014)

14. Anoop, S. K. M., Sarma, J.: On rotation distance of rank bounded trees. Fundamenta Informaticae **191**(2), 79–104 (2024)
15. Sleator, D.D., Tarjan, R.E., Thurston, W.P.: Rotation distance, triangulations, and hyperbolic geometry. In: Proceedings of the 18th Symposium on Theory of Computing (STOC 86), pp. 122–135 (1986)

On Saving Energy in Boolean Circuits via Negations

Jayalal Sarma[1(✉)] and Kei Uchizawa[2]

[1] Indian Institute of Technology Madras, Chennai, India
jayalal@cse.iitm.ac.in

[2] Graduate School of Science and Engineering, Yamagata University, Yamagata, Japan

Abstract. For a Boolean circuit C, the energy complexity of C is defined as the number of gates outputting ones in C, where the maximum is taken over all the input assignments. The energy complexity has been well-studied among various bases since 1960s.

Monotone circuits incur high energy in terms of their size. In order to understand how much energy can be saved by using negations in the circuit design, in this paper, we study the contrast between two infinite bases $\mathcal{B}_*$ and $\mathcal{N}_*$, where $\mathcal{B}_*$ contains the conjunction of unbounded fan-in and disjunction of unbounded fan-in, and the negation, while $\mathcal{N}_*$ contains the conjunction and disjunction with unbounded fan-in, where any input can be negated. Thus, a $\mathcal{B}_*$-circuit needs a single gate for negation, while $\mathcal{N}_*$-circuit can freely use negation without contributing to the energy of the circuit. We show that this "free" use of negations has a considerable impact on computational power of energy-bounded circuits, as follows:

- We observe that any Boolean function f can be computed by a $\mathcal{N}_*$-circuit of energy one. Thus, $\mathcal{N}_*$-circuits of constant energy are universal. In contrast, there is a Boolean function that cannot be computed by any $\mathcal{B}_*$-circuit of energy $o(\sqrt{n})$.
- There exists a DeMorgan $\mathcal{N}_*$-circuit of size n, energy e, and $(n/e+1)^e$ output patterns. This is in contrast with the known fact ([4]) that any $\mathcal{B}_*$-circuit has at most $2^{O(e \log e)}$ output patterns.
- If a Boolean function f is computable by a decision tree of size s and rank r, then f is also computable by a $\mathcal{N}_*$-circuit of size $O(s)$ and energy $O(r)$. In contrast, we show an explicit function which can be computed by rank 1 decision trees, but requires non-constant energy for any $\mathcal{B}_*$-circuit computing it.

The results above show equipping gates with negations are helpful for energy-bounded circuits. On the limitation of $\mathcal{N}_*$-circuits, we show:

- If C is $\mathcal{N}_*$-circuit that is either DeMorgan circuit or a layered circuit (with input variables connected only to gates in the bottom layer), computing the PARITY function on n variables with size s and depth d, then for large enough n, C must have energy at least $\frac{n}{d \log s}$.
- If a $\mathcal{N}_*$-circuit C on $2n$ variables computes a Boolean function of $f : \{0,1\}^n \times \{0,1\}^n \to \{0,1\}$, and has size s, depth d, then $\mathsf{EC}_{\mathcal{N}_*}(C) \geq \frac{\log(rk(M_f))}{d \log s}$. where M_f is the communication matrix corresponding to

A. Jeż and J. Otop (Eds.): FCT 2025, LNCS 16106, pp. 391–405, 2026.
https://doi.org/10.1007/978-3-032-04700-7_29

the Boolean function f. Thus, if C is a polynomial-size and constant-depth circuit computing DISJ_n, then C must have energy at least $\Omega(n/\log n)$. This lower bound is tight up to a constant factor. DISJ_n can indeed be computed by a circuit of size $O(n^2/\log n)$, depth 3 and energy $O(n/\log n)$.

1 Introduction

Energy complexity is a measure of Boolean functions well-studied in the 1960s, and then attracted more attention in the last decade. The measure was originally introduced by Vaintsvaig [12] where the main motivation was to understand the static power consumed in integrated circuits by dissipation of heat and maintaining a given potential at the nodes of a circuit. More recently, the energy complexity in the context of neural networks and threshold circuits has also been intensely studied (See [8]).

Let $\mathcal{B}$ be a set of Boolean functions. Let C be a logic circuit consisting of gates computing functions chosen from $\mathcal{B}$, and n be the number of input variables of C. For an input assignment $\mathbf{a} \in \{0,1\}^n$ to C, the energy complexity $\mathrm{EC}_{\mathcal{B}}(C, \mathbf{a})$ of C for $\mathbf{a}$ is defined as the number of the gates in C that evaluate to ones for $\mathbf{a}$. Then, for a Boolean function $f : \{0,1\}^n \to \{0,1\}$, we define $\mathrm{EC}_{\mathcal{B}}(f) = \min_C \max_{\mathbf{a}\in\{0,1\}^n} \mathrm{EC}_{\mathcal{B}}(C, \mathbf{a})$, where the minimum is taken over all the circuits C computing f, and $\mathrm{EC}_{\mathcal{B}}(n) = \max_f \mathrm{EC}_{\mathcal{B}}(f)$, where the maximum is taken over all the n-variate Boolean functions.

It is natural to expect the energy complexity to vary with the choice of the basis that we work with. Vaintsvaig [12] showed that for any finite basis $\mathcal{B}$, it holds that $\Omega(n) \le \mathrm{EC}_{\mathcal{B}}(f) \le O(2^n/n)$ (possibly with exponential size) where the lower bound was achieved for the basis $\mathcal{B} = \{\wedge_2, \vee_2, \neg\}$ and the upper bound was achieved by the basis $\mathcal{B} = \{\overline{\wedge_2}\}$ which is the NAND gate. Kasim-zade [2] demonstrated that for any constant d, and for a finite basis $\mathcal{B}$ with functions at the gates that has at most d input variables, the energy complexity over the basis $\mathcal{B}$ follows an interesting dichotomy : for every Boolean function f, either (1) $\Omega(n) \le \mathrm{EC}_{\mathcal{B}}(f) \le O(n^2)$ or (2) $2^{n/2d} \le \mathrm{EC}_{\mathcal{B}}(f) \le 2^n/n$.

The measures of $\mathrm{EC}_{\mathcal{B}}(n)$, $\mathrm{EC}_{\mathcal{B}}(f)$ has been recently investigated for several bases $\mathcal{B}$. For a basis $\mathcal{B}_{th}$ consisting of linear threshold functions, it is known that there exists explicit Boolean functions that any constant-depth threshold circuit needs even exponential size to compute if the energy is bounded, while these are computable by constant-depth and linear-size circuit [9–11]. For another basis $\mathcal{B}_\ell$ consisting of arbitrary Boolean functions of arity ℓ, Suzuki *et al.* [7]. showed that $\Omega((n - m_f)/\ell) \le \mathsf{EC}_{\mathcal{B}_\ell}(f) \le O(n/\ell)$ for any symmetric function f, where m_f is the maximum number of consecutive 0 s or 1 s in the value vector of f.

Coming to the standard basis for Boolean circuits, for the case of bounded fanin basis $\mathcal{B}_2 = \{\wedge_2, \vee_2, \neg\}$, where $\wedge_2$ and $\vee_2$ are a conjunction and a disjunction of arity two respectively, and $\neg$ is a negation, a series of works [1,6], established that $\mathrm{EC}_{\mathcal{B}_2}(f)$ and $\mathsf{DT}(f)$ are polynomially related, where $\mathsf{DT}(f)$ is the optimal decision tree depth of f.

The above arguments critically used the fact that the fanin of the gates is bounded. Indeed, it is unclear how to do a fanin reduction in circuits without increasing the energy. More recently, in [4], it was shown that energy restrictions affects the computational power of Boolean circuits even if unbounded fan-in and depth are allowed. In particular, for the standard basis $\mathcal{B}_* = \{\wedge, \vee, \neg\}$ where $\wedge$ and $\vee$ are the conjunction of unbounded fan-in and disjunction of unbounded fan-in, respectively, it was shown that if a Boolean function f can be computed by a circuit C of size s and energy e, then the number of output patterns that arise in C is independent of s, and bounded by $2^{e \log e}$. This upper bound implies that f can also be computed by a depth-3 circuit C' of size $s2^{e \log e + 4e}$. This was further used in [4] to establish lower bounds for the energy over the basis $\mathcal{B}_*$.

Motivation: Monotone circuits incur high energy relative to the size of the circuit. Indeed, if a circuit is monotone, the input assignment of all 1 s makes every gate in the circuit output one. So energy efficient circuits must be either monotone circuits of small size or non-monotone circuits which uses negations in a careful way to reduce the overall energy of the circuits. However, it is pertinent to note that every negation gate incurs energy since at least one among the input wire or output wire will carry a value of 1 for any input. This is true even if they are directly connected to input variables, because we can fix input variables to zeros to activate them. Thus, a circuit that uses a large number of negation gates must have large energy too. Despite this, it is conceivable that negation gates can be used carefully in the design to reduce the energy of the circuit. However, to begin with, it is even unclear *if negations ever help decrease the energy dissipated in the monotone gates of the circuit at all?*. This is the main question that we study in this paper.

To investigate the power of the negations in the context of energy complexity, we consider a basis $\mathcal{N}_*$ consisting of the conjunction and disjunction with unbounded fan-in, where any input variable can be negated, thus containing versions of the gates $\wedge$ and $\vee$. Note that any circuit C over the basis $\mathcal{B}_*$ can be converted to a circuit over the basis $\mathcal{N}_*$ by absorbing the negation gates to the gates that they feed into, choosing the appropriately negated version of the $\wedge$ and $\vee$ gates. Thus, a natural way to attack this question is by comparing the energy complexity with respect to the basis $\mathcal{N}_*$ and $\mathcal{B}_*$. Informally, does the availability of "free" negations help us decrease the energy of circuits?

Power of $\mathcal{N}_*$-Circuits: We begin with the observation that the energy over the new basis is remarkably different: even constant-energy circuits are universal.

Theorem 1. *For any function f, there exists a $\mathcal{N}_*$-circuit of size $|f^{-1}(0)|$ and energy one. Thus, $\mathsf{EC}_{\mathcal{N}_*}(n) = 1$.*

In contrast, we note that PAR_n (defined as $\bigoplus_{i\in[n]}$) function cannot be computed by a $\mathcal{B}_*$-circuit of energy $o(\sqrt{n})$, and MUX_n cannot be computed by a $\mathcal{B}_*$-circuit of energy $o(\sqrt{\log n})$, where the gates in the circuit are not with bounded fan-in (and hence the previous lower bounds do not apply).

We then demonstrate that the presence of "free" negations has an impact on the related parameters as well. For a Boolean circuit C of size s with gates $g_1, g_2, \ldots, g_s$, and an input assignment $\mathbf{a} \in \{0,1\}^n$, we define the s-dimensional vector $(g_1(\mathbf{a}), g_2(\mathbf{a}), \ldots, g_s(\mathbf{a})) \in \{0,1\}^s$ as an output pattern corresponding to $\mathbf{a} \in \{0,1\}^n$. For a circuit C, let $P(C)$ denote the set of possible output patterns of C. While it is known that any $\mathcal{B}_*$-circuit of energy e has at most $2^{O(e \log e)}$ output patterns [4], a $\mathcal{N}_*$-circuit can go beyond this upper bound. More precisely, we show that there exists a DeMorgan circuit over the basis $\mathcal{N}_*$ of size n, energy e, and $(n/e+1)^e$ output patterns.

Theorem 2. *There exists a DeMorgan circuit over the basis $\mathcal{N}_*$ of size n, energy e, and $(n/e+1)^e$ output patterns.*

We also relate $\mathcal{N}_*$-circuits to decision trees. Constructing energy efficient circuits over the basis $\mathsf{EC}_{\mathcal{B}_*}(f)$ was done by [6] in order to show that $\mathsf{EC}_{\mathcal{B}} \leq \mathsf{DT}(f)^2$ where $\mathsf{DT}(f)$ is the height of the smallest decision tree computing f. from decision tree.

We show that, in the new basis, the $\mathsf{EC}_{\mathcal{N}_*}(f)$ is related to the *rank* of decision trees computing f. The *rank* of a decision tree T is inductively defined as follows: if T consists of a single leaf, then $rank(T) = 0$; otherwise, the $rank(T)$ is $rank(T_\ell) + 1$, if $rank(T_\ell) = rank(T_r)$ or else it is $\max\{rank(T_\ell), rank(T_r)\}$, where T_ℓ and T_r are the left and right subtrees of the root of T, respectively. We show that, one can construct energy efficient circuits over the basis $\mathcal{N}_*$ from low rank decision trees.

Theorem 3. *If a Boolean function f is computable by a decision tree of size s and rank r, then f is also computable by a $\mathcal{N}_*$-circuit of size $O(s)$ and energy $O(r)$.*

In contrast, we show that such a construction is not possible for circuits over $\mathcal{B}_*$, by showing an explicit function which can be computed by rank 1 decision trees, but requires non-constant energy for any circuit computing it over $\mathcal{B}_*$.

Lower Bounds for Energy $\mathcal{N}_*$-circuits in Terms of Size and Depth: As mentioned above, the "free" negations enable exponential-size $\mathcal{N}_*$-circuits to have small energy, and can indeed be helpful even for polynomial-size circuits computing some Boolean functions. This leads to the natural question if there exists a Boolean function for which the exponential size is necessary when the energy is restricted to a constant. In particular, *are there explicit functions which require large energy over the basis $\mathcal{N}_*$ for any polynomial size (or low depth) circuit computing them?*

A natural approach towards such lower bound is to give a size-independent upper bound on the number of output patterns of an energy bounded circuit and use it to efficiently simulate the circuit using structurally different restrictions (like depth three circuits, as in the case of [4]). However, Theorem 2 shows that this approach fails. Thus, we need a new idea to obtain a lower bound.

Overcoming this difficulty, through different approaches, we show a lower bound on the energy in terms of the size and the depth of the circuit in two restricted cases: DeMorgan circuits and layered circuits. In addition, we also consider high-communication complexity Boolean functions.

Lower Bounds for DeMorgan Circuits: As our first result, we prove a lower bound for De-Morgan circuits[1] computing the PARITY function. For $x \in \{0,1\}^n$, recall that $\mathrm{PAR}_n(\mathbf{x}) = \bigoplus_{i \in [n]} x_i$.

Theorem 4. *Let C be a DeMorgan $\mathcal{N}_*$-circuit computing* PAR_n *with size s and depth d. Then,* $\mathsf{EC}_{\mathcal{N}_*}(C) \geq \frac{n}{d \log s}$.

The lower bounds for DeMorgan circuits show that even if we are allowed to use an unbounded number of negations connected to input variables, any constant-depth and polynomial-size DeMorgan circuit still needs almost linear energy for PAR_n.

Lower Bounds for Layered Circuits: We also consider the case where the circuit is layered and all input gates are connected to the gates in the bottom layer, and prove an energy lower bound.

Theorem 5. *Let C be a $\mathcal{N}_*$-circuit of size s and depth d that computes* $\mathrm{PAR}_{[n]}$. *If every input variable is connected to a gate in the bottom layer, then for sufficiently large n:* $\mathsf{EC}_{\mathcal{N}_*}(C) \geq \frac{n}{d \log s}$.

Lower Bounds from Rank of the Communication Matrix: As mentioned above, energy complexity for threshold circuits are well studied [8]. For threshold circuits, negation gates at the input can be absorbed into the threshold gate itself, and hence does not contribute to energy of the circuit. The techniques developed for showing lower bounds for threshold circuits, therefore, are possibly applicable to $\mathcal{N}_*$-circuits. In the next theorem, we adapt a connection that was recently established between rank of the communication matrix of a function and the energy complexity for threshold circuits [10], and refine it for our setting:

Theorem 6. *If a circuit C on $2n$ variables computes a Boolean function of $f : \{0,1\}^n \times \{0,1\}^n \to \{0,1\}$, and has size s, depth d, then* $\mathsf{EC}_{\mathcal{N}_*}(C) \geq \frac{\log(rk(M_f))}{d \log s}$ *where M_f is the communication matrix corresponding to the function f.*

Since $rk(M_{\mathrm{DISJ}_n}) = 2^n$, we can obtain an almost linear lower bound on energy using the above theorem. In particular, it implies that if a $\mathcal{N}_*$-circuit C is a polynomial-size and constant-depth circuit computing DISJ_n, then C must have energy at least $\Omega(n/\log n)$. We note that this lower bound is tight up to a constant factor by showing that DISJ_n can indeed be computed by a $\mathcal{N}_*$-circuit of size $O(n^2/\log n)$, depth 3 and energy $O(n/\log n)$.

[1] In this model, the DeMorgan circuits are those where the input-negated version of the gates $\wedge$ and $\vee$ can only have input variables feeding to them in the circuit.

2 Power of $\mathcal{N}_*$-Circuits

In this section, we show that gates equipped with negations are useful for decreasing energy.

2.1 $\mathcal{N}_*$-Circuits Are Universal

We first observe that restriction on energy can be critical for $\mathcal{B}_*$-circuits. In particular, the energy-restricted circuits over $\mathcal{B}_*$ can be simplified by fixing small number of variables to constants.

Lemma 1. *If a Boolean function f is computable by a $\mathcal{B}_*$-circuit of energy e, then we can force f to be constant by fixing at most $e^2 + e$ variables.*

Proof. We follow an argument given in [6]. Let C be a $\mathcal{B}_*$-circuit of energy e. Thus, C has at most e $\neg$-gates. Consider a bottommost $\neg$-gate g. Let C' be a subcircuit whose output is fed to g. Clearly, C' is monotone, and has at most e gates, since the all-1 assignment forces the gates in C' output ones. We then simply apply a gate elimination for C': If a $\wedge$-gate in C' directly receives an input variable x_i, we set $x_i = 0$; and if a $\vee$-gate in C' directly receives an input variable x_i, we set $x_i = 1$. Fixing a single input variable eliminates a gate, and hence the output of g is determined by setting at most e input variables.

Since C has at most e negations, after we repeatedly apply the above procedure at most e times, the resulting circuit must be monotone. Therefore, additional e input variables must force C compute constant function. □

The above proposition contrasts with the following claim (Theorem 1 from the introduction) showing that even energy-one $\mathcal{N}_*$-circuits are universal.

Theorem 1. *For any function f, there exists a $\mathcal{N}_*$-circuit of size $|f^{-1}(0)|$ and energy one. Thus,* $\mathsf{EC}_{\mathcal{N}_*}(n) = 1$.

Proof. Consider the negation $\bar{f}$ of f, and writing $\bar{f}$ in the sum of products form from the truth table. It can be verified that, other than the literals, for any input $\mathbf{a} \in \{0,1\}^n$ such that $f(\mathbf{a}) = 0$, exactly one $\wedge$ gate outputs 1 other than the root gate itself (which is an $\vee$-gate). We complete the construction simply by replacing the root $\vee$-gate with a negated $\vee$ gate. The resulting circuit computes f. In addition, for any input $\mathbf{a} \in \{0,1\}^n$ such that $f(\mathbf{a}) = 0$, exactly one $\wedge$ gate outputs 1 while for any input $\mathbf{a} \in \{0,1\}^n$ such that $f(\mathbf{a}) = 1$, only the root gate outputs 1. This establishes the proposition.

These two theorems immediately imply that the parity function is an example showing a distinction between $\mathcal{B}_*$-circuits and $\mathcal{N}_*$-circuits.

Corollary 1. PAR_n *cannot be computed by a $\mathcal{B}_*$-circuit of energy $o(\sqrt{n})$ irrespective of the size of the circuit, but can be computed by a $\mathcal{N}_*$-circuit of size 2^{n-1} and energy one.*

We can also observe that their computable functions are different for polynomial-size setting. We define the multiplexer MUX_n of $\log n + n$ input variables as follows: For $\mathbf{x} \in \{0,1\}^{\log n}$ and $\mathbf{y} \in \{0,1\}^n$, $\mathrm{MUX}_n(\mathbf{x}, \mathbf{y}) = y_i$, where $\mathbf{x}$ is the binary representation of i.

Theorem 7. MUX_n *cannot be computed by a* $\mathcal{B}_*$*-circuit of energy* $o(\sqrt{\log n})$*, but can be computed by a* $\mathcal{N}_*$*-circuit of size* $O(n)$*, depth two and energy two.*

Proof. Lemma 1 implies the first claim, since MUX_n cannot be constant no matter how we fix $\log n$ input variables.

Also, the standard construction for MUX_n gives the latter claim. Let $g_0, \ldots, g_{n-1}$ be gates in $\mathcal{B}$ such that g_j outputs one for $\mathbf{a} \in \{0,1\}^{\log n}$ and $\mathbf{y} \in \{0,1\}^n$ if and only if $\mathbf{a}$ is the binary representation of j and $y_j = 1$. Thus, g_j is a conjunction of $x_1, \ldots, x_{\log n}$ and y_j where some of $x_1, \ldots, x_{\log n}$ are negated. We complete the construction of C by taking a disjunction of $g_0, \ldots, g_{n-1}$. Clearly, C has size $n+1$. Since at most one of $g_0, \ldots, g_{n-1}$ together with the top gate output ones for any input variables, C has energy two. □

Number of Output Patterns: It is known that any $\mathcal{B}_*$-circuit of energy e has at most $2^{O(e \log e)}$ output patterns. Thus, the number of output patterns is determined independently of its size and the number of input variables. A $\mathcal{N}_*$-circuit can go beyond this upper bound, as we show in the following theorem from the introduction.

Theorem 2. *There exists a DeMorgan* $\mathcal{N}_*$*-circuit of size* n*, energy* e*, and* $(n/e+1)^e$ *output patterns.*

Proof. Consider the case where $n = ek$ for some k. Let $I_1, \ldots, I_e$ be a partition of $[n]$, where each of $I_1, \ldots I_e$ has the same size k. For each $j \in [e]$, we are going to make a circuit C_j of size k and energy one such that (i) C_j receives only the input variables in I_j; and (ii) C_j has $k+1$ output patterns, which implies the proposition.

We construct only C_1, because the other circuits are given in the same manner. The circuit C_1 receives $x_1, \ldots, x_k$, and has gates $g_1, \ldots, g_k$. For each $j \in [k]$, g_j receives x_i for $i \in [j-1]$ through a negation, and does x_j directly. Thus, g_j outputs one if and only if $x_1 = \cdots = x_{j-1} = 0$, but $x_j = 1$. C_1 has $k+1$ output patterns where no gate outputs one or only g_j outputs one.

2.2 Energy Upper Bounds from Rank-Bounded Decision Trees

In this section, we show that the $\mathsf{EC}_{\mathcal{N}_*}(f)$ is related to optimal rank of decision trees computing f. A decision tree T is a rooted binary tree in which each internal node is labeled by a variable and each leaf is labeled by zero or one. Given an input $\mathbf{a} \in \{0,1\}^n$ to T, the output $T(\mathbf{x})$ is determined by the following procedure starting from the root until reaching a leaf: if the input variable at the current node has the value zero in $\mathbf{a}$, then go to the left child; otherwise go to the right. If we reach a leaf labeled by $z \in \{0,1\}$, then $T(\mathbf{a}) = z$.

For a decision tree T of n input variables $x_1, \ldots, x_n$, we denote by a binary string a node (or a leaf) in T, as follows. We denote by the empty string ϵ the root of T. For a binary string $\mathbf{s} \in \{0,1\}^*$, we denote by $\mathbf{s}0$ and $\mathbf{s}1$ the left child and right child, respectively. Then we can identify T with a pair of sets S_T and L_T, where S_T is a set of strings representing the nodes in T, and L_T is a set of strings representing the leaves in T. For $\mathbf{s} \in S_T$, we write $x_{\mathbf{s}}$ for the query on the node $\mathbf{s}$. Similarly, we write $\ell_{\mathbf{s}} \in \{0,1\}$ for the value on the leaf $\mathbf{s}$. For an assignment $\mathbf{a} \in \{0,1\}^n$, we denote by $p(\mathbf{a})$ a string showing the computation path of T on $\mathbf{a}$.

For two strings $\mathbf{t} \in S_T$ and $\mathbf{s} \in S_T \cup L_T$, we write $\mathbf{t} \prec \mathbf{s}$ if $\mathbf{t}$ is a proper prefix of $\mathbf{s}$, implying that the node $\mathbf{t}$ is an ancestor of the node (or the leaf) $\mathbf{s}$. Note that the empty string ϵ is a proper prefix of any other string.

Recall the definition of size (number of leaves) and rank of a decision tree from the introduction. We first show that even functions computed by rank one decision trees may require high energy for $\mathcal{B}_*$-circuits to compute them.

Theorem 8. *There exists a n-variate Boolean function f such that a rank-1 decision tree can computes f, while any $\mathcal{B}_*$-circuit has energy $\Omega(\log n)$ to compute f.*

Proof. Consider the function defined as follows. $f(\mathbf{x}) = 0$ if $x_1 = 1$. Otherwise, $f(\mathbf{x}) = 0$ if there exists odd $i \in [n-1]$ such that $x_1 = \cdots = x_{i-1} = 0$ and $x_i = 1$, and otherwise $f(\mathbf{x}) = 1$. Clearly, f can be computed by a rank-1 decision tree. We below argue that this function requires energy $\Omega(\log n)$ for any circuit over the basis $\mathcal{B}_*$. It suffice to argue that any circuit computing f will require at least $\Omega(\log n)$ negation gates.

The *alternation* of a Boolean function is a measure of non-monotonicity of the Boolean function. More precisely, if we define a collection of distinct inputs $\alpha_0, \alpha_1, \alpha_2 \ldots, \alpha_n \in \{0,1\}^n$ satisfying $0^n = \alpha_0 \prec \alpha_1 \prec \alpha_2 \prec \cdots \prec \alpha_n = 1^n$ as a *chain* in the Boolean hypercube on $\{0,1\}^n$ then, alternation of f (denoted by $\mathsf{alt}(f)$) is defined as $\max\{\mathsf{alt}(f, \mathcal{C}) \mid \mathcal{C} \text{ is a chain in } \{0,1\}^n\}$ where $\mathsf{alt}(f, \mathcal{C}))$ is $|\{i \mid f(\alpha_{i-1}) \neq f(\alpha_i), x_i \in \mathcal{C}, i \in [n]\}|$. By a surprising result due to Markov [3], it was shown that for any Boolean function f, any circuit C computing f must have $\Omega(\log \mathsf{alt}(f))$ negations irrespective of the size of the circuit. Thus, for our purpose, it is sufficient to argue that the above function has $\mathsf{alt}(f) = n$. This can be seen by considering the chain of inputs $0^n \prec 0^{n-1}1 \prec 0^{n-2}11 \ldots \prec 1^n$. The value of the function will switch between 0 and 1 in this chain as per the definition and hence $\mathsf{alt}(f) = n$. Hence any Boolean circuit over the basis $\mathcal{B}_*$ must use $\Omega(\log n)$ negations and hence must have $\Omega(\log n)$ energy. □

On the other hand, an energy-bounded $\mathcal{N}_*$-circuit can simulate a decision tree T of rank r. The following theorem follows from an argument in [8].

Theorem 3. *If a Boolean function f is computable by a decision tree of size s and rank r, then f is also computable by a $\mathcal{N}_*$-circuit of size $O(s)$ and energy $O(r)$.*

Proof. Let T be a decision tree computing f. If $S_T = \emptyset$, the statement is true, since T computes a constant function. Below we thus assume that $S_T \neq \emptyset$.

It is known that rank is related to the number of ones in $\mathbf{s} \in L_T$.

Claim ([8]). If a Boolean function f can be computed by a decision tree T of size s and rank r, then f also can be computed by a decision tree T' of size s, where the number of ones in any string $\mathbf{s} \in L_{T'}$ is at most r.

Thus, there exists a decision tree T' of size s such that the number of ones in any string $\mathbf{s} \in L_{T'}$ is at most r. We construct from T' the desired $\mathcal{N}_*$-circuit C consisting of a gate $g_\mathbf{s}$ for every $\mathbf{s} \in S_{T'}$ together with an additional gate g whose output gives the circuit output.

For each $\mathbf{s} \in S_{T'}$, we make a $\wedge$-gate $g_\mathbf{s}$ with fan-in $|\mathbf{s}| + 1$, as follows. The gate $g_\mathbf{s}$ receives the outputs of $g_\mathbf{t}$ for each $\mathbf{t} \prec \mathbf{s}$, and it is negated if $\mathbf{t}0$ is a prefix of $\mathbf{s}$. The gate $g_\mathbf{s}$ also directly receives the input variable $x_\mathbf{s}$. The circuit output gate g receives the output of every gate $g_\mathbf{t}$ for $\mathbf{t}$ satisfying that either "$\mathbf{t}0 \in L_{T'}$ and $\ell_{\mathbf{t}0} = 1$" or "$\mathbf{t}1 \in L_{T'}$ and $\ell_{\mathbf{t}1} = 1$" holds, and the output is negated for the former case. This completes the construction.

By construction, the gate $g_\mathbf{t}$, $\mathbf{t} \in S_{T'}$, outputs 1 for a given assignment $\mathbf{a} \in \{0,1\}^n$ if and only if $\mathbf{t}1 \prec p(\mathbf{a})$, and g outputs one if and only if $\ell_{p(\mathbf{a})} = 1$. Thus, the resulting circuit has size $|S_{T'}| + 1$ and energy $r + 1$, as desired.

Bounding on the rank of the decision tree by logarithm of its size, we have:

Corollary 2. *Any Boolean function f that is computable by a decision tree of size s is also computable by a circuit over $\mathcal{N}_*$ of size $O(s)$ and energy $O(\log s)$.*

3 Limitations of $\mathcal{N}_*$-Circuits

In this section, we investigate if PAR_n is still hard for energy-bounded $\mathcal{N}_*$-circuit. As mentioned in the introduction, one of the successful approaches towards showing limitations of energy bounded circuits is by bounding the number of output patterns of the circuit [4]. However, as we have shown in the last section, a $\mathcal{N}_*$-circuit cave a large number of output patterns compared to $\mathcal{B}_*$-circuits. Therefore, we need to develop an argument avoiding bounds on output patterns.

We say that a gate g is open if g directly receives an input variable, and otherwise, we say that g is closed. We also define $J^C(g)$ as a set of the gates whose outputs are connected to g in C. For a Boolean function f of n variables, we denote by $S_0(f)$ (resp., $S_1(f)$) the number of assignments $\mathbf{a} \in \{0,1\}^n$ satisfying $f(\mathbf{a}) = 0$ (resp., $f(\mathbf{a}) = 1$). For a gate g in a $\mathcal{N}_*$-circuit C, we similarly define $S_0^C(g)$ (resp., $S_1^C(g)$) the number of assignments $\mathbf{a} \in \{0,1\}^n$ such that the output of g in C for $\mathbf{a}$ is zero (resp., one).

3.1 Lower Bounds for DeMorgan Circuits

For $I \subseteq [n]$, we define $\mathrm{PAR}_I(\mathbf{x}) = \bigoplus_{i \in I} x_i$. Thus, for any $I \subseteq [n]$, we have $S_0(\mathrm{PAR}_I) = 2^{|I|-1}$ and $S_1(\mathrm{PAR}_I) = 2^{|I|-1}$.

Theorem 4. *Let C be a DeMorgan $\mathcal{N}_*$-circuit computing* $\mathrm{PAR}_{[n]}$*. Then* $\mathsf{EC}_{\mathcal{N}_*}(C) \geq n/(d \log s)$.

Proof. Let C be a circuit computing $\mathrm{PAR}_{[n]}$, and s and d be the size and depth of C, respectively. We construct an input assignment $\mathbf{a}^*$ for which at least $n/(d \log s)$ gates output ones in C. Let $I = [n]$.

While $d \log s \leq |I|$, we repeatedly apply a procedure to find a partial assignment of the desired $\mathbf{a}^*$, and obtain a circuit C' computing $\mathrm{PAR}_{I'}$ for some $I' \subseteq I$ towards the next step (that is, we consider C' and I' as new C and I). We consider two cases (i) and (ii). In the case (i), we are going to set a single input variable, while, in the case (ii), we set $d \log s$ input variables.

Case (i): *There exists an open* $\vee$*-gate in* C :.
Let g^* be an open $\vee$-gate. Without loss of generality, let x_i be an input variable connected to g^*. We set $a_i^* = 1$ if x_i is directly connected to g^*. Otherwise (that is, x_i is connected to g^* through negation), we set $a_i^* = 0$. Note that g^* outputs one no matter how we fix the rest of the input variables, and the resulting circuit C' computes $\mathrm{PAR}_{I'}$ (or its complement) for $I' = I \backslash \{i\}$. We proceed to the next with these C' and I'.

Case (ii): *Every* $\vee$*-gate is closed in* C.
In this case, every gate in the bottom layer is an $\wedge$-gate. In addition, since C is a DeMorgan circuit, every negation appears at an input variable. Thus, C is an alternating circuit. For each $\ell \in [d]$, we define G_ℓ as a set of the gates in the l-th layer. Thus, G_ℓ consists of $\wedge$-gates if ℓ is odd, and of $\vee$-gates if ℓ is even. We now prove that there exists a gate $g^* \in G_1$ with fanin at most $d \log s$.
Suppose for the sake of contradiction that every $\wedge$-gate $g \in G_1$ has fanin larger than $d \log s$. Since PAR_I is of $|I|$ input variables, it holds that $S_1^C(g) \leq 2^{|I| - d \log s} = \frac{2^{|I|}}{s^d}$ for every $g \in G_1$. From this basis, we can inductively show that any gate in $g \in G_\ell$ satisfies that $S_1^C(g) \leq \frac{2^{|I|}}{s^{d-\ell+1}}$, as follows. Suppose for the induction hypothesis that, for any gate in $g \in G_{\ell-1}$, it holds that $S_1^C(g) \leq \frac{2^{|I|}}{s^{d-\ell+2}}$. If ℓ is odd, any gate $g \in G_\ell$ is an $\vee$-gate. Then g is closed and has fanin at most s. Thus, we have $|S_1^C(g)| \leq \sum_{g' \in J^C(g)} |S_1^C(g')| \leq s \cdot \frac{2^{|I|}}{s^{d-\ell+2}} = \frac{2^{|I|}}{s^{d-\ell+1}}$, as desired. If ℓ is even, any gate $g \in G_\ell$ is a $\wedge$-gate. Thus, $|S_1^C(g)| \leq \min_{g' \in J^C(g)} |S_1^C(g')| \leq \frac{2^{|I|}}{s^{d-\ell+2}}$, as desired. Consequently, $S_1(g)$ for the top gate g, that is, the number of the satisfying assignments of C, is bounded by $2^{|I|}/s$. Thus, $\frac{2^{|I|}}{s} < 2^{|I|}$, and hence this contradicts the fact that C computes PAR_I.
Therefore, there exists an $\wedge$-gate $g^* \in G_1$ such that $|J^C(g^*)| \leq d \log s$. Let K be a set of indices i such that x_i is connected to g^*. We fix a_i^* for every $i \in K$ so that we can make g^* output one no matter how the rest input variables are fixed. Let C' be a circuit obtained by fixing x_i for each $i \in K$ so that g^* outputs one. Let $I' = I \backslash K$. We continue the construction on C' and I'.

Thus, in either case, we fix at most $d \log s$ input variables. We repeat this procedure while $d \log s \leq |I|$, and lastly complete $\mathbf{a}^*$ by arbitrarily fixing the free variables. Since each application of the procedure above provides an unique gate g^* that outputs one for $\mathbf{a}^*$, we can repeat the procedure at most e times. Therefore, we have $n \le ed \log s$, as desired. □

Remark 1. Using a similar argument, we can obtain the lower bound not only for $\mathrm{PAR}_{[n]}$ but also for other Boolean functions. Let $f_{[n]}$ be a Boolean function over x_i, $i \in [n]$. For $I \subseteq [n]$ and a partial assignment $\mathbf{a} \in \{0,1\}^I$, let $f_I[\mathbf{a}]$ be a subfunction over x_i, $i \in [n]\backslash I$, obtained from $f_{[n]}$ by fixing x_i, $i \in I$, to $\mathbf{a}$. We can obtain the lower bound for $f_{[n]}$ if $f_I[\mathbf{a}]$ is not constant and it holds that $2^{\Omega(n)} \le |S_1^C(f_I[\mathbf{a}])|$ for every pair of $I \subseteq [n]$ and $\mathbf{a} \in \{0,1\}^I$.

3.2 Lower Bound for Layered Circuits

In the previous section, we considered DeMorgan $\mathcal{N}_*$-circuits, and show an upper bound by a bottom-up argument. In this section, we study layered $\mathcal{N}_*$-circuits where there may exist an internal gate with negations. We can obtain a similar lower bound in this case by a top-down argument, and prove Theorem 5 from the introduction.

Theorem 5. *Let C be a $\mathcal{N}_*$-circuit of size s and depth d that computes $\mathrm{PAR}_{[n]}$. If every input variable is connected to a gate in the bottom layer, then C has energy at least $n/(d \log s)$ for a sufficiently large n.*

Proof. Let C be a $\mathcal{N}_*$-circuit computing $\mathrm{PAR}_{[n]}$, and s, d and e be the size, depth and energy of C, respectively. We construct an input assignment $\mathbf{a}^*$ for which at least $n/(d \log s)$ gates output ones in C.

Let $I = [n]$. Similarly to the proof of Theorem 4, while $d \log s < |I|$, we repeatedly apply a procedure to find a partial assignment of the desired $\mathbf{a}^*$, and obtain a circuit C' computing $f_{I'}$ for some $I' \subseteq I$ for the next (that is, we consider C' and I' as new C and I). If C has a open $\vee$-gate, we can find such C' and I' by following the case (i) in the proof of Theorem 4. Below we assume that every $\vee$-gate is closed.

We now prove by induction on depth that, for every ℓ, there exists a gate g such that both $\frac{2^{|I|-1}}{s^{d-\ell}} \le |S_0^C(g)|$ and $\frac{2^{|I|-1}}{s^{d-\ell}} \le |S_1^C(g)|$ hold. Our induction proceeds from the top to the bottom.

Consider the base case where g is the top gate of C. In this case, since C computes $\mathrm{PAR}_{[n]}$, we have $2^{|I|-1} \le |S_0^C(g)|$ and $2^{|I|-1} \le |S_1^C(g)|$, as desired. Suppose for the induction hypothesis that there exists a gate $g \in G_{\ell+1}$ such that $\frac{2^{|I|-1}}{s^{d-\ell-1}} \le |S_0^C(g)|$ and $\frac{2^{|I|-1}}{s^{d-\ell-1}} \le |S_1^C(g)|$. We now give a procedure to choose the desired gate $g' \in G_\ell$ among the gates whose outputs are connected to g.

Consider the case where g is an $\vee$-gate. Then, for each $\mathbf{a} \in S_1^C(g)$, at least one of the gates connected to g outputs one for $\mathbf{a}$. Since the fan-in of g is at most s, we can choose a gate g' such that both $|S_0(g')|$ and $S_1(g')|$ are large, as follows. If the output of such a g' is directly connected to g, we have, $\frac{2^{|I|-1}}{s^{d-\ell-1}} \le |S_0^C(g)| \le |S_0^C(g')|$ and $\frac{2^{|I|-1}}{s^{d-\ell}} \le \frac{|S_1^C(g)|}{s} \le |S_1^C(g')|$. If the output of such a g'

is connected to g through negation, we have $\frac{2^{|I|-1}}{s^{d-\ell}} \leq \frac{|S_0^C(g)|}{s} \leq |S_0^C(g')|$ and $\frac{2^{|I|-1}}{s^{d-\ell-1}} \leq |S_1^C(g)| \leq |S_1^C(g')|$.

Consider the other case where g is an $\wedge$-gate. Similarly to the above, for each $\mathbf{a} \in S_0^C(g)$, at least one of the gates connected to g outputs zero for $\mathbf{a}$. Thus, we can choose a gate g' such that both $|S_0(g')|$ and $S_1(g')|$ are large. If the output of such a g' is directly connected to g, we have $\frac{2^{n-1}}{s^{d-\ell}} \leq \frac{|S_0^C(g)|}{s} \leq |S_0^C(g')|$ and $\frac{2^{n-1}}{s^{d-\ell-1}} \leq |S_1^C(g)| \leq |S_1^C(g')|$ If the output of such a g' is connected to g through negation, we have $\frac{2^{n-1}}{s^{d-\ell-1}} \leq |S_0^C(g)| \leq |S_0^C(g')|$ and $\frac{2^{n-1}}{s^{d-\ell}} \leq \frac{|S_1^C(g)|}{s} \leq |S_1^C(g')|$

Consequently, we can conclude that there exists a $\wedge$-gate $g^* \in G_1$ such that $\frac{2^{n-1}}{s^{d-1}} \leq |S_1^C(g^*)|$ which implies that $\frac{2^{n-1}}{s^{d-1}} \leq 2^{n-|J(g^*)|}$ Thus, the fan-in of g^* is at most $d \log s$. Let K be a set of indices i such that x_i is connected to g^*. We set a_i^* for every $i \in K$ so that we can make g^* output one no matter how the rest input variables are fixed.

Since each application of the procedure above provides an unique gate g^* that outputs one for $\mathbf{a}^*$, we can apply it at most e times. Therefore, we have $n \leq ed \log s$, as desired. □

Remark 2. We can obtain the lower bound not only for $\mathrm{PAR}_{[n]}$, but also for other Boolean functions: We can obtain the bound for $f_{[n]}$ if it holds that $2^{\Omega(n)} \leq |S_1^C(f_I[\mathbf{a}])|$ and $2^{\Omega(n)} \leq |S_0^C(f_I[\mathbf{a}])|$ for every pair of $I \subseteq [n]$ and $\mathbf{a} \in \{0,1\}^I$.

Theorem 5 implies the following corollary.

Corollary 3. *Let C be a layered circuit of size $(2^{\sqrt{n}}$ and constant depth that computes $\mathrm{PAR}_{[n]}$. Then $\mathsf{EC}_{\mathcal{N}_*}(C) = \Omega(\sqrt{n})$.*

The following theorem shows tightness of Theorem 5 for layered $\mathcal{N}_*$-circuits.

Theorem 9. $\mathrm{PAR}_{[n]}$ *is computable by a layered $\mathcal{N}_*$-circuit C of size $O(e2^{n/e} + 2^e)$, depth four, and energy $O(e)$. In particular,* $\mathrm{PAR}_{[n]}$ *is computable by a layered $\mathcal{N}_*$-circuit C of size $O(\sqrt{n}2^{\sqrt{n}})$, depth four, and energy $O(\sqrt{n})$.*

Proof. Suppose for simplicity that n is a multiple of e. Let $I_1, I_2, \ldots, I_e$ be a partition of $[n]$, where each of $I_1, I_2, \ldots, I_e$ has n/e integers. For each $j \in [e]$, by Theorem 1, we can construct a $\mathcal{N}_*$-circuit of size $O(2^{n/e})$, depth two, and energy one that computes $\mathrm{PAR}_{I_j}(\mathbf{a}) = \bigoplus_{i \in I_j} a_i = 1$ for every $\mathbf{a} \in \{0,1\}^n$. Since Theorem 1 also implies that the parity of the e outputs of $C_1, C_2, \ldots, C_e$ is computable by a circuit of size $O(2^e)$, depth two and energy one, $\mathrm{PAR}_{[n]}$ can be computed by a circuit of size $O(e2^{n/e})$, depth four and energy $O(e)$. We can verify the second statement by setting $e = \sqrt{n}$. □

3.3 Lower Bounds from the Rank of the Communication Matrix

In this section, we give almost linear lower bounds on energy complexity of $\mathcal{N}_*$-circuits. Let f be a Boolean function of $2n$ variables. A communication matrix M_f of f is a $2^n \times 2^n$ matrix, where each row and column are indexed by $\mathbf{a} \in$

$\{0,1\}^n$ and $\mathbf{b} \in \{0,1\}^n$, respectively, and each entry for $\mathbf{a} \in \{0,1\}^n$ and $\mathbf{b} \in \{0,1\}^n$, $M_f(\mathbf{a},\mathbf{b}) = f(\mathbf{a},\mathbf{b})$. We denote by $rk(M_f)$ the rank of M_f. If a circuit C (or a gate g) computes f, we may write M_C (or M_g) for the corresponding communication matrix.

Let C be a $\mathcal{N}_*$-circuit. Let G_C be a set of the gates in C. For every gate $g \in G_C$, let $I(g)$ be a set of input variables that are connected to g, and $I'(g)$ be a set of the gates in G_C whose outputs are connected to g. Thus, $I(g) \cup I'(g)$ is a set of the inputs of g. We then define $S(g)$ as a set of gates we can obtain from g by fixing the outputs of the gates in $I'(g)$: If g is a $\wedge$-gate, $S(g)$ contains a $\wedge$-gate over $I(g)$ and a gate computing constant zero; If g is a $\vee$-gate, $S(g)$ contains a $\vee$-gate over $I(g)$ and a gate computing constant one. We then define a set G_C^* as $G_C^* = \bigcup_{g \in G_C} S(g)$. We then define $r_C = \max_{g_1,\ldots,g_e \in G_C^*} rk(M_{\wedge[g_1,\ldots,g_e]})$ where $\wedge[g_1,\ldots,g_e]$ is a conjunction of $g_1,\ldots,g_e$.

Uchizawa and Abe [10] implicitly prove the following proposition:

Proposition 1 ([10]). *Let C be a circuit of size s, depth d and energy e. Then it holds that $rk(M_C) \leq (s^e \cdot r_C)^{O(d)}$.*

We give a proof of the proposition in Appendix for completeness. In the paper [10], they prove for threshold circuits that $r_C \leq (swn)^e$ holds. For $\mathcal{N}_*$-circuits, we can improve the bound to $r_C \leq 2^e$, which immediately implies an upper bound on $rk(M_C)$:

Theorem 10. *If a $\mathcal{N}_*$-circuit C computes a Boolean function of $2n$ variables, has size s, depth d and energy e, then it holds that $\log(rk(M_C)) = O(ed \log s)$.*

Proof. Proposition 1 implies that it suffices to give a bound on r_C. Clearly, G_C^* consists of $\wedge$-gate of $O(n)$ literals, $\vee$-gate of $O(n)$ literals and constant functions. Thus, the rank of M_g is at most two for any $g \in G_C^*$: If g is a $\wedge$-gate or a constant function, $rk(M_g) = 1$; and if g is a $\vee$-gate, $rk(M_g) = 2$. Note that, for any e tuples of $g_1,\ldots,g_e \in G_C^*$, we have $\wedge[g_1,\ldots,g_e] = M_{g_1} \circ M_{g_2} \circ \cdots \circ M_{g_e}$, where $A \circ B$ denotes the Hadamard product of two matrices A and B. Since $rk(A \circ B) \leq rk(A) \cdot rk(B)$, we have $rk(M_{f[g_1,\ldots,g_e]}) \leq 2^e$. Consequently, the proposition implies that $rk(M_C) \leq (s^e \cdot 2^e)^{O(d)}$. Taking its logarithm, we obtain the desired bound on $\log(rk(M_C))$. □

Let $f : \{0,1\}^n \to \{0,1\}$ be a Boolean function. Let $g : \{0,1\}^2 \to \{0,1\}$ by a 2-bit gadget function. Define $F_g : \{0,1\}^n \times \{0,1\}^n \to \{0,1\}$ defined as $F_g(x,y) = f(z_1,z_2,\ldots z_n)$ where $z_i = g(x_i,y_i)$. To simplify notation, we will write M_f^g for the matrix M_{F_g}. Let M_f^* be a matrix that satisfies $rk(M_f^*) = \max(rk(M_f^\wedge), rk(M_f^\vee))$. Shrestov [5] shows that for any Boolean function f, the $\deg(f)$ is upper bounded by $rk(M_f^*)$.

Corollary 4. *If a circuit C of size s, depth d, computes a Boolean function F such that $M_F = M_f^*$, then it holds that $\mathsf{EC}_{\mathcal{N}_*}(C) \geq \Omega\left(\frac{\deg(f)}{d \log s}\right)$*

Application to Disjointness Function: For $I \subseteq [n]$, the function DISJ_n is defined as $\mathrm{DISJ}_I = \bigwedge_{i \in I} \overline{x_i} \vee \overline{y_i}$. Consider f to be the $\vee$ function. Applying

Corollary 4, notice that $\mathrm{DISJ}[n]$ satisfies that $\mathsf{M}_{\mathrm{DISJ}_n} = rk(M_f^*)$ and hence, $\mathsf{EC}_{\mathcal{N}_*}(C)$ is at least $\Omega\left(\frac{\deg(\vee_n)}{d \log s}\right)$. Thus, we get that $\mathsf{EC}_{\mathcal{N}_*}(C)$ computing the disjointness must be at least $\Omega\left(\frac{n}{\log n}\right)$. Indeed, this can also be derived directly by observing that $rk(M_{\mathrm{DISJ}_n}) \geq 2^n$. In either case, we get the following:

Corollary 5. *Let C be a polynomial-size and constant-depth circuit computing* $\mathrm{DISJ}_{[n]}$. *Then* $\mathsf{EC}_{\mathcal{N}_*}(C) = \Omega\left(\frac{n}{\log n}\right)$.

The lower bound is tight up to a constant factor, as we show below

Theorem 11. $\mathrm{DISJ}_{[n]}$ *is computable by a DeMorgan $\mathcal{N}_*$-circuit C of size $O(e2^{2n/(e-1)})$, depth 3, and energy $O(e)$. In particular,* $\mathrm{DISJ}_{[n]}$ *is computable by a DeMorgan $\mathcal{N}_*$-circuit C of size $O(n^2/\log n)$, depth three, and energy $O(n/\log n)$*

Proof. Suppose for simplicity that n is a multiple of e. Let $I_1, I_2, \ldots, I_e$ be a partition of $[n]$, where each of $I_1, I_2, \ldots, I_e$ has n/e integers. By Theorem 1, for each $j \in [e]$, we can construct a depth-2 circuit D_j of size $2^{2n/e}$ and energy one that computes DISJ_{I_j}. Thus, taking a conjunction of $D_1, \ldots, D_e$, the resulting circuit has size $O((e-1)2^{n/(e-1)})$, depth 3 and energy $O(e)$. Setting $e = 1 + n/(2\log n)$, we have the desired circuit for the second statement. □

Acknowledgement. The authors thank the anonymous reviewers for the useful comments which improved the presentation and readability of the paper.

References

1. Dinesh, K., Otiv, S., Sarma, J.: New bounds for energy complexity of boolean functions. Theoret. Comput. Sci. **845**, 59–75 (2020)
2. Kasim-zade, O.M.: On a measure of active circuits of functional elements (Russian). Math. Probl. Cybern. Nauka **4**, 218–228 (1992)
3. Markov, A.A.: On the inversion complexity of a system of functions. J. ACM **5**(4), 331–334 (1958)
4. Sarma, J., Uchizawa, K.: Energy and output patterns in Boolean circuits. In: Chen, X., Li, B. (eds.) Theory and Applications of Models of Computation, pp. 185–196. Springer Nature Singapore, Singapore (2024)
5. Sherstov, A.A.: On quantum-classical equivalence for composed communication problems. Quantum Info. Comput. **10**(5), 435–455 (2010)
6. Sun, X., Sun, Y., Wu, K., Xia, Z.: On the relationship between energy complexity and other Boolean function measures. J. Comb. Optim. **43**, 1470–1492 (2022)
7. Suzuki, A., Uchizawa, K., Zhou, X.: Energy and fan-in of logic circuits computing symmetric Boolean functions. Theoret. Comput. Sci. **505**, 74–80 (2013)
8. Uchizawa, K., Douglas, R.J., Maass, W.: On the computational power of threshold circuits with sparse activity. Neural Comput. **18**(12), 2994–3008 (2008)
9. Uchizawa, K.: Size, depth and energy of threshold circuits computing parity function. In: Proceedings of 31st International Symposium on Algorithms and Computation (ISAAC 2020), vol. 181, pp. 1–13 (2020)

10. Uchizawa, K., Abe, H.: Exponential lower bounds for threshold circuits of sublinear depth and energy. In: 48th International Symposium on Mathematical Foundations of Computer Science (MFCS 2023), volume 272 of Leibniz International Proceedings in Informatics (LIPIcs), pp. 1–15 (2023)
11. Uchizawa, K., Takimoto, E.: Exponential lower bounds on the size of constant-depth threshold circuits with small energy complexity. Theoret. Comput. Sci. **407**(1–3), 474–487 (2008)
12. Vaintsvaig, M.N.: On the power of networks of functional elements (Russian). Doklady Akademii Nauk **139**(2), 320–323 (1961)

Cyclic Equalizability of Words and Its Application to Card-Based Cryptography

Kazumasa Shinagawa[1,3](✉) and Koji Nuida[2,3]

[1] University of Tsukuba, Ibaraki, Japan
shinagawa@cs.tsukuba.ac.jp
[2] Institute of Mathematics for Industry, Kyushu University, Fukuoka, Japan
[3] National Institute of Advanced Industrial Science and Technology, Tokyo, Japan

Abstract. Card-based cryptography is a research area to implement cryptographic procedures using a deck of physical cards. In recent years, it has been found to be related to finite group theory and algebraic combinatorics, and is becoming more and more closely connected to the field of mathematics. In this paper, we discuss the relationship between card-based cryptography and combinatorics on words for the first time. In particular, we focus on cyclic equality of words. We say that a set of words are cyclically equalizable if they can be transformed to be cyclically equal by repeated simultaneous insertion of letters. The main result of this paper is to show that two binary words of equal length and equal Hamming weight are cyclically equalizable. As applications of cyclic equalizability to card-based cryptography, we describe its applications to the information erasure problem and to single-cut full-open protocols.

Keywords: Card-based cryptography · Combinatorics on words

1 Introduction

1.1 Background

Suppose that n players $P_1, P_2, \ldots, P_n$ each having inputs $x_1, x_2, \ldots, x_n \in \{0, 1\}$, respectively, wish to compute the output value $f(x_1, x_2, \ldots, x_n) \in \{0, 1\}$ for some function $f\colon \{0, 1\}^n \to \{0, 1\}$. This is called distributed computing.

Secure computation is a kind of distributed computing where the inputs are not leaked from the protocol execution more than what is computed from the output value. It has many practical applications, e.g., a voting for two candidates: given a candidate 0 and a candidate 1, each player inputs one bit as a vote, and the output value is the index of the candidate who receives a majority of votes. Although secure computation is usually implemented on electronic computers, there is a line of research that implements secure computation using a deck of physical cards, and such a research area is called *card-based cryptography* [1,2, 11].

A. Jeż and J. Otop (Eds.): FCT 2025, LNCS 16106, pp. 406–419, 2026.
https://doi.org/10.1007/978-3-032-04700-7_30

In card-based cryptography, we usually use a deck of cards whose faces are either $\boxed{\clubsuit}$ or $\boxed{\heartsuit}$ and whose backs are the same symbol $\boxed{?}$. Here, the cards with the same symbol are assumed to be indistinguishable. We use a pair of cards to encode one bit information by $\boxed{\clubsuit}\boxed{\heartsuit} = 0$ and $\boxed{\heartsuit}\boxed{\clubsuit} = 1$, and a pair of face-down cards $\boxed{?}\boxed{?}$ encoding $x \in \{0,1\}$ is called a *commitment* to x. Given n commitments to $x_1, x_2, \dots, x_n \in \{0,1\}$, a protocol for $f : \{0,1\}^n \rightarrow \{0,1\}$ computes the output value $f(x_1, x_2, \dots, x_n)$ without revealing the inputs beyond the output by performing a sequence of physical operations such as shuffling.

The history of card-based cryptography began with the five-card trick proposed by den Boer [1] in 1989. The purpose of den Boer's paper itself was to construct a cryptographic protocol, which is assumed to be implemented on electronic computers, and physical cards were used to illustrate the idea of the protocol. In 1994, Crépeau and Kilian [2] showed the feasibility of card-based protocol for any function and developed the basic theory of card-based cryptography. The turning point in card-based cryptography was the proposal of a rigorous computational model of card-based protocols known as the Mizuki–Shizuya model by Mizuki and Shizuya [11]. After Mizuki–Shizuya's paper, this research field was revitalized as it became actively studied both in terms of protocol construction (e.g., [14]) and impossibility proofs regarding the number of cards and the number of operations (e.g., [5,7]). In recent years, the connection with mathematical fields has been deepening such as finite group theory [4,12], algebraic combinatorics [15], and formal method [6]. However, such an attempt has only just begun, and the development of new connections with other mathematical fields is important for the further development of this field.

In this paper, we focus on combinatorics on words [8–10] and study a new notion of *cyclic equalizability* of words. As discussed in Sect. 1.2, this concept appears naturally in card-based cryptography, but has never been studied. This is the first study to discuss card-based cryptography and combinatorics on words, and our main contribution is to bring these two fields together. We believe that this paper will lead to the development of the relation between card-based cryptography and combinatorics on words.

1.2 Cyclic Equalizability from Card-Based Cryptography

In this section, we see that cyclic equalizability appears naturally in card-based cryptography by observing a protocol called the *five-card trick* [1], which uses a *random cut.*

A random cut is a shuffling operation that cyclically shifts a sequence of face-down cards a random number of times. For a sequence of face-down cards $(c_0, c_1, \dots, c_{n-1})$, we obtain $(c_r, c_{r+1}, \dots, c_{r+n-1})$, where $r \in \mathbb{Z}_n$ is a secret number chosen uniformly at random and the subscript is modulo n. Here, the secret number r is hidden from all players and nobody can guess it. We use $\langle \cdot \rangle$ to denote the application of a random cut to a sequence of cards. Random cuts are considered to be the simplest shuffling operation proposed so far, since they can be easily performed by a well-known shuffling method called *Hindu shuffle.*

The five-card trick is a protocol for the AND function $a \wedge b$ using a random cut. A secure computation protocol for the AND function can be used in the following situation:

> *Alice and Bob are going to decide if they want to go on a date this weekend. However, Alice and Bob are very shy and embarrassed to tell each other directly whether they want to go on a date or not. This can be especially awkward if one of them wants to go on a date and the other does not. Is there any way to know just whether they both want to go on a date or not?*

If we assume that "1" means "I want to go on a date" and "0" means "I don't want to go on a date", and Alice and Bob each have inputs $a, b \in \{0, 1\}$, then this problem can be solved by secure computation for the AND function $a \wedge b$. The *five-card trick* computing the AND function proceeds as follows.

1. Make the input commitments to $a, b \in \{0, 1\}$.

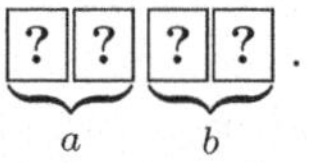

2. Swap the left and right cards of the commitment to a.

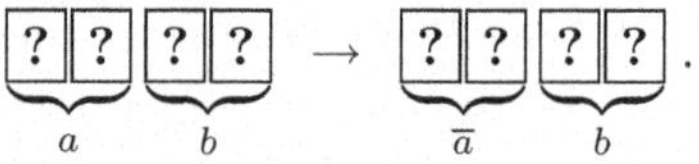

3. Insert a face-down card ♡ in the center.

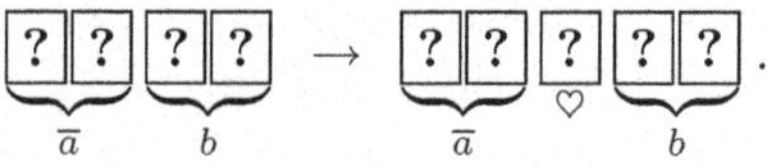

4. Apply a random cut.

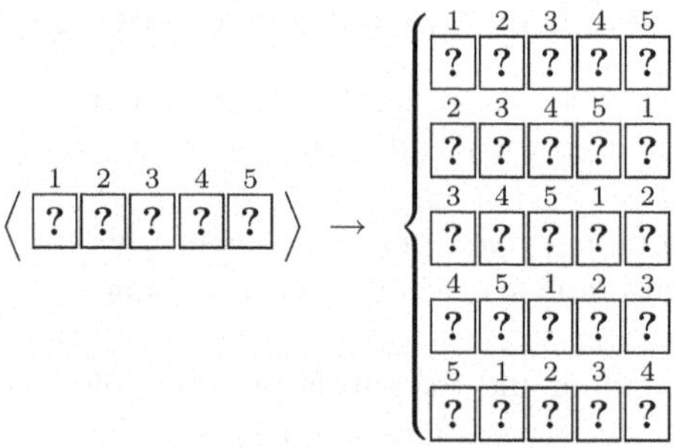

5. Open all cards. Output 0 if it is the left case and 1 if it is the right case.

♡	♣	♡	♣	♡		♣	♡	♡	♡	♣
♡	♡	♣	♡	♣		♣	♣	♡	♡	♡
♣	♡	♡	♣	♡		♡	♣	♣	♡	♡
♡	♣	♡	♡	♣		♡	♡	♣	♣	♡
♣	♡	♣	♡	♡		♡	♡	♡	♣	♣

The security of the five-card trick follows from the fact that for any given input (a, b) such that $a \wedge b = 0$, the resulting sequence is chosen uniformly at random from five patterns on the left side due to the effect of the random cut. Indeed, if $a \wedge b = 0$, a sequence of cards just after Step 2 is given as follows:

♡	♣	♣	♡	$(a, b) = (0, 0)$;
♡	♣	♡	♣	$(a, b) = (0, 1)$;
♣	♡	♣	♡	$(a, b) = (1, 0)$.

Then, a sequence of cards just after Step 3 is given as follows:

♡	♣	♡	♣	♡	$(a, b) = (0, 0)$;
♡	♣	♡	♡	♣	$(a, b) = (0, 1)$;
♣	♡	♡	♣	♡	$(a, b) = (1, 0)$.

Since these sequences are cyclically equal (see Definition 1), they become indistinguishable after a random cut is applied. We get three cyclically equal sequences by inserting ♡ in the middle. This is the key to understanding the structure of the five-card trick.

Let us reconsider the above as a problem of combinatorics on words. From now on, we consider words on Σ for a finite set of letters Σ. We say that k words $w_1, w_2, \ldots, w_k \in \Sigma^n$ are *cyclically equalizable* if all words can be transformed into cyclically equal words by repeated simultaneous insertion of letters, i.e., by inserting the same letters at the same positions (see Definition 3). The five-card trick corresponds to three binary words $1001, 1010, 0101$ with $\clubsuit = 0$ and $\heartsuit = 1$, which are cyclically equalizable by inserting 1 in the middle. In this paper, we consider a question of what is the necessary and sufficient condition for the cyclic equalizability. This is a problem of combinatorics on words that arises naturally from card-based cryptography.

1.3 Our Contribution

In this paper, we show that two binary words $w_1, w_2 \in \{0, 1\}^n$ are cyclically equalizable if and only if the Hamming weights of w_1 and w_2 are equal (Theorem 2). To show the main theorem, we first show that the cyclic equalizability does not change before and after simultaneous insertion (Theorem 1). Thanks to this, in the proof of Theorem 2, we can assume that w_1 and w_2 have no common letters on the same position. This makes the discussion easy to understand.

As applications of cyclic equalizability to card-based cryptography, we describe its applications to the information erasure problem and single-cut full-open protocols in Sect. 5. In particular, we discuss the fact that cyclic equalizability is equivalent to the possibility of information erasure (Sect. 5.1), and we show a lower bound on the number of cards in single-cut full-open protocols (Sect. 5.2).

2 Basic Notations

Let Σ be a finite set of *letters*. For a subset $\Delta \subseteq \Sigma$, we denote by Δ^* the set of all finite sequences of letters in Δ. Its elements $w \in \Delta^*$ are called *words* over Δ. The number of letters in w is called the *length* of w. For a letter $a \in \Sigma$ and a non-negative integer μ, a word $aa \cdots a$ that repeats a by μ times is denoted by a^μ. For a positive integer n and an integer m, the remainder of m divided by n (taken to be between 0 and $n-1$) is denoted by $m \bmod n$. For a special case, words over $\{0,1\}$ are called *binary words*. For a binary word $w \in \{0,1\}^*$, the number of 1's that appear in w is called the *Hamming weight* of w and is denoted by $\mathsf{wt}(w)$.

We identify k words of equal length $w_i = a_{i,0}a_{i,1} \cdots a_{i,n-1} \in \Sigma^n$ $(i = 1, \ldots, k)$ with a $k \times n$ matrix formed by placing the word w_i in the i-th row as follows:

$$\begin{pmatrix} w_1 \\ w_2 \\ \vdots \\ w_k \end{pmatrix} = \begin{pmatrix} a_{1,0} & a_{1,1} & a_{1,2} & \cdots & a_{1,n-1} \\ a_{2,0} & a_{2,1} & a_{2,2} & \cdots & a_{2,n-1} \\ \vdots & \vdots & \vdots & \ddots & \vdots \\ a_{k,0} & a_{k,1} & a_{k,2} & \cdots & a_{k,n-1} \end{pmatrix}.$$

We also denote a column vector consisting of k letters $c_1, \ldots, c_k$ from the top by $(c_1, \ldots, c_k)^{\mathrm{T}}$ or simply $(c_1 \cdots c_k)^{\mathrm{T}}$ without commas.

Definition 1 (Cyclically Equal). *Two words $w_i = a_{i,0}a_{i,1} \cdots a_{i,n-1} \in \Sigma^n$ $(i = 1, 2)$ of equal length are said to be* cyclically equal *if there exists an integer δ such that $a_{2,j} = a_{1,j+\delta \bmod n}$ for all $0 \le j \le n-1$. We denote this relation by $w_1 \sim w_2$. (Note that $\sim$ is an equivalence relation on Σ^n.) Similarly, k words of equal length $w_i \in \Sigma^n$ $(i = 1, 2, \ldots, k)$ are said to be* cyclically equal *if any pair of words w_i, w_j are cyclically equal.*

Definition 2 (Δ-Insertion/Deletion). *Let $\Delta \subseteq \Sigma$. Let $w_i \in \Sigma^n$ $(1 \le i \le k)$ be k words of length n. For $w_1, w_2, \ldots, w_k$, a Δ-*insertion *is an operation that transforms each $w_i = a_{i,0}a_{i,1} \cdots a_{i,n-1}$ into a word $u_0 a_{i,0} u_1 a_{i,1} \cdots u_{n-1} a_{i,n-1} u_n$, where $u_0, u_1, \ldots, u_n \in \Delta^*$ are $n+1$ words over Δ. We write $(w_1, \ldots, w_k) \xrightarrow{\Delta} (w'_1, \ldots, w'_k)$ if $(w'_1, \ldots, w'_k)$ is obtained from $(w_1, \ldots, w_k)$ by a Δ-insertion. The reverse operation of a Δ-insertion is called a Δ-*deletion.

We simply write "c-insertion" instead of "$\{c\}$-insertion". Note that the lengths of the words obtained by a Δ-insertion or a Δ-deletion are all equal.

Remark 1. A Δ-insertion means repetition of an insertion of a column vector $(cc\cdots c)^{\mathrm{T}}$ $(c \in \Delta)$ in the matrix representation of $w_1, w_2, \ldots, w_k$ as follows:

$$\begin{pmatrix} a_{1,0} \cdots a_{1,i}\ a_{1,i+1} \cdots a_{1,n-1} \\ a_{2,0} \cdots a_{2,i}\ a_{2,i+1} \cdots a_{2,n-1} \\ \vdots \quad \vdots \quad \vdots \quad \vdots \quad \ddots \quad \vdots \\ a_{k,0} \cdots a_{k,i}\ a_{k,i+1} \cdots a_{k,n-1} \end{pmatrix} \to \begin{pmatrix} a_{1,0} \cdots a_{1,i}\ c\ a_{1,i+1} \cdots a_{1,n-1} \\ a_{2,0} \cdots a_{2,i}\ c\ a_{2,i+1} \cdots a_{2,n-1} \\ \vdots \quad \vdots \quad \vdots \quad \vdots \quad \vdots \quad \ddots \quad \vdots \\ a_{k,0} \cdots a_{k,i}\ c\ a_{k,i+1} \cdots a_{k,n-1} \end{pmatrix}.$$

Conversely, a Δ-deletion means repetition of a deletion of a column vector $(cc\cdots c)^{\mathrm{T}}$ $(c \in \Delta)$ in the matrix representation of $w_1, w_2, \ldots, w_k$ as follows:

$$\begin{pmatrix} a_{1,0} \cdots a_{1,i}\ c\ a_{1,i+1} \cdots a_{1,n-1} \\ a_{2,0} \cdots a_{2,i}\ c\ a_{2,i+1} \cdots a_{2,n-1} \\ \vdots \quad \vdots \quad \vdots \quad \vdots \quad \vdots \quad \ddots \quad \vdots \\ a_{k,0} \cdots a_{k,i}\ c\ a_{k,i+1} \cdots a_{k,n-1} \end{pmatrix} \to \begin{pmatrix} a_{1,0} \cdots a_{1,i}\ a_{1,i+1} \cdots a_{1,n-1} \\ a_{2,0} \cdots a_{2,i}\ a_{2,i+1} \cdots a_{2,n-1} \\ \vdots \quad \vdots \quad \vdots \quad \vdots \quad \ddots \quad \vdots \\ a_{k,0} \cdots a_{k,i}\ a_{k,i+1} \cdots a_{k,n-1} \end{pmatrix}.$$

Definition 3 (Cyclically Equalizable). *Let $\Delta \subseteq \Sigma$. We say that k words $w_i \in \Sigma^n$ $(1 \le i \le k)$ are* Δ-cyclically equalizable *if there exist $w'_1, \ldots, w'_k \in \Sigma^*$ that are cyclically equal such that $(w_1, \ldots, w_k) \overset{\Delta}{\to} (w'_1, \ldots, w'_k)$. When $\Delta = \Sigma$, we simply say that they are cyclically equalizable.*

Remark 2. In Definition 2, we can assume without loss of generality that u_0 is an empty word. This is because whether $u_0 a_{i,0} u_1 a_{i,1} \cdots u_{n-1} a_{i,n-1} u_n$ $(1 \le i \le k)$ are cyclically equal is not changed by moving u_0 to the rightmost.

Given k words $w_i \in \Sigma^n$ $(1 \le i \le k)$, the *Δ-cyclic equalizability problem* asks whether they are Δ-cyclically equalizable or not, and how to insert if it is possible. In Sect. 5.1, we will see that the cyclic equalizability problem is equivalent to the *information erasure problem* in card-based cryptography as an application of cyclic equalizability.

3 Δ-Insertion Preserves Cyclic Equalizability

We prove the following theorem.

Theorem 1. *Let $\Delta \subseteq \Sigma$. Suppose that we have $(w_1, \ldots, w_k) \overset{\Delta}{\to} (w'_1, \ldots, w'_k)$ for $2k$ words $w_i, w'_i \in \Sigma^*$ $(1 \le i \le k)$. Then the following statements are equivalent.*

1. *$w_1, \ldots, w_k$ are Δ-cyclically equalizable.*
2. *$w'_1, \ldots, w'_k$ are Δ-cyclically equalizable.*

The following lemma is given for the proof of Theorem 1.

Lemma 1. *Suppose that $w_i = a_{i,0} a_{i,1} \cdots a_{i,n-1} \in \Sigma^n$ $(1 \le i \le k)$ are cyclically equal. Then for any $c \in \Sigma$ and $\mu \ge 0$, $\widetilde{w}_i = a_{i,0} c^\mu a_{i,1} c^\mu \cdots a_{i,n-1} c^\mu \in \Sigma^{(\mu+1)n}$ $(1 \le i \le k)$ are also cyclically equal.*

Proof. It is sufficient to show the case where $k = 2$. From the assumption, there exists an integer δ such that $a_{2,j} = a_{1,j+\delta \bmod n}$ for all $0 \leq j \leq n-1$. Let $\widetilde{w}_i = b_{i,0}b_{i,1}\cdots b_{i,(\mu+1)n-1}$ $(b_{i,j} \in \Sigma)$. Then $b_{i,j} = a_{i,j/(\mu+1)}$ if j is a multiple of $\mu+1$ and $b_{i,j} = c$ otherwise. Thus, for each $0 \leq j \leq (\mu+1)n-1$, $b_{2,j} = b_{1,j+(\mu+1)\delta \bmod (\mu+1)n}$. Therefore $\widetilde{w}_i$ are also cyclically equal. □

*Proof)**Theorem** 1).* Since it is trivial that the latter condition implies the former, we show the converse direction. It is sufficient to show that the latter condition follows from the former condition when $(w_1,\ldots,w_k) \xrightarrow{c} (w'_1,\ldots,w'_k)$ for some letter $c \in \Delta$ because any Δ-insertion can be decomposed into a series of such insertions.

Let $w_i = a_{i,0}a_{i,1}\cdots a_{i,n-1}$ $(a_{i,j} \in \Sigma)$. From Remark 2, we can assume that $w'_i = a_{i,0}c^{\mu_0}a_{i,1}c^{\mu_1}\cdots a_{i,n-1}c^{\mu_{n-1}}$ for some common non-negative integers $\mu_0,\mu_1,\ldots,\mu_{n-1}$. Let $\nu := \max\{\mu_0,\mu_1,\ldots,\mu_{n-1}\}$ and $\widehat{b} := bc^{\nu} \in \Sigma^{\nu+1}$ for each $b \in \Sigma$. Then for $(w'_1,\ldots,w'_k)$, a c-insertion yields $(u_1,\ldots,u_k)$ for $u_i = \widehat{a_{i,0}}\widehat{a_{i,1}}\cdots\widehat{a_{i,n-1}} \in \Sigma^{(\nu+1)n}$. From the assumption that $(w_1,\ldots,w_k)$ are Δ-cyclically equalizable and Remark 2, we obtain some common letters $d_{0,1},\ldots,d_{0,\ell_0},\ldots,d_{n-1,1},\ldots,d_{n-1,\ell_{n-1}} \in \Delta$ such that

$$a_{i,0}d_{0,1}\cdots d_{0,\ell_0}\ a_{i,1}d_{1,1}\cdots d_{1,\ell_1}\ \cdots\ a_{i,n-1}d_{n-1,1}\cdots d_{n-1,\ell_{n-1}}$$

are cyclically equal. Then, from Lemma 1,

$$\widetilde{u}_i := \widehat{a_{i,0}}\widehat{d_{0,1}}\cdots\widehat{d_{0,\ell_0}}\ \widehat{a_{i,1}}\widehat{d_{1,1}}\cdots\widehat{d_{1,\ell_1}}\ \cdots\ \widehat{a_{i,n-1}}\widehat{d_{n-1,1}}\cdots\widehat{d_{n-1,\ell_{n-1}}}$$

are also cyclically equal. Moreover, since $\widehat{d_{h,j}} \in \Delta^{\nu+1}$, we have $(u_1,\ldots,u_k) \xrightarrow{\Delta} (\widetilde{u}_1,\ldots,\widetilde{u}_k)$. As mentioned above, we have $(w'_1,\ldots,w'_k) \xrightarrow{\Delta} (u_1,\ldots,u_k)$, thus we also have $(w'_1,\ldots,w'_k) \xrightarrow{\Delta} (\widetilde{u}_1,\ldots,\widetilde{u}_k)$, therefore $(w'_1,\ldots,w'_k)$ are Δ-cyclically equalizable. □

The following corollary, which follows directly from Theorem 1, is useful to reduce the Δ-cyclic equalizability for a set of words to that of shorter words.

Corollary 1. *Suppose that $(w'_1,\ldots,w'_k) \in (\Sigma^{n'})^k$ is obtained from $(w_1,\ldots,w_k) \in (\Sigma^n)^k$ by a Δ-deletion for some subset $\Delta \subsetneq \Sigma$. Then the following statements are equivalent.*

1. *$w_1,\ldots,w_k$ are Δ-cyclically equalizable.*
2. *$w'_1,\ldots,w'_k$ are Δ-cyclically equalizable.*

4 Cyclic Equalizability of Two Binary Words

Throughout this section, we assume that $\Sigma = \{0,1\}$. The following theorem gives a complete characterization on cyclic equalizability of two binary words.

Theorem 2. *Two binary words $w_1, w_2 \in \{0,1\}^n$ are cyclically equalizable if and only if $\mathsf{wt}(w_1) = \mathsf{wt}(w_2)$.*

Proof. Since the only if part is trivial, we show the if part. Suppose that $\mathsf{wt}(w_1) = \mathsf{wt}(w_2)$. From Corollary 1, we can assume that no more $\{0,1\}$-deletion can be applied to w_1 and w_2, i.e., each column in the matrix representation of the pair (w_1, w_2) is either $(10)^{\mathrm{T}}$ or $(01)^{\mathrm{T}}$. Furthermore, since a cyclic shift of the columns of the matrix does not affect the claim, we can also assume that w_1 and w_2 are of the following form:

$$w_1 = 1^{\mu_0}0^{\mu_1}1^{\mu_2}0^{\mu_3}\cdots 1^{\mu_{2N-2}}0^{\mu_{2N-1}};$$
$$w_2 = 0^{\mu_0}1^{\mu_1}0^{\mu_2}1^{\mu_3}\cdots 0^{\mu_{2N-2}}1^{\mu_{2N-1}},$$

where each μ_j is a positive integer and it holds $\sum_{j=0}^{N-1}\mu_{2j} = \sum_{j=0}^{N-1}\mu_{2j+1}$ from the assumption $\mathsf{wt}(w_1) = \mathsf{wt}(w_2)$. We assume $N \geq 2$ since w_1 and w_2 are already cyclically equal if $N = 1$.

We say that a pair of binary words (w'_1, w'_2) is *admissible* if $\mathsf{wt}(w'_1) = \mathsf{wt}(w'_2)$ and they are of the following form:

$$w'_1 = 1^{\nu'_0}1^{\mu'_0}0^{\nu'_1}0^{\mu'_1}1^{\nu'_2}1^{\mu'_2}0^{\nu'_3}0^{\mu'_3}\cdots 0^{\nu'_{2N-1}}0^{\mu'_{2N-1}};$$
$$w'_2 = 1^{\nu'_0}0^{\mu'_0}0^{\nu'_1}1^{\mu'_1}1^{\nu'_2}0^{\mu'_2}0^{\nu'_3}1^{\mu'_3}\cdots 0^{\nu'_{2N-1}}1^{\mu'_{2N-1}},$$

for some parameters $\nu'_i \geq 0$ and $\mu'_i \geq 1$ $(0 \leq i \leq 2N-1)$. The pair (w_1, w_2) in the claim is admissible with $\nu'_i = 0$ and $\mu'_i = \mu_i$. For an admissible pair (w'_1, w'_2) and $0 \leq j \leq 2N-1$, we say that the pair is *consistent* up to the j-th slot if the following holds for all $0 \leq i \leq j$:

$$\nu'_i + \mu'_i = \mu'_{(i+1) \bmod 2N} + \nu'_{(i+2) \bmod 2N}.$$

For example, the following pair (w'_1, w'_2) is consistent up to the 1th slot if the underlined subsequences are equal:

$$w'_1 = \underline{1^{\nu'_0}1^{\mu'_0}0^{\nu'_1}0^{\mu'_1}}1^{\nu'_2}1^{\mu'_2}0^{\nu'_3}0^{\mu'_3}\cdots 0^{\nu'_{2N-1}}0^{\mu'_{2N-1}};$$
$$w'_2 = 1^{\nu'_0}0^{\mu'_0}0^{\nu'_1}\underline{1^{\mu'_1}1^{\nu'_2}0^{\mu'_2}0^{\nu'_3}}1^{\mu'_3}\cdots 0^{\nu'_{2N-1}}1^{\mu_{2N-1}}.$$

Note that when w'_1 and w'_2 are consistent up to the $(2N-3)$-th slot, they are cyclically equal from the following reason. If they are consistent up to the $(2N-3)$-th slot, they are also consistent up to the $(2N-2)$-th slot due to $\mathsf{wt}(w'_1) = \mathsf{wt}(w'_2)$, and furthermore, they are also consistent up to the $(2N-1)$-th slot since they are of equal length.

In the following, $\{0,1\}$-insertions for w_1 and w_2 are performed recursively to construct two cyclically equal binary words.

First, perform a $\{0,1\}$-insertion to be consistent up to the 0th slot. If w_1 and w_2 are already consistent up to the 0th slot, i.e., $\mu_0 = \mu_1$, then nothing needs to be done. So we assume that $\mu_0 \neq \mu_1$. If $\mu_0 > \mu_1$, then they will be consistent up to the 0th slot by the following 1-insertion:

$$\widetilde{w_1} = 1^{\mu_0}0^{\mu_1}\underline{1^{\mu_0-\mu_1}}1^{\mu_2}0^{\mu_3}\cdots 1^{\mu_{2N-2}}0^{\mu_{2N-1}};$$
$$\widetilde{w_2} = 0^{\mu_0}1^{\mu_1}\underline{1^{\mu_0-\mu_1}}0^{\mu_2}1^{\mu_3}\cdots 0^{\mu_{2N-2}}1^{\mu_{2N-1}}.$$

If $\mu_0 < \mu_1$, then it is enough to do the following 1-insertion:

$$\widetilde{w_1} = \underline{1^{\mu_1-\mu_0}}1^{\mu_0}0^{\mu_1}1^{\mu_2}0^{\mu_3}\cdots 1^{\mu_{2N-2}}0^{\mu_{2N-1}};$$
$$\widetilde{w_2} = \underline{1^{\mu_1-\mu_0}}0^{\mu_0}1^{\mu_1}0^{\mu_2}1^{\mu_3}\cdots 0^{\mu_{2N-2}}1^{\mu_{2N-1}}.$$

In either case, the current pair $(\widetilde{w_1}, \widetilde{w_2})$ is admissible and consistent up to the 0th slot.

Next, for $1 \le j \le 2N-3$, when the current pair (w_1', w_2') is admissible and consistent up to the $(j-1)$-th slot, we perform $\{0,1\}$-insertions to be consistent up to the j-th slot. Let $k_0 := \nu_j' + \mu_j' - (\mu_{j+1}' + \nu_{j+2}')$ and $k := |k_0|$. If w_1' and w_2' are already consistent up to the j-th slot, i.e., $k_0 = 0$, then nothing needs to be done. So we can assume $k_0 \neq 0$. If $k_0 > 0$, perform a 0-insertion of 0^k just before $0^{\nu_{j+2}'}$ when j is odd, and a 1-insertion of 1^k just before $1^{\nu_{j+2}'}$ when j is even. The following is an example when j is odd:

$$\widetilde{w_1} = \cdots 1^{\mu_{j-1}'}0^{\nu_j'}0^{\mu_j'}1^{\nu_{j+1}'}1^{\mu_{j+1}'}\underline{0^k}0^{\nu_{j+2}'}0^{\mu_{j+2}'}1^{\nu_{j+3}'}\cdots;$$
$$\widetilde{w_2} = \cdots 0^{\mu_{j-1}'}0^{\nu_j'}1^{\mu_j'}1^{\nu_{j+1}'}0^{\mu_{j+1}'}\underline{0^k}0^{\nu_{j+2}'}1^{\mu_{j+2}'}1^{\nu_{j+3}'}\cdots.$$

If $k_0 < 0$, for each $0 \le i \le j$ such that $i \equiv j \pmod 2$, a 0-insertion of 0^k just before $0^{\nu_i'}$ is performed when j is odd and a 1-insertion of 1^k just before $1^{\nu_i'}$ when j is even. The following is an example when j is odd:

$$\widetilde{w_1} = 1^{\nu_0'}1^{\mu_0'}\underline{0^k}0^{\nu_1'}0^{\mu_1'}1^{\nu_2'}1^{\mu_2'}\underline{0^k}0^{\nu_3'}0^{\mu_3'}\cdots 1^{\mu_{j-1}'}\underline{0^k}0^{\nu_j'}\cdots;$$
$$\widetilde{w_2} = 1^{\nu_0'}0^{\mu_0'}\underline{0^k}0^{\nu_1'}1^{\mu_1'}1^{\nu_2'}0^{\mu_2'}\underline{0^k}0^{\nu_3'}1^{\mu_3'}\cdots 0^{\mu_{j-1}'}\underline{0^k}0^{\nu_j'}\cdots.$$

In either case, the pair $(\widetilde{w_1}, \widetilde{w_2})$ is admissible and consistent up to the j-th slot.

By performing the above procedure up to $j = 2N-3$, we can make w_1, w_2 cyclically equal. This completes the proof of Theorem 2. □

5 Application to Card-Based Cryptography

5.1 Information Erasure

Suppose that we have a sequence of face-down cards $\boldsymbol{x} = (x_1, x_2, \ldots, x_n) \in \Sigma^n$. We know that $\boldsymbol{x} \in S \subseteq \Sigma^n$ but do not know which one. Consider the protocol in the following.

1. By inserting additional cards, arrange the following sequence:

$$\underset{x_1}{\boxed{?}}\underbrace{\boxed{?}\boxed{?}\cdots\boxed{?}}_{u_1}\underset{x_2}{\boxed{?}}\underbrace{\boxed{?}\boxed{?}\cdots\boxed{?}}_{u_2}\underset{x_3}{\boxed{?}}\underbrace{\boxed{?}\boxed{?}\cdots\boxed{?}}_{u_3}\cdots\underset{x_n}{\boxed{?}}\underbrace{\boxed{?}\boxed{?}\cdots\boxed{?}}_{u_n},$$

   where $u_1, u_2, \ldots, u_n \in \Sigma^*$.
2. Apply a random cut.
3. Open all cards.

We say that the above protocol is *information erasure* if the probability distribution of the opened symbols does not depend on $\boldsymbol{x} \in S$. Given $S \subseteq \Sigma^n$, the *information erasure problem* asks whether information erasure is possible, and how to insert if it is possible.

It is obvious that the information erasure problem can be solved if the cyclic equalizability problem for k words is solved for any k. From Theorem 2, the information erasure problem for $S \subseteq \{0,1\}^n$ with $|S| = 2$ is completely answered: for $S = \{w_1, w_2\}$, the information erasure problem is possible if and only if $\mathsf{wt}(w_1) = \mathsf{wt}(w_2)$ and how to insert is given in the proof of Theorem 2. The information erasure problems for $S \subseteq \{0,1\}^n$ with $|S| \geq 3$ and for $S \subseteq \Sigma^n$ with $|\Sigma| \geq 3$ are open problems.

5.2 Single-Cut Full-Open Protocols

Suppose that we have a sequence of face-down cards $\boldsymbol{x} = (x_1, \overline{x_1}, x_2, \overline{x_2}, \ldots, x_n, \overline{x_n}) \in \{0,1\}^{2n}$, i.e., n commitments to $x_1, x_2, \ldots, x_n \in \{0,1\}$. Let $f\colon \{0,1\}^n \to \{0,1\}$ be a function. Consider the protocol in the following.

1. Permute the order of the sequence by a known permutation $\pi \in S_{2n}$:

$$\underset{x_1}{\boxed{?}}\,\underset{\overline{x_1}}{\boxed{?}}\,\underset{x_2}{\boxed{?}}\,\underset{\overline{x_2}}{\boxed{?}}\cdots\underset{x_n}{\boxed{?}}\,\underset{\overline{x_n}}{\boxed{?}} \;\rightarrow\; \underset{y_1}{\boxed{?}}\,\underset{y_2}{\boxed{?}}\,\underset{y_3}{\boxed{?}}\,\underset{y_4}{\boxed{?}}\cdots\underset{y_{2n}}{\boxed{?}},$$

   where $(y_1, y_2, \ldots, y_{2n}) = \pi(\boldsymbol{x})$.
2. By inserting additional cards, arrange the following sequence:

$$\underset{y_1}{\boxed{?}}\,\underbrace{\boxed{?}\,\boxed{?}\cdots\boxed{?}}_{u_1}\,\underset{y_2}{\boxed{?}}\,\underbrace{\boxed{?}\,\boxed{?}\cdots\boxed{?}}_{u_2}\,\underset{y_3}{\boxed{?}}\,\underbrace{\boxed{?}\,\boxed{?}\cdots\boxed{?}}_{u_3}\cdots\underset{y_{2n}}{\boxed{?}}\,\underbrace{\boxed{?}\,\boxed{?}\cdots\boxed{?}}_{u_{2n}},$$

   where $u_1, u_2, \ldots, u_{2n} \in \{0,1\}^*$. Define $s(\boldsymbol{x}) := (y_1, u_1, y_2, u_2, \ldots, y_{2n}, u_{2n})$.
3. Apply a random cut.
4. Open all cards.

We say that the above protocol is a *single-cut full-open protocol* for f if there exist two binary words $z_0, z_1 \in \{0,1\}^*$ such that z_0, z_1 are not cyclically equal and two binary words $s(\boldsymbol{x})$ and $z_{f(\boldsymbol{x})}$ are cyclically equal. The number of cards in a protocol is defined as $|s(\boldsymbol{x})| = 2n + \sum_{i=1}^{2n} |u_i|$.

In 1989, den Boer [1] proposed a single-cut full-open protocol for the AND function $x_1 \wedge x_2$ called the five-card trick (see Sect. 1.2), which is the first card-based protocol in the history. In 2014, Heather, Schneider, and Teague [3] proposed a single-cut full-open protocol for the three-bit equality function, which outputs 1 if $x_1 = x_2 = x_3$ and 0 otherwise. In 2018, Shinagawa and Mizuki [13] rediscovered the Heather–Schneider–Teague's protocol and named it as the six-card trick. In addition, they designed a single-cut full-open protocol for the two-bit equality function, which is equivalent to a protocol for $x_1 \oplus x_2$, and posed an open problem of asking whether there exists a single-cut full-open protocol for an n-bit equality function for $n \geq 4$. So far, only these three single-cut full-open protocols are known. It is an open problem to construct single-cut full-open protocols for other interesting functions.

Based on the cyclic equalizability, we can derive a lower bound on the number of cards in single-cut full-open protocols. For $f\colon \{0,1\}^n \to \{0,1\}$ and $b \in \{0,1\}$, define $N_b(f) := |\{x \mid f(x) = b\}|$. From the definition, we have $N_0(f) + N_1(f) = 2^n$ for any function f. First, we have a simple lemma as follows.

Lemma 2. *If k different words $w_1, w_2, \ldots, w_k \in \Sigma^*$ are cyclically equal, then the length of the word $|w_i|$ is at least k.*

Proof. Any word $w \in \Sigma^*$ has at most $|w|$ words that are cyclically equal to w. □

From Lemma 2, we obtain a lower bound on the number of cards.

Theorem 3. *The number of cards in a single-cut full-open protocol for a function $f\colon \{0,1\}^n \to \{0,1\}$ is at least* $\max(N_0(f), N_1(f))$.

Proof. Let T be the truth table of f whose column corresponds to each symbol in $s(\boldsymbol{x})$, i.e., T has $|s(\boldsymbol{x})|$ columns and 2^n rows. From the definition of single-cut full-open protocols, all $N_b(f)$ rows are different binary words and cyclically equal for each $b \in \{0,1\}$. From Lemma 2, the number of cards $|s(\boldsymbol{x})|$ is at least $\max(N_0(f), N_1(f))$. □

6 Conclusion

In this paper, we show that two binary words $w_1, w_2 \in \{0,1\}^n$ are cyclically equalizable if and only if the Hamming weights of w_1 and w_2 are equal. We also give some applications of cyclic equalizability to the information erasure problem and single-cut full-open protocols.

Research on cyclic equalizability has just begun and there are many important open problems. What is the necessary and sufficient condition for k words $w_1, w_2, \ldots, w_k \in \Sigma^n$ to be cyclically equalizable? If they are cyclically equalizable, it is also important to minimize the length of the resulting words. These general problems are obviously important, but the following more specific problems are also of interest.

- What is the condition for two words $w_1, w_2 \in \{0,1,2\}^n$ to be cyclically equalizable?
- Are there two binary words $w_1, w_2 \in \{0,1\}^n$ of the same Hamming weight that are not 0-cyclically equalizable?
- Are there three binary words $w_1, w_2, w_3 \in \{0,1\}^n$ of the same Hamming weight that are not cyclically equalizable? (A counterexample exists for 0-cyclic equalizability. Indeed, 1001, 1010, and 1100 are not 0-cyclically equalizable.)

- Is there a single-cut full-open protocol for the three-bit AND function? In particular, is there a permutation $\pi \in S_6$ such that seven binary words $\{\pi(x_1\overline{x_1}x_2\overline{x_2}x_3\overline{x_3}) \mid (x_1, x_2, x_3) \in \{0,1\}^3 \setminus \{(1,1,1)\}\}$ are cyclically equalizable?
- Is there a single-cut full-open protocol for the four-bit equality function? In particular, is there a permutation $\pi \in S_8$ such that 14 binary words $\{\pi(x_1\overline{x_1}x_2\overline{x_2}x_3\overline{x_3}x_4\overline{x_4}) \mid (x_1, x_2, x_3, x_4) \in \{0,1\}^4 \setminus \{(0,0,0),(1,1,1)\}\}$ are cyclically equalizable?

Acknowledgement. This work was supported by JSPS KAKENHI Grant Numbers 21K17702, 23H00479, and 25K14994, JST CREST Grant Number MJCR22M1, and Institute of Mathematics for Industry, Joint Usage/Research Center in Kyushu University: Short-term Visiting Researcher 2022a006, 2023a009, and 2024a015.

A Concrete Example of Theorem 2 and Corollary 1

1. Let $w_1, w_2 \in \{0,1\}^{20}$ be two binary words of the same Hamming weight as follows:
$$w_1 = 1\dot{0}10\ddot{1}\ 001\dot{0}\ddot{1}\ \dot{0}1101\ \dot{0}1000;$$
$$w_2 = 00011\ 11001\ 00010\ 00111.$$
where they are separated by 5 bits for ease of reading, and the columns of $(00)^{\mathrm{T}}$ are dotted at the top and the columns of $(11)^{\mathrm{T}}$ are double-dotted at the top.
2. By applying a 1-deletion, we can remove the double-dotted columns as follows:
$$w_1' = 1\dot{0}100\ 01\dot{0}\dot{0}1\ 101\dot{0}1\ 000;$$
$$w_2' = 00011\ 10000\ 01000\ 111.$$
3. By applying a 0-deletion, we can remove the dotted columns as follows:
$$w_1'' = 11000\ 11101\ 1000;$$
$$w_2'' = 00111\ 00010\ 0111.$$
4. Now we are ready to apply the proof of Theorem 2. By applying a 1-insertion, we can make it consistent up to the 0th slot:
$$\underline{\overline{\mathbf{1}}11}00\ 01110\ 11000;$$
$$\mathbf{1}00\underline{11\ 1}0001\ 00111,$$
where the bold columns are the columns inserted so far (including this time), the overlined columns are the columns inserted this time, and the underlined subsequences are the matching patterns. Hereafter, we use these notations.

5. We notice that they are already consistent up to 1th slot since the next sequences just after the underlined sequences are both 000. The underlines are updated as:

$$\underline{\mathbf{1}1100\ 0}1110\ 11000;$$
$$\mathbf{1}00\underline{11\ 1000}1\ 00111.$$

6. By applying a 1-insertion, we can make it consistent up to the 2th slot:

$$\underline{\mathbf{1}1100\ 0111}0\ \overline{\mathbf{11}}110\ 00;$$
$$\mathbf{1}00\underline{11\ 10001\ \mathbf{11}}001\ 11.$$

7. By applying a 0-insertion, we can make it consistent up to the 3th slot:

$$u_1 = \underline{\mathbf{1}11\overline{\mathbf{0}}0\ 00111\ \overline{\mathbf{0}}0}\mathbf{11}1\ 1000;$$
$$u_2 = \mathbf{1}000\mathbf{0}\underline{1\ 11000\ \mathbf{0}1\mathbf{11}0\ 0}111.$$

   Now u_1, u_2 are cyclically equal. Thus we can confirm that w_1'', w_2'' are cyclically equalizable. This corresponds to the end of the proof of Theorem 2.
8. Next we apply Corollary 1. Let u_1', u_2' be the binary words obtained by replacing 0 and 1 of w_1', w_2' by $\hat{0} := 01$ and $\hat{1} := 11$, respectively, as follows:

$$u_1' = \hat{1}\hat{0}\hat{1}\hat{0}\hat{0}\ \hat{0}\hat{1}\hat{0}\hat{0}\hat{1}\ \hat{1}\hat{0}\hat{1}\hat{0}\hat{1}\ \hat{0}\hat{0}\hat{0};$$
$$u_2' = \hat{0}\hat{0}\hat{0}\hat{1}\hat{1}\ \hat{1}\hat{0}\hat{0}\hat{0}\hat{0}\ \hat{0}\hat{1}\hat{0}\hat{0}\hat{0}\ \hat{1}\hat{1}\hat{1}.$$

   Then we have $(w_1, w_2) \xrightarrow{1} (u_1', u_2')$ because (w_1, w_2) does not have two or more consecutive double-dotted columns $(11)^{\mathrm{T}}$.
9. Let u_1'', u_2'' be the binary words obtained by replacing 0 and 1 of w_1'', w_2'' by $\tilde{0} := \hat{0}\hat{0}\hat{0} = 010101$ and $\tilde{1} := \hat{1}\hat{0}\hat{0} = 110101$, respectively, as follows:

$$u_1'' = \tilde{1}\tilde{1}\tilde{0}\tilde{0}\tilde{0}\ \tilde{1}\tilde{1}\tilde{1}\tilde{0}\tilde{1}\ \tilde{1}\tilde{0}\tilde{0}\tilde{0};$$
$$u_2'' = \tilde{0}\tilde{0}\tilde{1}\tilde{1}\tilde{1}\ \tilde{0}\tilde{0}\tilde{0}\tilde{1}\tilde{0}\ \tilde{0}\tilde{1}\tilde{1}\tilde{1}.$$

   Now we notice that u_1'' and u_2'' are obtained by inserting $\hat{0}$'s in u_1' and u_2' because (u_1', u_2') has at most two consecutive columns $(\hat{0}\hat{0})^{\mathrm{T}}$. By combining this with $(w_1, w_2) \xrightarrow{1} (u_1', u_2')$ in Step 8, we have $(w_1, w_2) \xrightarrow{\{0,1\}} (u_1'', u_2'')$.
10. Finally, in the same operations to obtain u_1 and u_2 from w_1'' and w_2'', the following binary words are obtained by inserting $\tilde{0}$'s and $\tilde{1}$'s in u_1'' and u_2'' as follows:

$$u_1''' = \tilde{\mathbf{1}}\tilde{1}\tilde{1}\tilde{\mathbf{0}}\tilde{0}\ \tilde{0}\tilde{0}\tilde{1}\tilde{1}\tilde{1}\ \tilde{\mathbf{0}}\tilde{0}\tilde{\mathbf{1}}\tilde{\mathbf{1}}\tilde{1}\ \tilde{1}\tilde{0}\tilde{0}\tilde{0};$$
$$u_2''' = \tilde{\mathbf{1}}\tilde{0}\tilde{0}\tilde{\mathbf{0}}\tilde{1}\ \tilde{1}\tilde{1}\tilde{0}\tilde{0}\tilde{0}\ \tilde{\mathbf{0}}\tilde{1}\tilde{\mathbf{1}}\tilde{\mathbf{1}}\tilde{0}\ \tilde{0}\tilde{1}\tilde{1}\tilde{1}.$$

   Now u_1''', u_2''' are cyclically equal of length $19 \times 6 = 114$. Thus we can confirm that w_1, w_2 are cyclically equalizable.

The length of the resulting binary words in the above construction depends on the order of the characters to be deleted. In fact, if we apply a 0-deletion first and then a 1-deletion, we have the resulting binary words of length $19 \times 4 = 76$.

References

1. Boer, B.: More efficient match-making and satisfiability *the five card trick.* In: Quisquater, J.-J., Vandewalle, J. (eds.) EUROCRYPT 1989. LNCS, vol. 434, pp. 208–217. Springer, Heidelberg (1990). https://doi.org/10.1007/3-540-46885-4_23
2. Crépeau, C., Kilian, J.: Discreet solitary games. In: Stinson, D.R. (ed.) CRYPTO 1993. LNCS, vol. 773, pp. 319–330. Springer, Heidelberg (1994). https://doi.org/10.1007/3-540-48329-2_27
3. Heather, J., Schneider, S., Teague, V.: Cryptographic protocols with everyday objects. Formal Aspects Comput. **26**(1), 37–62 (2014)
4. Kanai, K., Miyamoto, K., Nuida, K., Shinagawa, K.: Uniform cyclic group factorizations of finite groups. Comm. Algebra **52**(5), 2174–2184 (2024)
5. Kastner, J., Koch, A., Walzer, S., Miyahara, D., Hayashi, Y., Mizuki, T., Sone, H.: The minimum number of cards in practical card-based protocols. In: Takagi, T., Peyrin, T. (eds.) ASIACRYPT 2017. LNCS, vol. 10626, pp. 126–155. Springer, Cham (2017). https://doi.org/10.1007/978-3-319-70700-6_5
6. Koch, A., Schrempp, M., Kirsten, M.: Card-based cryptography meets formal verification. In: Galbraith, S.D., Moriai, S. (eds.) ASIACRYPT 2019. LNCS, vol. 11921, pp. 488–517. Springer, Cham (2019). https://doi.org/10.1007/978-3-030-34578-5_18
7. Koch, A., Walzer, S., Härtel, K.: Card-based cryptographic protocols using a minimal number of cards. In: Iwata, T., Cheon, J.H. (eds.) ASIACRYPT 2015. LNCS, vol. 9452, pp. 783–807. Springer, Heidelberg (2015). https://doi.org/10.1007/978-3-662-48797-6_32
8. Lothaire, M.: Combinatorics on Words, vol. 17. Addison Wesley, Boston (1983)
9. Lothaire, M.: Algebraic Combinatorics on Words, vol. 90. Cambridge University Press, Cambridge (2002)
10. Lothaire, M.: Applied Combinatorics on Words, vol. 105. Cambridge University Press, Cambridge (2005)
11. Mizuki, T., Shizuya, H.: A formalization of card-based cryptographic protocols via abstract machine. Int. J. Inf. Secur. **13**(1), 15–23 (2014)
12. Shinagawa, K., Kanai, K., Miyamoto, K., Nuida, K.: How to covertly and uniformly scramble the 15 puzzle and rubik's cube. In: Broder, A.Z., Tamir, T. (eds.) Fun with Algorithms, vol. 291 of LIPIcs, pp. 30:1–30:15. Schloss Dagstuhl, Dagstuhl (2024)
13. Shinagawa, K., Mizuki, T.: The six-card trick: secure computation of three-input equality. In: Lee, K. (ed.) Information Security and Cryptology. LNCS, vol. 11396, pp. 123–131. Springer, Cham (2018)
14. Shinagawa, K., Nuida, K.: A single shuffle is enough for secure card-based computation of any Boolean circuit. Disc. Appl. Math. **289**, 248–261 (2021)
15. Suga, Y.: A classification proof for commutative three-element semigroups with local AND structure and its application to card-based protocols. In: 2022 IEEE International Conference on Consumer Electronics - Taiwan, pp. 171–172. IEEE (2022)

Quantum First-Order Logics and Quantum Natural Deduction

Tomoyuki Yamakami(✉)

Faculty of Engineering, University of Fukui, 3-9-1 Bunkyo, Fukui 910-8507, Japan
tomoyukiyamakami@gmail.com

Abstract. A quantum analogue of the classical first-order logic (FO) was discussed lately in [CiE 2024, LNCS 14773, pp. 311–323] from a semantic viewpoint in order to characterize logarithmic-time/space quantum computability. From a syntactic perspective, we further study this new, intriguing quantum logic by introducing a quantum analogue of Gentzen's tree-like proof system of classical natural deduction and further analyzing the expressing power of this new quantum natural deduction system. In particular, we prove the fundamental features of the soundness and completeness of this quantum proof system and, as a consequence of them, we further show the compactness of the system. Quantum natural deduction is expected to pave a road to automated theorem proving in a quantum-mechanical fashion.

Keywords: quantum first-order logic · quantum computability · quantum proof system · quantum natural deduction · QND

1 Background, Motivations, and Challenges

1.1 Quantum Computing and Quantum Logic

Quantum physics has opened a door to vast new perspectives of quantum computing. Two of the most significant features of quantum computation are *entanglement* of quantum states and *reversibility* of the computation (unless quantum measurement is performed during the computation). To logically "express" quantum mechanical behaviors, quantum logical frameworks have been sought out in the literature from an early study of quantum logic by Birkhoff and von Neumann [2] to recent notions of quantum dynamic logic [4], quantum predicate logic [15], quantum Hoare logic [21], etc.

In the classical setting, nevertheless, *classical first-order logic* (FO) of [1,6,10–12] was studied for the characterization of time/space-bounded computability from a semantic viewpoint. In this line of study, a quantum analogue of FO was introduced lately in [20] under the name of *quantum first-order logic* (QFO) to logically express a fundamental nature of quantum computation. The QFO formalism of [20] heavily relies on the earlier works on recursion-theoretical schematic definitions [17,19], which were used to define quantum

A. Jeż and J. Otop (Eds.): FCT 2025, LNCS 16106, pp. 420–433, 2026.
https://doi.org/10.1007/978-3-032-04700-7_31

polynomial-time and polylogarithmic-time/space computabilities of quantum functions, which map from quantum states to quantum states. A primary purpose of the introduction of QFO was to precisely characterize time- and space-bounded quantum computabilities.

As a plausible trait of quantum recursion schemes, the syntax of QFO includes quantum-AND ($\wedge$), quantum-OR ($\|$), and quantum-NOT ($\neg^Q$), which reflect the specific behaviors of quantum computation and thus their actions are quite different from the classical counterparts. Moreover, the construction of each quantum formula in QFO roughly describes an "algorithmic process" of quantum computation rather than a simple description of quantum physical phenomena. Moreover, the special quantum predicate symbol $\simeq_\eta$ relates to a measurement of a quantum bit (or a qubit, for short). See Sect. 3 for a quick overview of these notations. It was proven in [20] that QFO possesses sufficient expressing power to characterize logarithmic-time/space computabilities.

From a syntactic viewpoint, on the contrary, in the classical setting, a logical calculus of classical "natural deduction", coined by Gentzen [7,8], has been used to visualize the process of verifying the validity of a given logical formula. A proof system of *natural deduction* thus provides us a constructive "proof" of a valid formula, in which each intermediate formula is deduced from a series of premises by applying a finite set of inference rules. Early proof systems of natural deduction were sought by Gentzen [7] and Jaśkowski [13] to model mathematical reasoning. Such proof systems then have provided a logical framework to mathematical proofs and also in general they have provided a practical means of automated verification procedures. Gentzen's proof system, NK, is composed of inference rules (introduction and elimination) of logical connectives and quantifications and it further allows us to discuss the *proof complexity* of given target formulas. The field of proof complexity has been well-developed over the years (see textbooks, e.g., [5]).

Based upon various benefits of the introduction of a proof system of natural deduction into first-order logics, it is natural to seek for a similar proof system of "quantum natural deduction" in the hope that quantum proof systems will allow us to discuss the *quantum proof complexity* of target quantum formulas.

1.2 Our Challenges and Main Results

From a semantic perspective, the QFO formalism of [20] was used to logically "express" the process of logarithmic-time/space quantum computations. Further taking a syntactic approach, we wish to introduce a quantum analogue of proof systems of natural deduction for QFO, succinctly called *quantum natural deduction*. In this work, we particularly intend to look into Gentzen's tree-like proof systems and to develop similar proof systems for QFO. Our quantum natural deduction reflects Yamakami's schematic approach toward quantum functions [17,19]. A fundamental idea of introducing our quantum inference rules is to mimic each step of quantum computation produced by a quantum machine. In Sect. 4, we will formally introduce axioms and quantum inference rules to define this quantum proof system. For simplicity, we here abbreviate our proof system of quantum natural deduction as QND.

In Sect. 5, as the main contribution of this work, we will present fundamental features, known as the *soundness* and *completeness*, of the proof system QND, which establish a close connection between semantical meaning and syntactical manipulation. Such syntactical manipulation leads us to automated theorem proving in a quantum-mechanical fashion. In Sect. 6, we further show the *compactness* of QND by an application of the soundness and completeness.

2 Basic Definitions

We formally provide the fundamental definitions of quantum first-order logic (QFO) and its relevant notions. In what follows, we begin with the syntax and semantics of QFO.

2.1 Numbers, Languages, and Promise Problems

All the *integers* form the set $\mathbb{Z}$ and all *natural numbers* (i.e., nonnegative integers) do the set $\mathbb{N}$. Moreover, the notation $\mathbb{N}^+$ is used to denote $\mathbb{N} - \{0\}$. The sets $\mathbb{R}$ and $\mathbb{C}$ consist of all *real numbers* and of all *complex numbers*, respectively. For two numbers $m, n \in \mathbb{Z}$ with $m \leq n$, $[m, n]_{\mathbb{Z}}$ indicates the *integer interval* $\{m, m+1, m+2, \ldots, n\}$. This is compared to a real interval, e.g., $[0, 1]$. For each number $n \in \mathbb{N}^+$, $[1, n]_{\mathbb{Z}}$ is abbreviated as $[n]$. Given a natural number n, $\mathrm{ilog}(n)$ denotes $\lceil \log(n) \rceil$, provided that $\mathrm{ilog}(0)$ is set to be 0. An *alphabet* Σ is a nonempty finite set of "letters" or "symbols". Given an alphabet Σ, Σ^* denotes the set of all *strings over* Σ (i.e., finite series of symbols of Σ) and the *length* of a string w over Σ is denoted $|w|$. The special notation λ is reserved for the *empty string* of length 0. Given a string w of length n and a number $i \in [n]$, the notation $w_{(i)}$ indicates the ith symbol of w. Thus, w matches the series $w_{(1)} w_{(2)} \cdots w_{(n)}$. A *language over* Σ is a subset of Σ^*. A *promise (decision) problem* over Σ is a pair (A, R) of disjoint subsets of Σ^*, where elements in A are considered to be "positive" or "accepted" and elements in R are "negative" or "rejected". When $R = \Sigma^* - A$ (denoted $\overline{A}$), (A, R) is identified with A itself.

2.2 Quantum States and Quantum Operations

Throughout this work, we consider only finite-dimensional Hilbert spaces. We use Dirac's notation $|\cdot\rangle$ to express a quantum state and $\langle\cdot|$ for its dual notation. In particular, we use $|0\rangle$ and $|1\rangle$ as the computational basis. A *quantum bit* (or a *qubit*, for short) is a unit-norm vector of the 2-dimensional Hilbert space spanned by $\{|0\rangle, |1\rangle\}$. In other words, a qubit $|\phi\rangle$ has the form $\alpha|0\rangle + \beta|1\rangle$ for two coefficients $\alpha, \beta \in \mathbb{C}$ with $|\alpha|^2 + |\beta|^2 = 1$. More generally, we consider the Hilbert space of dimension 2^n spanned by the basis vectors of $\{|s\rangle : s \in \{0,1\}^n\}$. For each number $n \in \mathbb{N}$, the special notation $\mathcal{H}_{2^n}$ is used to denote this 2^n-dimensional Hilbert space. In contrast, the notation Φ_{2^n} refers to the set of all unit-norm quantum states in $\mathcal{H}_{2^n}$ and each element of Φ_{2^n} is called a *quantum string* (or a *qustring*) of length n. The *null vector* is denoted $\mathbf{0}$, which is treated as a unique element of Φ_0 although $\mathbf{0}$ in fact belongs to every $\mathcal{H}_i$. When the first qubit of $|\phi\rangle$ is observed (or measured) in the computational basis, we instantly obtain the quantum state $\langle 0|\phi\rangle$ and $\langle 1|\phi\rangle$ with the probability $\|\langle 0|\phi\rangle\|^2$ and $\|\langle 1|\phi\rangle\|^2$, respectively, where $\|\cdot\|$ indicates the *norm*, i.e., $\||\phi\rangle\| = \sqrt{\langle\phi|\phi\rangle}$. For example, when $|\phi\rangle = \frac{1}{\sqrt{2}}|011\rangle + \frac{1}{\sqrt{2}}|110\rangle$ is observed, we obtain $\langle 0|\phi\rangle = \frac{1}{\sqrt{2}}|11\rangle$ and $\langle 1|\phi\rangle = \frac{1}{\sqrt{2}}|10\rangle$. For other special uses of $|\cdot\rangle$, see [17,19].

A *quantum transform* is in general expressed as a unitary matrix on a finite-dimensional Hilbert space. We use the following special quantum transforms: $I(|\phi\rangle)$ (identity) and $ROT_\theta(|\phi\rangle)$ (rotation at angle θ) for any constant $\theta \in [0, 2\pi)$. The special operator $SWAP(i, |\phi\rangle)$ moves the ith qubit of $|\phi\rangle$ to the front. More formally, for any quantum state $|\phi\rangle = \sum_{u\in\{0,1\}^n} \alpha_u|u\rangle$, we define $SWAP(i, |\phi\rangle) = \sum_{u\in\{0,1\}^n} \alpha_u|u^{(i)}\rangle$, where $u^{(i)} = u_{(i)}u_{(1)}u_{(2)}\cdots u_{(i-1)}u_{(i+1)}\cdots u_{(n)}$.

The notion of promise problems can be expanded within a quantum framework to *quantum promise problems*, each of which is a pair (A, R) of disjoint sets composed of quantum states in the same dimensional Hilbert space.

3 Quick Review of Quantum First-Order Logic or QFO

We quickly review the syntax and semantics of the quantum first-order logic (QFO) given in [20]. Unfortunately, the precise definition of them are quite lengthy, and thus we here provide only "informal" descriptions of them and refer to [20] for the precise definitions and the underlying ideas of introducing them.

Analogously to the formalism of classical first-order logic (FO), we call $\mathcal{T}$ a *vocabulary* (or an *alphabet*) if $\mathcal{T}$ consists of quantum predicate symbols of fixed arities, and quantum function symbols of fixed arities, where quantum function symbols of arity 0 are called constant symbols. A *structure* (or a *model*) $\mathcal{M}$ is a set of "objects" assigned to elements of $\mathcal{T}$.

We use two categories of variables for classical numbers and quantum states. In addition to classical terms (built from the basis symbols $0, 1, n, \mathrm{ilog}(n)$), we use *quantum terms*, which are built inductively from quantum variables (called *pure-state quantum variable*) and the function symbols ($QBIT$ and MES). The notation $QBIT(i, y)$, abbreviated as $y[i]$, intuitively means the ith qubit (or the ith component) of y, $|y|$ means the *qubit size* (i.e., the number of qubits) of y, and $MES(b, y)$, abbreviated as $y\lceil_b$, means the result obtained by measuring the first qubit of y with respect to a bit b.

We then introduce two crucial notions: *interpretation* (or object assignment) and *evaluation* (or truth assignment). An interpretation assigns an "object" of $\mathcal{M}$ to each element of $\mathcal{T}$. For example, a quantum state $|\phi_y\rangle$ is assigned to a pure-state quantum variable y. For a classical term i expressing a natural number $\underline{i}$, $QBIT(i, y)$ is interpreted as $SWAP(\underline{i}, |\phi_y\rangle)$. For a quantum term s

expressing a pure quantum state $|\phi_s\rangle = \sum_{j\in[2^{|s|}]} \beta_j|j\rangle$, $QBIT(s,y)$ is interpreted as $SWAP(|\phi_s\rangle, |\phi_y\rangle)$, which equals $\sum_{j\in[2^{|s|}]} \beta_j SWAP(j, |\phi_y\rangle)$.

In FO, the special notation X is used as a predicate symbol, expressing an input given to an underlying classical computation. In contrast, in QFO, X is treated as a function symbol, called the *instance function symbol* whose value $X(i)$ expresses the ith qubit of a pure quantum state assigned to X.

The expression $z \simeq_\eta b$ for the quantum predicate symbol $\simeq_\eta$ (with $\eta \in [0,1]$) means that b is observed with probability at least η when measuring z. The predicate symbols, I and ROT_θ, are used in the form $R(x:y)$ (where $R \in \{I, ROT_\theta \mid \theta \in [0,2\pi)\}$), which means that the first argument x is transformed to the second argument y by the quantum operator R. Quantum formulas are constructed from quantum predicates with classical/quantum terms by applying $(\exists i \le s)$, $(\forall i \le s)$, $(\exists^Q y, |y|=s)$, $(\forall^Q |y|=s)$, $\wedge$ (quantum AND), $\|$ (quantum OR), $\neg^Q$ (quantum NOT), where $\forall^Q$ and $\exists^Q$ are used only for pure-state quantum variables y. Quantum AND and quantum OR in particular have the forms $P \wedge Q$ and $(z)[P(z) \| Q(z)]$, respectively. Here, we remark that an application of quantum connectives $\{\wedge, \|, \neg^Q\}$ is intended to "mimic" each step of a quantum machine. Based on a given interpretation, we evaluate each quantum formula A to be either 0 or 1; that is, $eval[A](\mathcal{M}) = b$ for a given structure $\mathcal{M}$ and a bit b. For example, we set $eval[P \wedge Q](\mathcal{M}) = 1$ exactly when $eval[P](\mathcal{M}) = eval[Q](\mathcal{M}) = 1$. Moreover, $(z)[P(z) \| Q(z)]$ is evaluated to be 1 iff, when z is measured to be 1 (resp., 0), $P(z)$ (resp., $Q(z)$) is evaluated to be 1 (resp., 0). Unlike classical NOT, quantum NOT affects only the measurement predicate (not $\wedge$ or $\|$) and quantum quantifiers. For example, the evaluations of $\neg^Q(z \simeq_\eta b)$ and $\neg^Q(\exists^Q y, |y|=t)P(y)$ are the same as those of $z \simeq_\eta \bar{b}$ and $(\forall^Q y, |y|=t)(\neg^Q P(y))$, respectively, where $\bar{b} = 1-b$ for a bit b. In our system, we do not have the classical inequality ($\neq$) because it is possible to express "$x \neq y$" between two natural numbers x and y by the quantum formula of the form $(\exists^Q z, |z|=1)[(z)[z \simeq_1 0 \wedge x < y \| z \simeq_1 1 \wedge y < x]]$.

Unfortunately, not all quantum formulas are admissible in quantum mechanics. We thus need to restrict the above quantum formulas by imposing the "well-formedness" conditions on the quantum formulas. A quantum variable y (as well as its component $y[i]$) in a given quantum formula ϕ is called *predecessor-dependent* if it appears in the second argument place of a certain predicate symbol of ϕ; otherwise, it is called *predecessor-independent*. For each quantum formula R, the *variable connection graph* of R is a directed graph $G_R = (V,E)$, where E consists of all pairs (y,z) for two quantum variables y and z such that there exists a quantum predicate symbol in a given quantum formula for which either y or its component $y[i]$ is the first argument of R and either z or its component $z[j]$ is the second argument of R.

Unlike classical quantifiers, we acknowledge two different roles played by quantum quantifiers. Given a quantum formula ϕ of the form $(\exists^Q y, |y|=s)R(y)$, if y is predecessor-dependent in ϕ, then $\exists^Q y$ is called *consequential quantum existential quantifier*. Otherwise, we call it an *introductory quantum existential quantifier*. Notice that quantum universal quantifiers are always treated as introductory quantum quantifiers. The introductory quantum existential quantifiers together with quantum universal quantifiers are collectively called *introductory quantum quantifiers*.

A structure $\mathcal{M}$ is said to *satisfy* a quantum formula ϕ (or ϕ is *satisfiable* by $\mathcal{M}$) if ϕ is evaluated to be 1 by $\mathcal{M}$ (i.e., $eval[\phi](\mathcal{M}) = 1$). We use the notation $\mathcal{M} \models \phi$ to express that $\mathcal{M}$ satisfies ϕ. A quantum formula P is said to be *quantumly valid* (or simply *valid*) if it is satisfiable by all structures. Moreover, a set Ψ of quantum formulas is *satisfiable* if there exists a structure $\mathcal{M}$ and an evaluation such that $eval[A](\mathcal{M},\xi) = 1$ for all $A \in \Psi$. Otherwise, Ψ is *unsatisfiable*.

The notation $Err(\phi)$ denotes the sum of all values $1-\eta$ that appear in the form of $y \simeq_\eta b$ in ϕ. A promise problem $(L^{(+)}, L^{(-)})$ over the binary alphabet $\{0,1\}$ is said to be *(syntactically) expressible* by a quantum sentence ϕ if there exists a fixed constant $\varepsilon_0 \in [0,1/2)$ such that, for any given string $x \in \{0,1\}^*$, (1) $Err(\phi) \leq \varepsilon_0$ holds, and (2) assuming that the length of x evaluates the constant symbol n and x itself evaluates the instance function symbol X, if $x \in L^{(+)}$, then these values make ϕ true, and if $x \in L^{(-)}$, then they make $\neg^Q \phi$ true. We write QFO to denote the class of all promise problems that are expressible by quantum sentences of quantum first-order logic.

4 A Proof System of Quantum Natural Deduction: QND

In what follows, we will introduce a key notion of quantum proof system of quantum natural deduction. Remember that Gentzen's natural deduction aims at proving tautologies by a series of applications of inference rules whereas quantum natural deduction tries to implement quantum computation starting with given inputs by applications of quantum inference rules. Since quantum computation is reversible, we cannot freely discard quantum formulas that appear in the derivations.

4.1 Axioms and Inference Rules

Firstly, we list a set of axioms. Note that the evaluations of $X(y)$ and n depend on a given input to a target promise problem $\mathcal{L}$. Fix $b \in \{0,1\}$ and $\theta \in [0,2\pi)$.

Axioms: *Axioms* are of the form:

(1) $|X| = n$, $|y| = \mathrm{ilog}(n) \wedge |X(y)| = 1$, $t \leq \mathrm{ilog}(n) \wedge |x[t]| = 1$, $t \leq \mathrm{ilog}(n) \wedge t < suc(t)$, and $t = t$.
(2) $|y| = 1 \wedge \mathcal{P}_I(b:y) \wedge y \simeq_1 b$, $|y| = 1 \wedge \mathcal{P}_I(b:y) \wedge y \simeq_0 \bar{b}$, $|y| = 1 \wedge y\lceil_b \simeq_1 b$, and $|y| = 1 \wedge y\lceil_b \simeq_0 \bar{b}$.
(3) $|y| = 1 \wedge \mathcal{P}_{ROT_\theta}(0:y) \wedge y \simeq_{\cos^2\theta} 0$, $|y| = 1 \wedge \mathcal{P}_{ROT_\theta}(0:y) \wedge y \simeq_{\sin^2\theta} 1$, $|y| = 1 \wedge \mathcal{P}_{ROT_\theta}(1:y) \wedge y \simeq_{\sin^2\theta} 0$, and $|y| = 1 \wedge \mathcal{P}_{ROT_\theta}(1:y) \wedge y \simeq_{\cos^2\theta} 1$.
(4) $(z\lceil_b)\lceil_b \simeq_1 b$, $(z\lceil_b)\lceil_{\bar{b}} \simeq_0 b$, and $(z\lceil_b)\lceil_{\bar{b}} \simeq_0 \bar{b}$.

Our proof system of quantum natural deduction is composed of introduction rules and elimination rules. In some inference rules, quantum formulas are expanded by attaching auxiliary "assumptions" in the form of the pair $[\cdot]$ of brackets, such as $P(i,x,y)$ $[i \leq \mathrm{ilog}(n)]$, $P(x,y)$ $[|y| = \mathrm{ilog}(n)]$, and $P(i,x,y)$ $[i \leq |y|]$. Moreover, for a string $c \in \{0,1\}^{\mathrm{ilog}(n)}$ and for any quantum term y of qubit size $\mathrm{ilog}(n)$, an abbreviation "$y \simeq_\eta c$" is used to denote the quantum formula $(\forall i \leq \mathrm{ilog}(n))[y[i] \simeq_{\eta_i} c_{(i)}]$, where $c_{(i)}$ is the ith bit of c and η_i's satisfy $\eta = \prod_{i \leq \mathrm{ilog}(n)} \eta_i$. A *double horizontal line* used in some of the inference rules indicates an omission of a derivation consisting of quantum formulas manipulated by a series of inference rules.

In each inference rule, a formula at the bottom of the inference rule is called a *principal quantum formula* and formula(s) in the top of the inference rule are called *auxiliary quantum formulas*.

Equality, Inequality, and Successor: We first discuss inference rules regarding classical/quantum terms.

[Numeral Inference Rules]. Let f and ϕ denote arbitrary function and predicate symbols.

(1) $\frac{x_1=y_1 \wedge \cdots \wedge x_k=y_k}{f(x_1,\ldots,x_k)=f(y_1,\ldots,y_k)}$ (f-change)

(1') $\frac{\phi(x_1,\ldots,x_k) \wedge x_1=y_1 \wedge \cdots \wedge x_k=y_k}{\phi(y_1,\ldots,y_k)}$ (ϕ-change)

(2) $\frac{x \leq y \wedge y \leq x}{x=y}$ (=-intro) (2') $\frac{x=y}{x \leq y \wedge y \leq x}$ (=-elimi)

(3) $\frac{x=y}{y=x}$ (=-symmet) (3') $\frac{x=y \wedge y=z}{x=z}$ (= -transi)

(4) $\frac{x \leq y}{x < suc(y)}$ (suc-right-intro) (4') $\frac{x<y}{suc(x) \leq y}$ (suc – left-intro)

[Measuremental Inference Rules]. The following inference rules concern quantum measurement.

(1) $\frac{y[s] \simeq_\eta b}{\neg^Q(y[s] \simeq_\eta 1-b)}$ ($\neg^Q$-intro) (1') $\frac{\neg^Q(y[s] \simeq_\eta b)}{y[s] \simeq_\eta 1-b}$ ($\neg^Q$-elimi)

Remark: rules (1) and (1') are only meaningful whenever the interpretation $\xi(y[s]) = \rho_{y[s]}$ satisfies either $\|\langle 0|\rho_{y[s]}\|^2 \geq \eta$ or $\|\langle 1|\rho_{y[s]}\|^2 \geq \eta$.

(2) $\frac{y[s] \simeq_{\eta'} b}{y[s] \simeq_\eta b}$ ($\simeq$-change), provided that $\eta \leq \eta'$.

(3) $\frac{|z|=1 \wedge z \simeq_\eta b}{|z|=1 \wedge z\lceil_b \simeq_\eta b}$ ($\lceil\simeq$-intro) (3') $\frac{|z|=1 \wedge z\lceil_b \simeq_\eta b}{|z|=1 \wedge z \simeq_\eta b}$ ($\lceil\simeq$-elimi)

(4) $\frac{P_F(x[s]:y[t]) \wedge x[s] \simeq_{\eta'} b'}{P_F(x[s]:y[t]) \wedge y[t] \simeq_\eta b}$ ($\simeq$-replace), provided that the following condition (*) holds.

Condition (*): when $F = I$, (η', b') equals (η, b). When $F = ROT_\theta$, (η', b') must satisfy that $\|\langle b|\rho_{y[t]}\|^2 \geq \eta$ iff $\|\langle b'|\rho_{x[s]}\|^2 \geq \eta'$.

[Structural Inference Rules]. The next two inference rules regard structural properties.

(1) $\dfrac{P \quad P}{P}$ (contra) (2) $\dfrac{R_1 \wedge P \wedge Q \wedge R_2}{R_1 \wedge Q \wedge P \wedge R_2}$ ($\wedge$-excha) (3)

$$\dfrac{P \qquad \dfrac{\overline{P}}{Q \wedge R}}{P \wedge R} \text{ (cut)}$$

Remark: the rule of "contra" can be derived from the rules of $\wedge$-intro and $\wedge$-elimi. In the rule of $\wedge$-excha, R_1 and R_2 may be empty. The top horizontal line above P in the cut rule indicates that this is no more inference rule applicable to this P.

[Conditional Inference Rules]

(1) $$\dfrac{\dfrac{\overline{i \le t}}{P(i)}}{P(i)\ [i \le t]}$$ ([·]-intro) (2) $\dfrac{s \le t \wedge P(i)\ [i \le t]}{s \le t \wedge P(s)}$ ([·]-elimi)

Remark: the [·]-elimi rule is redundant because it can be derived by the rules of $\forall$-intro and $\forall$-elimi. Since the assumption $i \le t$ in the [·]-intro is "absorbed" into the quantum formula $P(x, y)\ [i \le t]$, the first horizontal line above $i \le t$ indicates that $i \le t$ is no longer an assumption (or a hypothesis). In this case, we say that the assumption is *discharged*.

Other Quantum Inference Rules: Next, we introduce the remaining inference rules concerning quantum logical connectives and quantum quantifiers. Our treatment of quantum OR and quantum NOT are particularly different from their classical counterparts.

[Logical Inference Rules]. The following condition is applied: all quantum formulas in each inference rule must be well-formed. Let b denote any constant in $\{0, 1\}$.

(1) $\frac{P \quad Q}{P \wedge Q}$ ($\wedge$-intro) (1′) $\frac{P \wedge Q}{P}$ $\frac{P \wedge Q}{Q}$ ($\wedge$-elimi), where both rules must be applied together at once.

(2) $\frac{|z|=1 \wedge P(z\lceil_0, y) \qquad |z|=1 \wedge Q(z\lceil_1, y)}{|z|=1 \wedge (z)[P(z,y) \| Q(z,y)]}$ ($\|$-intro)

(2′) $\frac{|z|=1 \wedge P(z\lceil_0, y)}{|z|=1 \wedge (z\lceil_0)[P(z,y) \| Q(z,y)]}$ ($\|\lceil_0$-intro)

(2″) $\frac{|z|=1 \wedge Q(z\lceil_1, y)}{|z|=1 \wedge (z\lceil_1)[P(z,y) \| Q(z,y)]}$ ($\|\lceil_1$-intro)

(2‴) $\frac{|z|=1 \wedge (z)[P(z,y) \| P(z,y)]}{|z|=1 \wedge P(z,y)}$ ($\|_I$-elimi)

(2⁗) $$\dfrac{|z| = 1 \wedge (z)[P(z, y) \parallel Q(z, y)] \qquad \dfrac{\overline{|z| = 1 \wedge P(z\lceil_0, y)}}{R} \qquad \dfrac{\overline{|z| = 1 \wedge Q(z\lceil_1, y)}}{R}}{|z| = 1 \wedge R}$$

($\|$-elimi).

Remark: the top horizontal lines above $|z| = 1 \wedge P(z\lceil_0, y)$ and $|z| = 1 \wedge Q(z\lceil_1, y)$ indicate that those assumptions are discharged, that is, no more inference rules are applicable to these quantum formulas. The rule of $\|_I$-elimi is actually a special case of $\|$-elimi.

(3) $\frac{P(x[j], y)\ [j \le t]}{(\forall i \le t)P(x[i], y)}$ ($\forall$-intro), where j is a fresh free classical variable and t is a classical term.

(3′) $\dfrac{(\forall i \leq t)P(x[i], y)}{s \leq t \wedge P(x[s], y)}$ ($\forall$-elimi), where s, t are classical terms.

Remark: this inference rule looks quite different from the corresponding classical one because we have no logical connective expressing "implication".

(4) $\dfrac{s \leq t \wedge P(x[s], y)}{(\exists i \leq t)P(x[i], y)}$ ($\exists$-intro), where s and t are classical terms.

(4′) $$\dfrac{(\exists i \leq t)P(x[i], y) \qquad \dfrac{\overline{j \leq t \wedge P(x[j], y)}}{Q}}{Q}$$ ($\exists^Q$-elimi), where j is a fresh classical variable and Q does not include j. The top horizontal line above $j \leq t \wedge P(x[j], y)$ means that this quantum formula is discharged, and thus no more inference rule is applicable to this quantum formula.

(5) $\dfrac{P(x, u) \; [|u| = \ell]}{(\forall^Q y, |y| = \ell)P(x, y)}$ ($\forall^Q$-intro), where u is a fresh free variable and predecessor dependent in P.

(5′) $\dfrac{(\forall^Q y, |y| = \ell)P(x, y)}{|t| = \ell \wedge P(x, t)}$ ($\forall^Q$-elimi), where t is a quantum term and y is predecessor dependent in P.

Remark: a similar condition imposed on the $\forall$-elimi rule must be applied to this rule.

(6) $\dfrac{|t| = \ell \wedge P(x, t)}{(\exists^Q y, |y| = \ell)P(x, y)}$ ($\exists^Q$-intro), where t is a quantum term and predecessor dependent in P.

(6′) $$\dfrac{(\exists^Q y, |y| = \ell)P(x, y) \qquad \dfrac{\overline{|z| = \ell \wedge P(x, z)}}{Q}}{Q}$$ ($\exists^Q$-elimi), where z is a fresh free variable and predecessor dependent in P, and Q does not include z. Here, the quantum formula $|z| = \ell \wedge P(x, z)$ is discharged as stated before.

(7) $\dfrac{P}{\neg^Q\neg^Q P}$ ($\neg^Q\neg^Q$-intro) (7′) $\dfrac{\neg^Q\neg^Q P}{P}$ ($\neg^Q\neg^Q$-elimi)

(8) $\dfrac{\neg^Q P \wedge \neg^Q Q}{\neg^Q(P \wedge Q)}$ ($\neg^Q\wedge$-intro) (8′) $\dfrac{\neg^Q(P \wedge Q)}{\neg^Q P \wedge \neg^Q Q}$ ($\neg^Q\wedge$-elimi)

(9) $\dfrac{|z| = 1 \wedge (z)[\neg^Q P(z, y) \,\|\, \neg^Q Q(z, y)]}{|z| = 1 \wedge \neg^Q(z)[P(z, y) \,\|\, Q(z, y)]}$ ($\neg^Q\|$-intro)

(9′) $\dfrac{|z| = 1 \wedge \neg^Q(z)[P(z, y) \,\|\, Q(z, y)]}{|z| = 1 \wedge (z)[\neg^Q P(z, y) \,\|\, \neg^Q Q(z, y)]}$ ($\neg^Q\|$-elimi)

(10) $\dfrac{(\exists^Q y, |y| = \ell)(\neg^Q P(x, y))}{\neg^Q(\forall^Q y, |y| = \ell)P(x, y)}$ ($\neg^Q\forall^Q$-intro)

(10′) $\dfrac{\neg^Q(\forall^Q y, |y| = \ell)P(x, y)}{(\exists^Q y, |y| = \ell)(\neg^Q P(x, y))}$ ($\neg^Q\forall^Q$-elimi)

(11) $\dfrac{(\forall^Q y, |y| = \ell)(\neg^Q P(x, y))}{\neg^Q(\exists^Q y, |y| = \ell)P(x, y)}$ ($\neg^Q\exists^Q$-intro)

(11′) $\dfrac{\neg^Q(\exists^Q y, |y| = \ell)P(x, y)}{(\forall^Q y, |y| = \ell)(\neg^Q P(x, y))}$ ($\neg^Q\exists^Q$-elimi)

Treatment of X for Classical Inputs: The instance function symbol X should be treated in a quite different way. We deal with only the case where inputs to

$\mathcal{L}$ are classical strings. Since X represents such an input, we write $X^{-1}(b)$ to mean the set $\{z \in \{0,1\}^{\mathrm{ilog}(n)} \mid X(z) = b\}$.

(1) $\dfrac{|y| = \mathrm{ilog}(n) \wedge y \simeq_{\eta_c} c\ [c \in X^{-1}(b)]}{|y| = \mathrm{ilog}(n) \wedge X(y) \simeq_{\eta} b}$ (X $\simeq$-intro), provided that $\eta = \sum_{c \in X^{-1}(b)} \eta_c$.

Remark: the notation in the top of the auxiliary quantum formula means the series of the quantum formulas of the form $|y| = \mathrm{ilog}(n) \wedge y \simeq_{\eta_c} c$ for all choices of c in $X^{-1}(b)$.

(2) $\dfrac{|y| = \mathrm{ilog}(n) \wedge X(y) \simeq_{\eta'} b'}{|y| = \mathrm{ilog}(n) \wedge |z| = 1 \wedge \mathcal{P}_F(X(y) : z) \wedge z \simeq_{\eta} b}$ ($X\mathcal{P}_F$-intro), provided that if $F = I$, then (η', b') equals (η, b); if $F = ROT_\theta$, then (η', b') satisfies that $\|\langle b' | \phi_{X(y)} \rangle\|^2 \geq \eta'$ iff $\|\langle b | \rho_z \|^2 \geq \eta$.

(3) $\dfrac{|y| = \mathrm{ilog}(n) \wedge P(X(y)\lceil_0, z) \quad |y| = \mathrm{ilog}(n) \wedge Q(X(y)\lceil_1, z)}{|y| = \mathrm{ilog}(n) \wedge (X(y))[P(X(y), z) \parallel Q(X(y), z)]}$ ($X \parallel$ -intro).

4.2 Quantum Proofs

Now, we introduce the notion of QND quantum proofs.

Definition 1 (Quantum proofs). *(1) A* QND quantum proof *(or a QND-proof, for short) of a quantum formula ϕ is a finite tree whose nodes are labeled with quantum formulas such that the root is labeled by ϕ, all leaves are labeled by axioms, and each non-leaf node labeled by a formula ψ follows from a node labeled by another formula ψ' obtained directly by applying one of the inference rules. A formula ϕ is said to be* provable in QND *if there exists a QND quantum proof of ϕ. (2) Given a set Ψ of quantum formulas, if we allow the leaves of a proof tree to be labeled by elements of Ψ, then we call such a proof a* QND-Ψ-proof. *If we have a QND-Ψ-proof of a quantum formula A, then we succinctly write $\Psi \vdash A$. (3) A promise problem $(L^{(+)}, L^{(-)})$ is said to be* provable in QND *if there exists a quantum sentence $\phi_{n,X}$ (i.e., containing no free variable) such that, for any $n \in \mathbb{N}$, (i) for any $x \in L^{(n)} \cap \{0,1\}^n$, $\phi_{n,X}$ is provable in QND, (ii) for any $x \in L^{(-)} \cap \{0,1\}^n$, $\neg^Q \phi_{n,X}$ is provable in QND, and (iii) $\phi_{n,X}$ syntactically expresses $(L_{(+)}, L^{(-)})$*

Here, we present a simple example of quantum proof. The target promise problem $\mathcal{L} = (L^{(+)}, L^{(-)})$ is composed of $L^{(+)} = \{x \in \{0,1\}^* \mid \#_1(x) \geq (2/3) \cdot 2^{|x|}\}$ and $L^{(+)} = \{x \in \{0,1\}^* \mid \#_1(x) \leq (1/3) \cdot 2^{|x|}\}$ (where $\#_b(x)$ denotes the total number of b in x). It is not difficult to see that $\mathcal{L}$ is semantically expressible in QFO. In what follows, we assign a number $n \in \mathbb{N}$ and a string in $L^{(+)} \cap \{0,1\}^n$ to the symbols n and X, respectively. The corresponding quantum sentence $\phi_{n,X}$ is defined to be $(\exists^Q y, |y| = \mathrm{ilog}(n))[(\forall i \leq |y|)(\mathcal{P}_{ROT_{\pi/4}}(0 : y[i])) \wedge X(y) \simeq_{\eta'} 1]$, where $\eta' = \sum_{c \in X^{-1}(1)} \eta(c) \geq 2/3$, $\eta(c) = \prod_{i \leq \mathrm{ilog}(n)} \eta(c, i)$, and $\eta(c, i) = 1/2$ for any $c \in \{0,1\}^{\mathrm{ilog}(n)}$. We also choose an interpretation ξ satisfying $\xi(u) = 2^{2\mathrm{ilog}(n)} \sum_{c \in \{0,1\}^n} |c\rangle$. To simplify the rendering of a proof tree, we slightly relax its inference rules by allowing the presence of other irrelevant quantum formulas as in this example. For example, the original $\forall^Q$-intro rule is relaxed to:

$$\frac{R \wedge P(x,u)\ [|u| = \ell]}{R \wedge (\forall^Q y, |y| = \ell) P(x,y)}\ (\forall^Q\text{-intro})$$

for any quantum formula R. The dashed horizontal line below corresponds to the abbreviation of "$u \simeq_{\eta(c)} c$".

$$\dfrac{\dfrac{\dfrac{\dfrac{\dfrac{(\forall i \le |u|)(\mathcal{P}_{ROT_{\pi/4}}(0:u[i])) \qquad \dfrac{\dfrac{\dfrac{(\forall i \le |u|)(\mathcal{P}_{ROT_{\pi/4}}(0:u[i]))}{\mathcal{P}_{ROT_{\pi/4}}(0:u[j])} \qquad \dfrac{\overline{j \le |u|} \qquad |u| = \mathrm{ilog}(n) \wedge (\mathcal{P}_{ROT_{\pi/4}}(0:u[j]) \wedge u[j] \simeq_{\eta(c,j)} c_{(j)})\ [c \in X^{-1}(1)]}{|u| = \mathrm{ilog}(n) \wedge (\mathcal{P}_{ROT_{\pi/4}}(0:u[j]) \wedge u[j] \simeq_{\eta(c,j)} c_{(j)})\ [j \le |u|, c \in X^{-1}(1)]}}{\mathcal{P}_{ROT_{\pi/4}}(0:u[j]) \wedge |u| = \mathrm{ilog}(n) \wedge (\mathcal{P}_{ROT_{\pi/4}}(0:u[j]) \wedge u[j] \simeq_{\eta(c,j)} c_{(j)})\ [j \le |u|, c \in X^{-1}(1)]}}{|u| = \mathrm{ilog}(n) \wedge (\mathcal{P}_{ROT_{\pi/4}}(0:u[j]) \wedge u[j] \simeq_{\eta(c,j)} c_{(j)})\ [j \le |u|, c \in X^{-1}(1)]}}{(\forall i \le |u|)(\mathcal{P}_{ROT_{\pi/4}}(0:u[i])) \wedge |u| = \mathrm{ilog}(n) \wedge u[j] \simeq_{\eta(c,j)} c_{(j)}\ [j \le |u|, c \in X^{(-1)}(1)]}}{|u| = \mathrm{ilog}(n) \wedge (\forall i \le \mathrm{ilog}(n))(\mathcal{P}_{ROT_{\pi/4}}(0:u[i])) \wedge u[j] \simeq_{\eta(c,j)} c_{(j)})\ [j \le |u|, c \in X^{(-1)}(1)]}}{|u| = \mathrm{ilog}(n) \wedge (\forall i \le \mathrm{ilog}(n))(\mathcal{P}_{ROT_{\pi/4}}(0:u[i])) \wedge (\forall i \le |u|)(u[i] \simeq_{\eta(c,i)} c_{(i)})\ [c \in X^{(-1)}(1)]}}{|u| = \mathrm{ilog}(n) \wedge (\forall i \le |u|)(\mathcal{P}_{ROT_{\pi/4}}(0:u[i])) \wedge u \simeq_{\eta(c)} c\ [c \in X^{(-1)}(1)]}$$

$$\dfrac{\dfrac{|u| = \mathrm{ilog}(n) \wedge (\forall i \le |u|)(\mathcal{P}_{ROT_{\pi/4}}(0:u[i])) \wedge u \simeq_{\eta(c)} c\ [c \in X^{(-1)}(1)]}{|u| = \mathrm{ilog}(n) \wedge (\forall i \le |u|)(\mathcal{P}_{ROT_{\pi/4}}(0:u[i])) \wedge X(u) \simeq_{\eta} 1}}{(\exists^Q y, |y| = \mathrm{ilog}(n))[(\forall i \le |y|)(\mathcal{P}_{ROT_{\pi/4}}(0:y[i])) \wedge X(y) \simeq_{\eta} 1]}$$

The negation $\neg^Q \phi_{n,X}$ is similarly proven for X that represents a string in $L^{(-)} \cap \{0,1\}^n$. Therefore, $\mathcal{L}$ is provable in QND.

5 Fundamental Properties of the QND Proof System

The most fundamental properties of a logical system include (local) soundness and (local) completeness. These properties establish a bridge between semantic meaning and syntactic manipulation.

5.1 Soundness of QND

We begin with demonstrating the soundness of QND, which states that every quantum formula provable in QND is semantically valid. For this purpose, we explain conventional terminology. A quantum formula ψ is said to be a *logical consequence* of a set Ψ of quantum formulas if every evaluation (induced by any interpretation) that satisfies Ψ also satisfies ψ. To express this situation, we write $\Psi \models \psi$. A *universal closure* of a formula $\phi(a_1, a_2, \ldots, a_n)$ involving with free variables $a_1, a_2, \ldots, a_n$ is of the form $(\forall x_1)(\forall x_2) \cdots (\forall x_n)\phi(x_1, x_2, \ldots, x_n)$, where $x_1, x_2, \ldots, x_n$ are bound variables. For convenience, we denote by $\forall\phi$ for the universal closure of a quantum formula ϕ. For a set $\mathcal{C}$ of quantum formulas, $\forall\mathcal{C}$ denotes the set of all universal closures of any quantum formulas in $\mathcal{C}$.

Proposition 2 (Soundness). *Let Ψ denote a set of quantum formulas. For any quantum formula ψ, if there is a QND quantum proof Π of ψ from assumptions, which are a finite subset Ξ of Ψ, then ψ is a logical consequence of $\forall\Psi$ (i.e., the universal closure of Ψ).*

Proof Sketch. For the direct use of mathematical induction, it is more convenient to prove a slightly stronger statement described below.

Claim 3. *If Π is a QND-Ψ-proof of ψ, then $\forall\psi$ is a logical consequence of $\forall\Psi$.*

In what follows, we sketch the proof of this claim. The following argument proceeds by induction on the number of quantum formulas in Π. If ψ is an axiom, say, B, then its universal closure $\forall B$ is semantically valid, and thus it is a logical consequence of $\forall \Psi$. Assuming otherwise, we pick up inference rules one by one and verify that, for each inference rule, the universal closure of its bottom quantum formula is a logical consequence of the universal closure of the top quantum formula(s). Here, we show only one significant case.

(**$\|$-intro**) Assume that $\forall \Psi \models \forall(|z| = 1 \wedge (z)[P(x,z) \| Q(x,z)])$. Let us consider any structure $\mathcal{M}$ that satisfies $\forall \Psi$; namely, $eval[|z| = 1 \wedge (z)[P(x,z) \| Q(x,z)](\mathcal{M}, \rho_{a_1}, \ldots, \rho_{a_m}) = 1$ holds for all free variables $a_1, \ldots, a_m$, where ρ_a is an object assigned to a. From $eval[(z)[P(x,z) \parallel Q(x,z)](\mathcal{M}, \rho_{a_1}, \ldots, \rho_{a_m}) = 1$, it follows that $eval[P(x,z)](\mathcal{M}, \rho_{a_1}, \ldots, \rho_{a_m}, \rho_{z_0}) = 1$ and $eval[P(x,z)](\mathcal{M}, \rho_{a_1}, \ldots, \rho_{a_m}, \rho_{z_1}) = 1$, where $\xi(z\lceil_0) = \rho_{z_0}$ and $\xi(z\lceil_1) = \rho_{z_1}$ by an appropriate interpretation ξ. Hence, we conclude that $\forall \Psi \models \forall(P(x, z\lceil_0))$ and $\forall \Psi \models \forall(Q(x, z\lceil_1))$. □

5.2 Completeness of QND

The completeness of QND roughly states that every semantically valid quantum formula is provable in QND. We remark that, in our formalism, the vocabulary $\mathcal{T}$ contains only countably many function symbols and predicate symbols.

Proposition 4 (Completeness). *Given a semantically valid quantum formula ϕ, there exists a set Ψ of quantum formulas satisfying $\Psi \models \phi$ and there is also a finite subset Ψ' of Ψ for which $\Psi' \vdash \phi$ holds with no use of "elimination" rules or the "cut" rule.*

Proof Sketch. We intend to show that, according to the definition of semantical validity, all valid quantum formulas are indeed provable. The validity of a quantum formula is determined by the inductive definitions of evaluation. Hence, it suffices to verify that each inductive definition can be syntactically realized by a certain proof subtree.

We loosely follow an argument used to prove the completeness of the classical predicate calculus. See, e.g., [5]. We first enumerate all quantum formulas $P_1, P_2, \ldots$, all classical terms $t_1, t_2, \ldots$, and all quantum terms $s_1, s_2, \ldots$ We remark that these enumerations contain free variables that appear in Ψ. For a later use, however, we assume that the enumerations further contain a countably infinite set of free variables not in Ψ. We further enumerate all pairs of the form (P_i, t_j) and (P_i, s_j).

The following construction proceeds according to the choice of the next pair, either (P_i, t_j) or (P_i, s_j). If P_i does not appear as some of the leaves of Π, then we skip this step and move to the next one. In what follows, we assume that P_i appears in some of the leaves of Π.

(1) If $P_i \in \Psi$, then we do nothing. (2) If P_i is atomic, then we skip this step and move on to the next step. (3) Assume otherwise. Depending on the form of P_i, we modify Π by replacing P_i with an associated derivation as follows. Here, we show only two important cases.

(quantum $\exists^Q$) Assume that P_i has the form $(\exists^Q y, |y| = \ell)P(x, y)$. Assuming that (P_i, t_j) is selected, if $P(x, t_j)$ is not satisfied, then we skip this step and move to the next step. Assume otherwise. In this case, we replace P_i by the following derivation, update Π, and move to the next step.

$$\frac{|t_j| = \ell \wedge P(x, t_j)}{(\exists^Q y, |y| = \ell)P(x, y)}\,.$$

(quantum $\forall^Q$) If P_i has the form $(\forall^Q y, |y| = \ell)P(x, y)$, then we choose a fresh free variable u and replace P_i with the following derivation.

$$\frac{P(x, u)\ [|u| = \ell]}{(\forall^Q y, |y| = \ell)P(x, y)}\,.$$ □

6 Applications of the Soundness and Completeness

We investigate other important properties of QND, which will be useful in a future study of quantum logics. One of such properties is the "compactness" of QND.

Proposition 5 (Compactness). *Let Ψ be an unsatisfiable set of quantum formulas. There exists a finite subset of Ψ that is unsatisfiable.*

Proof. Assume that Ψ is unsatisfiable. This means that there is no evaluation satisfying all quantum formulas in Ψ. We first replace all free variables appearing in Ψ by different constant symbols so that all quantum formulas in Ψ become quantum sentences. Even by this modification of Ψ, the notion of unsatisfiability does not change. We thus use the same notation Ψ to denote the resulting set.

Next, we define the quantum sentence $CONTRA$ to be $(\forall^Q z, |z| = 1)(\mathcal{P}_I(0 : z) \wedge z \simeq_1 1)$. It then follows that $CONTRA$ is unsatisfiable, and thus it is not provable in QND by the soundness and completeness of QND. We then assert the following claim.

Claim 6. *Ψ is unsatisfiable iff $\forall\Psi \vdash CONTRA$.*

Next, we prove the proposition with the use of Claim 6. Since $\forall\Psi$ is unsatisfiable, by the claim, we obtain $\forall\Psi \vdash CONTRA$. Since every QND-$\forall\Psi$-proof is a finite tree, we can take a finite subset Ξ of $\forall\Psi$ such that $CONTRA$ has a QND-Ξ-proof. By the claim again, we conclude that Ξ is unsatisfiable. □

References

1. Barrington, D.A.M., Immerman, N., Straubing, H.: On uniformity within NC^1. J. Comput. System Sci. **41**, 274–306 (1990)
2. Birkhoff, G., von Neumann, J.: The logic of quantum mechanics. Ann. Math. **37**, 823–843 (1936)
3. Bernstein, E., Vazirani, U.: Quantum complexity theory. SIAM J. Comput. **26**, 1411–1473 (1997)

4. Baltag, A., Smets, S.: Quantum logic as a dynamic logic. Synthese **179**, 285–306 (2011)
5. Cook, S., Nguyen, P.: Logical Fondations of Proof Complexity. Cambridge University Press, Cambridge (2010)
6. Fagin, R.: Gneralized first-order spectra and polynomial-time recognizable sets. In: Karp, R. (ed.) Complexity of Computation. SIAM-AMS Proceedings, vol. 7, pp. 27–41 (1974)
7. Gentzen, G.K.E.: Untersuchungen über das logische Schließen. I. Math. Zeitschrift **39**, 176–210 (1934)
8. Gentzen, G.K.E.: Untersuchungen über das logische Schließen. II. Math. Zeitschrift **39**, 405–431 (1935)
9. Gurevich, Y., Lewis, H.R.: A logic for constant-depth circuits. Inform. Control **61**, 65–74 (1984)
10. Immerman, N.: Relational queries computable in polynomial time. Inform. Control **68**, 86–104 (1986)
11. Immerman, N.: Languages that capture complexity classes. SIAM J. Comput. **16**, 760–778 (1987)
12. Immerman, N.: Expressibility and parallel complexity. SIAM J. Comput. **18**, 625–638 (1989)
13. Jaśkowski, S.: On the rules of suppositions in formal logic. Stud. Log. **1**, 6–32 (1934)
14. Kitaev, A.Y., Shen, A.H., Vyalyi, M.N.: Classical and Quantumn Computation. Graduate Studies in Mathematics. Americal Mathematical Society (2002)
15. Kornell, A.: Discrete quantum structures. arXiv:2004.04377v6 (2022)
16. Yamakami, T.: Quantum NP and a quantum hierarchy. In: Proceedings of the 2nd IFIP International Conference on Theoretical Computer Science (TCS 2002), Kluwer Academic Press (under the title of Foundations of Information Technology in the Era of Network and Mobile Computing), The International Federation for Information Processing, vol. 96 (Track 1), pp. 323–336 (2002)
17. Yamakami, T.: A schematic definition of quantum polynomial time computability. J. Symb. Log. **85**, 1546–1587 (2020)
18. Yamakami, T.: Nonuniform families of polynomial-size quantum finite automata and quantum logarithmic-space computation with polynomial-size advice. Inf. Comput. **286**, article 104783 (2022)
19. Yamakami, T.: Elementary quantum recursion schemes that capture quantum polylogaritmic-time/space computability of quantum functions. Math. Struct. Comput. Sci. **34**, 710–745 (2024)
20. Yamakami, T.: Quantum first-order logics that capture logarithmic-time/space quantum computability. In: Proceedings of the 20th Conference on Computability in Europe (CiE 2024). Lecture Notes in Computer Science, vol. 14773, pp. 311–323. Springer (2024). A complete and corrected version is available at arXiv:2501.12007 (2025)
21. Ying, M.: Hoare logic for quantum programs. arXiv:0906.4586v1 (2009)

Minimizers in Semi-dynamic Strings

Wiktor Zuba[1,2](✉), Oded Lachish[3], and Solon P. Pissis[2,4]

[1] University of Warsaw, Warsaw, Poland
w.zuba@mimuw.edu.pl
[2] CWI, Amsterdam, The Netherlands
solon.pissis@cwi.nl
[3] Birkbeck, University of London, London, UK
o.lachish@bbk.ac.uk
[4] Vrije Universiteit, Amsterdam, The Netherlands

Abstract. Minimizers sampling is one of the most widely-used mechanisms for sampling strings. Let $S = S[0] \dots S[n-1]$ be a string over an alphabet Σ. Further, let $w \geq 2$ and $k \geq 1$ be two integers and $\rho = (\Sigma^k, \leq)$ be a total order on Σ^k. The minimizer of window $X = S[i \mathinner{.\,.} i+w+k-2]$ is the smallest position in $[i, i+w-1]$ where the smallest length-k substring of X based on ρ starts. The set of minimizers for all $i \in [0, n-w-k+1]$ is the set $\mathcal{M}_{w,k,\rho}(S)$ of the minimizers of S. The set $\mathcal{M}_{w,k,\rho}(S)$ can be computed in $\mathcal{O}(n)$ time. The folklore algorithm computes the minimizer of every window in $\mathcal{O}(1)$ amortized time using $\mathcal{O}(w)$ working space.

It is thus natural to pose the following two questions:

1. *Can we efficiently support other dynamic updates on the window?*
2. *Can we improve on the $\mathcal{O}(w)$ working space?*

We answer both questions in the affirmative:

1. We term a string X *semi-dynamic* when one is allowed to insert or delete a letter at any of its ends. We show a data structure that maintains a semi-dynamic string X and supports minimizer queries in X in $\mathcal{O}(1)$ time with $\mathcal{O}(1)$ amortized time per update operation.
2. We show that this data structure can be modified to occupy strongly sublinear space without increasing the time complexity of its operations. To the best of our knowledge, this yields the first algorithm for computing $\mathcal{M}_{w,k,\rho}(S)$ in $\mathcal{O}(n)$ time *using $\mathcal{O}(\sqrt{w})$ working space.*

Keywords: String algorithms · Sampling · Minimizers · Dynamic strings

1 Introduction

Minimizers is a mechanism for string sampling that has been introduced independently by Schleimer et al. [30] and by Roberts et al. [28]. Since then, it has been used ubiquitously in modern sequence analysis applications, underlying some of the most widely used tools [22,23,34]. Minimizers also form the basis of more involved sampling schemes, such as syncmers [10], strobemers [29], mod-minimizers [13], and bd-anchors [24]. For more information, see [6,35].

We start by providing the standard definition of minimizers [28,30].

A. Jeż and J. Otop (Eds.): FCT 2025, LNCS 16106, pp. 434–447, 2026.
https://doi.org/10.1007/978-3-032-04700-7_32

Definition 1 (Minimizers). *Let $S = S[0] \ldots S[n-1]$ be a string over an alphabet Σ, $w \geq 2$ and $k \geq 1$ be two integers, and $\rho = (\Sigma^k, \leq)$ be a total order on Σ^k. The* minimizer *of the window $X = S[i \mathinner{.\,.} i + w + k - 2]$ of S is the smallest position in $[i, i + w - 1]$, where the smallest length-k substring of X starts based on ρ. The set of minimizers for all $i \in [0, n - w - k + 1]$ is the set $\mathcal{M}_{w,k,\rho}(S)$ of the minimizers of S.*

Two standard instantiations of the ρ order are the lexicographic order (e.g., when Σ is a totally ordered alphabet [33]) or the total order implied by the Karp-Rabin (KR) rolling hash of the elements of Σ^k [18]. For efficiency, practitioners mostly use such rolling hash functions [14,24,30].[1] We will henceforth assume the KR order, unless explicitly stated otherwise. In particular, this implies oracle access to some order ρ. We can compute the KR fingerprint (KRF) of a length-k substring of S in $\mathcal{O}(1)$ amortized time [18], and apply a folklore algorithm[2] computing the minimum elements in a sliding window of size w in $\mathcal{O}(1)$ amortized time per window using $\mathcal{O}(w)$ working space (i.e., the space on top of storing S). We summarize the above discussion into the following folklore result.

Proposition 1 (cf. [24]). *Given a string S of length n over an alphabet Σ, two integers $w \geq 2$ and $k \geq 1$, and oracle access to an order $\rho = (\Sigma^k, \leq)$, the set $\mathcal{M}_{w,k,\rho}(S)$ of the minimizers of S can be computed in $\mathcal{O}(n)$ time.*

It is thus natural to pose the following two questions:

1. *Can we efficiently support other dynamic updates on the window?*
2. *Can we improve on the $\mathcal{O}(w)$ working space?*

Problem Formulation. Let X be a string of length ℓ over alphabet Σ. The string X is, in general, called *dynamic* in the literature [2,7,15] when it can be modified through three different types of *edit* operations: letter insertion, deletion, or substitution. Here, we consider a *semi-dynamic* model, where one is allowed to insert or delete a letter at any of the ends of X. More formally, we define the four following types of modifications that are allowed on X:

- $\mathsf{Prepend}(a)$ that updates the string X into aX, for $a \in \Sigma$;
- $\mathsf{Append}(a)$ that updates the string X into Xa, for $a \in \Sigma$;
- $\mathsf{DeleteFirst}()$ that updates the string $X = aX'$ into X', for $a \in \Sigma$;
- $\mathsf{DeleteLast}()$ that updates the string $X = X''a$ into X'', for $a \in \Sigma$.

We will collectively call the above four types of modifications *border modifications* and such a string X a *semi-dynamic* string. A *semi-dynamic minimizer* data structure, for a fixed integer $k \geq 1$ and a semi-dynamic string X, is then a data structure that maintains X to support queries asking for the minimizer of X; a minimizer query must return the *position* (and the KRF *value* if needed) of

[1] Note that (unlikely but possible) hash collisions can only influence the number of minimizers and do not affect the correctness of the approach.

[2] For example, see https://codeforces.com/blog/entry/71687 or Sect. 2.3.

the corresponding substring. Such a data structure should be much more efficient than the naive comparison of all its length-k substrings. Note that border modifications form a subset of the most commonly used edits, and as such, problems concerning those can be answered using more general solutions (see Sect. 2.2). On the other hand, the sliding-window mechanism, used in the context of minimizers over string S (see Definition 1), is a special case of a semi-dynamic string. In particular, using a semi-dynamic minimizer data structure, we can simulate the sliding window and also output the minimizer of the current window X.

Our Results. Interestingly, while the problem of a *dynamic minimizer* data structure (i.e., the problem for a *dynamic* string X) does not seem to be solvable faster than $\mathcal{O}(\log \ell)$ time per edit, in the case of border modifications, a faster solution is indeed possible. At the same time, a semi-dynamic minimizer data structure is already quite powerful, as it generalizes the most commonly used sliding-window approach (cf. Proposition 1) and the computation of minimizers on a trie [12], for which a standard approach would use Proposition 1 on every leaf-to-root path. In particular, we make the following contributions:

1. In Sect. 2, we show a warm-up dynamic minimizer data structure that supports $\mathcal{O}(1)$-time minimizer queries with: (i) $\mathcal{O}(\log \ell)$ time per border modification; and (ii) $\mathcal{O}(k \log \ell)$ time per general edit. We also provide evidence as to why any improvement over the time complexity for general edits is unlikely (in fact, impossible in the comparison model of computation).
2. In Sect. 3, we show a semi-dynamic minimizer data structure that supports $\mathcal{O}(1)$-time minimizer queries with $\mathcal{O}(1)$ amortized time per update operation.
3. In Sect. 4, we show that the above semi-dynamic minimizer data structure can be modified to occupy strongly sublinear space in $|X|$, the length of X, without increasing the time complexity of its operations. To the best of our knowledge, this yields the first algorithm for computing the set $\mathcal{M}_{w,k,\rho}(S)$ of the minimizers of S in $\mathcal{O}(n)$ time *using* $\mathcal{O}(\sqrt{w})$ *working space.*

Computational Model. We operate within the standard word RAM model, where machine words are composed of $\Omega(\log n)$ bits, with n representing the input size. Basic operations on these words – like indirect addressing or arithmetic operations – are assumed to take $\mathcal{O}(1)$ time. When assessing the space complexity of an algorithm, we quantify it in terms of the number of words utilized.

Related Work. The sliding-window and the semi-dynamic models have been considered for maintaining the *suffix tree* [5,11,21], the *directed acyclic word graph* [16,31], and for numerous other string-processing tasks [1,3,9,25–27]. In fact, the first papers on dynamic variants of the classical longest common subsequence and edit distance problems considered updates that were a subset of prepend, append, and delete the first or last letter [17,19,20,32].

2 Preliminaries

An *alphabet* Σ is a finite set of elements called *letters*. A sequence $X[0]\ldots X[n-1]$ of letters from Σ is called a *string* X of *length* $n = |X|$. By $X[i\,.\,.\,j] = X[i]\ldots X[j]$, we denote a *fragment* of X. Strings that are fragments of X are called *substrings* of X.

In Sect. 2.1, we detail how the substrings of X can be efficiently compared and how the starting positions of the substrings of X are maintained. In Sect. 2.2, we present a simple dynamic minimizer data structure and explain why it is unlikely to be improved. In Sect. 2.3, we present the sliding-window approach for computing the set $\mathcal{M}_{w,k,\rho}(S)$ of the minimizers of S.

2.1 Substring Comparison and Starting Positions of the Substrings

The cost of the modifications of our data structures is determined by the number of length-k string comparisons. The techniques used in the solutions presented in the paper do not impose restrictions on the alphabet Σ, other than the fact that we are given oracle access to an order $\rho = (\Sigma^k, \leq)$. A naive comparison of length-k strings imposes a multiplicative $\mathcal{O}(k)$ factor in the time complexities. To avoid this problem, we define $\rho = (\Sigma^k, \leq)$ through the corresponding KRF values. In our setting, the KRF of the new, removed or modified length-k fragment of X can be computed in $\mathcal{O}(1)$ time by modifying the stored KRF of its neighbouring length-k fragment (occurring one position earlier or later). Throughout the paper, we store the handle to substring comparison (either a pointer to its position in the string or the KRF) as a *value* which occupies $\mathcal{O}(1)$ space and can be computed in $\mathcal{O}(1)$ time. In the complexity analysis of the algorithms, we assume that the $\mathcal{O}(1)$-time comparison of the KRFs is used.

By modifying X with $\mathsf{Prepend}(a)$ or $\mathsf{DeleteFirst}()$ operations, we shift the starting positions (relative to the beginning of the string) of all fragments without actually altering them in any way. Since data structures store information about the positioning of substrings and updating all of those would be costly, we always refer to the starting position relative to the starting position of the string in its initial state (the starting position of a substring can be negative). The relative position of the smallest substring (the minimizer) is obtained by subtracting from the answer the position of the first letter of X (which is stored along the structure and updated in $\mathcal{O}(1)$ time) for border modifications or by more complicated data structures for edits. Henceforth, we assume that this approach is shared by all our structures and always refer to this immutable starting position of a substring.

2.2 General Priority-Queue Approach

Upon inserting a letter to string X, we produce up to k length-k fragments (when $|X| \geq k$), and upon asking a minimizer query, we want to find the length-k substring with the smallest value-this alone suggests that the dynamic minimizer data structure is some sort of a priority queue storing pairs (*position*, *value*) sorted first by value and then by position in the case of fragment ties: two or more fragments of X can correspond to one substring of X.

Standard implementations of priority queues [8], such as a *heap* data structure, allow efficient insert, delete, and find-min operations. In such data structures, the two modification operations are performed in $\mathcal{O}(\log \ell)$ time (where $w = \ell - k + 1 \leq \ell$ is the number of elements stored in the structure); the smallest element is then found in $\mathcal{O}(1)$ time. This gives the following fact.

Fact 1. *The general priority-queue implementation of a dynamic minimizer data structure for a fixed integer $k \geq 1$ and a dynamic string X of length ℓ supports $\mathcal{O}(1)$-time minimizer queries and performs border modifications in $\mathcal{O}(\log \ell)$ time and general edits in $\mathcal{O}(k \log \ell)$ time.*

Proof. Recall that at any moment we store most $w = \ell - k + 1 \leq \ell$ elements. Upon inserting (resp. deleting) a new length-k fragment, we can compute its value in $\mathcal{O}(1)$ time, and thus we know exactly what (*position*, *value*) pairs should be inserted (resp. deleted).

For border modifications, exactly one such pair is inserted (resp. deleted), hence such an operation can be performed in $\mathcal{O}(\log \ell)$ time in a heap implementation of priority queue.

The case of general edits is a little more complicated-a single letter modification can affect up to k substrings of X, hence the modification of the structure requires deleting and inserting up to k pairs, which gives us $\mathcal{O}(k \log \ell)$ time per modification. Additionally, insertions or deletions in the middle of X change the positioning of the substrings in a non-consistent way. However, this problem can be solved with a data structure (e.g., a segment tree) that stores these position shifts and supports $\mathcal{O}(\log \ell)$ time interval updates [4].

In both cases, finding the minimizer can be performed by simply asking for the smallest pair (possibly stored along the structure). □

As mentioned in the proof of Fact 1, dealing with internal changes is more involved and costly; in particular, in some settings, the $\mathcal{O}(\log \ell)$ time per update cannot be avoided (even if we do not need to care about the uneven positioning), as illustrated by the following lemma.

Lemma 1. *Any dynamic minimizer data structure, for an integer $k \geq 1$ and a dynamic string of length ℓ over a general ordered alphabet, cannot support substitution modifications in $o(\log \ell)$ amortized time in the comparison model.*

Proof. We restrict our focus to the case of $k = 1$ and naive string comparison, which is equivalent to KRF for $k = 1$ (for most orderings the first letter of a fragment is the most important, and thus the problem for larger k reduces directly to the considered one). Given a sequence of ℓ numbers, we can construct a string X representing this sequence: the ith letter of X is the ith number in the sequence. With the dynamic minimizer structure for substitution operations (that is, a subset of edit operations), we can repeat ℓ times a procedure of finding the position of the smallest length-1 substring (the number) and replacing the number at that position with the maximum of the whole sequence. Such a procedure sorts the sequence of numbers using $\mathcal{O}(\ell)$ modifications of the sequence, hence those operations cannot cost less than $\Omega(\log \ell)$ comparisons [8]. □

For a constant-size alphabet Σ and for small values of k, one can resort to bucketing (for $k = 1$ one can simply store the queue as a set of $|\Sigma|$ buckets); this approach is generally less practical, however, and does not solve additional problems such as supporting positional shifts.

In the following, we restrict the operations on X to border modifications in order to design an efficient, yet simple to implement, data structure.

2.3 Sliding-Window Approach

Minimizers are most commonly used for string sampling-a set of string *anchors* is defined as the set of minimizers of all length-ℓ fragments of the input string S for some fixed w, k. Note that by Definition 1, we have $\ell = w + k - 1$.

In this setting, a folklore method to compute minimizers is the sliding-window approach-the length-ℓ fragment starting one position later differs by very little, and hence the update of the variables kept by the algorithms is usually cheap. From the perspective of minimizers, usually the position of the minimizer stays the same; it can change in two cases: either the new length-k fragment becomes the smallest one; or the current minimizer falls out of the window under consideration. The standard technique exploits this setting of very restricted changes to store only information about the substrings that can become minimizers in the future. Formally, the set of the substrings kept consists of only the position-value pairs (p, v) such that there exists no length-k substring in the current window with a representing pair (p', v') such that $p' > p$ and $v' < v$ (the pair (p, v) does not *dominate* any other pair). Thus, the deque (in this case a bidirectional linked list is enough) storing those substrings is sorted non-decreasingly forming a Manhattan skyline (although reversed as we look for the smallest values instead of the largest ones), which allows updates in $\mathcal{O}(1)$ amortized time; see Fig. 1. Observe that, if the sequence of *all* values is monotonically increasing, the space used by this approach is $\Omega(w)$. In Sect. 3, we provide more details and generalize this approach.

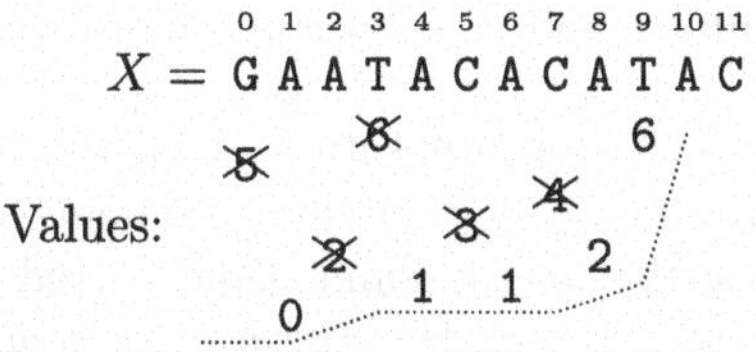

Fig. 1. Window X (slid over string S) with the values for each length-3 fragment (for legibility matching their lexicographic order) starting at each position of X. Values for which there exists a smaller value to the right are crossed out-they cannot become a minimizer, as this smaller value will leave the window later. The remaining values form a non-decreasing sequence; notice that the last value cannot be crossed out.

3 Constant-Time Update Data Structure

In Sect. 3.1, we show how the sliding-window approach (see Sect. 2.3) can be generalized to support *a subset* of the border modifications. In Sect. 3.2, we present our data structure for the *complete* set of border modifications. We implement the semi-dynamic string X using a bidirectional linked list.

3.1 Generalizing the Sliding-Window Approach

We show that the sliding-window approach with storing information about only the substrings that can become minimizers in the future can be applied directly to obtain a minimizer data structure with a *limited set of border modifications*: $\mathsf{Prepend}(a), \mathsf{Append}(a), \mathsf{DeleteFirst}()$. This then becomes a building block of our data structure for all border modifications.

In our deque data structure (or in a bidirectional linked list), we store the (*position*, *value*) pairs that represent the length-k fragments that can become minimizers of X after some modifications. A pair (p, v) dominates a pair (p', v') if $p < p'$ and $v > v'$. We keep the pairs that do not dominate any other one (from the representation of all the length-k fragments of X) in a deque sorted increasingly by their first element (position), and (because of this domination property) at the same time sorted non-decreasingly by their second element (value). We call this the *limited semi-dynamic minimizer* data structure.

Lemma 2. *The limited semi-dynamic minimizer data structure supports* $\mathcal{O}(1)$*-time minimizer queries and limited border modifications in* $\mathcal{O}(1)$ *amortized time.*

Proof. Let p be the position of the minimizer, v its value, and let (p_1, v_1) be the first element of the deque (retrieved trivially in $\mathcal{O}(1)$ time); by the definition of the minimizer $v_1 \geq v$. If $p_1 > p$, then the pair (p, v) would appear in the deque before (p_1, v_1)-it clearly does not dominate any pair since it has the smallest value. If $p_1 < p$ and $v_1 > v$, then the pair (p_1, v_1) would dominate the pair (p, v) and therefore would not appear in the deque. This shows that $p_1 = p$ is the position of the minimizer and that it can be retrieved in $\mathcal{O}(1)$ time. When updating the structure we have to consider the three possible modifications:

- $\mathsf{Append}(a)$: we insert a new pair (p, v) representing the new fragment to the structure; as the one with the largest position it will have to be stored in the deque (as its last element). We also need to remove from the deque the pairs with a value greater than v, which is done in time linear in the number of removals since the pairs are stored ordered by those values.
- $\mathsf{DeleteFirst}()$: we remove a single fragment from the structure. If its position is equal to p_1, then the pair (p_1, v_1) is removed-the second element stored in the deque represents the new minimizer, as every fragment in between has a strictly larger value.
- $\mathsf{Prepend}(a)$: we only want to add the new pair (p, v) if it does not dominate any pair currently stored in the structure. To check this property, it suffices to compare the value v with v_1: if $v \leq v_1$, then (p, v) should be added to the

structure and represent the new minimizer (the other stored pairs cannot be dominated by this pair, as their values are greater or equal to v_1); otherwise, no modification of the structure is required.

The cost of operation $\mathsf{Prepend}(a)$ (resp. $\mathsf{DeleteFirst}()$) is $\mathcal{O}(1)$: a single comparison and at most one insertion (resp. deletion) of a pair. The cost of $\mathsf{Append}(a)$ is amortized by the number of elements removed from the deque; namely, by the number of insertions and, in turn, the number of $\mathsf{Prepend}(a)$ and $\mathsf{Append}(a)$ operations performed earlier. □

By symmetry, the technique underlying Lemma 2 can be adapted to support the set of operations $\mathsf{Append}(a), \mathsf{Prepend}(a)$, and $\mathsf{DeleteLast}()$, and with some modifications to solve the problem of a window sliding both sides over string S. Unfortunately, however, this technique does not extend to the full set of border modifications because of the following fact.

Fact 2. *When both* $\mathsf{DeleteFirst}()$ *and* $\mathsf{DeleteLast}()$ *operations are allowed on string* X*, every length-*k *fragment of* X *can become the minimizer in the future as one can perform subsequent* $\mathsf{DeleteFirst}()$ *or subsequent* $\mathsf{DeleteLast}()$ *operations so that* X *contains only this single length-*k *fragment.*

This shows that keeping the information about all the fragments that can become minimizers in the future brings us back to the general approach with storing information about all length-k fragments; in the next section, we show how to work around this problem.

3.2 The Semi-dynamic Minimizer Data Structure

Our approach consists of relaxing the property of storing the (*position*, *value*) pairs for all fragments that can become minimizers to storing them only for the fragments that can become minimizers in the near future. When this "time period" expires (when we are no longer sure that the current minimizer is stored), we resort to rebuilding the whole structure from scratch.

Recall that X is our semi-dynamic string of length ℓ. We distinguish its single position x and partition the set of length-k fragments of X into those that start before position x and those that start at or after position x. For each of those subsets, we store a separate queue structure represented by a stack (or a one-directional linked list, since this time such a structure is enough).

The *left stack* represents the pairs (p, v) such that there exists no pair (p', v') with $p' > p$ and $v' < v$ among the left subset of length-k fragments (with (p_1, v_1) stored on top)-that is exactly the same as in Sect. 3.1. Symmetrically, the *right stack* represents the pairs (p, v) such that there exists no pair (p', v') with $p' < p$ and $v' \leq v$ among the right subset of length-k fragments. The represented subsequence is *strictly* decreasing, as we do not need to store multiple pairs with the same value (only the one with the leftmost position is important). Again, pairs are stored in the stack in the order of their positions-the one with the rightmost position (hence again with the smallest value) is at the top. See Fig. 2.

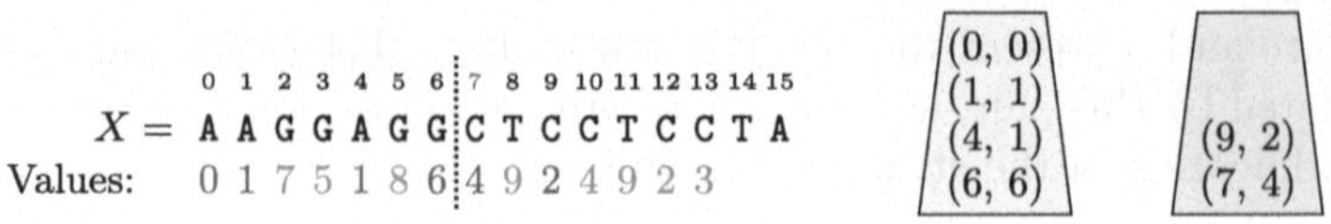

Fig. 2. A graphical representation of the data structure. On the left, we present string X with the values for each length-3 fragment (for simplicity matching their lexicographic order) starting at each position of X. Position x drawn in red divides X into two parts-for each one, the "interesting" values are drawn in blue; the ones in gray are not represented but they may be computed again during the full structure rebuild. One can notice that the distinguished values can repeat in the left part but not in the right one. On the right, the "interesting" (*position*, *value*) pairs are stored in two stacks. Note that, in particular, the positions x and $x - 1$ are always represented (the stacks are nonempty). (Color figure online)

Recall that for efficiency, the values are computed as KRFs. Thus, we also store the KRF values of the leftmost and rightmost length-k fragments of X to allow for fast computation of the KRF of the length-k fragment that is inserted or deleted during the modification of X. These take up $\mathcal{O}(1)$ space.

We call this the *two-stack semi-dynamic minimizer* data structure. Each of the stacks stores as its top element the only candidate for the minimizer over its subset of length-k fragments; hence, the minimizer can be obtained using a single comparison (or stored along the structure to support $\mathcal{O}(1)$-time queries in case of brute-force comparison). The modifications to the left stack are performed in the same way as in Sect. 3.1 without the Append(a) operation, which is the only one whose cost was amortized (we never remove multiple pairs in a single operation). By symmetry, the right stack is handled analogously (with the simple change of handling the ties upon insertion to not store two pairs with the same value). However, this time, a single exceptional case can happen: if we try to remove a pair from an empty stack, the behaviour is undefined; we would like to remove something from the other stack, but as mentioned in Fact 2 this causes problems. In such a situation, we restore the "basic state" of the structure by "dividing the single stack into two stacks" as follows:

1. We recompute (*position*, *value*) pairs for all the length-k fragments of X;[3]
2. We set the value of x to the starting position of X plus $\lceil (|X| - k)/2 \rceil$;
3. We divide the set of length-k fragments according to this new value of x;
4. We populate the left and right stacks by inserting the new pairs.

The above four steps can be clearly implemented in $\mathcal{O}(\ell)$ total time. Note that the cost of rebuilding is amortized by the number of insertions plus the number of deletions performed since the last rebuild. Indeed, assume that after the last rebuild both stacks had size h, and now one stack is of size 0, while the other one is of size ℓ. This means that we had to perform at least h operations DeleteFirst() and at least $\ell - h$ operations Append(a).

[3] We do not compute all of the pairs at once; instead we simply perform ℓ insertions on an empty structure. Such an approach requires only $\mathcal{O}(1)$ extra construction space.

The correctness of the approach in the standard case follows from Lemma 2. In the special case, we rebuild everything from scratch; hence we cannot miss any (*position*, *value*) pair needed. Upon recomputing the two stacks, we fall back to the standard case: all values that may be needed before the next rebuild are stored. We have thus arrived at the main result.

Theorem 3. *The two-stack semi-dynamic minimizer data structure supports* $\mathcal{O}(1)$*-time minimizer queries and border modifications in* $\mathcal{O}(1)$ *amortized time.*

4 Space-Efficient Data Structure

In Sect. 3, we showed how to obtain a fast data structure by storing only the pairs that may become important in the near future. Still, those pairs translate into storing the representation of all the $\ell - k + 1$ length-k fragments of X if the sequences of their values before and after position x are monotonous. In this section, we show how to extend the two main ideas of our data structure-storing only the pairs that may become useful in the near future and re-computation-to further reduce the space occupied by the data structure. We implement the semi-dynamic string X using a bidirectional linked list.

4.1 The High-Level Idea

We partition X into blocks; each block is represented by a single (*position*, *value*) pair for the length-k fragment with the smallest value among the ones starting within this block. We treat the blocks as if they were just single length-k fragments and store information about those in stacks. Additionally, we represent all the length-k fragments of the first block that may correspond to minimizers in the future (before the stack is emptied or the whole structure is rebuilt) just as in Sect. 3.2 (the first block is treated like the whole left part of X therein). We also represent the last block of X analogously. The total number of pairs stored is bounded by the number of blocks plus the total size of a constant number of blocks. Conceptually, we have two layers: the *first layer* corresponds to the length-k fragments of X; and the *second layer* corresponds to the blocks of X.

4.2 Fixed-Length Blocks

We start with the simplest block division design, where each block has the same fixed length, to show the technical details of the approach in a possibly clear setting. Let $c > 1$ be a fixed integer and let $B_\kappa = \{x + \kappa \cdot c, x + \kappa \cdot c + 1, \cdots, x - 1 + (\kappa + 1) \cdot c\}$, for some integer κ. Recall that x is a position on the string X. This divides the length-k fragments of X into blocks of length c; i.e., length-k fragments starting at the positions in B_κ form a *block*.

Assume that we want to build the structure for a nonempty string X; for example, this happens when the full structure is rebuilt. For the first and last nonempty blocks (containing some length-k fragment), we build stacks in the

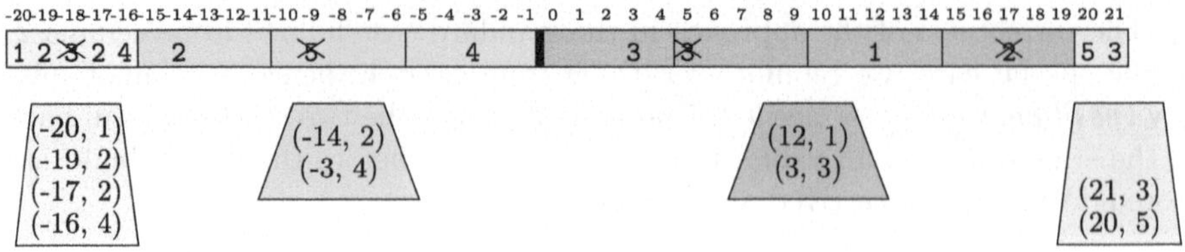

Fig. 3. A graphical representation of the two-layer data structure for $c = 5$ and $x = 0$. The "internal" blocks (with darker colour in the figure) are represented by one fragment each (the one with the smallest value). We store one second-layer stack representing the internal blocks on each side, and one first-layer stack for each border block. Note that in the process of updating the structure, there are either one or two such blocks for each side of the string.

same way as in Sect. 3.2. Additionally, for each "internal" block (if any) we find the fragment with the smallest value (breaking ties for the one with the smallest position) and build two stacks for only those fragments, again just as in Sect. 3.2. The blocks B_κ with $\kappa < 0$ are represented with the left second-layer stack, and the blocks B_κ with $\kappa \geq 0$ are represented with the right second-layer stack; see Fig. 3. We call this the *two-layer semi-dynamic minimizer* data structure. We prove the following result.

Lemma 3. *Suppose that we have read-only access to a semi-dynamic string X of length ℓ. The two-layer semi-dynamic minimizer data structure with blocks of length c, for any integer $c > 1$, supports $\mathcal{O}(1)$-time minimizer queries, performs border modifications in $\mathcal{O}(1)$ amortized time, and occupies $\mathcal{O}(c + \frac{\ell}{c})$ space.*

Proof. The minimizer is stored as the top element of one of the stacks. If it comes from a "border" block, then the correctness follows from Theorem 3. Similarly, if it comes from one of the "internal" blocks, then it must be its representing element, thus again it is stored as the top element of the second layer stack.

When the structure is updated, and adding a new fragment does not require starting a new block (respectively, we do not attempt to remove an element from an empty first-layer stack), then the operation takes $\mathcal{O}(1)$ time like in Sect. 3.2.

If a new block gets started upon inserting a letter, we simply create an empty stack for the new block and insert the pair representing the new fragment there. This means that we can have more than one first-layer stacks for this side of X. For space reasons if there are three such stacks for a single side of X, we remove the most internal one; i.e., insert its top element into the second-layer stack (if it satisfies the conditions of not dominating any pair) and simply discard the rest of the stack-this takes $\mathcal{O}(c)$ time. Symmetrically, when we want to remove an element from an empty first-layer stack, we use the first-layer stack for the next block, and if such a stack is not in the memory, we recompute it from scratch in $\mathcal{O}(c)$ time (plus possibly remove the top element of the second-layer stack).[4]

[4] Notice that we cannot simply store exactly one first-layer stack at all times. Such an approach could require this stack to be discarded and rebuilt every other operation, which in turns would cost $\mathcal{O}(c)$ amortized time per operation.

Like in the vanilla version of the two-stack semi-dynamic minimizer structure, if we pass the position x, the whole structure gets recomputed from scratch.

The complexity analysis is similar to the one of Sect. 3.2. The only costly operations are rebuilds. In the case of a full rebuild, the costs are covered by the number of insertion and deletion operations since the last full rebuild.

The re-computation of a first-layer stack takes $\mathcal{O}(c)$ time, yet, for such an operation to happen, at least c letters had to be deleted from this side of X since the previous re-computation or stack removal. Indeed, whether the previous such operation was a re-computation or a removal, one full stack was in the memory afterwards. Analogously, in case of removal of a first-layer stack, one had to insert at least c letters to this side of X. In case of the first re-computation or removal after the full rebuild, we charge its cost to this full rebuild (the cost is correctly covered even if $\ell < c$). In total, this shows that the amortized cost of operations is $\mathcal{O}(1)$.

The space follows from the fact that there are at most four first-layer stacks at a time, each of size $\mathcal{O}(c)$, plus two second-layer stacks, each of size $\mathcal{O}(\ell/c)$. □

4.3 The $\mathcal{O}(\sqrt{\ell})$-Space Data Structure

Previously, we never actually used the fact that the blocks are of fixed length; the complexity analysis relied only on the property that the more "internal" block (for which the stack is removed or recomputed) is up to a constant factor larger (or actually smaller) than the block that is further outside (whose completion or emptying initializes another block rebuild or removal).

Let $B'_\kappa = \{x + \kappa^2, x + \kappa^2 + 2, \cdots, x + (\kappa + 1)^2 - 1\}$, for $\kappa \geq 0$, and $B'_\kappa = \{x - \kappa^2, x - \kappa^2 + 1, \cdots, x - (\kappa + 1)^2 - 1\}$, for $\kappa < 0$. The blocks on each side have lengths $1, 3, 5, 7, \cdots$ from the middle (from position x) outwards. Using this design, we have $\mathcal{O}(\sqrt{\ell})$ blocks on each side of x, and the size of each block is bounded by $\mathcal{O}(\sqrt{\ell})$ for $\ell = |X|$ (even when changing the value of ℓ dynamically). We call this way of partitioning string X *progressing blocks*, and from Lemma 3, we obtain the following result.

Theorem 4. *Suppose that we have read-only access to a semi-dynamic string X of length ℓ. The two-layer semi-dynamic minimizer data structure with progressing blocks supports $\mathcal{O}(1)$-time minimizer queries, performs border modifications in $\mathcal{O}(1)$ amortized time, and occupies $\mathcal{O}(\sqrt{\ell})$ space.*

Corollary 1. *Given read-only access to a string S of length n over an alphabet Σ, two integers $w \geq 2$ and $k \geq 1$, and oracle access to an order $\rho = (\Sigma^k, \leq)$, the set $\mathcal{M}_{w,k,\rho}(S)$ can be computed in $\mathcal{O}(n)$ time using $\mathcal{O}(\sqrt{w})$ working space.*

4.4 Multi-layer Data Structure

The two-layer data structure naturally generalizes to multiple layers; blocks can be aggregated into second-layer blocks, just as fragments were grouped into blocks. In this way, for a three-layer semi-dynamic minimizer data structure, the space becomes $\mathcal{O}(\sqrt[3]{\ell})$. When using h layers, the space becomes $\mathcal{O}(h \cdot \ell^{1/h})$. At the same time, the cost of operations actually grows-there are at most $4h - 2$ stacks at the same time, hence the cost of finding the minimizer grows to $\mathcal{O}(h)$ (although it can be stored along the structure and have its cost covered by the modifications). The cost of border modifications also increases to amortized $\mathcal{O}(h)$; a single modification pays for the amortization of every layer.

In the limit, each block of level h consists of a constant number of blocks of level $h-1$, and new layers are introduced when needed, that is, the blocks form a regular balanced tree over the length-k fragments of X. The size of the data structure becomes $\mathcal{O}(\log \ell)$, while operations require $\mathcal{O}(\log \ell)$ amortized time.

Acknowledgments. A research visit during which part of the presented ideas were conceived was funded by a Royal Society International Exchanges Award.

Wiktor Zuba was supported in part by the Netherlands Organisation for Scientific Research (NWO) through Gravitation-grant NETWORKS that has received funding from the European Union's Horizon 2020 research and innovation programme under the Marie Skłodowska-Curie grant agreement No 101034253.

Solon P. Pissis was supported in part by the PANGAIA and ALPACA projects that have received funding from the European Union's Horizon 2020 research and innovation programme under the Marie Skłodowska-Curie grant agreements No 872539 and 956229, respectively.

References

1. Akagi, T., et al.: Combinatorics of minimal absent words for a sliding window. Theor. Comput. Sci. **927**, 109–119 (2022)
2. Amir, A., Landau, G.M., Lewenstein, M., Sokol, D.: Dynamic text and static pattern matching. ACM Trans. Algorithms **3**(2), 19 (2007)
3. Bannai, H., Charalampopoulos, P., Radoszewski, J.: Maintaining the size of LZ77 on semi-dynamic strings. In: CPM 2024, pp. 1–20 (2024)
4. de Berg, M., Cheong, O., van Kreveld, M.J., Overmars, M.H.: Computational Geometry: Algorithms and Applications, 3rd Edition. Springer (2008)
5. Brodnik, A., Jekovec, M.: Sliding suffix tree. Algorithms **11**(8), 118 (2018)
6. Chikhi, R., Holub, J., Medvedev, P.: Data structures to represent a set of k-long DNA sequences. ACM Comput. Surv. **54**(1), 1–22 (2022)
7. Clifford, R., Gawrychowski, P., Kociumaka, T., Martin, D.P., Uznanski, P.: The dynamic k-mismatch problem. In: CPM 2022, pp. 1–15 (2022)
8. Cormen, T.H., Leiserson, C.E., Rivest, R.L., Stein, C.: Introduction to Algorithms, Second Edition. The MIT Press and McGraw-Hill Book Company (2001)
9. Crochemore, M., Héliou, A., Kucherov, G., Mouchard, L., Pissis, S.P., Ramusat, Y.: Absent words in a sliding window with applications. Inf. Comput. **270** (2020)
10. Edgar, R.: Syncmers are more sensitive than minimizers for selecting conserved k-mers in biological sequences. PeerJ **9**, e10805 (2021)
11. Fiala, E.R., Greene, D.H.: Data compression with finite windows. Commun. ACM **32**(4), 490–505 (1989)
12. Gabory, E., Liu, C., Loukides, G., Pissis, S.P., Zuba, W.: Space-efficient indexes for uncertain strings. In: ICDE 2024, pp. 4828–4842. IEEE (2024)
13. Groot Koerkamp, R., Liu, D., Pibiri, G.E.: The open-closed mod-minimizer algorithm. Algorithms Mol. Biol. **20**(1), 4 (2025)

14. Groot Koerkamp, R., Martayan, I.: SimdMinimizers: computing random minimizers, fast. bioRxiv (2025). https://doi.org/10.1101/2025.01.27.634998
15. Gu, M., Farach, M., Beigel, R.: An efficient algorithm for dynamic text indexing. In: SODA 1994, pp. 697–704 (1994)
16. Inenaga, S., Shinohara, A., Takeda, M., Arikawa, S.: Compact directed acyclic word graphs for a sliding window. J. Discrete Algorithms **2**(1), 33–51 (2004)
17. Ishida, Y., Inenaga, S., Shinohara, A., Takeda, M.: Fully incremental LCS computation. In: Liśkiewicz, M., Reischuk, R. (eds.) FCT 2005. LNCS, vol. 3623, pp. 563–574. Springer, Heidelberg (2005). https://doi.org/10.1007/11537311_49
18. Karp, R.M., Rabin, M.O.: Efficient randomized pattern-matching algorithms. IBM J. Res. Dev. **31**(2), 249–260 (1987)
19. Kim, S., Park, K.: A dynamic edit distance table. J. Discrete Algorithms **2**(2), 303–312 (2004)
20. Landau, G.M., Myers, E.W., Schmidt, J.P.: Incremental string comparison. SIAM J. Comput. **27**(2), 557–582 (1998)
21. Larsson, N.J.: Extended application of suffix trees to data compression. In: DCC 1996, pp. 190–199 (1996)
22. Li, H.: Minimap and miniasm: fast mapping and de novo assembly for noisy long sequences. Bioinform. **32**(14), 2103–2110 (2016)
23. Li, H.: Minimap2: pairwise alignment for nucleotide sequences. Bioinform. **34**(18), 3094–3100 (2018)
24. Loukides, G., Pissis, S.P., Sweering, M.: Bidirectional string anchors for improved text indexing and top-k similarity search. IEEE Trans. Knowl. Data Eng. **35**(11), 11093–11111 (2023)
25. Mieno, T., Fujishige, Y., Nakashima, Y., Inenaga, S., Bannai, H., Takeda, M.: Computing minimal unique substrings for a sliding window. Algorithmica **84**(3), 670–693 (2022)
26. Mieno, T., Funakoshi, M.: Shortest unique palindromic substring queries in semi-dynamic settings. In: IWOCA 2022, pp. 425–438 (2022)
27. Mieno, T., Watanabe, K., Nakashima, Y., Inenaga, S., Bannai, H., Takeda, M.: Palindromic trees for a sliding window and its applications. Inf. Process. Lett. **173**, 106174 (2022)
28. Roberts, M., Hayes, W.B., Hunt, B.R., Mount, S.M., Yorke, J.A.: Reducing storage requirements for biological sequence comparison. Bioinform. **20**(18), 3363–3369 (2004)
29. Sahlin, K.: Effective sequence similarity detection with strobemers. Genome Res. **31**(11), 2080–2094 (2021)
30. Schleimer, S., Wilkerson, D.S., Aiken, A.: Winnowing: local algorithms for document fingerprinting. In: SIGMOD 2003, pp. 76–85 (2003)
31. Senft, M., Dvořák, T.: Sliding CDAWG perfection. In: Amir, A., Turpin, A., Moffat, A. (eds.) SPIRE 2008. LNCS, vol. 5280, pp. 109–120. Springer, Heidelberg (2008). https://doi.org/10.1007/978-3-540-89097-3_12
32. Thorup, M.: Equivalence between priority queues and sorting. J. ACM **54**(6), 28 (2007)
33. Verbeek, H., Ayad, L.A.K., Loukides, G., Pissis, S.P.: Minimizing the minimizers via alphabet reordering. In: CPM 2024, pp. 1–13 (2024)
34. Wood, D.E., Salzberg, S.L.: Kraken: ultrafast metagenomic sequence classification using exact alignments. Genome Biol. **15**(3), R46 (2014)
35. Zheng, H., Marçais, G., Kingsford, C.: Creating and using minimizer sketches in computational genomics. J. Comput. Biol. **30**(12), 1251–1276 (2023)

Author Index

A. Jeż and J. Otop (Eds.): FCT 2025, LNCS 16106, pp. 449–450, 2026.
https://doi.org/10.1007/978-3-032-04700-7